Best wishes,

Mike Stimmann

A GUIDEBOOK TO

California Agriculture

A GUIDEBOOK TO

California Agriculture

By Faculty and Staff of the
University of California

Edited by
Ann Foley Scheuring

UNIVERSITY OF CALIFORNIA PRESS
Berkeley • Los Angeles • London

University of California Press
Berkeley and Los Angeles, California
University of California Press, Ltd.
London, England
©1983 by
The Regents of the University of California

Library of Congress Cataloging in Publication Data

Main entry under title:
A Guidebook to California agriculture:

    Includes index.
    1. Agriculture—California. 1. Scheuring, Ann
Foley.
S451.C2G84   338.1'09794   82-2669
ISBN 0-520-04709-5   AACR2

Printed in the United States of America

1 2 3 4 5 6 7 8 9

Contents

Foreword

California's agriculture is somewhat of a paradox. Demographically, the state has an extremely urban nonagrarian population. Its national and international reputation is grounded on motion pictures, aerospace industries, sandy ocean beaches, spectacular mountain and ocean scenery, cable cars, out of the ordinary politicians, and radical political movements. Undergirding this popularly held view of California and of equal or greater fame among the people who know it, is an agricultural complex that contributes one-tenth of the national gross value of all agricultural crops, employs one out of every five employed people, and uses less than three percent of the state's total population in the production process.

How this agricultural prominence has come about is the subject of this book. It will become apparent to the reader that no single event or organization is alone responsible for the size and rank of California's agriculture today. It is a product of its location and the evolution of many forces operating together in generally favorable economic, political, natural, and social environments.

Stresses and strains, conflicts and tensions have constantly influenced the development of California's agriculture, yet it seems to emerge from each so-called crisis stronger and more vital than when it initially encountered seemingly unsurmountable difficulties.

I want to draw attention to the role that publicly supported teaching, research, and extension, as exemplified in the Land Grant College and University concept, has had in the progress of U.S. agriculture. The University of California, as California's Land Grant institution, has had a significant influence on the state's agriculture. By instructing undergraduate and graduate students, researching and solving many varied problems, and seeing that solutions are applied where problems exist, the University has helped agriculture develop in spite of a multitude of real and potential adversities.

Linking agricultural teaching, research, and extension together under a single system of management in the stimulating environments of institutions of higher education is a concept that has taken hold only in the United States. The validity of this concept is nowhere more evident than here in California, where a highly productive agriculture relies on educated participants, research-based knowledge, and rapid adoption of new and innovative practices.

That this concept was ingenious and of inestimable value to the well-being of the public at large is exemplified by a simple economic comparison. For the fiscal year 1980–81, the California taxpayers invested 88 million dollars to support University of California agricultural research and extension programs. During this same period, California's agricultural products, before processing, packaging, distribution, and marketing, had a "farm gate" value of about 15.3 *billion* dollars. The total stimulus to the state's economy is estimated to be two to two-and-one-half times the farm gate figure, or nearly 35 *billion* dollars. The public investment in agricultural research and extension, therefore, amounted to only six-tenths of one percent of the farm value of products, and less than three-tenths of one percent of the economic contribution of agriculture to the state. This investment ensures not only employment opportunities, but also abundant and varied food, fiber, and ornamental products at low costs.

While, in the minds of many, agriculture in this country is a resounding success story and a tribute to individuals, organizations, and government working together, some people raise justifiable concerns about its future. For an activity as dynamic as agriculture, the past is truly just a prologue. The problems ahead are as critical as, and perhaps more pervasive than, any that have been faced before. Education, research, and the extension of knowledge can and should continue to focus attention on issues and problems of the future.

Not all of agriculture's future problems lend themselves to technological solutions. In a society of people separated by several generations from their agrarian-oriented forebears and living among an abundance of food and fiber products, lack of understanding and appreciation for the agriculture of America can lead quickly to complacency and neglect. This book is dedicated to closing the gaps between ignorance and knowledge and to restoring the appreciation for a magnificent period of agricultural development in California. We hope more "nonagricultural" than "agricultural" people will read this volume and gain an increased understanding of this complex activity that contributes so significantly to their individual lives.

To Ann Scheuring and her many collaborators I offer my congratulations and appreciation for conceiving and completing a monumental task. We all will feel rewarded if you, the reader, find things between these covers that you never knew existed.

J. B. KENDRICK, JR.
Vice President
Agriculture and University Services
University of California, Berkeley

Preface

It is a privilege to have had the opportunity to provide part of the support for the project that has consummated in the publication of *A Guidebook to California Agriculture*. As the reader will see, the project is a multidisciplinary approach to California agriculture. The California food and fiber system is so large and complex that no one individual is an authority for all the many facets that are represented in its operation. However, the collective knowledge that has been incorporated in the Guidebook represents an authoritative, up-to-date analysis of California agriculture's past, present, and future. It is hoped that the Guidebook will provide fuller appreciation of the total California agricultural system to those already familiar with some of its parts, and to provide individuals only peripherally involved with agriculture with a comprehensive resource that explains the greatness as well as the complexities and challenges that make up California's number one industry.

CHARLES E. HESS
Dean, College of Agricultural and Environmental Sciences
University of California, Davis

Acknowledgments

Preparing a volume of this scope and size has been a massive undertaking involving many individual efforts. My task as editor was made the less laborious and the more rewarding because so many individuals willingly lent their support and cooperation to a project they thought worthwhile. Authors have been credited elsewhere; here I wish to make special thanks to some whose participation was also crucial. Gratitude and appreciation to: Harry Troughton for his artwork on the figures; Ray Coppock for being a sympathetic and helpful back-up editor; Joan Learned, Guy Whitlow, and Dick Wayman for hundreds of patient hours on the word processor; T.J. Burnham for numerous excellent uncredited photographs; Hays Fisher and the photographers of Cooperative Extension at UC Davis; and many dozens of unnamed individuals and organizations around the state who supplied information, answered questions and made suggestions. Special thanks also to those University administrators who enabled the book to be done by giving vital encouragement and financial support—James Kendrick, Charles Hess, Loy Sammet, Glenn Hawkes, and Warren Johnston. Without the collaboration of such individuals the project could never have been completed. *A Guidebook to California Agriculture* is the product of all their efforts, and it has been a pleasure for me to be associated with so many fine contributors.

ANN FOLEY SCHEURING
Editor
University of California, Davis
January 8, 1982

Tables

Figures

An aerial view of the San Joaquin Valley south of Tracy shows Interstate 5 hugging the base of the Diablo Range. The Delta-Mendota Canal and the California Aqueduct carry irrigation water south through almond orchards in bloom. *Photo by Jack Clark.*

Introduction:
A Special Place

A relief map tells some of the story of California agriculture; a century and a half of western history tells the rest.

The United States as a nation possesses astonishing physical diversity, including some of the most fertile lands in the world. But there is something special about California. No other single state has such a startling range of subregions varying in climate, topography, soils, and native vegetation. California extends through 10 degrees of latitude, encompassing extremes of elevation, temperature, and natural precipitation. The state has over a thousand miles of rugged coastline along the Pacific Ocean; the immense undulating interior desert called the Mojave; the spectacular chain of the Sierra Nevada forming a backbone along much of its eastern border; a crumpled mass of high volcanic uplands to the northeast; the densely forested, moist Cascades and Klamath Mountains of the northwest; and, in its very heart, the great Central Valley, a giant, verdant hollow lying within circling rings of mountains.

Nature, though generally kind, also presents grim legacies from California's geologic development—rough terrain, threat of earthquake and volcanic eruption, arid desert, natural marshlands, the threat of wildfire and recurrent floods. Much of the state is still neither agricultural nor urban; maps show sparsely settled uncultivated areas in large tracts of desert, mountain, forest or brushland. Though 20th century transportation technology has minimized many of the physical barriers which separate California from the rest of the continent, vast spaces still remain as reminders of why the state was settled late and developed its own unique way of doing things.

In some ways the great Central Valley—450 miles long, averaging 50 miles wide—dominates most of California. From an airplane flying overhead in summer we see the colorful geometry of a hugely productive modern agriculture in sharp contrast with the brown and gold foothills. Broad, flat fields of meticulously groomed crops are dissected by meandering rivers and bullet-straight canals—both carrying water from the mountain uplands.

Here in the heartland is much of the wealth of the state, where human effort has developed what nature began eons ago. The Central Valley, once ebbing and flowing with seasonal life as the winter rains and melting mountain snows provided means, has been channeled and cultivated nearly year-round. Abundant sunshine and alluvial soils are the basis of a remarkable agricultural system. Yet even here there is surprising variation. The encircling ranges form a unique symbiosis with the valley. Water supplies and soil formation depend on the mountains; even the multitude of micro-climates and vegetative zones relate to the wind and precipitation patterns which have evolved around them. Though in most of the valley the growing season is nine or ten months long, there are pockets of variation in crop suitability, making its agriculture a complex mosaic rather than a broad homogeneity.

Other areas of California also contribute to a varied food and fiber system. Midway through the Central Valley, the major rivers flowing from north and south drain into the Delta and thence to San Francisco Bay and the Pacific Ocean. The Delta, home to great flocks of migratory waterfowl for centuries, now additionally supports a diversified agriculture on its leveed islands, though the fragility of its peat soils is underscored by frequent threats of flood. Coastal valleys supply many of the nation's specialty vegetables, premium wines, and exotic crops such as artichokes and avocados. Even the deserts of California, barren as they were once thought, now burgeon with irrigated crops. On the millions of acres of foothill rangeland, a healthy livestock industry survives, while the state's forests are managed to sustain a timber yield.

Commercial California agriculture sends farm commodities all over the world. The state supplies more than 250 agricultural products, some of them produced nowhere else in the nation and in few other places in the world. Introduced species of plants and animals from many origins have adapted and now thrive in the California environment. Though much of California's farm output comes from a relatively small acreage, less than 15 percent of the state, the blending of soil, water, weather and technology results in yields for some crops far above those of other areas. California is not the breadbasket of America—it lacks the enormous expanses of prairie soil that produce most of the nation's grain—but it is the salad bowl and fruitplate. Up to 25 percent of the nation's table food comes from California alone.

Land Developers and Irrigation

It was not always thus. California's astonishing productivity dates back only a few decades. Much of what we see today has taken place, in fact, since World War II—the span of less than two generations.

The history of California agriculture is the development of an enormously promising but challenging land by the restless sons and daughters of a pragmatic culture. Certainly the state's farm scene is a product of a unique historical past as well as a rare combination of topographical and climatic conditions. California was a mecca for dreamers and the dispossessed from the days of the Gold Rush onward. The state was open for settlement during the beginnings of an era of revolutionary technological change, and agricultural expansion took place in an opportunistic and nontraditional political and social climate.

California represents in microcosm the evolution of agriculture from pastoral grazing through extensive dryland grain farming into today's complex, intensive specialty cropping with irrigation. Even now all three types of agriculture are carried on, sometimes within the same counties. In most mountain and foothill areas livestock ranching often remains the major agricultural activity, while dryland farming is still found in the areas peripheral to valley floors. Depending on soil type and water availability, however, irrigation in the valleys has gradually brought more specialization in higher-value crops. In fact, modern California agriculture is largely the story of irrigation.

Irrigation is nearly as old as agriculture itself, but its scale in the American West was something new. Early schemes in California were often privately financed by large landholders, early ethnic or religious colonies, or land developers speculating on the future. The Wright Act of 1887 was a landmark piece of state legislation enabling landowners to form public irrigation districts with power to take over private companies, issue bonds and levy taxes for canals and water delivery systems. The Modesto and Turlock Irrigation Districts, tapping the Tuolumne River, were the first such successful public irrigation projects in California, with the Turlock main canal completed in 1901 and the Modesto in 1904. Other early irrigation districts, however, suffered from lack of design and management skills—many floundered and failed, embroiled in litigation, bankruptcy, and fraud. Nevertheless, the concept of water as a "public utility" had been established. Dozens of new districts were formed between 1910 and 1921. Later, the Colorado River Project, the Central Valley Project and the State Water Project invested enormous amounts of federal and state money in massive water supply systems, bringing, as well, centralized supervision and political controversy over allocations, water pricing, and other matters.

Though nonagricultural use has become increasingly important, 85 percent of California's developed water supply is currently used in agriculture. Through the early decades of the 20th century and continuing even today, irrigation projects have brought thousands of acres into new agricultural production, changing the face of the state. Vast acreages of former semiarid scrubland or grazing land now are planted to cotton, alfalfa, row crops, vineyards, and orchards.

Another Ingredient: Technology

None of this would have been possible without the combination of high technology with abundant natural resources. Perhaps more than anywhere else, California reflects the successes—and the problems—of manipulating nature to serve man's purposes. Moving the water to the place where it was needed was only one way the California farmer manipulated his environment. Labor shortages and the vastness of early California grain fields encouraged farmers and tinkerers to develop various kinds of machinery for plowing and harvesting. This early bent for mechanization has continued throughout the years, extending to more and more crops.

Other technological developments have also been quickly adopted by Californians. Much time and effort have gone into breeding of plant and animal varieties for California conditions. Chemicals have been widely used to protect crops from numerous pests that thrive in the state, while immense amounts of fertilizers have been applied to coax the yields necessary for profit. New cultural methods for a number of crops have been pioneered in California. Although some of these technologies have become increasingly controversial, their adoption has contributed to the volume and efficiency of production which is characteristic of the state's farming today.

Paralleling agricultural development over the years has been that of allied marketing and service industries. California agriculture has been a dynamic, changing activity partly as a result of the industrialization of the state. Banking, processing, and transportation have grown along with and supported greater agricultural intensification, while public education and research and government policy have likewise encouraged it. The present agriculture of California is a hugely complex system not only because of geographical factors but because of the social, economic, and political fabric in which it is embedded.

More Changes Ahead

California agriculture in the 1980s is—as it has always been—in a period of transition. Precisely because it reflects so directly the successes and problems of a technological society, it brings to focus the concerns of many over the effects of industrialization in agriculture. Trends which have been clear in American agriculture for a long time—the declining number of farms, their greater size, reduction in employment, consolidation of firms, increasing use of chemicals, the growth of specialized services, among others—are even more apparent in California. Many observers are uneasy about changes in the economic and social structure of agriculture. Rural-urban polarization appears clearly in California too, where 90 percent of the population is concentrated heavily along the coast, particularly in southern California. Urbanization has engulfed many of the small farms that used to ring the metropolitan areas, and many urbanites are very far removed, physically and psychologically, from the life of rural California. Increasing concern about environmental problems has brought much attention to agriculture as it affects the environment. Value conflicts come into sharper definition as interest groups plead their causes before government or in the media. Agricultural versus recreational or residential land use; water for cities or for crops; farm development versus soil conservation; dilemmas of pesticide use; the allocation and costs of energy; unemployment in the face of mechanization; the social and economic discrepancies between farm operators and laborers; the costs and benefits of agricultural research—these are some of the issues engendering controversy over California agriculture today. The 1980s and beyond will inevitably bring some profound changes, for agriculture is a dynamic system which has always responded to complex signals.

A UC Project

The purpose of this book is, primarily, to describe California agriculture as it has evolved since World War II. As a public service project of the University of California, it is intended for the public. Over 50 members of the UC faculty and staff have written about their specialties for a general audience, while several others from outside the University also have contributed their expertise.

The book has been designed to set the state's farm scene in overall perspective. It is divided into five major sections. The first describes human resources and lays out some of the statistical dimensions of California agriculture. The second describes natural resources and how they are used for agriculture. The third looks closely at the actual production and significance of most major food, fiber, and ornamental crops; a chapter on timber, rangeland and wildlife is also included. The fourth section—interestingly, the longest—focuses on the social, economic, and political infrastructure which has grown up around the production system. The fifth and last section takes an occasionally challenging look at directions and trends, identifying issues which will ultimately have to be resolved.

California is so complex and varied that it would take a much larger book than this to cover all its agricultural resources and practices exhaustively. Of necessity, many subjects here have been delineated with a broad brush, or barely mentioned. Uniformity of approach and style has not always been possible with the large number of contributing authors. Though statistical obsolescence is a danger, some use of numerical data has been unavoidable; in most cases, however, trends rather than static figures have been shown. Much of the material is a synthesis from many sources, but citations in the text have purposely been kept to a minimum. Sources of information are given at the ends of most chapters and readers should be able to find out more about the intricacies of California agriculture by consulting them.

Preparation of this volume has been a challenging but fascinating task. For many of the contributors, the necessity to condense huge amounts of material into reasonably concise and coherent form has been both frustrating and educational. But if readers can take away with them some sense of the richness of California agriculture, as well as a clearer understanding of its components, our work will be well rewarded.

ANN FOLEY SCHEURING
Editor
Davis, California

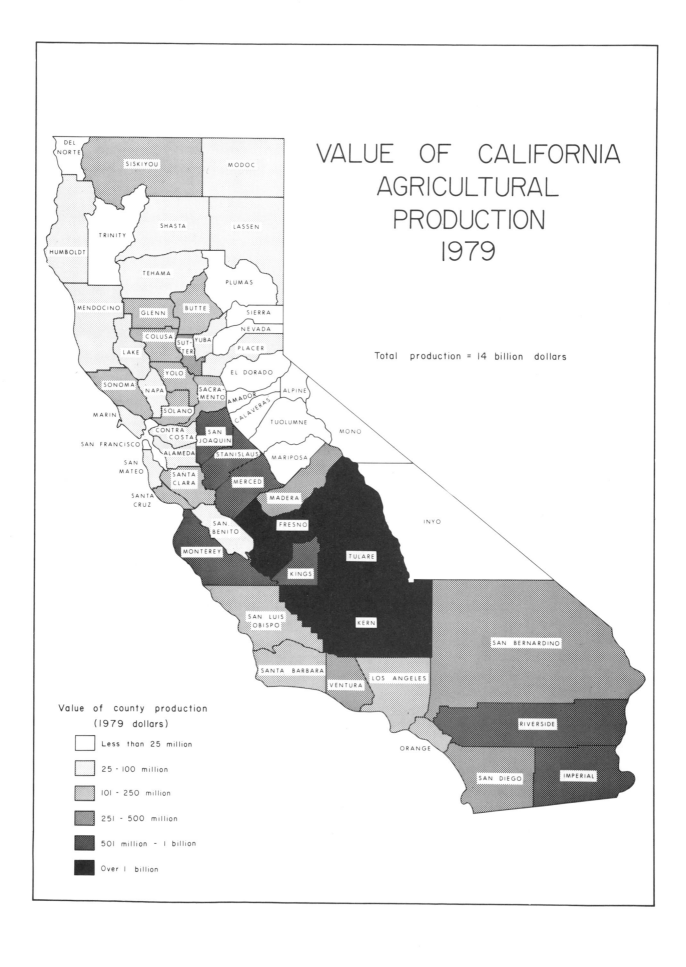

VALUE OF CALIFORNIA AGRICULTURAL PRODUCTION 1979

Total production = 14 billion dollars

Value of county production
(1979 dollars)

Less than 25 million

25 - 100 million

101 - 250 million

251 - 500 million

501 million - 1 billion

Over 1 billion

PART I 🌿 Agriculture
in California:
An Overview

A young farm couple checks conditions in their almond orchard in early spring. In the foreground are bee hives moved into the orchard for pollination of the blossoms.

1

California Agriculture: The Human Story

Carole Frank Nuckton, Refugio I. Rochin, and Ann Foley Scheuring

Though the climate and other natural resources of California have proven to be highly suitable for intensive agricultural development, it has been the combination of these resources with historical circumstance that has resulted in the state's food and fiber system. The scattered Indian tribes of prehistory were primitive people who gathered what nature offered but did little to change their environment. Although early European explorers saw forbidding difficulties, they recognized the bounty which the land might yield. The colonizing of California by Spanish missionaries and soldiers from Mexico began two centuries of increasingly intensive agricultural development, culminating in the massive publicly-built water projects and other technological wonders of the mid-20th century. Hundreds of thousands of persons, lured to this once-isolated coastal state by perceived opportunity, have played roles in the remarkable evolution of agriculture here. This chapter describes some of the human diversity in California farming. The following chapter presents a statistical view of the overall dimensions and structure of the system.

The California Farmer

It is nearly impossible to describe a "typical" California farmer. With well over 200 commodities being produced commercially in California, the term "farm" covers an astonishing array of operations. The term "farmer" can be used only loosely to designate farm operators who may be in fact distinctly different from each other in background and outlook. One California farmer harvests dates grown in the desert while another cuts hay in an alpine valley. The scope of operations ranges from enormous cattle feedlots,

where animal manure sales alone bring in over a million dollars annually, to a retired person's mini-orchard where supplemental income is earned by direct sales of fruit at a roadside stand. A Monterey area artichoke grower looking for harvest help has very different concerns than does a Mendocino sheepman battling coyote predation on his flock.

A few brief sketches give some idea of the variety of people engaged in California agriculture, and their many motives and methods:

- A ranch along the Sacramento River was established by a German-born grandfather in the 1890s. His son and grandsons now form a family corporation producing rice, sugarbeets, and walnuts on 1500 acres of rich alluvial soil.
- Although a middle-aged rancher of Irish descent owns nearly two thousand acres in the rugged northeastern part of the state, his wife supplements the income from their modest cow-calf operation with her in-town job.
- A young college graduate leasing 400 acres of valley land to grow wheat, beans, corn, and milo is now seriously considering pursuing a more lucrative occupation.
- A hard-working Japanese truck farmer in the Delta markets asparagus and other vegetables at several independent grocery and roadside enterprises.
- Although his first-generation Italian father ran a small Grade B dairy, a Stockton-area tomato grower has recently turned his attention toward real estate development. He now divides his time between supervision of farming activities and his financial transactions.

- A former research chemist, dissatisfied with urban life, has parlayed his modest family inheritance and his interest in fine wines into a small but excellent vineyard in the Napa Valley.
- Three brothers of Armenian ancestry running a fig and raisin operation near Fresno combine production with packing for some of their neighbors with smaller operations. They are looking for more land to expand the family holdings so that several sons can be taken into the farming corporation.
- A large-scale grower-shipper who sends lettuce, broccoli, carrots, and celery to the fresh market in the East manages 6,000 acres in the Salinas Valley. To reduce his risk, he contracts some fields to a local freezer-processor. To extend his shipping season through the winter, he leases acreage in southern California and Arizona.
- Formerly migrant workers, a Mexican family with six children now shares earnings with 11 other families on a 48-acre cooperative strawberry farm near Oxnard, initially funded by a federal economic opportunity grant.
- With a recent degree in agricultural economics, a young farm manager for a large land company in the southern San Joaquin Valley executes top management decisions, supervises field work, and keeps crop records. The company grows cotton, alfalfa, fruits, and nuts on several thousand acres, contracting its hay and by-products to a large Imperial County feedlot.
- In the Los Angeles basin, a greenhouse operator has a contract to supply potted plants to the largest grocery chain in the area.
- An airline pilot in his late 40s owns and operates a 20-acre avocado grove north of San Diego. The extra income currently helps with his children's college expenses; he plans to expand his farming venture when he retires from flying.

Perhaps in no other sector of our economy do the actors differ so from one another. Family background, degree of affluence, complexity of business arrangements, and attitudes of farm operators all vary, as does the length of their commitment to the land. Some farm primarily for the country life-style; others are business persons differing little from managers of nonfarm enterprises.

Having stressed the diversity in California farm operations, however, it must also be said that the popular image of the California "business farmer" has much basis in fact. Chapter 2 details census and economic data to develop an overall picture of California farms which varies significantly from national averages. Farms in California have tended to be larger, more specialized, more productive, more prosperous, more technologically developed (and dependent), more financially leveraged, and somewhat more likely to be corporately owned. A brief historical review is helpful in comprehending why the state's agriculture developed as it did.

Evolution of California Farming

Farming began on a relatively large scale in California and long-term economic forces have not substantially changed the initial pattern in spite of some recent social pressures to do so. In the eastern part of the nation, farms had to be carved from the forest wilderness and so by necessity started small. The vast open spaces of the west, however, and the typically semi-arid climate led to farming much larger acreages. In California several other factors converged to favor the development of agriculture as we know it today.

Between 1760 and 1822 the Spanish established 21 missions and four military settlements (presidios) along California's El Camino Real (now U.S. Highway 101). They also made 25 land grants ranging in acreage from 4,000 to 300,000 acres (Beck and Haase). During the Mexican occupation of California (1822–1848) another 428 landgrants were conferred, most between 5,000 and 25,000 acres. Thus, California's agricultural development was instituted on an extremely large scale. It was understood at the time of the Treaty of Guadalupe Hildalgo in 1848 that existing Mexican property rights would be guaranteed by the U.S. government, but in the confusion that ensued after statehood was conferred, extremely large tracts of land were sold to the wealthiest of Anglo settlers. Thus, some of the richest and most productive lands were immediately removed from the public domain and did not fall under the Pre-Emption Act of 1841 (later the Homestead Act of 1862). Much good quality land, therefore, was precluded from sale to settlers in traditional 160 acre parcels.

Furthermore, there were just not that many settlers around. The long distance from eastern population centers presented a formidable barrier to the early establishment of a network of family

Both business opportunity and social event, the annual California Woolgrowers' Association ram sale is the biggest in the West, bringing together farmers and ranchers from all parts of the state.

farms in California. The fact that one or two growing seasons had to be sacrificed just to get to California kept many would-be settlers back in the Middle West (Gregor).

Besides distance, another discouraging factor was the winter-wet, summer-dry climate. Accustomed to planting in the spring so that crops could flourish moistened by summer showers, many new migrants had great difficulty adapting to California's normal seasonal drought and accordingly sent discouraging reports back east. Those who arrived in midsummer were confirmed in their suspicion that the entire country west of the Missouri River was a great desert. The agriculture of early California consisted chiefly of winter grains harvested before the heat of summer, and rangeland cattle. Cattle hides, cured and shipped to eastern markets, were a major product of early ranchos.

By the mid-1860s most of the state's arable land belonged to only a few thousand people. Early speculators took advantage of many loopholes in federal law to amass large holdings from the public domain even when the land seemed less than desirable for agriculture. Ownership

patterns became firmly fixed in the decade between 1860 and 1870 so that even in the 1930s, only 516 landholders in California owned 8,685,439 acres of land (McWilliams). Sixteen controlled at least 84 square miles of land each. In Fresno County, 48 owners had more than 79,000 acres each. To this day, the largest landowner in the state, with over 20 million acres, is the Southern Pacific Railroad. This is because the federal government granted 12,800 acres (20 sections) of land from the public domain for every mile of track laid, in a checkerboard pattern alternating mile by mile, extending out on either side of the tracks. That checkerboard pattern can yet be clearly seen on federal public lands maps where railroad holdings alternate in some national forest and desert areas.

While landholdings were large, labor was scarce in early California. Mechanization of grain production after the Civil War coincided with the development of the West. California was particularly attractive for machine adoption since, although the long seasonal drought had its disadvantages, grains could do well under dryland cultivation, and there were few interruptions in machine operation during harvest.

By the late 1880s, eight to twenty teams of horses or mules were hauling combines as big as houses over the wheat fields of California....Still later steam traction engines began to replace the animals, and the combines grew even larger, some cutting a swath forty-two feet wide (Gregor, p. 88).

Wheat was the crop that first brought California into world trade since it could withstand the long journey in ocean-going vessels, and was profitable enough when farmed on large acreages. The decades between 1870 and 1900 were known as the bonanza wheat years. Huge amounts of grain were shipped out of California ports, particularly to England.

Towards the end of the 19th century, however, the character of California agriculture began to change. The international grain market became less profitable for California growers, while wheat yields began to diminish within the state as the land's natural fertility declined after years of cropping. Many incoming farmers began to experiment with new crops. As irrigation developed in local areas, it became possible to grow orchard, row, and vine crops—taking advantage of the long summer days while maintaining a dependable supply of moisture. The railroad and refrigerated cars meanwhile opened up eastern markets for California-grown perishables. Successful experiments with new varieties and new processing techniques further encouraged the development of California's now-famous fruit, grape, nut, and vegetable crops.[1]

The decades of the 20th century have seen a steady evolution in California agriculture, both in scope and in complexity, as other economic and social factors buttressed its expansion. Immigration, government policy, the formation of marketing associations, the increasing supply of credit, the development of allied supporting industries: all have contributed to a synergistic combination of circumstances encouraging growth.

Today's patterns are very different from those of a century ago. Much has changed even since World War II. One such pattern, the distinction between farm ownership and operation, is worth discussing here.

1. Much of the early history of these crops is contained in *California Agriculture*, edited by C. B. Hutchison (University of California Press, 1946). Chapters in Section III of this book detail further development since World War II.

Farm Owners and Farm Operators

Not all farm operators own their land; conversely, not all farm owners operate their farms. One national estimate suggests that of about 900 million acres in farmland in the U.S., currently about 60 percent is actually operated by owners (Schertz). The rest is operated (or "managed") by tenants or lessees. Although in American political ideology there has always been an emphasis on owner-operated farms, recent years have favored an evolution away from farming as a "way of life" to farming as a business venture. Given the demands of technologically sophisticated modern farming methods and low-margin returns on many crops, this has resulted in an increasing emphasis on management, and a growing distinction between owner and operator.

Investors and Absentee Landlords

Only about 3 percent of America's farmland changes hands each year, and most transfers are farmer to farmer. According to the USDA, purchases to expand farming operations accounted for 63 percent of farmland transfers in 1978. Thus, only a small portion of land is available for sale to nonagriculturalist investors; much of this may be upon the retirement or death of a farm operator. Investment by outsiders, nevertheless, has had, historically and currently, an impact on the structure of agriculture—perhaps more in California than elsewhere. (For a discussion of tax-shelter agricultural investments, see Chapter 20.)

Active investment in farmland by nonagriculturalists is only one reason, however, for the increasing growth of tenancy in U.S. agriculture. Another is ownership by widows of former operator-owners, by heirs who do not desire to farm, or by those who have retired from the operation side of farming. Like the investor, they seek someone else to manage their land holdings on a day-to-day basis. In addition, some of the extremely large land holdings in California are simply too large to be managed by one operator but are farmed instead by many, each on a leased farm.

It is generally true that there are many more ownership units than there are farm operators. In the nation as a whole, in 1978, some 2.3 million operators farmed 6.2 million ownership units (Boxley). Under the old "agricultural ladder" concept, tenancy was an important step

Irrigation water is the lifeblood of much of California agriculture. This irrigator wields a shovel as furrows fill with water in a row crop field.

from hired farm hand to owner-operator. Many of today's tenant farmers, however, no longer have the goal of full ownership; they simply use other people's land for their farming operations. Leasing can lower the barrier to entry erected by high land prices. Specialty crop agriculture also has contributed to the importance of tenancy in California farming. Specialists in tomatoes, for example, often move their crop to different parcels of land from year to year in order to control disease or take advantage of rotation crops that enhance fertility of soils. Tenancy may also be a device for risk sharing. (See Chapter 2 for more detail on farm tenure in California.)

Some Current Patterns of Farm Management
Whenever the owner-operator functions are separated in these or other situations, the management function may be performed either jointly by the owner and operator, by the operator solely, by a third party, or in a number of other different arrangements.

In most cases where the owner may still share some decision-making with the actual operator, the farming is done under lease. Leasing agreements usually have some provision about the crop or type of crop to be grown. Both share-crop and cash-rent arrangements are used on leased land; share-crop is probably more prevalent in field crops and orchards, cash rent in vegetables. For example, a grain grower who may lease ground from his neighbors in addition to that which he himself owns, may use his own equipment, arrange for his own labor needs, and make his own decisions on planting, cultivation, pest and weed control. At year's end he will split the gross income from his crop with the landlord, depending on local custom and the specific agreement. Another operator, perhaps in row crops, may simply agree to pay the landlord a fixed amount for each acre. His profit will then vary according to the year's expenses (including the rent), his yield, and market returns. Leases of these types are usually on an annual basis, subject to yearly renewal.

In California, where so many acres are devoted to perennial crops, a type of management has evolved that is quite different from that of leased annual crop acreage. Permanent crops, which must be tended with a long-range future in mind, do not lend themselves well to traditional leasing arrangements. An entrepreneur-

manager who owns his own expensive equipment and perhaps has his own permanent work crew may become a specialist in one or a few perennial crops. He then manages orchards, under contracts of some duration, for several landlords. San Diego's avocado orchards, for example, are often managed in this manner. Another case is that of citrus cooperative associations, which for years have arranged production details for their members. Other crops in many areas of the state are also managed under such contracts.

There is also a wide spectrum of "third party" management firms, ranging from small one-man businesses to large organizations employing dozens of people. These firms generally assume total responsibility for the farms which are contracted to them. They may obtain a tenant for the operation—perhaps a neighboring farmer—who will then individually supervise cropping programs, management practices, marketing, and other matters which might affect the landlord; or they may hire a resident manager to operate under the firm's direct supervision. Management firms may also serve as consultants, giving advice on specific farming problems, or may perform rural appraisal services. Some trust departments in banks may perform some broker-management functions, although this is less prevalent in California than elsewhere.

Corporate landholdings such as those of Standard Oil, Getty Oil, Tenneco, and Southern Pacific are often leased out for cash rent to individual operators who then make their own management decisions. Other large land companies hire their own management staffs who make all the main decisions including the hiring and firing of employees, as in any other large business. Smaller, but still large, landowners may employ a resident manager on a salaried basis. Such a person must know techniques of recruiting, supervising, and training personnel, negotiating marketing and labor contracts, accounting, and other business management skills.

California farm managers, like California farmers, operate under a wide variety of conditions and arrangements, but it is their tens of thousands of individual decisions which shape the face of agriculture as we know it today. Besides farm owners and farm managers, however, the human side of farming includes the laborers who work in the fields. Farm workers in California have a particularly interesting history, some of which is recounted here.

California Farm Workers

A Brief Social History

At the time when California farming was beginning to undergo a major metamorphosis and needed a large pool of low-cost labor, some 10,000 Chinese laborers became available, having completed work on the transcontinental railroad. Fifty thousand other Chinese were in California working in mines and in light manufacturing in the cities. White worker competition drove many of these to join the ex-railroad workers in farm work. Both on the railroad and in the mines the Chinese had been accustomed to working in gangs under contract, a system which was readily adopted by growers. By the 1880s more than half of California's farm labor workforce was Chinese.

Most of the Chinese had been tenant farmers in their home country and were highly skilled. They taught their employers the intricacies of fruit and vegetable growing, and did much of the work of building levees, reclaiming marshlands, and developing irrigation systems that opened large new areas of the state to farming.... Growers were so pleased with their labor that they imported thousands more workers directly from China, with the help of merchants and others who also profited from the immigration. (Meister and Loftis, pp. 6–7)

In spite of fruit and vegetable growers' satisfaction with the performance of Chinese workers, political agitation about "cheap" Chinese labor from small growers, white unionists, and small manufacturers resulted in 1882 in the Exclusion Act, prohibiting further immigration of Chinese labor. Another law passed in 1894 during an economic depression was intended to force the deportation of many Chinese, but was largely ignored by the growers. During this period, bands of unemployed white workers actually invaded farm labor camps forcing Chinese to flee to the cities where many took jobs unwanted by white workers and established their own separate communities. From 1900 on there was a rapid decline in the importance of Chinese in California agriculture as their numbers dwindled to a relatively insignificant portion of the farm labor workforce.

Concerned about their labor supply, growers talked of importing Portuguese, Italians, Mexicans, blacks from the South or of introducing more capital-intensive, labor-saving farming methods. In 1885, Japanese emigration had been legalized by that government, and in time it was the Japanese who filled the labor vacuum left by the Chinese.

At first, during the difficult economic times of the 1890s the Japanese resorted to wage cutting in order to get established in farm work. They were well accepted by growers because they, like the Chinese, organized themselves into hard-working gangs, provided their own food and housing, and lived apart in their own communities (Fuller). After 1900, the employment of Japanese increased rapidly until they represented a majority of the workforce in labor-intensive crops.

The story of the Japanese in California agriculture, however, did not end there. Once established in agricultural employment, many industrious and ambitious workers began taking advantage of the labor contract system to move up and out of the seasonal farm worker class. The beginnings of farm worker trade unionism can be traced to their collective bargaining tactics such as "quickie" strikes when fruit was ripe; the refusal to scab against fellow nationals; the provision of fewer men than needed, to drive wages up; and boycotts of certain growers. Growers found that leasing land to Japanese on a share-crop system was more advantageous to both parties than labor contracts which could not be counted on when most urgently needed. By 1910, one-fifth of the Japanese in California were farming their own land, as owners, lessees or share-croppers (Meister and Loftis). By 1920, the Japanese owned 74,769 acres of farmland and leased another 383,287 acres. They controlled 91 percent of the berry acreage, 81 percent of the onions, 65 percent of the asparagus, 58 percent of the green vegetables, 53 percent of the celery, and substantial portions of the acreage in several other commodities (Nagai).

Japanese farm workers worked intensively for Japanese growers. Ethnic cooperation paid off handsomely until they were subjected to even stronger discrimination than that endured by the Chinese. The build-up of anti-Japanese pressure resulted in the state Alien Land Law of 1913 preventing immigrant Japanese from owning or leasing land more than three years—a law which was strengthened in 1920. In 1924, further Japanese immigration was cut off by national legislation. By 1930, the Japanese had lost their position of dominance in the California hired farm labor force, although they continued as tenants on farms vested in the names of others.

Under the pressure of real or threatened labor scarcity, other foreign farm labor groups were welcomed by California growers. The Hindustanis (Pakistanis or Indians), and smaller numbers of Greeks, Armenians, Portuguese, Italians, and other Europeans were hired for farm work. Hindustanis were subjected to much of the same discrimination applied to Orientals in general. The Europeans were not long in becoming farmers themselves or in migrating to the cities. The Portuguese made important contributions in dairying; the Italians in viticulture; the Armenians in marketing. Between 1920 and 1930 the number of Filipinos in California increased tenfold from 2,700 to 30,500. These immigrants were predominantly young males, experienced in agriculture; they, like the Chinese and Japanese, worked in a gang under a contracting boss.

Before 1900, relatively few Mexicans came to work in the United States, but their numbers increased substantially between 1900 and the Great Depression of the 1930s. A revolution in Mexico had created economic depression and unemployment there; World War I cut off immigration from Europe; and representatives of growers' associations and industrial companies actively recruited in Mexico for workers (Hoffman). For these and other reasons, in the 1920s the Mexican population in California tripled from 121,000 in 1920 to 368,000 in 1930 (Fisher).

The Depression and the "dust bowl" drought conditions in Oklahoma and Arkansas brought an influx of poor whites to California, reducing employment opportunities for Mexicans in California and discouraging further immigration. Unemployed Mexicans in the United States meant an additional burden on already strained social agencies and pressure was brought upon the Mexicans to return to their home country. By the end of the 1930s over 400,000 Mexican aliens had been repatriated to Mexico, leaving the farmlands to be worked largely by white "refugees" (Hoffman).

The Bracero Program

At the beginning of World War II when national employment rates rose, absorbing most domestic unemployed, urgent demands were voiced by growers for the renewed importation of Mexican farm workers. Among the first petitioners were sugarbeet growers who sought approval from the U.S. Department of Agriculture to recruit contract workers in Mexico. After Mexico declared war on the Axis powers in June 1942, the provision of contract Mexican agricultural labor was considered part of the war effort. Soon the workers came to be known as *braceros*, which translated means "arms"—those who do hand labor. Although at the conclusion of the war there was some talk of ending the *bracero* program, both agriculturalists and workers liked the arrangement and wanted it continued. Thus, in 1950, Public Law 78 was passed giving the Secretary of Labor the authority to recruit Mexican workers for employment, including those in the U.S. illegally, provided they had been here five years.

In time, the costs of the *bracero* program began to be resented, and it was claimed that a relatively few large growers—those who could provide labor camps, transportation, and other facilities—benefited most. Furthermore, once again the claim was made that the presence of foreign workers was detrimental to domestic ones, depressing wages and taking away jobs (Galarza). Under pressure from organized labor, P.L. 78 was officially ended on December 31, 1964, but the *bracero* program continued until August 1968 under emergency provision of the Immigration and Nationality Act, Public Law 414. In its 26 years of existence, from 1942 to 1968, nearly five million *braceros* had been contracted.

Legal and Illegal Foreign Farm Workers

It has frequently been stated that the termination of the *bracero* program would have spelled disaster for the California processing tomato industry had it not been for the fortuitous development of the tomato harvester and a variety of tomato capable of being mechanically harvested. Controversial as this may be, masses of ex-*braceros* returned to California, legally or illegally, and continue coming to this day.

For those wishing to enter legally there are provisions of "green cards" (officially called the U.S. Alien Registration Receipt Card), entitling the holder to immigrant status. Valid indef-

Though mechanization has reduced many labor needs, cherries and some other fruit crops are still harvested by hand. This picking crew has a woman crew leader.

initely, the "green card" permits the bearer to reside, travel, and work anywhere in the U.S. and to exit and reenter the country without loss of these rights, provided the holder does not stay away longer than a year. Many Mexican holders of such cards, perhaps as many as 60,000, live in Mexico near the border and commute to the U.S. to work. The issuance of "green cards" by the Immigration and Naturalization Service (INS), however, is a slow, selective process, reaching relatively few of the Mexican workers wanting employment across the border. Hence, the great bulk of Mexican workers return to the U. S. surreptitiously.

The INS has estimated the number of Mexicans living illegally today in the U.S. to be between eight and twelve million. More conservative recent surveys place the number between three to six million, given that many return to Mexico during off-season. Because entries are illegal there is no way to obtain accurate data

on the phenomenon, but there may be as many as three million illegal border crossings in a single year. In 1979, according to the INS, 998,761 illegal entrants were returned to Mexico, a number which may be inflated by repeaters who return and then try again. Many illegals are employed in nonagricultural jobs such as in the garment industry, but, nevertheless, a very large labor pool is still available to California agriculture today, as in the past.

Seasonal Labor and Mechanization

Much of the labor on California's farms is highly seasonal. Herein lie many of the problems confronting workers, such as difficulties in maintaining a steady income stream, in finding permanent employment, and in having assurance of employment season to season. At peak-season, employment in Fresno County is more than ten times the employment in the off-season.

Only 15 percent of all farm workers can be considered year-round employees of a single farm. Many workers are locked into seasonal farm work because of serious obstacles between them and alternative employment: inability to speak English, low educational attainment, lack of money, lack of housing for large families in areas where employment may be available, and finally lack of jobs for people with these limitations (Sosnick).

The existing size and structure of farming in California and the labor-intensive nature of many crops also has resulted in dilemmas for growers. Huge seasonal peak labor supplies generally cannot be satisfied with local residents. Rather, the major part of supply must come from a migrant stream of workers, over half of whom are interstate transients whose employment may range in distance from California to as far away as Texas. If the stream is interrupted or stopped, farmers may lose their crops. Alternatives to such dependence on the labor stream are to mechanize, which may be possible in some crops but not in others, or to shift to less labor-intensive crops.

The decision to plant other crops may mean the loss of the comparative advantage that California has achieved in the current crop mix. Growers have tended to adopt labor-displacing technology as it is developed, not only because it may save on wages but also because management problems and uncertainty are reduced. Mechanization research done by the University of California, however, has been challenged in court by California Rural Legal Assistance for its alleged impact on farm employment. It is difficult to determine the real impact of mechanization, however, for as demand for California specialty crops has strengthened in world and domestic markets, new acreages have been planted, employment on some California farms has increased, and the total effect of labor-saving machinery in some crops has been obscured.

The economic situation of California's farm workers has gradually improved over the years as agricultural labor has become better organized and more specialized in skills; protective legislation and social welfare programs have also contributed toward the improvement of farm labor working and living conditions. (See Chapters 12, 19 and 20 for information on farm worker unionization and state and federal government programs and agencies dealing with farm labor.)

The Other Side of Agriculture

At the production level, over 600,000 individuals work on California farms in the course of a year. Yet these farmers, managers, and farm workers make up but one part of the state's vast food and fiber system. Behind them stand those in farm machinery manufacture and sales, agricultural chemicals, seeds, custom services, banking and finance, and University research and extension. After the harvest comes the participation of those in transportation, processing, packing, and marketing. Thousands of government officials are employed in the protection and regulation of agriculture, while engineers and workers construct and maintain water projects essential to farming in this semiarid state. Finally, all of us as consumers are an integral part of the system; as a recent bumper sticker declaimed, "Agriculture is everybody's bread and butter." Subsequent chapters will explore some of these other components.

References

Beck, W.A. and Haase,Y.D. *Historical Atlas of California*. Norman, Oklahoma: University of Oklahoma Press, 1974.

Fisher, L.H. *The Harvest Labor Market in California*. Cambridge: Harvard University Press, 1953.

Fuller, V. *The Supply of Agricultural Labor as a Factor in the Evolution of Farm Organizations in California*. Unpublished Ph.D. dissertation, University of California, Berkeley, 1939. Reprinted in Senate Hearings, Subcommittee of Education and Labor Committee, 76th Congress, 3rd Session, 1940, Part 54, commonly referred to as the LaFollette Committee Hearings.

Galarza, E. *Farm Workers and Agri-Business in California, 1947–1960*, University of Notre Dame Press, 1977.

Gregor, H.F. "The Industrial Farm as a Western Institution." *Journal of the West* 9(January 1970):78–92.

Hoffman, A. *Unwanted Mexican Americans in the Great Depression: Repatriation Pressures, 1929–1939*. Tucson: University of Arizona Press, 1974.

McWilliams, C. *Factories in the Fields, The Story of Migratory Farm Labor in California*. 1935. Reprinted by Peregrine Smith, Inc., Layton, Utah, 1971.

Meister, D. and Loftis, A. *A Long Time Coming: The Struggle to Unionize America's Farmworkers*. New York: The MacMillan Publishing Company, 1977.

Nagai, N. "Japanese in California Agriculture." Unpublished paper, University of California, Davis, 1977.

Schertz, L.P. et al. *Another Revolution in Farming?*. U. S. Department of Agriculture, Economics, Statistics, and Cooperative Services, Report No. 441, December 1979.

Sosnick, S.H. *Hired Hands: Seasonal Workers in the United States*. Santa Barbara: McNally and Loftin, 1978.

Reminiscent of the state's earliest patterns of agriculture, sheep are silhouetted against the sky as they browse the stubble of a dryland grain farm in the Sacramento Valley. *Photo by Jack Clark.*

2

The Dimensions of California Agriculture

C. O. McCorkle and Carole Frank Nuckton

The previous chapter gives a historical and human perspective on California agriculture, emphasizing the contributions of ethnic groups as well as the evolution of ownership and management arrangements. To help the reader gain a sense of the overall scope and structure of the agricultural economy, this chapter presents selected statistics bearing on regional variations within the state. Additionally, California is contrasted in a number of ways with other states.

The scope and complexity of California's agriculture mirrors the richness and variability of its natural and human resources. Early ranchers and settlers were challenged to adapt their livestock and farming systems to the highly seasonal pattern of winter rain and summer drought, but they early saw opportunities to modify as well as adapt to the environment. California's challenge over the years has been to structure an agriculture—with its supporting processing, marketing, financing and other supply industries—that could take full advantage of the potential offered by the unique combination of resources within the state.

It is the purpose of this chapter to review the complex nature and the broad dimensions of California's agriculture as it exists today. Beginning with an overview of the major production regions, highlighting their characteristics and differences, the discussion moves on to productivity and input use. A short examination of financial characteristics is followed by a look into the nature of ownership and organization of California's farms and ranches.

Regional Distinctions

Agricultural production possibilities in any of the various regions or provinces of the state are determined primarily by climate, soils, and water conditions. To these are added various economic factors such as availability and cost of inputs, transportation costs to market, and market prices. These major determinants influence yields and interseasonal variation in yields and quality, as well as production practices, costs, and choice of enterprises. Among regions there is extensive variability in soils, climate, and water conditions, as detailed in Chapters 3 and 4. Agriculturalists have learned to take full advantage of these differences in developing cropping and livestock systems, locating enterprises, and developing farm businesses which may involve several geographically dispersed tracts of land. They have also learned to modify climatic influences through such means as planting tree rows to serve as windbreaks, providing heat or wind machines to protect against freezing temperatures at critical times in the growing season, using plastic mulches to raise soil temperatures as well as control weeds, and constructing greenhouses for total climate control.

Irrigation in the state is the greatest modification of regional climatic conditions. Movement of water hundreds of miles from areas of relatively high rainfall to areas of low, and development of deep wells to reach underground reservoirs, have changed the size and shape of California agriculture completely. Highly developed techniques of drainage, land leveling, and soil management in some areas are among the most refined in the world today. The relatively flat terrain in most cropped areas, the ability to control soil moisture, and the specialized agricultural production that has developed have encouraged mechanization and, in turn, the expansion of farm size. The state's unique environment has served as a laboratory for sophisticated research in plant genetics and nutrition, adding further to California productivity.

In viewing California's agriculture, it is customary to divide the state into six production regions. A rough approximation of these six regions is shown in Figure 1 following county lines. Counties included in each region are listed in Table 1. Even this division masks important variations within regions, however, as will become apparent. There are slightly more than 100 million acres of land in California, roughly one-third of which is in farms and ranches. Of this, less than 12 million acres are cropped and approximately 8.6 million are irrigated (Table 2). These data indicate the vastness of California's mountain and desert regions which account for the remaining two-thirds of the land area of the state.

North Coast and Mountain Regions

The North Coast and Mountain regions comprise a little more than one-third of the state, but less than 6 million acres are in farms and under 1 million are cropped. These regions contain less than 10 percent of the cropland and approximately 7 percent of the irrigated acreage. Farm numbers are relatively small, and average size of farm is the highest of any region—more than 800 acres. The percentage of land cropped per farm is 10.6 and 19.1 in the North Coast and Mountain regions respectively, reflecting the rough character of much of the terrain. Not surprisingly, sales per farm and per acre are the lowest by far of any of the six regions, given the less advantageous climate, soil, and topography.

Cattle and sheep operations dominate agriculture in these two regions, utilizing a combination of owned land, a portion of which is typically devoted to hay or irrigated pasture production, and leased public rangeland. Some highly productive farming areas are found, such as the Tulelake district in the northeast part of the Mountain region and the grape growing areas of Mendocino County in the North Coast region. Some dairying is found on the coastal plains of Humboldt County, benefiting from

TABLE 2.1
Production Regions of California*

NORTH COAST	SACRAMENTO VALLEY	MOUNTAIN	SOUTHERN CALIFORNIA
Del Norte	Tehama	Siskiyou	Santa Barbara
Humboldt	Glenn	Modoc	Ventura
Mendocino	Butte	Trinity	Los Angeles
	Colusa	Shasta	Orange
CENTRAL COAST	Sutter	Lassen	San Bernardino
Sonoma	Yuba	Plumas	Riverside
Lake	Yolo	Sierra	Imperial
Napa	Solano	Nevada	San Diego
Marin	Sacramento	Placer	
Contra Costa		El Dorado	
Alameda	SAN JOAQUIN VALLEY	Amador	
San Mateo	San Joaquin	Calaveras	
Santa Clara	Stanislaus	Tuolumne	
Santa Cruz	Merced	Mariposa	
San Benito	Madera	Mono	
Monterey	Fresno	Inyo	
San Luis Obispo	Kings		
	Tulare		
	Kern		

*San Francisco County is in the Central Coast Region and Alpine County is in the Mountain Region; neither county is included in the Census of Agriculture.

FIGURE 2.1 Production regions of California. Regions are divided along county lines for broad purposes of statistical comparison. These are the usual divisions of the Crop and Livestock Reporting Service, with the exception of the mountain region which here combines three smaller ones.

proximity to the ocean, high rainfall, productive native pastures, and level terrain.

Central Coast Region

The Central Coast region consists of a number of highly productive valleys lying between predominantly north-south mountain ridges of the Coast Range. The climate in these areas is modified by the coastal influence and the soils are typically fertile and easily tilled. The region contains approximately 10 percent of the land area in California, 17 percent of the land in farms, 12 percent of the cropland and 13 percent of the farms. Only one-third of the cropland in this re-

gion is irrigated, indicating substantial acreages of dryland grain and nonirrigated orchards and vineyards. The noncropped land in farms is usually grazed by livestock, including dairy cattle, beef cattle, and sheep. In recent years substantial increases in irrigated acres have accompanied the expansion of vineyards and other high-value crops, particularly in the more southern counties. The Napa Valley, famous for its fine wine grapes, and the Salinas Valley, long-known for its vegetable crops, are in this region. Of the six regions, the farms of the Central Coast have the third highest value of farm products sold per farm.

TABLE 2.2

Farming Characteristics of Six Production Regions of California

		North Coast	Mountain	Central Coast	Sacramento Valley	San Joaquin Valley	Southern California	Total for California[a]
Land area	1000 acres	5,187	30,716	10,067	7,143	17,506	28,957	100,069
Land in farms	1000 acres	1,451	4,341	5,673	4,574	11,055	5,626	33,130
Percentage of total land in farms	percent	28.0	14.1	56.4	64.0	63.1	19.4	33.1
Number of farms	number	1,697	5,374	11,821	9,657	28,142	16,418	81,706
Average size farm	acres	855	808	480	474	393	343	405
Cropland	1000 acres	153	829	1,422	2,231	5,394	1,424	11,721
Percentage of land in farms that is cropland	percent	10.6	19.1	25.1	48.8	48.8	25.3	35.4
Irrigated land	1000 acres	47	555	472	1,588	4,819	1,021	8,604
Percentage of land in farms that is irrigated	percent	3.3	12.8	8.2	34.7	43.6	18.2	26.0
Average per acre value of farm products sold[b]	dollars	47	40	234	216	380	446	281
Average per farm value of farm products sold	dollars	40,134	32,705	112,152	102,457	149,359	152,938	114,121

[a] Land area totals do not add since Alpine and San Francisco counties are excluded from the County Census of Agriculture reports. Other totals do not add since county data are based on the mail survey alone, whereas the state figures are supplemented by an estimate of farms missed by the mail survey.

[b] Total gross sales value of farm products sold, divided by land in farms.

SOURCE: U.S. Bureau of the Census, *Census of Agriculture, 1978.*

The Central Valley

The rich Central Valley of California is the agricultural heart of the state. Nearly half of the state's farmland, two-thirds of the cropland, and almost 75 percent of the irrigated land is found in the Sacramento Valley, the Delta and the San Joaquin Valley which together comprise the Central Valley.

THE SACRAMENTO VALLEY. With its cooler winters and higher rainfall, the Sacramento Valley produces small grain crops and excellent seasonal grazing on its nonirrigated acreage. Rice is the predominant irrigated crop in the areas of relatively inexpensive water and impervious soils. A variety of fruit and nut crops are produced on the deeper, better-drained and more fertile soils, while row crops such as tomatoes, beans, corn, sugar beets, milo, and sunflowers are grown extensively. Alfalfa and a variety of crops for seed are also found. Where air drainage modifies winter climates, some citrus fruits are produced. The foothills of the Sacramento Valley provide seasonal grazing for sheep and cattle, and irrigated pastures are developed on some of the poorer upland soils where water can be provided at reasonable cost.

The region is drained by the Sacramento River and its tributaries. Impounding water in this region provides for much of the irrigation not only in the Sacramento Valley, but also in the San Joaquin and Southern California regions which are heavily dependent on water from the north for multiple agricultural, domestic and industrial uses.

THE DELTA. The confluence of the San Joaquin and Sacramento rivers is located in a most interesting agricultural area known as the Delta. Drainage rather than irrigation is the primary challenge here, since much of the area lies below the surrounding waterways. A series of man-made levees keeps the water from invading large tracts of highly fertile peat soils. However, these "islands" continue to subside as the natural organic matter in the soil is oxidized and eroded by wind. The climate, available moisture, and highly productive soils support a rich but uncertain agriculture in the Delta, calling for exceptional management skills and a willingness to assume substantial risks.

THE SAN JOAQUIN VALLEY. The southern portion of the great Central Valley is the most extensive and productive of the agricultural regions of California. A third of the land in farms in the state is in this valley and more than a third of the farms. Nearly half of the cropland and more than half of the irrigated acreage in California lie in this region. Without irrigation most of the San Joaquin Valley would be little more than a desert—but water, together with a long warm growing season and reasonably good agricultural soils, has turned it into the state's agricultural showcase. The proportion of land in farms that is irrigated—44 percent—is higher in this region than in any other. There are few crops or livestock known to man (with the exception of those that grow only in tropical climates) that cannot thrive at some location in the valley, and most will yield more abundantly in this region. Farms here on the average are smaller (393 acres) than in other regions, yet the average per farm value of farm products sold is far higher ($149,359) than in any of the regions to the north, about equal with the southern region, and 30 percent greater than the state average ($114,121).

Many high-value crops are grown here. Deciduous tree fruits and nuts, grapes and citrus are major cash crops, in addition to cotton, alfalfa and a broad spectrum of vegetable and other field crops. Dairying is important throughout the valley, as are poultry enterprises. Beef cattle and sheep production is also carried on in the foothills, on irrigated pasture developed on the shallower and bench soils, and on some of the heavier bottom soils in the valley trough.

Agriculture on the east side of the valley and along the San Joaquin River developed early largely because of the availability of water and the excellent soil quality. As deep well turbine pumping developed, cultivation spread to other parts of the valley. Finally, when the State Water Project brought water to the parched western reaches, the San Joaquin's development as an agricultural colossus was complete. The size of holdings and cropping patterns reflect this history. The older and intensively cultivated farms to the east are smaller and concentrate on tree fruit, nut, grape, and other high value crop production. The farms to the west are much larger and have concentrated on irrigated field crops, early vegetables and, more recently, various types of nuts and some deciduous fruits. Much of the west side of the valley has been held for a long time by very large corporations which acquired land early in conjunction with development of oil resources and transportation in the west.

Thousands of acres in southern California produce citrus nearly year-round. Hand harvesters empty fruit from picking bags into bins which will be picked up with a forklift. *Photo courtesy of Sunkist Growers.*

Southern California

As an agricultural area, Southern California is the most difficult to describe—because of the enormous variation in growing conditions and the complex interactions between agricultural and urban development. The region encompasses nearly 30 percent of the state, but less than 20 percent of the area is in farms. Farms are smaller in size on the average than in other parts of California, but the average value of farm products sold per acre and per farm exceed all other regions.

Along the south coast where the climate is strongly moderated by the Pacific Ocean, a number of high value crops such as lemons, avocados, vegetables and flowers are grown. In the hot interior valley, including the distinctive and highly productive Coachella and Imperial Valleys which lie to the east of the coastal mountains, citrus, dates, winter vegetables, cotton, small grains and alfalfa predominate. Irrigation is essential to production in this region. Farms here are generally larger than along the coast, given the enterprises which are adapted to the more severe climatic conditions of the interior. Livestock feeding is an important industry in the Imperial Valley and also along the Colorado River in the easternmost portion of the region, and some large poultry and egg operations are found in the higher desert areas. Drylot dairying is extensive and important east of Los Angeles. The latter industry developed in response to a large local demand for fluid milk; feed supplies were readily available from surrounding areas, including the Imperial Valley and the southern San Joaquin. Both dairying and the once extensive citrus industry in the Los Angeles basin, however, have been largely displaced and forced to relocate by expanding urban areas. Both industries continue to move northward into the San Joaquin Valley where production conditions are highly favorable and the pressures of urban growth are far less.

Statewide Scope

Having emphasized the regional diversity in California's agriculture, some comments on the nature of agriculture in the state as a whole can be offered with less danger of reading too much into the overall analysis.

Characteristics of California Farms

Average farm sizes in California have increased steadily for the past 40 years in keeping with U.S. trends, though the percentage increase is less since the average size farm in California has always been greater than U.S. averages (Table 3).

TABLE 2.3
Farm Numbers, Size; U.S. and California[a]

Year	United States		California	
	NUMBER OF FARMS	AVERAGE SIZE (ACRES)	NUMBER OF FARMS	AVERAGE SIZE (ACRES)
1940	6,102,417	175	132,658	230
1945	5,859,169[b]	195	138,917	252
1950	5,388,437	216	137,168	267
1954	4,782,416[b]	242	123,075	307
1959	3,710,503	303	99,274	372
1964	3,154,857	352	80,852	458
1969	2,730,250	389	77,875	454
1974	2,314,013	440	67,674	493
1978[c]	2,257,775	449	73,194	447

[a] All the figures except those for 1974 and 1978 reflect the 1959 definition of farm as a place with at least 10 acres having gross sales of at least $50 or if less than 10 acres, gross sales of at least $250. The 1974 definition dropped the acreage requirement entirely but raised the sales criterion to at least $1,000. Such definitional changes make year to year comparisons difficult.

[b] Excludes Hawaii and Alaska.

[c] In order to make the 1978 figures comparable to those for 1974, the estimated number of farms missed by the 1978 mail survey—220,687 for the U.S. and 8,512 for California—was subtracted from the totals reported for 1978 since the 1974 numbers were derived from a mail survey alone. The average size farm reported was also recalculated accordingly.

SOURCE: U.S. Bureau of the Census, *Census of Agriculture, 1969, 1974,* and *1978.*

At least three contributing factors are: (1) original settlement of much of the state in large Spanish land grants rather than by homesteadig, (2) extensive corporate land holdings originally for other than agricultural purposes, and (3) the large acreages required to provide a livelihood when the terrain was unsuited to irrigation or water was not available.

It is well known that the average increase in size has been accompanied by a decrease in number of farms, both in California and in the United States. Between 1940 and 1978 the number of farms in California dropped by 50 percent; in the U.S. the decline was over 60 percent. Land holdings have continued to be consolidated throughout this period. (Recent data on changes in farm size and numbers suggest a reversal of long time trends for California. There is no doubt an increase in small farms, as many persons have chosen to live in more rural areas and to engage in farming on a small scale, either as a way of life or as a supplement to urban employment. Recent changes in counting methods for census purposes have led to the inclusion of large numbers of farms which would not have been counted earlier. Also, inflation has meant that more farms now meet the census definition of a farm: a place with at least $1,000 of agricultural commodity sales.)

While roughly one-third of California's land is in farms and ranches, nationally the proportion approaches one-half (Table 4). Less than 12 percent of California's acreage is farmed cropland, in contrast to the U.S. figure of over 20 percent. Harvested cropland in California is slightly higher than the national average—76 percent as contrasted to 70 percent. The sharpest differences are, not surprisingly, in the proportion of land that is irrigated. Twenty-six percent of the land in farms in California is irrigated, compared to

TABLE 2.4
National and State Comparisons in Agricultural Land Use (1,000 acres, except for percentages)

	United States	California
Total land area (acres)	2,263,591	100,069
Land in farms and ranches	1,029,695	33,130
Percent of total land area in farms and ranches	45.5	33.1
Total cropland, 1,000 acres	461,341	11,721
Percent of total land area that is cropland	20.4	11.7
Percent of land in farms and ranches that is cropland	44.8	35.4
Harvested cropland	320,666	8,899
Percentage of cropland that is harvested	69.5	75.9
Woodland including woodland pasture	94,892	1,379
Percent of land in farms that is woodland and woodland pasture	9.2	4.2
Other pasture and rangeland	436,729	18,831
Percent of total land area in pasture and rangeland	19.3	18.8
Percent of land in farms and ranches that is pasture and rangeland	42.4	56.8
All other land in farms[a]	36,733	1,199
Percent of land in farms	3.6	3.6
Irrigated land, includes irrigated pasture	50,838	8,604
Percent of total land area that is irrigated	2.2	8.6
Percent of land in farms that is irrigated	4.9	26.0

[a] Land in houselots, ponds, roads, wasteland, etc.
SOURCE: U.S. Bureau of the Census, Census of Agriculture, 1978.

Truck farming is still viable on the fringes of some urban areas, as here along the river near Sacramento. Many vegetables are raised in sequence for sale to local fresh markets.

less than 5 percent for the nation. (The national trend is for more acres to receive supplemental water when rainfall is inadequate, particularly east of the Mississippi River; hence, the low national figure is expected to rise signficantly in the next decade.)

Ranking of Commodities

Because of the extensive acreage and high production of the many commercial crops in California, one is tempted to conclude that crop farming dominates the agricultural picture. In terms of farm value of sales, tree crops, vegetables, and field crops account for roughly two-thirds each year while livestock and livestock products account for the remaining third. In value of marketings by commodities, cattle and calves,[1] and milk and cream, rank first and second in California, with a combined value in 1980 of $3.2 billion (Table 5). Cotton and grapes, which follow, had a combined value of $2.6 billion. The extensive range and pasture acreages in California provide a base for cow-calf and feeder cattle operations which supply some of the animals fed in the many feedlots in the state. California's nearly 23 million people provide a ready market for finished beef, as well as a large in-state market for both fluid milk and manufactured dairy products. Hay production, predominantly alfalfa, is the fifth highest value crop, indicating the importance of cattle and dairy operations, which provide the market for the hay produced in California. Additional hay is purchased seasonally from surrounding states.

Significance of Exports

California's agricultural exports valued at port amounted to nearly $4 billion in 1980—a 43 percent increase over 1979. The two largest percentage increases over the year were poultry products at 182 percent and rice at 117 percent. Since 1962, when exports were first estimated by the California Crop and Livestock Reporting Service, the total value of the state's international marketings has increased 800 percent.

About 36 percent of California's harvested cropland is now used to produce exports, and 23 percent of the value of total farm marketings is sold abroad. It is likely, given these figures, that

1. Cash receipts from sales of cattle and calves includes the value of large numbers of stockers and feeders shipped into the state, fed, and resold.

TABLE 2.5
California's 20 Leading Commodities in 1980

Farm Product	Ranking	Value[a]
		Million Dollars
Milk and cream	1	1,771
Cattle and calves	2	1,439
Cotton	3	1,389
Grapes	4	1,216
Hay	5	723
Nursery products	6	498
Almonds	7	473
Rice	8	424
Flowers and foliage	9	399
Lettuce	10	383
Chicken eggs	11	370
Wheat	12	358
Tomatoes, processing	13	327
Chickens	14	229
Oranges	15	224
Strawberries	16	201
Sugarbeets	17	183
Turkeys	18	179
Peaches	19	176
Walnuts	20	168

[a] Based on value of quantity harvested for crops and on value of quantity marketed for livestock and poultry products.

SOURCE: California Department of Food and Agriculture, *California Agriculture, 1980.*

California farmers would have difficulty selling all they produce on the domestic market alone.

Cotton remains the leading California agricultural export with seven out of eight bales going abroad; 28 percent of California's total export value for all products is accounted for by cotton lint. Almonds are second, at 11 percent of the total, and rice third (Table 6).

Productivity and Input Use

The last 50 years have been a time of tremendous increase in agricultural output in the United States and in California. Four factors are typically cited: (1) increases in area cropped; (2) increases in yield per acre; (3) increases in number of crops per year; and (4) replacement of lower by higher yielding crops. All have placed a role in growth of production, though the major increases nationally have come primarily from other than increasing the land area cropped. In California, extension of irrigation in the great valleys has

TABLE 2.6

Estimated Value of California's Ten Leading Agricultural Exports, Valued at Port of Exportation, 1980

Rank	Commodity	Value	Percentage of Total Export Value
		1000 Dollars	
1	Cotton lint	1,135,559	28.4
2	Almonds	429,817	10.8
3	Rice	318,808	7.8
4	Wheat	283,145	7.1
5	Grapes (fresh, raisins and crushed)	230,864	5.8
6	Oranges	177,094	4.4
7	Lemons	93,210	2.3
8	Cattle, calves and products	91,762	2.3
9	Walnuts	82,957	2.1
10	Peaches (including canned)	75,372	1.9

SOURCE: California Crop and Livestock Reporting Service, *Exports of Agriculture Commodities Produced in California, Calendar Year 1980,* August 1981.

brought not only tens of thousands of acres of new land into crop production, but has fostered substitution of a wide variety of irrigated crops for dryland grain or pasture on land that was already being tilled. One prominent agricultural economist states that while the 1979 level of national grain production (316.2 million tons) required 162.1 million acres, production of that same amount at 1910 yields would have required 509.9 million acres. Similar dramatic changes have occurred nationally in the productivity of labor. From 1920 to 1979 the labor input in terms of hours used for farmwork in American agriculture dropped by about 80 percent. Inputs of fertilizers, pesticides, and mechanical power and machinery meanwhile rose sharply during this period. The net result, viewed in terms of crop production per unit of labor nationally, is that labor productivity has risen by 13 times between 1920 and 1979.

Given the broader spectrum of agricultural commodities produced in California, the more certain weather during the growing season, and better control over such important factors as water, the pattern of growth in crop productivity in the West, and in California particularly, is somewhat different. In yields per harvested acre of some important national commodities that are also important in California, the leadership of this state is clearly evident (Table 7). For crops in which California supplies relatively less of total U.S. production, such as in corn and cotton, the margins are significantly greater between national and state averages. Of course, in such

TABLE 2.7

Yield Per Harvested Acre, Representative Crops, United States and California, 1979

Commodity	Units	U.S.	California
Corn, for grain	bushels	109.4	117.0
Cotton, upland	pounds	548	1000
Lettuce, summer	hundredweight	275	305
Rice, medium grain	pounds	5327	6500
Strawberries	hundredweight	188	425
Sugarbeets	tons	19.6	26.8
Tomatoes, processing	tons	23.5	25.4
Wheat, winter	bushels	36.9	70

SOURCE: U.S. Department of Agriculture, *Agricultural Statistics, 1980.*

TABLE 2.8
Agricultural Employment in California by Type of Worker

Year	Total	Farmers and Unpaid Family	Hired Domestic			Contract Foreign[c]
			TOTAL	REGULAR[a]	SEASONAL[b]	
1950	357,300	132,100	217,800	108,600	109,200	7,400
1960	333,700	99,000	192,000	93,500	98,500	42,700
1970	289,100	78,100	211,000	97,100	113,900	–0–
1980	288,300	64,200	224,100	104,600	119,500	–0–

[a] Workers employed by the same employer for 150 or more consecutive days.
[b] Workers employed less than 150 consecutive days by the same employer.
[c] Agricultural workers brought to California under contract from outside the United States.
SOURCE: California Employment Development Department, *Agricultural Employment Estimates,* Report 881-X, January 1981.

California-dominated commodities as summer lettuce and processing tomatoes, the margins are much smaller as is to be expected.

Agricultural Inputs

FARM LABOR. According to census figures, California paid over 20 percent of the nation's hired farm wage bill in 1978. It is estimated that California's farmers hire an annual average of 200,000 workers (seasonal and year round), nearly three times as many as the second state, Texas. In the peak harvest months of September and October, over 300,000 farm workers are employed in California. These figures may represent jobs not workers, however; the number in the farm work force is probably much larger. The number of jobs for hired workers on California farms has held relatively constant over the past 30 years, while the number of farmers and unpaid family workers employed on farms has steadily decreased (Table 8). The continuing large number of hired workers despite increased mechanization in many crops reflects the continuing expansion of high-value, relatively labor-intensive crops in the state.

FERTILIZERS AND CHEMICALS. Inputs of fertilizers and other chemicals have grown markedly in American agriculture and particularly in the production of high value crops in areas where available soil moisture can be controlled (Table 9). Such controlled conditions are found extensively in California which accounts for much of the yield differential in specific crops between California and the rest of the nation. Not only are average yields higher, but variations from season to season are lower, because of greater control and less climatic variability during the growing and harvesting season. Data for the Pacific region (Washington, Oregon, California), the smallest geographic area for which systematic data on productivity are reported, indicate that between 1940 and 1979 the use of chemicals in agricultural production rose nearly 15 times, while nationally it rose ten times. In the Pacific region, nitrogen usage alone expanded from 33,000 tons in 1940 to 855,000 tons in 1979, an increase of 26 times.

MECHANIZATION. The use of mechanical power and machinery in the West has closely paralleled the increase nationally, though the data fail to reflect the full contribution since the indices do not allow adequately for upgrading in quality. Since 1940, these inputs have increased approximately four times nationally, 3.5 times in the Pacific region. The slight difference in rate of

TABLE 2.9
Indices of the Use of Agricultural Chemicals and Mechanical Power and Machinery, U.S. and Pacific Region, Selected Years (1967=100)

	Agricultural Chemicals		Mechanical Power and Machinery	
	U.S.	PACIFIC REGION	U.S.	PACIFIC REGION
1940	13	10	42	44
1950	29	24	84	88
1960	49	55	97	100
1970	115	120	100	100
1975	113	104	127	144
1979	129	149	182	152

SOURCE: U.S. Department of Agriculture, *Economic Indicators of the Farm Sector, Production and Efficiency Statistics, 1979,* February 1981.

growth probably reflects the fact that mechanization of specialty crops on the Pacific coast was encouraged early, because of the heavy dependence on seasonal hired labor.

Outlook for the Future

Since the profitability of further mechanization and the use of chemicals and pumped water are all affected by energy costs, the expected long term upward trend in these costs raises serious questions about the future amounts of these critical elements that might be used. There is already growing evidence that the amounts of agricultural chemicals applied per acre may be decreasing, partly because of rising costs. There are other reasons as well. More careful timing and placement of application allows plants to utilize fertilizer more efficiently. Insect and disease control methods are being modified to reduce environmental loading of undesirable chemicals. Water application is becoming more efficient, with wider adoption of water conserving practices such as drip irrigation systems. Reducing tillage and increasing the use of herbicides is another practice designed to conserve petroleum-derived energy. The long term outlook is for reduced chemical inputs per acre, but greater total use in the U.S.—as their application expands with more intensive management of forests and greater intensification of agricultural production in areas of the country where usage is now relatively low. Machinery use is being curtailed in many regions of the country, and to some extent California, as farmers seek to control rising costs and reduce soil erosion.

The future yields of major crops are of growing concern to scientists and policy makers. Rising productivity has been responsible for continued low costs of food to Americans and, increasingly, has provided foodstuffs to meet growing demand abroad. California has been a significant contributor. Despite record yields in the last several years, there are several reasons why future agricultural productivity has become an important issue: (1) the decline in productivity already in evidence in other sectors of the economy; (2) the change in relative factor prices, especially of fossil fuels and fossil fuel-based products; (3) decline in recent years in the real funds committed to agricultural research and development; (4) some evidence that observed yields are approaching experimental levels, suggesting that the gap between available technology and present practice is closing. For all

these reasons it is important that trends in agricultural productivity be more thoroughly understood if public policy is to guide effectively the development of our agricultural resources in the face of new constraints, including the strong commitment to preservation of the quality of our natural resource base.

Financial Dimensions

Trends in Farmland Values

Trends toward larger and fewer farms have been accompanied more recently by rapid escalation in farmland values both in normal and real terms (Table 10). The 48-state average value per acre of farm real estate increased 2.7 times from 1970 to 1980, the corn-belt states leading the escalation. In California, farmland values have historically been much higher. Though they only doubled in the past decade, they remain considerably above the national averages. These increases in value reflect the complex interaction of a number of economic forces related to inflation, taxation, and the value of additional parcels when added to an existing farm operation.

The data in Table 10 represent averages of farm real estate values per acre for all types of agriculture. Consequently, they mean less in a state such as California, where an acre of undeveloped rangeland and one of highly productive irrigated farmland are weighted equally in computing the average. Values by region and type of use illustrate more accurately the variations in California farm real estate values (Table 11), underscoring the enormous variability in California agriculture as reflected in anticipated

TABLE 2.10

Average Per Acre Value of Land and Buildings, United States and California, 1970 and 1980 (dollars per acre)

	United States		California	
	NOMINAL	REAL[a]	NOMINAL	REAL[a]
1970	196	214	479	524
1980	725	409	1426	804

[a] To obtain "real" values, nominal values were divided by the GNP Implicit Price Deflator (U.S. Council of Economic Advisers, 1981).

SOURCE: U.S. Department of Agriculture, *Farm Real Estate Market Developments, Outlook and Situation*, August 1981.

net returns per acre. The richness and diversity of agriculture in the Central Coast, the Central Valley, and Southern California is again evident as contrasted with the North Coast and Mountain regions.

Farm Income

California's gross cash receipts from farm marketing totaled $13.5 billion in 1980, representing nearly 10 percent of total U.S. cash receipts—a proportion that has consistently approached 10 percent since 1960. No other state is close to this level; the second state in 1980 was Iowa at $10 billion, followed by Texas at $9 billion. California holds this position by consistently ranking at the top in cash receipts from sale of crops, and second or third in livestock and livestock products. In the major field crop categories, California ranks high in cotton, hay, and sugarbeets. In fruits and vegetables California is consistently at or near the top in the majority of commodities in this category. In gross income and net income per farm, California ranks second to Arizona, which has led the nation for 20 years. (This is ex-

TABLE 2.11

California Per Acre Farm Real Estate Values by Region by Type of Crop, 1979 (dollars per acre)

	North Coast & Mountain	Central Coast	Sacramento Valley	San Joaquin Valley	Southern California
IRRIGATED					
Truck and vegetables	—	4200	2280	2650	4200
Intensive field crops[a]	1400	2440	1900	2260	2400
Extensive field crops[b]	1050	1580	1660	1970	2300
Pasture	830	1750	1500	1550	2000
NONIRRIGATED					
Cropland	600	790	880	960	1800
Pasture	540	680	700	775	—
Rangeland	300	460	430	520	—
ORCHARDS[c]					
English walnuts	—	4300	4400	5200	—
Almonds	—	—	3850	5600	—
Apples	—	7950	—	—	—
Peaches	—	—	4050	5750	—
Pears	—	4850	4200	—	—
Apricots	—	—	—	4350	—
Prunes	—	5050	3650	5950[d]	—
Plums	—	—	—	6000	—
Avocados	—	—	—	8600[d]	12700
Olives	—	—	—	3800	—
Vineyards	—	—	—	—	—
Raisin varieties	—	—	—	6900	—
Wine varieties	—	8900	—	6400	—
Table varieties	—	—	—	6500	—
Citrus	—	—	—	—	—
Valencia oranges	—	—	—	4650	7550
Navel oranges	—	—	—	4850	7200
Lemons	—	—	—	4250	8600
Grapefruit	—	—	—	6100[d]	6200

[a] Includes land used for cotton, sugarbeets, rice, etc.
[b] Includes land used for barley, beans, corn, and sorghum.
[c] Observations lying outside the range of plus or minus one standard deviation of the mean—assuming a normal distribution, about one-third of the observations—were dropped. The figures reported are the means of the retained observations.
[d] 1980 values.

SOURCE: U.S. Department of Agriculture. *Farm Real Estate Market Developments, Outlook and Situation*, August 1981.

Flat, regular fields make geometric patterns in the Central Valley as they stretch for miles with few interruptions. *Photo by Robert Campbell.*

plained by the small number of farms in Arizona and their relatively large size.) California's net farm income per farm has exceeded by roughly three times the national average in each of the last 30 years.

To supplement income derived from farming many smaller farmers earn additional income in off-farm activities. In recent years, approximately half of the personal income of the nation's farm population has come from nonfarm sources. In 1978, 59 percent of all California farm operators reported some off-farm work. Most of these were small farm operators who worked part-time for other farmers or at other types of jobs for part of the year. Many farmers derive income from other off-farm sources—interests and dividend income, rental income from farm or nonfarm properties, or income from farm-related activities such as trucking, custom feeding of livestock, or sales commissions on farm inputs, to name a few.

Farm Assets and Liabilities

California agriculture reports a 1980 total farm asset value of nearly $43 billion, exceeded only by Iowa at $70 billion, Texas at $67 billion and Illinois at $63 billion. In all cases real estate accounts for approximately 80 percent of the total asset value. Total farm debt as a percent of total farm assets in California in recent years has been running over 25 percent—in sharp contrast to the 12 to 15 percent in the other three high-asset states, and the national average of approximately 17 percent. In all cases the value of farm real estate has been climbing sharply and the absolute

debt load has been moving up as well. While real estate debt has risen, nonreal estate debt has risen much faster, indicating the requirement for greater borrowing by farmers to finance current production expenses. Rising farm land values have made many otherwise unattractive production loans appear more favorable to lenders.

These data, aggregated at the state level, can only suggest in a general way some differences that exist in California. The primary difference is that farmers in California, because of the types of crops they grow and the high cash production expenses they must meet, tend to borrow more money seasonally to finance production. Furthermore, with the high degree of commodity specialization of many farms, the opportunity to meet late season expenses for one farm enterprise by sales from an early season enterprise is more limited. Since variation in financial status among the state's farms is marked, any comparisons and conclusions drawn above must be carefully interpreted.

Farm Ownership Patterns

As total farm numbers have declined, the proportion of farms in California fully owned by farm operators has remained relatively constant at around 70 percent for the past 40 years (Table 12). Nationally the trend has been slowly upward from 50 percent toward 60 percent over the same time period.

Nationally, the percent of farm operators who are part owners has climbed steadily over the same 40 year period, rising from 10 to nearly 30

TABLE 2.12
Tenure of Farm Operators, United States and California

	United States					California				
	1940	1950	1959	1969	1978	1940	1950	1959	1969	1978
Percent of farm operators who are full owners of the farms they operate	50.6	57.5	57.1	62.4	58.6	67.7	73.5	69.0	68.9	71.6
Percent of farm operators who are part owners of the farms they operate	10.1	15.6	21.8	24.5	28.8	10.6	12.7	17.9	18.4	16.3
Percent of farm operators who are tenants or managers	39.3	26.9	21.1	12.9	12.7	21.7	13.8	13.1	12.5	12.1

SOURCE: Nuckton and McCorkle, 1980, and *Census of Agriculture, 1978*.

percent—while in California the comparable figures are from 10 to between 15 and 20 percent. This reflects two types of actions by farmers. First, borrowing funds has enabled many farm operators to purchase additional acres to add to an existing operating unit, thus changing from full- to part-owner status. Second, leasing additional acreage to add to an existing fully owned unit is an increasingly popular means of increasing farm size with minimum capital outlay. This type of expansion, by definition, also shifts full owners to part-owner status. The increased percentages in part ownerhsip over time have generally been at the expense of the tenant and manager class—those who lease (or manage) all the acres they operate. Tenant farming nationally has dropped sharply since 1940, from approximately 40 percent to 12 percent, where it appears to be stabilizing. In California, the drop in tenant farms was sharpest between 1940 and 1950 and has been very stable at between 12 and 14 percent for 30 years. The decline in tenant farming is attributed to tenants' becoming part owners or owners by purchasing some or all of the acreage they are farming.

The predominant legal form of ownership of farms, both nationally and in California, remains the individual or family type organization (Table 13). Eighty-eight percent of the nation's farms are so controlled; in California the figure is approximately 80 percent. Partnerships account for about 10 percent of the farm ownerships nationally, and 14 percent in California.

INCORPORATED FARMS. Increasingly prevalent, and a highly publicized type of organization for farming, is the corporation—though the numbers are still relatively small. The data available on both partnerships and corporations are somewhat misleading, since many farms so classified are operated as "family" farms. In 1978, family corporations, in which 51 percent or more of the stock is owned by persons related by blood or marriage, constituted 89 percent of the total number of farming corporations in the nation, 84 percent in California.

The decision of the family farmer to incorporate his operation is generally the result of careful business planning with the income tax structure and estate arrangements in mind. Fringe benefits can be offered employees of the corporation, including the farmer himself. Double taxation usually associated with corporations—once on corporate income, once on dividends—can be avoided if the farm firm qualifies under subchapter S of the Internal Revenue Code. Possibly more important is the fact that the corporate farm can elect to retain earnings within the corporation to reinvest in the business and be taxed on these earnings at corporate income tax rates which are generally lower than individual income tax rates. This advantage may give the corporate family farmer an edge over unincorporated farmers in bidding for additional land or in purchasing machinery and equipment.

Incorporation also provides a useful tool in estate planning. For example, annual tax-free gifts of corporate shares can be given to potential heirs by an individual and spouse and thus pass

TABLE 2.13

Type of Organization of the Farming Operation, United States and California, 1978

	United States	Percent	California	Percent
Total number of farm operations	2,478,642	100.0	81,863	100.00
Individual or family type organization	2,175,437	87.8	65,981	80.6
Partnership	240,290	9.7	11,340	13.9
Corporation	51,270	2.1	3,926	4.8
Family held	45,418	—	3,287	—
Other than family held	5,852	—	639	—
Other—cooperatives, estates or trusts, institutional, etc.	10,645	0.4	616	0.8

SOURCE: U.S. Bureau of the Census, *Census of Agriculture, 1978.*

outside the estate, avoiding estate taxes. By incorporating, a farmer may gain certain advantages that will help preserve the "family farm." The benefits of incorporation, of course, must be weighed against the additional costs involved in the organizational and administrative time required and in legal and accounting assistance needed.

Conclusion

The great diversity in California's soils, water availability, climatic conditions, topography, and markets has created a complex and unique agriculture. Individual farmers and their farms vary greatly between and within the major agricultural regions, reflecting their adaptation to the physical, biological, economic, and social environment as well as their personal preferences for agricultural enterprises and practices. General and highly aggregated statistics can convey a sense of the dimensions and structure of this agriculture—but its rich mosaic, its enormous vitality, and its potential fragility can only be sensed through deeper understanding of the many forces that shape the individual farmer's decisions. The remainder of this book attempts to convey that understanding.

References

Archibald, S.O., and McCorkle, C.O. "Trends in Productivity of American Agriculture." University of California, Davis, Department of Agricultural Economics, Working Paper No. 81-1.

California Crop and Livestock Reporting Service. *Exports of Agricultural Commodities Produced in California, Calendar Year 1980*. California Department of Food and Agriculture: August 1981.

California Department of Food and Agriculture. *California Agriculture, 1980*.

California Employment Development Department. *Agricultural Employment Estimates*. Report 881-X, January 1978; Report 881-X, January 1981.

Carman, H.F. "Coming: More Corporate Farms in California." *California Agriculture*, 34 (January 1980):9–10.

Carter, H.O., and Johnston, W.E. "Agricultural Productivity and Technological Change: Some Concepts, Measures and Implications," in *Technological Change, Farm Mechanization, and Agricultural Employment*. University of California, Division of Agricultural Sciences, Priced Publication No. 4085, July 1978.

Carter, H.O., and Johnston, W.E., Principal Investigators. *Farm Size Relationships, with an Emphasis on California*. University of California, Department of Agricultural Economics, Davis, Giannini Foundation Project Report, December 1980.

Nuckton, C.F., and McCorkle, C.O. *A Statistical Picture of California Agriculture*. University of California, Division of Agricultural Sciences, Leaflet No. 2292, April 1980.

Penn, J.B. "The Structure of Agriculture: An Overview of the Issues," in *Structure Issues of American Agriculture*. U.S. Department of Agriculture, Economics Statistics and Cooperatives Service, Agricultural Economics Report No. 438, pp. 2–23, November 1979.

U.S. Bureau of the Census. *Census of Agriculture*. Washington, D.C.: U.S. Government Printing Office, 1969, 1974, and 1978.

U.S. Council of Economic Advisers. *Economic Report of the President*. Washington, D.C.: U.S. Government Printing Office, 1981.

U.S. Department of Agriculture. *Agricultural Statistics*. Washington, D.C.: U.S. Government Printing Office, 1980.

US. Department of Agriculture. *Economic Indicators of the Farm Sector, Production and Efficiency Statistics, 1979*. Economics and Statistics Service, Statistical Bulletin No. 657, February 1981.

U.S. Department of Agriculture. *Economic Indicators of the Farm Sector, State Income and Balance Sheet, 1979*. Economics and Statistics Service, Statistical Bulletin No. 661, March 1981.

U.S. Department of Agriculture. *Farm Real Estate Market Developments, Outlook and Situation*. Economic Research Service, CD 86, August 1981.

U.S. Department of Agriculture. *Economic Indicators of the Farm Sector, Income and Balance Sheet Statistics, 1980*. Economic Research Service, Statistical Bulletin No. 674, September 1981.

Though many landholdings in the state are huge, an increasing number of very small farms was noted in some areas in the 1978 Census of Agriculture. *Photo by Jack Clark.*

PART II Resources:
The Natural
Base

From 570 miles high, space imagery from NASA's Landsat system reveals this panoramic view of northern California in October. A portion of the Cascade Mountains is seen at the top of the picture, with snow capped Mt. Lassen to the right; lower center and right show the northern end of the Sacramento Valley, with the Coast Range to the left. Shasta Reservoir, top center, and the foothills and agricultural fields of the valley can be clearly distinguished. Dark areas are forested mountains, with a light dusting of snow. *Photo from the Remote Sensing Research Program, University of California Space Sciences Laboratory, Berkeley.*

3

Soils and Climate

Paul J. Zinke and Constant C. Delwiche

Mountains, foothills, valleys, deserts: landscapes in California are broad and varied. The soils deposited during eons of geologic time form a complex mosaic throughout the state. The climate, too, shaped by oceanic and continental influences, varies dramatically from place to place. For agriculture this means a range of growing seasons, from twelve months in a few parts of the state down to a two-month season in others. This chapter describes the land resources of California by region from a basic geological perspective.

The Geologic Background for the Regions of California

The natural regional variability of California gives rise to the remarkable diversity and productivity of its agriculture. This can be best understood by reference to the geologic history of the state. Many of the state's farming regions are defined by mountain ranges and their various watershed and valleys.

A large portion of California represents accretions to the continental mass as North America moved across and above the ocean floor to the west. This process, much like pushing one rug underneath another, ruffled the western edge of the continent as the deep folding and bending produced a series of mountain ranges and valleys.

Two basic consequences of these geologic processes have particularly affected California agriculture. The mountains and valleys have created compartments of climate which affect the rates of soil formation, the natural vegetation, and agri-

cultural productivity. In addition, the varied rocks in California influence the types of soils, their inherent productivity, and ultimately their plant cover—whether natural vegetation or cultivated crops.

The processes which form a soil depend upon the parent material (the rock or other material which weathered to form soil); the climate, which affects the nature and rate of the weathering processes; the slope of the land and its exposure to sunlight and wind; the vegetation or other biologic factors which produce humus and fertility elements; and finally, the period of time or age of the surface on which the soil has formed.

All of these geologic processes have divided California into a number of compartments or "provinces" where the nature of agricultural activity is determined mainly by the combination of soil and climate—although there are other variables, including economic ones. The geologic processes which formed the major provinces of California agriculture were long term, with dimensions of tens of millions of years. On an intermediate time scale—millions of years—has been the gradual uplift of the terraces and foothills surrounding the present valley areas of California. As a consequence, although parts of the valleys continue to receive new sediment to form new soils, there are frequently terraces of older alluvium surrounding the valleys. Some of these are very old, with lowered productivity for most intensive agriculture, although they may be valuable as range land.

On a much shorter time scale, at the peak of

the latest glaciation period—thirty to fifty thousand years ago—enough water was tied up in ice shields and caps in cold parts of the world, so that sea level was lower by 91 meters (300 feet) than it is now. As this ice melted and water was released, the sea level began to rise and areas of low elevation in California as elsewhere were flooded and became marshlands. The water-logged soils inhibited the penetration of oxygen necessary for decomposition. Consequently, as the sea level slowly continued to rise, un-decomposed plant material accumulated in the basin which is now the Sacramento-San Joaquin Delta. This process eventually built up a thick layer of peat soils containing largely organic plant remains, but also some of the silty material from the river system feeding the Delta. Mean-while, other parts of the central valleys were re-ceiving new sediment from the erosion created by mountain glaciers in the Sierra Nevada. At present these are young sandy soils.

The various agricultural "provinces" that come out of this geologic process are shown in Figure 1. Each has its own special characteristics.

The principal mountain range complexes of California—including the Sierra Nevada, the Cas-cades, the Coast range, the transverse ranges to the south including the Tehachapi mountains, and those to the north including the Siskiyous and the Klamath Mountains—are all large-scale features which strongly affect the overall climate and geography of California. These features de-termine the availability of water in the various California provinces.

Within these large provinces are smaller units with their own unique climates, soils, and agri-culture. Thus, for example, we find the high rainfall areas of the north coast supporting prod-uctive redwood and Douglas fir forests and rangeland with high carrying capacity for grazing animals; the Napa and Sonoma Valleys north of San Francisco with climates particularly suited to varietal wine grapes; the Salinas Valley, inland from Monterey Bay, with cool summer climates adapted to certain truck crops; and the Imperial Valley with its year-round productive desert cli-mate. Some areas, such as the latter, may be nat-urally unsuited for agriculture because of aridity, but given a sufficient water supply they can be made productive.

The distribution of water, in fact, largely deter-mines agricultural productivity in the state. The natural distribution is governed by the latitude,

proximity to the coast, and the effects of moun-tain ranges which increase precipitation on their western slopes but create arid "rain shadows" on their easterly inland slopes. Figure 2 shows the distribution of water yield in the state.

It is apparent that 80 percent of the available water yield for agricultural use comes from the northern forest and mountainous regions of the state. However, most of the curent agricultural production, whether in northern or southern California, is located where local water supply is insufficient to meet crop demand. A primary fea-tures of agriculture in California, therefore, is the need to transport water from watershed supply areas in the forest and mountains to arable areas in the valleys. Thus, there are strong political and economic pressures to export water from areas of high water yield to those of high water demand.

The mountains and valleys of California give rise to another aspect of climate that is peculiar to the state. Mountain ranges tend to limit air circulation, so it is possible for an entire valley or large parts of it to commonly experience a tem-perature inversion with the accumulation of stag-nant air and smog problems. In this respect, the Sacramento-San Joaquin Valley repeats on a large scale conditions at the southern end of the San Francisco Bay area and in the Los Angeles-San Bernardino Basin. This phenomenon is particu-larly significant because agriculture depends on sunlight and clean air. Particulates in the atmo-sphere cut down light intensity, and oxidants from automobile or industrial exhaust gases in-jure plants. Furthermore, the soils of a region be-come the final storage place for many of these pollutants. A study by researchers at the Univer-sity of California at Riverside has indicated that a large proportion of the lead from automobile ex-haust for the past 50 years in southern California can be found in the surface soils of that region.

Regional Characteristics and Related Agriculture

Our consideration of the various regions of Cali-fornia will begin with the deserts of the eastern portion of the state; proceed to the southern coastal plains and the coastal ranges and valleys of central California; to the central valleys of the San Joaquin and Sacramento rivers and their com-mon delta region; to the Sierra Nevada region

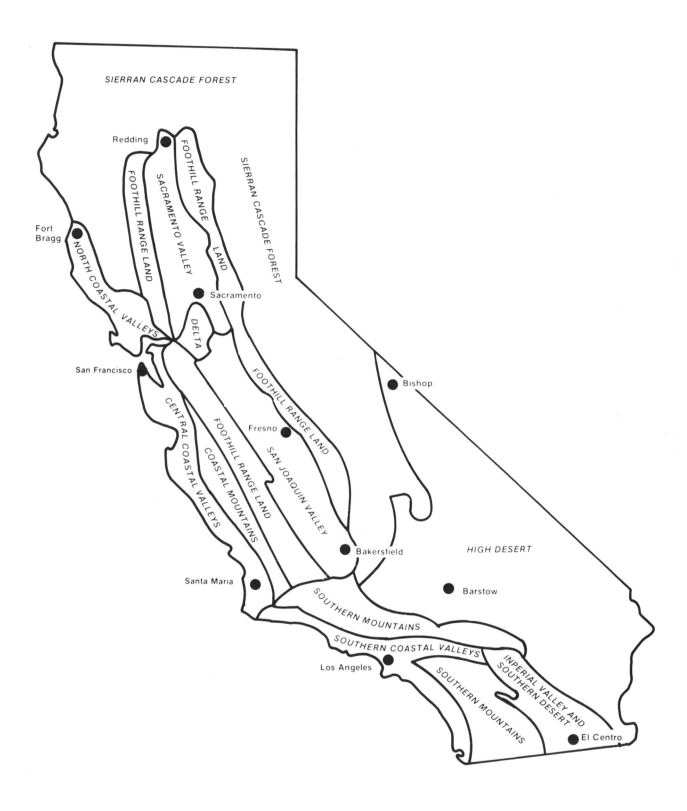

FIGURE 3.1 Agricultural provinces of California. Divisions are based on the nature of agricultural activity, which in turn is determined largely by climate, soils, and topographical features. *Figure by C. Delwiche.*

FIGURE 3.2 Water resources in principal rivers of California. Thickness of line represents average discharge in cubic feet per second. *Adapted from "Water Resources Investigations in California," U.S. Geological Survey, January 1962.*

and the Cascade mountains; and finally to the northern coastal and transverse ranges and valleys. These regions can be seen in Figure 1.

The Deserts

Those portions of California which are characterized as desert usually have less than 150 mm (6″) of annual rainfall, and very high evaporation, measuring as much as 1000 mm (40″) per year or more.[1] These evaporation rates are highest in the southern deserts, and drop to around 635 mm (25″) in the desert areas of northeastern California. The natural vegetation, in response to this climate, is usually desert shrub at lower elevations and rainfall. Desert woodlands of pinyon pine and junipers scattered in a denser cover of desert shrubs occur at higher elevations where precipitation is somewhat higher. Agricultural development is dominated by these climatic factors and the associated needs for supplemental irrigation water.

The soils that have formed under a desert climate are usually calcareous or high in salts due to lack of leaching by rain. They are valley soils developed from alluvium deposited by intermittent streams fed by thunderstorms in adjacent mountain ranges. These soils are coarser in texture (sandier) near these sources, and finer in texture toward the lower parts of the valleys. Some areas of the Coachella and Imperial Valleys are dominated by soils such as the MYOMA and the ROSITAS soil series developed from windblown sands. The east mesa of the Imperial Valley has very sandy alluvial deposits typified by the SUPERSTITION soil series[2] which requires excessive amounts of water, where irrigated, because of its high permeability. Progressing to lower elevations in the Imperial Valley, the soils are of increasingly finer texture and tend to have poorer drainage with more clay content. Representative soils are the MELOLAND, the IMPERIAL and the HOLTVILLE soil series. These soils

---

1. "Evaporation" means the rate of loss from a free water surface, usually measured from a pan of water.

2. Soils are examined on a pit face, and those with similar properties of texture, color, consistency, structure, thickness of layers, reaction or pH, and similar sequences of change of properties with depth are designated as Soil Series and given a name, usually derived from the locality where first observed. The names given can then be used as a reference to the soil characteristics which are usually published in a report accompanying the pertinent survey. Although the names may be changed in subsequent surveys, the original survey is still valid as a map and report description of this Soil Series and its occurrence.

accumulate salt, and because of their low permeability to water present a problem of management so that salts do not continue to accumulate through additions from irrigation water and from rising ground water. Thus excessive salts in the soil and those added by irrigation water must be removed by artificial drainage. In some areas up to five feet of irrigation water are needed per year to meet the combined water requirements for crop use and salt removal from soil. Drain tiles are also necessary because of poor percolation in the heavier textured soils. In the area east of Holtville a canal is required for drainage in each quarter square mile; up to 32,000 km. (20,000 miles) of tile drains lead into these canals, draining salts from 48,600 hectares (120,000 acres). Drainage water may remove as much as 100 tons of salts per hectare, mainly as calcium and sodium chlorides and sulphates. These problems and their costly solutions require high investment in the farming enterprise even though the average farm size is small (about 45 hectares or 100 acres).

Irrigation in this naturally arid area has made it possible to capitalize on a year-round growing season with many cloud-free days. As a consequence, record productivity has been obtained with such crops as irrigated alfalfa, row crops, fruits, and vegetables. As many as several crops per year of some commodities can be obtained from these areas. The fact that many of the crops can be harvested for marketing at times of the year when agricultural production in other areas is at a low ebb makes this crop region one of very high economic value.

The other desert areas in this region are not nearly so productive as the Imperial and Coachella Valleys. Limitations of water supply constrain their potential productivity. Although some "dryland" crops can be grown on an average of ten inches or less of precipitation properly timed, or managed from soil moisture carried over from fallow years, many crops reach their maximum productivity with from three to ten feet of water per year applied in irrigation.

Many long term problems based on the climatic and soil characteristics of this desert region remain to be solved. Many cultures that have depended upon irrigated agriculture in the past have eventually collapsed because of accumulation of salts, lack of maintenance of water sources or other detrimental soil-related processes.

We now know that to solve the agricultural problems of the desert regions, three basic requirements need to be met:

1. A means for resupplying to the soil the plant nutrients which are exported with the crop.
2. A means of disposing of drainage waters which contain the unwanted salts from the soil and the irrigation water, concentrated by evaporation.
3. An abundant supply of high quality water for irrigation. This means water free from deleterious quantities of salts, or of other materials that may accumulate in the soil.

Although the inherent problems of agriculture in this desert region are formidable, all of them are subject to solution—although with methods sometimes costly or socially controversial. An adequate water supply for the desert soils may depend upon importation from other regions. Export of salty drainage waters may mean pollution for another region. The need for food and fiber production may nevertheless force the acceptance of otherwise unattractive technological solutions.

The northern deserts of eastern California differ from those of the south in that they are at higher elevations and thus have cold winters and shorter growing seasons. The soils used for agriculture are mainly found on old lake basins and associated terraces, or on alluvial fans, and tend to be gray colored and calcareous. They are used mainly for grain and pasture. An example of this northern type of desert is in the Susanville area of the Honey Lake Valley.

The Southern Coastal Plain and Adjacent Mountains

The south coastal portion of California is like a half valley, bounded on one side by high mountains and the other by the ocean. As a result, the topography is dominated by the mountains, alluvial fans and flood plains originating where rivers and streams debouch from the mountains. Along the ocean margins there are coastal terrace deposits at various elevations above sea level representing past ocean levels or coastal uplifts. The climate varies mainly with distance from the sea, and with rising elevation into the mountains. It is subtropical at lower elevations, but alpine on the mountain peaks. Precipitation ranges from 300 mm (12″) per year at the lower elevations, to 1000 mm (40″) on the higher peaks,

much in the form of snow. Record years of up to 2300 mm (90″) precipitation have occurred at the high elevations. There is a considerable deficit of water in the summer, and annual water need for evaporation and vegetation use is around 760 mm to 840 mm (30″ to 33″). The natural adjustment of vegetation to this results in annual grassland which is green during the wet winter and golden brown during the dry summer. Increasing elevation and accompanying wetness is reflected in increasing amounts of woody vegetation, first grass-oak woodland, then chaparral,[3] and finally pine and fir forests at high elevations.

Soils formed in place on the steep mountain slopes are usually shallow and subject to removal during occasional intensive storm years, being deposited downstream on the coastal plain or interior valleys. This alluvial material is usually cobbly and sandy near the mountains, forming (for example) the SOBOBA soil series. With increasing distance into the valley, as in a transect down the fan of San Antonio Creek along the border of western San Bernardino County, the soils have decreasing rock fragments and increasing clay content resulting in the sequence of soil series: SOBOBA-TUJUNGA-GRANGEVILLE-CHINO. Agriculture has developed in adaptation to this sequence with the sandier soils used for orchards or vineyards, and the finer textured clayey soils used for pasture and dairy farms. Where the deposition is from mountains of sedimentary rocks (such as sandstones and claystones), the entire sequence of soils is finer textured, represented by the sequence of series REIFF-CARRETSON-MOCHO-CROPLEY. Most of the interior valleys of southern and central California have similar soil type sequences on the alluvial fans.

The mountains from which these deposits are derived are usually covered with shallow stony and sandy soils held in place on steep slopes by the chaparral (brush) and forest vegetation. This protective cover is removed at intervals by catastrophic wildfires. When this denudation is combined with a flood year, the slopes are eroded to form the sediment which forms the valley soils. Since urbanization has occurred in these areas, these periodic events become costly and trau-

---

3. Chaparral is a Spanish term for a brushy plant formation containing a large amount of scrub oak. The term has been expanded to include most brushy cover on steep mountain slopes. This vegetation gives the characteristically dark color to southern California mountains.

Citrus orchards thrive in the frost-free climate of southern California. As many acres of older orchards have been absorbed by urbanization in Riverside, Los Angeles, and Orange counties, new orchards have been planted in the San Joaquin Valley. This view shows the dramatic contrast between irrigated agriculture and the natural landscape. *Photo by Max Clover.*

matic to city dwellers whose homes are threatened by this centuries-old natural process.

Terraces occur around some of the interior valleys and along the coastal region extending inland for 20 kilometers or more. The older the terrace, the redder the soil, beginning with grey or brown soils on the younger terraces. Thus, just north of San Diego on the Kearney mesa are some very old red soils with iron hardpans called the REDDING soil series. Some of the coastal terraces are formed on old ocean bottom deposits, and the soils have sandy to loamy surfaces and clayey subsoils. They are susceptible to erosion and slipping. The LAS FLORES soil series near Carlsbad develops deep gullies when eroded by concentrations of drainage water from fields or adjacent urban areas.

A succession of agricultural crop types has occurred in this area since the original Spanish settlement in 1769 at San Diego, progressing from rangeland to grain to grapes to apricots, walnuts, and finally citrus. Much of this land has now been engulfed by urban use. Currently avocados are the major remaining orchard crop in the southern part, occupying soils from the valley bottoms to steep mountain slopes.

The Central Coast Ranges and Valleys
A mild climate with winter rain and occasional winter frosts, but with an overall water deficit for crop growth, is characteristic of the area extending from Santa Barbara County to the San Francisco Bay area, and inland to the great Central Valley. Precipitation ranges from slightly less

The coastline of Santa Cruz County is both spectacular and highly productive. Coastal areas of California produce many varieties of cool season vegetables for national consumption throughout the year. *Photo by Robert Campbell.*

than 380 mm (15") per year in some of the inland valleys, to more than 1000 mm (40") on some of the higher coastal mountains. The potential water loss by evaporation or use by fully watered crop plants varies from 760 mm (30") in more inland areas to about 685 mm (27") per year adjacent to the coast. The natural vegetation adapted to this climate is annual grassland, grass-oak woodland, and chaparral in the areas of greatest soil water deficit, Douglas fir and redwood forests occur in this region.

This is a rough mountainous area with peaks of moderate height and long narrow valleys. The soils of the area are mainly alluvial deposits along the river valleys, derived from the sediment from the natural erosion of shallow mountain soils on steep slopes. The nature of the alluvial soils is determined by the source of the sediment from which they were formed, and the age of the deposit. Where the upland areas are dominated by sandstones, shallow brush covered soils such as the GAVIOTA and RELIZ series develop. Where parent material is soft calcareous sedimentary rock, soils favoring good grass growth such as the LINNE and the ZACA soil series form. Southeast of Atascadero is an area of low granite hills giving rise to young, shallow granitic soils whose eroded sandy materials find their way down the Salinas River, giving its bed a characteristic broad white sandy wash appearance.

The valley soils such as those in the Salinas River Valley range from various recent flood deposits on the immediate banks of the stream to older, often red-colored soils on the higher terraces. For example, in the vicinity of King City sandy soils such as the ELDER and the ARROYO SECO series occur on the recent alluvium. The METZ series, a highly productive agricultural soil, also is found in this area. It is on recent alluvium, and in contrast to the other soils it is calcareous[4] throughout its depth. With distance downstream or laterally from the river channel, the alluvial deposits become finer textured. The SALINAS and CROPLEY soil series represent this stage. Soils on the adjacent terraces are represented by the LOCKWOOD and CHUALAR series. This complex of soils forms the basis for

one of the most intensive vegetable farming areas in the state.

Agriculture in the central coast range area is concentrated on the recent alluvial soils where water is available. The mountain slopes are devoted to rangeland use on the many good grassland soils, and with clearing of some of the brushy chapparal the grass cover has been expanded—as in the vicinity of San Ardo along the upper Salinas River. Specialty agriculture flourishes where the long valleys debouch upon the coastal plain and have the benefit of the cool summer maritime climate. Flower seed production and artichokes are examples of the crops making use of the unique combinations of climate and soil.

The Great Central Valley Regions

The "Great Central Valley" of California usually gives rise to an impression of one large valley with seemingly uniform agriculture and land use. In reality, this area, like the entire state, can best be understood if subdivided into regions— each with its characteristic agriculture reflecting differing soils and climate. But some generalizations are possible:

1. The soils of the eastern portion of the San Joaquin Valley, originating from the granitic rocks of the Sierra Nevada, tend to be coarser in texture. They generally have a greater sand content than those of the eastern portion of the Sacramento Valley which come from a mixed parent material higher in detritus from igneous and metamorphic rocks.
2. Soils on the western side of both the Sacramento and San Joaquin Valleys are generally finer textured, originating from parent material of the highly metamorphosed rocks and sedimentary rocks of the coast ranges. Since this area is in the rain shadow of the coast range, the soils develop characteristics typical of arid areas. In the southern portion of the San Joaquin Valley this is particularly the case with some alkaline soils[5] containing appreciable quantities of calcium carbonate and gypsum due to limited leaching. Such soils may have peculiar fertility problems related to the tie-up of essential elements such as phosphorus, or trace elements such as zinc.

4. A calcareous soil has calcium carbonate present, indicated in the field by fizzing (releasing carbon dioxide) when a mild acid solution (such as lemon juice or 0.1 N acid) is applied to it.

5. Alkaline soils are high in salts usually of calcium and sodium, and thus have a pH higher than 7.0, similar to baking soda.

Also, the presence of excess boron in some of these soils limits certain sensitive crops such as citrus.

3. Approaching the foothills, in the northern portion of the Sacramento Valley and in the eastern side of both the Sacramento and San Joaquin valleys, one encounters old valley terraces. Many of these soils are very old and characterized by such features as claypans or hardpans at varying depths form the surface. Their fertility is generally lower than the soils of more recent origin in the lower portions of the valley. The lower portion of the Sacramento Valley, from Sacramento northward to Marysville has fine textured soils with comparatively poor drainage. This disadvantage for most crops has been turned into an asset in using these soils for rice crops.

4. The delta of the Sacramento-San Joaquin rivers, as discussed above, has distinct soil properties. Some of the most intensive agricultural production in the state occurs on more than 300,000 acres here.

Because of their unique characteristics, these regions of the Great Central Valley will be discussed separately.

The San Joaquin Valley

The San Joaquin Valley, the southern portion of the Great Central Valley, has a warm climate, relatively little frost, low rainfall and a large annual water deficit. A large part of the valley, particularly the southern third and the westside, receives 250 mm (10″) or less precipitation while the loss of water either through evaporation from open water surfaces or from irrigated soil is as much as 1000 mm (39″) per year. Thus, if precipitation is the only water source, there is a deficit of nearly 760 mm (30″). The result is that natural vegetation in the valley is grassland composed of species that are annuals—green only during the rainy season, and yellow-brown and dry the remainder of the year. Where groundwater occurs near the surface, in the vicinity of rivers or old lake basins, the natural vegetation is tule swamps or riparian willow and shrub growth. Occasional valley oak groves occur on eastside alluvial fans where groundwater is available at moderate depth.

The San Joaquin Valley has a great variety of soils. The topography of the valley bottom—

basins, alluvial fans, flood plains, stream ridges or terraces—results in various soil types with agricultural practices and crops adapted to them.

Soils on the west side of the valley are finer textured then on the east side and nearly all are alkaline or calcareous. On the lower parts of the west side alluvial fans the soils are commonly saline. Typical soil series are the PANOCHE, PANHILL, ORESTIMBA, and VERNALIS. Coarser textured basin rim soils on the east side are typified by the FRESNO, POND, TRAVER, and MILHAM soil series. Sandy deposits, windblown from dry stream and river courses during the summer, occur over some parts of young river fans on the east side, giving rise to the CALHI, DELLO, and the DELHI soil series in association with HANFORD, GREENFIELD, and DINUBA soils formed from the alluvium. Frequently, abrupt changes in crop types mark the boundaries between these contrasting groups of soil.

Recent alluvial soils of the flood plains and alluvial fans on the east side of the valley are the TUJUNGA, HANFORD, HONCUT, and the YETTEM soil series. They are coarse textured. Outcrops of non-granitic rocks occur in the Sierra foothills bordering the valley, such as diorite or basalt and andesitic volcanic tuff. Materials from these give rise to heavy textured valley soils such as the PORTERVILLE and the CENTERVILLE soils in the south, and the PENTZ and PETERS series in the northeast portion of the San Joaquin Valley.

Elevated older terraces in the San Joaquin Valley are characterized by redder soils, with fine subsoil textures frequently underlain by hardpans. These include the REDDING, SAN JOAQUIN, MADERA, and the YOKOHL soil series. Some old stream deposits occur on these terraces and the deep, sandy material gives rise to the reddish RAMONA soil series, which where reworked through wind movement and deposition, has subsequently become the ATWATER soil series.

This wide variety of soils gives rise to a mosaic of agricultural crops and uses. The sandy soils are used for grapes and melons and the finer textured terrace soils for fruit orchards. Where soils are older and have developed hardpans they may be used for grain crops or rangeland grazing. Some of the terraces offer a climatic advantage in being elevated above the winter and springtime temperature inversions of the valley floor that result from cold air drainage and radiation cooling. Such "thermal belts" may be planted to citrus.

The practice of irrigation and soil modification has greatly influenced both cropping practice and productivity. Where irrigation is available in this area of long growing seasons, high light intensities, and warm temperatures, yields have been spectacular. Grapes, particularly the high sugar table, raisin and varietal wine types deserve special mention as one of the more important crops from an economic standpoint, but poultry, cotton, cattle, and a whole spectrum of products all represent significant output from this intensively managed region. Statistics on annual crops and livestock tend to be exceedingly variable and responsive to market demands. On the other hand, fruit crops requiring several years to reach productive levels are large and stable sources of productivity of the San Joaquin Valley.

The future of such a complex and intricate production structure is subject to many variables which are difficult to evaluate. From a long-term standpoint, however, perhaps the most significant problems are those dealing with water management and drainage of waste waters. Most of San Joaquin Valley agriculture depends upon a large supply of irrigation water, much of which comes from the Sierra. Although limited, it is of high quality. These waters, coupled with pumped ground water, meet most of the need. In recent years, additional water has been imported from the north by way of the California Aqueduct. The use of added fertilizers, however, and extensive use of pesticides and herbicides pose problems which demand attention. A major problem is the maintenance of irrigated agriculture as a permanent activity without degradation of soil or water quality. The high productivity of the region makes solution of these problems highly important. The agricultural output of Fresno County alone exceeds that of many of the states of the Union, and it is but a small part of this region.

The Delta of the
San Joaquin-Sacramento Rivers

The Delta region is by no means an area of uniform soil, climate and crop production. It consists of a number of "islands" of several thousand or more acres, each surrounded by the meandering channels of the Sacramento, San Joaquin and other tributary rivers. Near the channels, natural levees were formed by the periodic overflowing of the river system with the main body of each island being somewhat lower in level and in most cases, near sea level. As an extensive marsh, prior to settlement, the Delta supported a large population of wildlife and served as a major center and way point for migratory water fowl. Even today, the income from portions of the Delta where drainage is temporarily suspended to attract wildfowl during the hunting season rivals the income which might be obtained from agriculture on the same area. Starting in the latter part of the 19th century, these islands were successfully drained after the natural levees along the channels and around the islands were improved. This process required massive engineering efforts. Although a small region, the Delta is important in agricultural production because of its unique soil characteristics, mild climate, and ready availability of water. The rainfall averages 380 mm (15") per year while the yearly evaporation or plant water use potential is around 760 mm (30") per year. Although this would indicate a water deficit, the low elevation, delta location, and resulting high water table created naturally marshy conditions resulting in a luxuriant growth of tules. Originating in this accumulation of plant detritus over millenia, delta soils are rich in organic matter, with variable amounts of inwash of mineral sediment from the adjacent river channels. Typical soil series in such a topographic sequence of decreasing organic matter are the RINDGE, KINGILE, EGBERT, and SYCAMORE. The SYCAMORE soil series was formed from earthy mineral deposits along the Sacramento River, some of which originated from hydraulic mining in the Sierra Nevada gold fields. The mineral soils support orchards, while the organic soils such as the RINDGE soil series are used for corn, tomatoes, and other field crops.

The delta soils are highly productive, but each has its own management problems, depending on the depth to groundwater, permeability, and need for drainage. The irrigation method of the organic soils is unique; water flows laterally through them from small, shallow ditches. Drainage of these soils by pumping has exposed the organic materials to loss through oxidation, wind erosion and compaction. The removal of

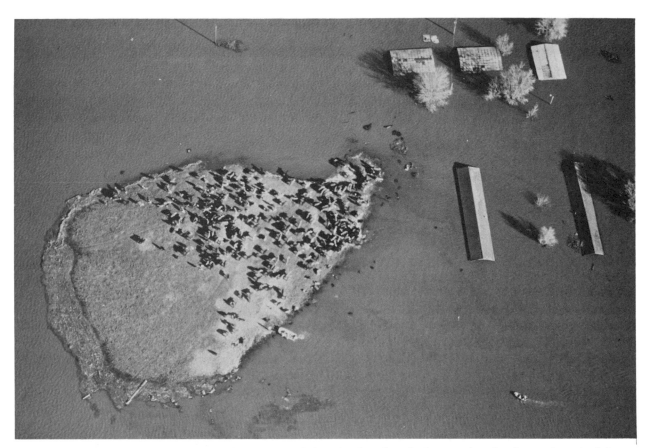

Periodic floods endanger islands in the Delta. Here cattle have gathered on high ground as floodwater breached the levees on some tracts in 1980. *Photo by William Wildman.*

organic matter by oxidation[6] has resulted in a relative increase in quantity of silt and clay present. This improves the surface texture and other features of the soil, so that from the standpoint of management, the soils are perhaps better than when they were first drained. Because the organic matter also contained nitrogen and other minerals, as these soils were oxidized each year, additional fertility was released and made available for the growing crops. Delta soils have been particularly fertile because of this.

However, since the soil material itself has been oxidized, the result has been subsidence, or lowering of the soil surface. As a consequence these islands are now 6 meters (20') or more below sea level in many places. Their protection from flooding becomes an increasingly difficult problem. The levees, particularly where man-

6. Oxidation is a biochemical change whereby oxygen or its chemical equivalent is utilized by soil microorganisms to decompose the organic matter of the peaty soil, causing its loss through transformation to carbon dioxide released to the atmosphere.

improved, rest on the surface of organic material, an unstable foundation. Their weight is sufficient sometimes to displace the organic foundation beneath and they collapse, leaving the islands open to massive flooding such as that which affected Brannan and Andrus islands in the mid-1970s and the Jones tract in the 1980s.

Simply to abandon the Delta to flooding will not eliminate the problems. The soil levels have lowered so much through oxidation that the islands would now becomes inland bodies of shallow water quite different from the marshlands which they once were.

Wind erosion has been another serious problem. The organic soils are very light when dry, and subject to erosion by the high winds of the area. This problem can be handled by interrow planting of taller crops.

The extremely high productivity of delta soils and their convenient location near markets, emphasizes the need for some solution to the problems. Maintenance of the levees to protect these soils from flooding will require an increasing investment. The dilemma is that continued cropping accelerates their subsidence.

In addition to these problems, water quality changes are occurring. Salt water intrusion due to tidal influence is now affecting the ground water of the area. A proposed canal around the periphery of the Delta has been designed that could increase fresh water flow through the area.

All of these problems seem to indicate a need for a regional plan designed to address the complexities of delta agriculture. An integrated systems approach rather than the present piecemeal, problem by problem approach is needed to maintain the integrity of the area.

The Sacramento Valley

Like many other regions of California, the Sacramento Valley also has a deficit of water need over natural precipitation supply. The precipitation ranges from 380 mm (15″) at its southern end to nearly 1000 mm (40″) in its northern end. The annual water need for vegetation, or loss to evaporation, ranges from 760 mm (30″) at the southern end of the valley to more than 900 mm (36″) at its northern end. Fortunately, in most areas near the Sacramento River and its tributaries, groundwater supplies are available. However, irrigation water supplied by canals is necessary for many of the areas away from the river. Temperatures are mild through most of the year in the southern part of the valley but the northern third is subject to frequent frost in winter, and the summers are much warmer.

The relationships between Sacramento Valley topography and the formation of its soils are similar to the San Joaquin Valley. Almost all of the soils are formed from alluvial material deposited in the past by creeks and rivers. The location of soils in basins, along river flood plains, or on various levels of terraces representing older valley bottoms or fans, determines their general nature.

On the west side of the valley where the climate is drier, soils which are alkaline, such as the WILLOWS and the MARVIN soil series, have formed in fine textured materials in basin areas. On the east side of the valley the same types of basins contain the finer textured STOCKTON and SACRAMENTO soil series. The mountains surrounding the northeastern side of the Sacramento Valley are dominated by the volcanic rocks of the Cascades. Here one finds a sequence of soils occurring on the stream deposits of eroded materials. For example, the Butte Creek fan near Chico has a sequence from its upper to lower end of the MOLINOS, VINA, ANITA, and finally the clayey STOCKTON soil series in the furthest downstream basins.

The river floodplain soils bordering the Sacramento River and its tributary streams occur at a few feet higher elevations than the basin soils along the center of the valley. The alluvium is a mixture of sediment from sources in the coast range to the west, the Siskiyou mountains to the north, and the Cascades to the north and east. The COLUMBIA soil series, a moderately coarse textured soil, occurs in the valley along the upper and middle reaches of the Sacramento while further to the south the SYCAMORE soils are dominant. The Feather River, a major tributary, is flanked by HONCUT and WYMAN soils. Along the western side of the valley the soils are formed on alluvial deposits from the erosion of sedimentary rocks with some interbedded serpentine and metamorphic rocks in the Coast Range. On the most recent of these one finds the YOLO and the CORTINA soil series. As these sediments age, the soils develop finer textured subsoils, as in the ZAMORA soil series. In the basins between streams, the clay textured MYERS soil series is found.

There are terraces along both the eastern and western margins of the Sacramento Valley which are only a few miles wide from its southern end to the vicinity of Orland. They widen toward the north to form most of the valley floor north of Red Bluff. The low terraces of the southern part are occupied by the HILLGATE and the TEHAMA soil series, which generally have smooth to undulating surfaces and are suited for irrigation if carefully managed. Further north, the terraces are older and higher, and the older reddish soils such as CORNING, RED BLUFF, and REDDING soil series form. These have fine textured subsoils, the REDDING having a silica iron hardpan and numerous fertility problems. Old volcanic mudflows cover wide areas of the terraces along the northeastern Sacramento Valley bottom, giving rise to shallow stony soils called the TUSCAN soil series.

This wide range of soil characteristics mainly determined by topographic situation brings about a variety of agricultural crop potentials. The southern portion of the valley has many level alluvial fields suited to intensively managed crops such as tomatoes, maize, orchards, and irrigated grain. Most of the finer textured basin soils are used for rice production and, depending upon the market, sorghum, safflower, and other crops. Further to the north on the older terrace soils there is dryland grain farming. The foothills on either side of the valley support extensive grazing and both irrigated and dryland hay crops. Fruit and nut crops including almonds, pears, and others, occur on the more recent soils on the upper areas of the stream fans, or along the low ridges of recent sediments deposited along the ivers. In areas relatively free from frost, such as in the northwest portion near Orland and Corning, olives and citrus fruit are grown.

Level alluvial soils in the Sacramento Valley are managed so intensively that total production is high. The older, more strongly developed soils are comparatively difficult to manage, but share the same climate and are potentially highly productive providing they can be supplied with adequate water, drainage, and fertilizers. In the context of present day agriculture, the concept of a "fertile" soil becomes somewhat meaningless, since all intensive agricultural management depends upon the supply of the elements necessary for fertility from some external source, and the proper physical manipulation of the soil. Problems of drainage in hardpan or clay pans of some of these older soils can be corrected, or in many cases even exploited for the conservation of water and mineral nutrients.

The climatic factors—length of the growing season and yearly total of solar energy—may well become the more important determinants dictating the nature of future agriculture in the area. Soils which are not now being intensively managed may take on greater importance in the future. It is even possible that new, man-made soils might renovate the old terrace surfaces. A common practice in Europe during the Middle Ages was a process called "colmatage," in which soil washed from the hills was used to create new fields. An inadvertent example of this can be seen on tens of thousands of acres of the outwash from the old Cherokee Hill Gold Mine north of Oroville. The upstream portions of the debris from this hydraulic mine were too coarse and are an infertile sandy wash, but along the lower reaches of the deposit a medium textured soil called the RAMADA soil series has been cultivated and successfully planted to almonds. Most of California's valley soils were formed by nature in similar processes.

Major Watershed Regions

The regions of California which are water sources for agriculture are the major forested regions of the state which also provide timber crops as well as summer and winter recreation. In the Sierra Nevada, Cascade ranges and the North Coast ranges and valleys, forest and man have operated as a team to furnish a crop that is truly natural to the region—a contrast to agriculture which usually depends upon exotic crop species. The following regions of the state have this natural form of agriculture, utilizing native forest tree species as the crop—or in the case of the grassland and shrub areas, the natural vegetation which is harvested by grazing animals.

The Sierra Nevada and the Cascade Range
The climate of the mountain regions of California presents a contrast to most of the regions discussed earlier. Precipitation totals in the Sierra through the year result in a large annual water surplus. At higher elevations it occurs mainly as one of the deepest snow packs in the United States—frequently storing more than 1000 mm (40") of water, which melts quickly in the spring. Precipitation increases with elevation, from a minimum of about 380 mm (15") in the lower foothills to between 1000 mm (40") and 2030 mm (80") at the middle elevations of 1500-2750 meters (5,000' to 8000'), decreasing in amount above this. Precipitation may vary widely from year to year—from twice the mean in very wet years to half the mean in dry years, while the water used by vegetation or lost by evaporation varies little. Much of the excess pours out of the Sierra Nevada as a springtime snowmelt, which if unobstructed, would flood the great valley below before running to the sea. To turn this natural sequence of destructive flooding to use, the snowmelt flood is stored in reservoirs and released during the long period of summer heat to meet irrigation needs. The large water surplus in the mountain region results in soils which are leached of salts and thus tend to be more acidic than those of the other regions. These acidic soils favor the coniferous forests typical of the area.

The Feather River separates the basalt and andesitic volcanic rocks of the Cascade range to the north from the predominantly granitic and metamorphic rocks of the Sierra Nevada to the south. This difference in geology distinguishes two major soil regions.

The Coast Ranges flank the broad, flat grain fields of Colusa County in the Sacramento Valley. *Photo by Jack Clark.*

Forested regions of northern California receive high rainfall and serve as water producing areas for regions to the south. Mt. Shasta caps this view of Shasta County. *Photo by Robert Campbell.*

The soils of the Sierra Nevada are often shallow and stony at high elevation. Glaciers eroded previous soils from the high mountains, leaving barren rock surfaces. These have weathered very slowly in the cold climate to form very shallow soils or none at all where bare granite slopes remain. The CHIQUITO soil series is an example. It is a shallow sandy soil at elevations of about 2750 m to 3000 m (9,000' to 10,000') in the San Joaquin, Kings, and Kern river drainages.

At lower, relatively warmer elevations outside the influence of glaciation, older, deeper, and reddish brown colored soils of the HOLLAND series occur. At still lower elevations, where long time weathering of the rock has occurred due to large amounts of moisture and warmer climate, the reddish brown MUSICK soil series is found just within the lower edge of the main coniferous forest belt of the Sierra Nevada. Below this forest belt, the foothill climate is drier and warmer, giving rise to soils such as the AUBERRY and the sandy VISTA soil series. These soils support annual grassland and oak woodland vegetation.

The Sierra Nevada granitic core has a wide belt of metamorphic rocks along its westerly edge, north of the vicinity of Mariposa and the Merced River. Shallow to moderately deep reddish soils have formed on these lower elevation unglaciated areas. The AUBURN series is a widespread example while to the south, the similar MARIPOSA soil series occurs.

The Cascade Range to the north of the Feather River has soils derived from basalt and andesite lava flows and associated ash, pumice deposits, and volcanic mud flows. Some of these also cap the granitic ridges in the northern third of the Sierra Nevada. Depending upon the age of these volcanic formations, the soils range from the older, red AIKEN series, through the intermediate age brown COHASSET series, to the younger stony grayish brown and dark brown MC CARTHY, WINDY, and WACA series. The latter are at higher elevations where soil formation is slower. Toward the northeasterly portion of California across the crest of the Cascades, precipitation decreases in the rain shadow of the mountains, and the forests end where the soils become much less acid due to lower rainfall. In the area around Alturas, the LASSEN soil series, a fine textured calcareous soil on basalt, is found. It supports a vegetation of sagebrush, Juniper, and grass.

The higher slopes of the Sierra Nevada and Cascade mountains were severely eroded by glaciation during the ice age. Locally, numerous bouldery, sandy soils are found on glacial moraines and glacial outwash deposits resulting from this glacial erosion. The vegetation of the Sierra Nevada and the Cascades ranges is generally coniferous forest, with local areas of brush on shallow and stony soils, and grassland in meadows where groundwater is available. The coniferous forests are usually of mixtures dominated by ponderosa pine in the lower elevations, increasing amounts of white fir and then red fir at higher elevations. Finally, the highest elevations are dominated by lodgepole pine and scrubby whitebark pines and mountain hemlocks nearest the timberline. The areas where soils were eroded away by glaciers in the past have generally remained bare rock or with patches of lodgepole and scrubby whitebark pines, and hemlocks or Sierra Junipers in isolated soil pockets. The lower elevation mixed conifer forests of ponderosa pine, Douglas fir, incense cedar, sugar pine and white fir form the main commercial timber harvest areas.

The main use of the mountain soils is for timber production. Timber harvest, however, is partially contingent upon watershed management requirements, since these areas also represent the water supply regions of the rest of the state. Forest cover is essential in the prevention of erosion, and in the maintenance of water quality. Recreational use has preempted some of the more scenic areas. Fortunately, these are usually at higher elevations where soils are shallower and timber site productivity is lower. The productivity of the soils for timber depends upon sufficient rain and snow (usually more than 635 mm (25") precipitation) plus adequate soil depth to store moisture for the summer drought period. Thus the deeper soils at the mid elevations of the western slopes of the Cascade range and Sierra Nevada are most productive.

A large subregion of the Cascades occurs as the east side lava plateau of Modoc and Lassen and eastern Siskiyou counties. These areas on the lava plateau have low precipitation, warm summers and long cold winters. Topographic position in relation to the lava flows and intervening valleys is a primary determinant in the types of soil and subsequent agriculture. Occasional influence of pumice and volcanic ash fallout affect the soils. The flat lava surfaces are dominated by a brown forest soil called the

TOURNQUIST soil series. Volcanic scablands bound broad basins where inwash of volcanic ash has occurred. Large open meadow or sagebrush flats occur in these basins. Where the groundwater table is close to the surface, meadow soils favor the main agricultural pursuit of pasture management which supports extensive range use of adajcent slopes. The vegetation of this eastside lava plateau country is mainly ponderosa and Jeffrey pine forest with occasional Sierra Juniper, interspersed with broad grassland meadows and sagebrush and bitter brush flats.

The agricultural use of soils on this plateau is dominated by the short growing season and the susceptibility to frost at any time of the year. The crops are usually grass or grain on the basin and alluvial soils, with grazing on the upland areas and some timber production on the higher ridge areas. Potatoes are a specialty crop in the northern basins of this region along the Oregon border. They are grown in broad valleys such as the Tule Lake Valley south of Klamath Falls, and in the Butte Valley to the west.

The North Coast Ranges and Valleys

The climate of this region produces a large annual water surplus due to precipitation which ranges from 635 mm (25") to more than 2500 mm (100") per year, mostly as winter rainfall. Potential water use by vegetation or loss through evaporation is about 680 mm (27"). This produces a large water surplus amounting to more than 1750 mm (70") of runoff in some watersheds. This occurs as wintertime peaks in streamflow, with floods in very wet years. These occur at intervals of from 20 to 50 years, devastating the agricultural and urban areas in the narrow alluvial valleys. Most of the water surplus from this region flows into the sea, forming scenic rivers and excellent fisheries of salmon and steelhead for recreational and commercial use. The flow to the sea in this region amounts to more than one-half of California's potential water supply. The Mattole River, for example, yields as much water to the sea as all the coastal California rivers south of San Francisco Bay. This surplus water in the north has produced a strong drive from water deficient agricultural regions of California to divert at least a portion of this flow to the Sacramento River. In addition, all the urban regions of California are in areas of water deficit, and thus their populations largely depend upon importation of water from such water surplus areas.

As in the southern Coast Ranges, the north coast region is dominated by a topography of long narrow river valleys usually trending northwesterly, spreading out into broad valleys where the rivers discharge into the ocean.

The soils are either mountain soils derived mainly from sedimentary sandstones and shales, and occasionally from intrusions of serpentine and basalts, or are valley soils formed in alluvium from the natural erosion of the mountain slopes. The uplands in this area have some of the highest natural erosion rates of the world and thus there is rapid alluvial soil formation in the valleys. Deposits of one or two feet of new soil are not unusual in the valley bottom areas during the periodic floods, and these deposits quickly become productive soils unless they are too sandy or stony. These natural high rates of erosion impose constraints on land use of the mountain slpes. Thus, road construction, timber harvest, and grazing need special care with regard to erosion control measures in this region.

Depending upon their age and rates of formation, the mountain soils range from the shallow brush-covered MAYMEN soil series, to the grey brown HUGO, reddish brown JOSEPHINE, or the red SITES soil series. The redder soils indicate older topographic surfaces in the area. Frequently these are old residual plateau tops that have not yet eroded. In southern Lake County, an area of volcanic rocks occurs on which soils somewhat like those of the Cascade Range have formed, the COHASSET and the GUENOC soil series being dominant. Along the coast near Fort Bragg and inland for twenty miles are soils of the coastal terraces. Some of them are infertile sands derived from ancient dune and ocean shoreline deposits—the BLACKLOCK soil series. These support pygmy forests of cypresses and pines that contrast with the very tall redwood and Douglas fir forests on adjacent soil types. (Some people have transplanted pygmy cypress seedlings from this area to their home gardens expecting to have miniature tress, but have later found that such seedlings may grow more than one hundred feet tall on fertile soil.) Inland from these terraces are deeply weathered soft sedimentary rocks giving rise to the reddish brown MENDOCINO soil series, one of the most productive timber soils in the north coast area.

Alluvial materials derived from the mountain soils give rise to the productive valley soils of the region. The LOLETA, SOQUEL, and FERNDALE soil series are the immature soils of the coastal valleys, with YOLO soils forming in valleys further inland. Finer textured alluvium in less well drained parts of some interior valleys give rise to the clayey CLEAR LAKE soil series widely used for pear orchards. Each of the valleys has its own set of local terraces which give rise to older and more strongly developed soils. For example, the SAN YSIDRO and the PINOLE soils series are found on terraces in the Anderson Valley near Boonville and in the upper Russian River Valley near Ukiah.

The vegetation of this region is mainly coniferous forest. Redwood and Douglas fir are the main timber species. Hardwood trees such as tan-oak and madrone are associated with them. The magnificent redwood groves, some of the world's tallest trees, are confined to deep alluvial soils in the valleys. Most of the commercial timber forests are on hill and mountain slopes. However, a stand of redwood on the valley bottom of the Big River east of Mendocino City has one of the world's record rates of growth for timber at more than 5,000 board feet per acre per year.[7] Redwood extends inland about as far as the influence of marine air or where geologic changes result in soils that do not meet its fertility requirements. Further inland, ponderosa pine, as in the Sierra and Cascade ranges, becomes a component of the forest along with Douglas fir. A belt of grasslands occurs along the coastal margin, often extending inland as wind-swept ridgetop prairies with very dark organic soils such as the KNEELAND series. These have a very high carrying capacity for grazing animals.

Agriculture in this north coastal region is concentrated on recent alluvial soils, and the adjacent terraces and lower hill slopes bounding the valleys. Orchards, alfalfa, and irrigated pastures are found on the lower valley soils, while vineyards and rangelands occupy the terraces and lower hillsides. The upland soils are used mainly for timber production where depth is adequate to maintain trees through the summer drought period, and where the rainfall is higher. These are generally soils deeper than one meter (39"). Shallow soils may be occupied by chaparral brush, or, if the rock contains sufficient calcium, by grassland glades. Brush fields are hazardous

wildfire areas subject to periodic fires if left to re-grow naturally. Efforts to convert the brush to productive and less fire-hazardous range grass-land have been successful on some soil types in western Lake County. Naturally brush-covered deep soils derived from volcanic rocks south of Clear Lake have been cleared for planting to walnut orchards.

The prevailing cool summer climate and usual occurrence of late spring rains allow especially good prolonged grass production and high carrying capacities to support range cattle in coastal areas of northern Mendocino and Humboldt County. Similar climate gives rise to a unique lily-bulb industry on the wide coastal terraces of Del Norte County north of the Smith River.

Regional Issues Affecting Production Agriculture

In each region of California there are problems relating to the soil, the climate, and the crops and vegetation of the area. These generally require a societal solution, since they extend beyond the boundaries of any one landownership. Examples are water supply allocation and distribution, the provision of adequate drainage, sand dune control, erosion control, and air pollution control.

Landowners are often unaware of the effect of their use of the land in the total landscape context. As a consequence, houses and cities are built in locations which invite disaster from brush and forest fires, or subsequent floods and landslides. Land use problems of this type become massive, testing the wisdom of planners and the courage of legislators and others. Too often, decisions are made on the basis of short-term economic needs of individuals or small groups, and not on the basis of long-term public need.

The Central Valley

The soils of this vast area, among the most valuable in the world, are exposed to a number of hazards.

INAPPROPRIATE LAND USE. The conversion of prime agricultural land to other uses is, from the long range view, both wasteful and short-sighted.

ACCUMULATION OF SALT. This problem is difficult to control, sometimes difficult to recognize and

7. Growth rates measured by Professor Emanuel Fritz on a flat along the Big River east of Mendocino in Mendocino County and reported in the *Journal of Forestry*, January 1945.

Vineyards are terraced into the hills surrounding Spring Mountain in the Napa Valley. The microclimates of such undulating landscapes give rise to special qualities in production of grapes and other tree and vine crops. *Photo by Jack Clark.*

increasingly likely as water supplies become more limited. Soils in the lower portion of both the Sacramento and San Joaquin Valleys are particularly subject to this threat.

HAZARD FROM RESIDUES. We have come a long way since the use of arsenic, copper and other heavy metal pesticides was largely discontinued. However, other potential hazardous herbicides and pesticides will continue to be necessary. Their use will have to be continually monitored to keep the risk to the soil resource within tolerable limits.

The Delta

Particularly significant to the soils of the Delta are wind erosion, salt water intrusion, and subsidence. In some areas, subsidence is the overriding problem; in others, wind erosion or salt water are of more concern.

Forest Regions

These lands, too, have overall problems that threaten their productivity.

INAPPROPRIATE LAND USE. A forest that has been growing to harvestable size for 50 years may suddenly be preempted as a scenic backdrop or second-home tract. Large areas of Califorinia's most productive forest land are currently set aside for exclusive recreational use, under the "single-use" concept. The solution requires demonstrated timber harvest techniques that are compatible with other uses.

LOSS OF SITE FERTILITY. A large proportion of the fertility of a forest is often stored in the tree biomass, with greater proportionate amounts in the branches, twigs and leaves. Today's emphasis on total harvest—for example, to augment fuel supplies—leads to removal of these smaller but more fertile materials, which previously were returned to the soil. Consideration of fertility balance in each forest site may be needed.

SOIL EROSION. This is a particular problem with road construction for highways, subdivisions and timber harvest access. Timber harvest effects on soil erosion are most serious where the soils are shallow and held in place on steep slopes mainly by the vegetation. Most vulnerable to erosion are the immature soils derived from granite as in the Sierra Nevada, and the Siskiyou and Trinity Alps. Some of these soils may require a very constrained timber harvest.

CATASTROPHIC FIRE. With human encroachment on forest lands, the potential for wildfire has in-creased. A regionally integrated fuel management program is needed to solve the problem. Such a program may require very judicious use of "controlled burning," with all the inherent liabilities of that procedure. Adequate removal of residues (slash) following timber harvest—either by utilization or by controlled burning—also is necessary. A similar problem is that of brush control on range land.

WATERSHED STREAMFLOW AND WATER QUALITY. Production and protection of water supplies is a prime objective for most forest areas of California. Trees, however, use water; timber harvest increases water yield. A problem is how to maintain this increase while growing a new timber crop. Excessive removal of forest over too large an area of watershed may result in faster runoff and hgher flood peaks. Also, much of the high quality of California water is due to the filtering action of organic, well-vegetated forest soils.

Desert Regions

Soils of the desert are subject to many of the hazards that threaten other regions, such as erosion, inappropriate use, and damage due to poor salt balance. An additional threat is the possibility of loss of the desert itself, with its clear dry air and unique natural vegetation, through intensive recreational use.

A central problem pertains to all these locales—maintenance of the natural regional quality that comprises the heritage of California. Thus, in the deserts there should be areas that still elicit the boundless feelings of looking across broad, undisturbed desert valleys; in the forest there should be areas which still have the majestic trees that are the wonder of people throughout the world; and there should still be areas of delta vegetation where waterfowl may rest in numbers that past generations of Californians have enjoyed. Our use of our California heritage must be wise enough to protect the splendor of the land itself.

Sources of Additional Information

For any of these regions there is still considerable local variation which has not been described here. Much of this information is in the form of maps and reports of surveys for geologic, vegetation, and soils information, and as climatic data. The major sources for this information are described below.

Climatological Data

The weather measurements from more than 600 observation stations form the basis for the climatological data of California. These data are published monthly in "Climatological Data— California" by the National Oceanic and Atmospheric Administration, Environmental Data and Information Service, National Climatic Center, Federal Building, Asheville, North Carolina, 28801. The current subscription price is $5.85 per year ($13.00 foreign) for these monthly reports, including an annual summary. The data are organized by regions which are practically the same as those used in this report. Stations are indexed on a map for quick regional reference.

Geologic Data

The geologic data useful for a more detailed understanding of California's regions are in the form of published maps and reports available from various agencies:

> U.S. Geological Survey
> 345 Middlefield Road
> Menlo Park, California 94025

or:

> California State Division of Mines and
>   Geology
> The Resources Agency
> 1416 Ninth Street
> Sacramento, CA 95814

In addition to the detailed surveys of local areas, there are some broad regional summaries of geology, such as the *Geologic Guidebook of San Francisco Bay Counties* (California Department of Natural Resources, Division of Mines), a 392 page book describing the local geology in detail. It offers many self-guided tour routes to major geologic features. A book which deals with the entire state is *Geology of California*, by R. Norris (Wiley, 1976).

Most soil surveys may be obtained free of charge at the local Soil Conservation Services (USDA) office. Some soil surveys and a list of available soil surveys may be obtained from the following:

> Department of Land, Air and Water
>   Resources
> University of California
> Davis, CA 95616

Maps of the soil and vegetation of wildland areas in private ownership are published in quadrangle form by the California Department of Forestry. Information as to their availability can be obtained from the following:

> Resource Inventory Maps
> Soil-Vegetation Survey, Room 1342-5
> 1416 Ninth Street
> Sacramento, California 95814

Vegetation Information

The available information on vegetation in California is either in survey map and report publication, or as other published literature.

Mapping of California vegetation began in the days of Spanish California when some of the missions, such as San Jose, had vegetation and land use maps made of their holdings. The California State Board of Forestry and State Division of Forestry have been involved in such mapping since 1886. Since 1926, this interest of the State Division of Forestry has been supplemented by the vegetation mapping of the U.S. Forest Service, Pacific Southwest Forest and Range Experiment Station. These efforts have led to a nearly complete vegetation map of the state, quadrangle by quadrangle, with a current update occurring in connection with the Soil-Vegetation Survey mentioned above. A list of the various types of vegetation maps can be obtained from the Resource Inventory Maps address.

The most comprehensive book concerning California vegetation is *Terrestrial Vegetation of California*, edited by M. Barbour and J. Major (John Wiley & Sons, 1977). This 1002 page book covers each of the major regions of the state in descriptive detail. It also has an excellent chapter on the climate of the state in relation to distribution of vegetation.

Specific local information on soils and vegetation can be obtained from the local county office of the University of California Cooperative Extension Service.

4

Water and Agriculture

J. Herbert Snyder

Equally as important to agriculture as the land resources are the water resources of California. Like the land, water in California is so complex a subject that it almost defies condensation into one chapter. Unlike land, however, water in California is *in motion* and the uncertain nature of its supply has led to complicated development of delivery systems, involving many economic and social issues. The objectives of this chapter are to give an improved appreciation for the complexities of California water resources, an awareness of the need for increasingly careful evaluation of future alternative use and development strategies, and an understanding of some of the political process that shapes and ultimately determines policies for this vital resource.

The importance of water to an efficient and productive California agricultural industry can hardly be overstated. Prior to the introduction of tilled crops and controlled grazing by Spanish and Mexican settlers, the Indian populations of the region depended upon the productivity of the landscape nurtured by natural precipitation. Native plants produced grains, grasses, bulbs, roots, fruits, and leaves that were used directly for human food and also supported large native animal and bird populations, which, along with fish in the streams, also contributed to the diet of the Indian peoples. These food sources depended on seasonal and geographic features that are still characteristic of certain aspects of California agriculture.

Though California is fortunate to have abundant water within its borders, the supply has unique aspects of seasonality and location. Winter storms bring significant amounts of precipitation to most areas, and snowfall in the Sierra Nevada creates a snowpack that retains large volumes of water until the spring snowmelt. Major rivers move water through the northern and central parts of the state, and permeable soils have allowed the development of substantial groundwater reserves. While generally plentiful, California's water is not always present at the right time and in the right place; so sophisticated water storage and diversion systems have evolved. Although the natural flow of water in California has thus been greatly modified, physical factors still clearly influence its availability, storage, and movement.

The history of the various efforts to manage California water resources reveals a range of increasingly complex issues. In most instances in the past, simple problems were solved by relatively simple approaches: surface flow from streams was diverted for use, wells were dug, then small dams and short conveyance structures were built. As agricultural activity grew, the variety and size of water development projects as well as management and use schemes also expanded. The resulting complex interaction among legal, economic, social, and institutional factors has given California major challenges in developing water use strategies. The following pages review these factors, particularly as they relate to California agriculture.

California Water and Its Origin

Much of California has a two-season climate. The winter months from November through February typically are wettest, during which nearly 75 percent of the state's average annual precipitation occurs. The period of April through October is typically much drier. Although average temperatures generally increase from north to south, the amount of precipitation and its distribution varies throughout California largely as a function of topography. However, there is a definite decrease in amount from north to south in all regions. A description of topography, the water cycle and weather relationships provides the basis for discussing water supply in California.

Geology and Landforms

As discussed in the previous chapter, California is a land of contrast containing many distinct regions with different rock types, soils and landforms. Mountain ranges cover more than one-half of the state's surface.

The Sierra Nevada occupies approximately one-fifth of the total land area of the state. Varying in width from 40 to 80 miles with elevations up to 14,495 feet (Mt. Whitney), it extends in length some 400 miles. The southern Cascade Range, consisting largely of volcanic peaks and lava flows, extends through north central California to the northern edge of the Sierra Nevada.

The Coast Ranges, the second of the two great mountain systems in California, extend the length of the state, trending roughly along the coast as parallel ridges broken only at the Carquinez Straits, where the Sacramento and San Joaquin rivers drain into San Francisco Bay. In southern California, the Coast Ranges merge with the Tehachapi Mountains which join with the San Gabriel and San Bernardino Mountains to form the Transverse Ranges. The Tehachapi Mountains, which form a short connecting link between the Sierra Nevada and the Coast Ranges, separate the fertile Central Valley from the Mojave Desert.

The Central Valley, between the Coast Ranges and the Sierra Nevada, extends north to south nearly 450 miles from the Klamath/Cascades to the Tehachapis. This alluvial plain contains the largest irrigated agricultural area west of the Rocky Mountains. The arid Mojave Desert, a land of broad flat basins and low mountains, lies to the southeast of the Valley. To the north of the Mojave and bordered on the west by the abrupt east walls of the Sierra Nevada are the north-south mountain ranges and valleys known as the Trans Sierra, which extend eastward into the Great Basin. The Owens Valley and Mono Basin of this area provide one major source of water for the city of Los Angeles. In the southeast corner of California, the Colorado Desert lies in a trough that continues southward into the Gulf of California. In this arid desert, an area that receives less than three inches of rainfall annually, lie the Coachella and Imperial Valleys, two man-made oases made productive by water from the Colorado River.

The Central Valley's two largest rivers, the Sacramento and the San Joaquin, after draining more than one-third of the state, flow into a complex network of interconnecting channels known as the Delta. The Delta, consisting of 738,000 acres of organic soils largely below sea level, is interlaced with 700 miles of meandering waterways surrounding 60 islands and tracts made possible by 1100 miles of levees. The Delta is a focal point of California's intricate water diversion systems. Thus, almost any controversy concerning water development and management in California involves the Delta.

The Water Cycle

Precipitation, as abundant as it seems at certain times and locations, comprises only a minute portion of water occurring in nature. Water may be classified or separated into different types, depending on where it is found—i.e., atmospheric, surface, or below the surface. The terms "water cycle" or "hydrologic cycle" are used to describe and analyze the relationships among the various forms of naturally occurring water.

Approximately 97 percent of the water to be found in nature is located in the salt water oceans covering more than four-fifths of the earth's surface. Nearly 2.5 percent is entrapped in polar caps. Radiant energy from the sun causes evaporation from the oceanic water surfaces, and naturally pure (salt-free) water is transformed to atmospheric form. Depending upon a variety of atmospheric and orographic conditions, the vapor in clouds condenses; when heavy enough, drops aggregate and fall in the form of rain, snow, sleet, or hail. Precipitation may fall directly back into the ocean, into lakes or streams, or onto land and plant surfaces. Some water is evaporated from land surfaces and some is transpired by plants back into the atmosphere. If the volume of precipitation is sufficient, an excess runs off the land surfaces, coalesces into streams, may be stored temporarily in lakes, or returns naturally to the oceans. Some water enters the ground under various conditions and may accumulate in aquifers—zones of groundwater storage—for varying periods of time from very long (up to thousands of years) to very short (less than one year).

The hydrologic cycle is of great importance, for it supplies continually renewed fresh supplies of water by moving it from the oceans to inland locations. Thus, man makes use of the water cycle to obtain the water needed for domestic, agricultural and industrial consumption. While non-irrigated agriculture makes direct use of the natural precipitation, the development of technology to store and transport water or obtain it from underground reservoirs has permitted the development of California's highly productive irrigated agriculture, as well as the urban and industrial economy.

Weather

California's weather is as diverse as its topography. The most obvious variations occur with changes in elevation from sea level to mountain peaks, with distance from the Pacific Ocean—the major moisture source—and with changes in latitude from subtropical to temperate locations. These factors result in a great variety of weather conditions throughout the state.

California generally has dry summers and wet winters. In summer an offshore high pressure cell forces the mid-latitude storms far to the north and gives the state relatively dry, clear weather. Winter precipitation comes almost completely from the mid-latitude storms when the high pressure cell has moved southward, permitting these cyclonic storms to traverse the state. Because California lies to the south of the center track of many of these storms, there is a heavier and longer rainy season in the north state than in the south.

As moist air from the Pacific Ocean moves east, it encounters California's two major barriers, the Coast Ranges and the Sierra Nevada. The air flow is lifted and cooled by the western

and southern slopes of the mountains, resulting in heavier precipitation on these sides than in the "rain shadow" on the lee side of the barriers. The much higher Sierras tend to catch the moisture that has evaded the lower Coast Range. The resulting precipitation pattern is shown in Table 1 and by Figure 1.

TABLE 4.1
Average Annual Precipitation at Selected California Stations*

Station	Elevation	Average Annual Precipitation
	In Feet	*In Inches*
Salinas	74	13.74
Fresno	331	11.14
Huntington Lake	7020	32.45
Bishop	4108	4.84

*These stations are similar in latitude but change in longitude from near the coast eastward. Salinas, located in a valley of the Coast Range, receives more rain than Fresno, which is in the Central Valley and in the "rain shadow" of the Coast Ranges. Huntington Lake in the western Sierra is on the windward side of that major topographic barrier. Bishop is in the "rain shadow" on the east side of the Sierra.

SOURCE: *The California Water Atlas*, State of California, Sacramento, 1979.

Mid-summer precipitation is limited almost completely to the southeastern deserts, the Sierra Nevada, and the eastern slopes of the Transverse and Peninsular ranges. Infrequent invading moist air masses from the Gulf of Mexico or the tropical eastern Pacific Ocean provide the source of precipitation for these arid regions and the only variation to California's almost universal winter-wet/summer-dry pattern.

Temperature, influenced by latitude, air flow and topography, also varies widely across the state. In desert regions the daily peak temperature remains above 90°F for over 150 days annually while in the Sierra Nevada and the Modoc Plateau, low temperatures averaging below 32°F for a similar period of time during the winter months are normal. Another notable feature is the large variation in summer temperature between the coast and interior. Moderated by sea breezes, temperatures at the coast may be as much as 40°F cooler than in the Central Valley. Conversely in winter, the moderating effect of the ocean produces higher maximum temperatures on the coast than the interior.

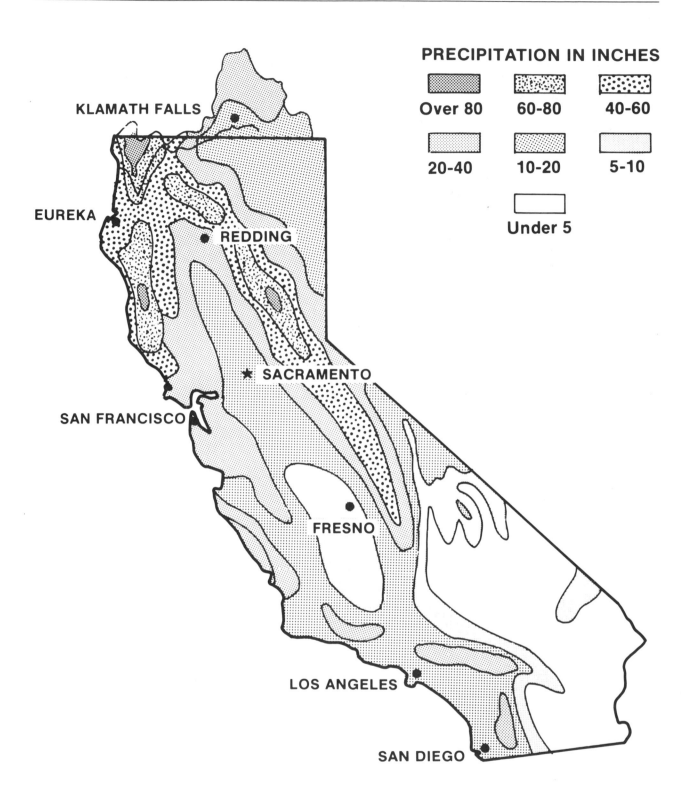

FIGURE 4.1 Mean annual average precipitation in California. The range of precipitation varies from less than 5 inches in the southern interior desert to more than 80 inches in parts of the northern coastal mountains. *Adapted from "California Water: A Report on Critical Problems and Proposed Actions," California Westside Study Team, U.S. Bureau of Reclamation, 1974.*

Rainfall and California Agriculture

The amount of rainfall and its seasonal distribution are both important factors in determining the type and range of agriculture possible in an area. Under natural conditions—i.e., without any effort to augment or manage the seasonally available moisture in the soil—crop production limits are set by the amount of water available for plant growth in the root zone during the growing season. Without irrigation, since there is little or no effective precipitation during the typical California summer, production is naturally restricted to amounts of water stored in the soil during the winter months. The preceding chapter discussed some of the natural vegetative and agricultural diversity that is found in California under non-irrigated conditions.

The mountainous regions of the Coast Range and the Sierra Nevada, although characterized by large amounts of precipitation (30 to more than 100 inches mean annual rainfall), do not support significant crop production because of limiting factors such as poor soil and shortness of growing season. The 18–30 inch mean annual rainfall of the Sacramento Valley can support a variety of fruit, vegetable, nut and vine crops as well as seasonal pasture, but the absence of precipitation to replace depleted soil moisture late in the season limits productivity for most nonirrigated crops and increases the risk of crop failure for others. The more arid and desertlike environments of the southern San Joaquin Valley and the Mojave-Colorado desert basin, with rainfall amounts averaging 5–10 inches per year, can support without irrigation only marginal yields of grain and seasonal pasture crop production, and that with varying degree of risk. The foothills of the western Sierra Nevada and the coastal plains of southern California, with mean annual precipitation of 10–30 inches, have supported a great variety of nonirrigated fruit, vegetable, vine and nut crops but population pressures and higher competing yield in irrigated areas have created serious inroads on agricultural land use in these regions.

Hydrology

Surface runoff provides about 60 percent of the water used in California while the remainder comes from groundwater supplies. The mean annual runoff from all streams in the state amounts to approximately 71 million acre-feet (MAF), but there are wide variations by month and geo-graphical location. The volume of runoff and the temporal patterns of flow depend on the type of precipitation—winter rain, summer convective storms or spring snowmelt. For example, along the Coast Ranges, precipitation is generally in the form of rain or, at times, snow that melts rapidly. Surface runoff occurs soon after a storm begins and diminishes after it ceases. In this area most runoff occurs between the months of November and March. In contrast, the much colder Sierra Nevada receives most of its precipitation in the form of snow which accumulates until the spring thaw. Consequently, runoff usually occurs after March, peaks in May and can continue to be strong through July.

Variation in streamflow magnitude and seasonality can be illustrated by comparing streams in different sections of the State as shown in the following table (Table 2). Where possible, the data reflect conditions prior to flow modification by dams and other water resources projects. As can be seen in the table, sustained and uniform streamflow is rare in California. The greater variability between high and low flows in the San Joaquin River than in the Sacramento River is due in large part to the fact that the Sacramento station drains a larger basin. In addition, the San Joaquin depends more on spring snowmelt for total flow, and the two basins have different drainage pattern characteristics. Many of the Sierra rivers show a later peak than the coastal rivers due to dependence on spring meltwater for peak flow. Coastal rivers show a general decline in water availability from north to south.

In many ways, spatial distribution of runoff presents a greater problem to California than does seasonal flow variation. Streamflow in many locations has been regulated by impoundments with a pattern of release that tends to smooth out the peaks and troughs in the natural streamflow regime. However, roughly two-thirds of the annual precipitation occurs in the northern one-third of the state with large amounts of runoff flowing away from areas with the greatest consumptive need. This natural distribution imbalance has been greatly overcome by construction of extensive water-transfer projects. The imbalance is further mitigated by the existence of huge groundwater reservoirs.

TABLE 4.2
Flow Characteristics of Selected Streams in California

Area Drained/River	Basin Size	Annual Average Flow	Peak Flow	Percent of Annual Average Flow	Minimum Flow	Percent of Average Flow
	Square Miles	*Cubic ft / Second*	*Cubic ft / Second*		*Cubic ft / Second*	
MODOC PLATEAU						
Pit	4711	2931	4449 *March*	126	2100 *August*	60
SIERRA AND CENTRAL VALLEY						
San Joaquin	1676	2381	6324 *June*	221	596 *November*	21
Feather	3624	5834	12120 *April*	172	1792 *September*	26
Sacramento	9022	11790	25030 *February*	176	4343 *September*	30
EASTERN SIERRA						
East Fork, Carson	341	356	1124 *May*	263	78 *October*	18
NORTH COAST						
Eel	3113	6896	19780 *February*	237	128 *September*	1.5
CENTRAL COAST						
Russian	793	1473	4211 *February*	236	146 *August*	8.2
SOUTH COAST						
Los Angeles	832	144	380 *March*	219	20 *August*	11

SOURCE: *California Streamflow Characteristics*, Vols. 1 & 2, U.S. Geological Survey, Menlo Park, California, 1971.

Groundwater in California

Given the extreme variation in precipitation and resultant surface water supply in California, settlers and farmers learned early that water stored in underground reservoirs provides a year around source of water for agricultural and domestic use. Earliest reliance was upon simple dug wells that penetrated only the upper water table, plus a few artesian wells. Development of drilled wells, internal combustion and electrically powered engines to drive pit centrifugal pumps, and later, deep well turbine pumps expanded the use of groundwater supplies in the state. By the mid-1920s there was widespread dependence on water from wells several hundred feet in depth. Groundwater use has expanded until it now supplies 40 percent of the total applied water used in California (about 16.5 MAF per year). This is approximately 25 percent of all groundwater pumped in the United States.

The bulk of California's groundwater is found below inland and coastal valleys in alluvial materials deposited by streams. Water in this younger alluvium is usually contained in sand and gravel layers known as aquifers. Water occupies the small spaces between individual particles. Clay and fine silt layers are usually interspersed with the aquifers. Although these are also saturated with water, the spaces between the individual silt and clay particles are so small that the layers, known as aquicludes or confining layers, generally bar water movement.

The largest groundwater reservoir is located in the Central Valley and has an estimated total volume of 100 MAF at depths ranging from 400 to 4,000 feet below sea level. This reservoir is approximately equal in total volume to the 50 other groundwater reservoirs of significant size in the state. Approximately 40 of the 50 developed aquifers in California are located in the north-south valleys of the Coastal Ranges. Smaller reserves of groundwater are found in some of the extensive areas of volcanic material in the Modoc Plateau and the Cascade Range of the north central and northeastern portions of the tate, in parts of the Owens valley and nearby ranges, and in parts of the south coast. In the southeastern desert, groundwater reservoirs contain usable water but

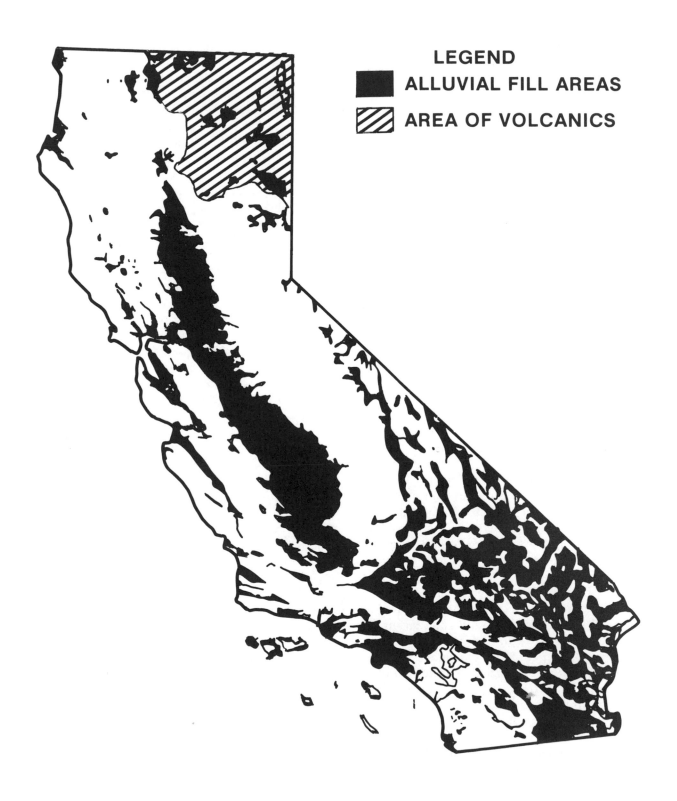

FIGURE 4.2 Areas of groundwater occurrence in California. Dark areas indicate alluvial fill basins, striped area is volcanic-origin basin. *Adapted from "California Water: A Report on Critical Problems and Proposed Actions," California Westside Study Team, U.S. Bureau of Reclamation, 1974.*

have not yet been extensively exploited. Smaller accumulations of groundwater occur in some glacial moraines at higher elevations and some usable supplies are found in a few coastal areas in thin marine terraces. Fractured rock formations in many upland areas yield small amounts of water for individual domestic and stock-watering purposes, but the volume is not sufficient for large-scale development. Some of the deeper sediments in the Central Valley were deposited in sea water and these marine sediments often contain salt water, usually at depths of 1,000 feet or more. In the Sacramento-San Joaquin Delta and at the mouths of coastal valleys, salt water may be encountered at as little as 100 feet below ground surface. Where physical conditions permit, the salt water may be flushed out from these shallow aquifers by percolating fresh water, thus creating usable fresh water aquifers for local water supply needs, as in coastal Sonoma and Santa Cruz counties.

The total storage capacity of California's groundwater basins has been estimated to be approximately 1.3 billion acre-feet of water. The gross storage capacity of surface reservoirs in the state is 77 million acre-feet, or only about one-twentieth of that amount. Because of physical and economic limitations, however, not all of the groundwater storage capacity is usable. It is estimated that usable groundwater storage is approximately 200 million acre-feet.

The current average annual withdrawal rate exceeds overall recharge by more than 2 million acre-feet. During the extreme drought years of 1976 and 1977 the estimated overdraft exceeded 5 million acre-feet per year. Recent studies by the California Department of Water Resources have identified conditions of critical overdraft in 16 out of some 200 inventoried groundwater basins. Unfortunately, these overdrafted basins account for approximately 20 percent of the state's usable groundwater storage capacity. A condition of critical overdraft is defined by the Department of Water Resources as conditions that cause or threaten to cause significant adverse environmental, social, or economic impacts at the local, regional or state level. Such impacts may include land subsidence, seawater intrusion, or water quality degradation. Ultimately, continued overdrafting depletes the groundwater and land has to be either abandoned or dry farmed. For the near future, groundwater overdrafting will undoubtedly require the use of more energy to pump water from increasingly greater depths.

Under the current conditions of rising energy costs, it is reasonably certain that future management strategies for California groundwater will require an overall balance of groundwater pumping that will not perpetuate pumping at or beyond critical overdrafting levels. If, however—as in many of the state's groundwater basins—the net return realized by a farmer from pumping an extra acre-foot of water exceeds the cost of additional pumping, it makes economic sense to go ahead and extract it. In the absence of environmental and water-quality degradation, overdraft can result in either beneficial or adverse social and economic impacts. However, although overdrafting groundwater in many cases may be socially beneficial in the initial stages, it will eventually reach a point where the long-term costs of overdraft exceed the short-term benefits.

Groundwater levels and pumping costs in parts of California—particularly the identified critical overdrafted basins of the San Joaquin Valley—have reached the stage when it is essential to define workable management plans to determine when and if overdraft will become too costly to water users.

Water Storage and Distribution Systems

As important as groundwater is, the development of surface water supplies for irrigation and other water uses has historically received the most emphasis. This development has remained the central focus of many private, public and quasi-public efforts that have produced a vast network of man-made watercourses and storage facilities making possible controlled delivery and application of water throughout the length and breadth of California.

The earliest mission efforts of the Franciscan friars required modification of natural rivers and streams at most locations to provide water needed for permanent tree and vine planting. The Gold Rush brought simple hydraulic engineering to California's rivers—dams, flumes, ditches and other devices moved water from one place to another to permit more effective working of the gold-bearing earth and rock. The experience gained in the control of surface water supplies was transferred to the rapidly expanding agricultural settlements, launching the transition from a

Pine Flat Dam, on the Kings River in Fresno County, holds back spring runoff from the Sierras, releasing water at planned intervals for summer irrigation. *Photo by Jack Clark.*

livestock-grazing agricultural economy to one of mixed crop production. In some localities, large acreages devoted to grain production developed during the mid to late 1800s, and significant acreages are still dry-farmed. However, in much of California, natural precipitation did not provide satisfactory crop yield. Individuals and groups, on a small and local basis, began organizing to impound and divert water for mid-season application to trees, vines, and other crops. Private water companies, mutual water companies and irrigation districts were important in providing stable and reliable water supplies for local farm and urban water requirements throughout the late 1800s and early 1900s.

Satisfactory water supplies for further agricultural development, however, could not be developed by local organizations. Similarly, although water requirements for early urban populations were satisfied by local storage and diversion facilities, forecasts throughout California for major expansions of cities brought realization that major and far-reaching surface water developments would be needed.

Although some planning work was done as early as the late 1800s, it was not until well into the 20th century that construction of significant large-scale storage and conveyance systems was undertaken in California. The earliest large-scale water development projects were for urban water supply systems for such areas such as Los Angeles, San Francisco and Oakland. Later came the Central Valley Project, essentially for agricultural use, while the State Water Project, initiated in the 1960s, is an integrated agricultural-urban waterproject.

Urban Supply Systems

Los Angeles-Owens River Aqueduct

Los Angeles obtains its water from five sources: the Los Angeles Owens River Aqueduct, the Second Los Angeles Aqueduct, the Colorado River Aqueduct, the State Water Project, and groundwater (Figure 3). About 80 percent of the supply comes from the first two aqueducts which run parallel to bring water from Owens Valley and Mono Basin.

The concept of transporting Owens River water over 230 miles to Los Angeles was first envisioned as early as 1892, but not until 1905, after a series of dry years, did the voters of Los Angeles approve a $1.5 million bond issue authorizing

acquisition of land and water rights in the Owens Valley. Two years later, they approved a $23 million bond issue and construction of the aqueduct began. In 1930, the aqueduct was lengthened by 105 miles to reach Mono Basin water supplies. The second Los Angeles Aqueduct, completed in 1970, originates at Haiwee Reservoir, south of Owens Lake. The City of Los Angeles currently exports its maximum entitlement from the Owens-Mono area.

Metropolitan Water District of Southern California

In 1924, anticipating a need for water supplies additional to the Owens Valley Aqueduct, Los Angeles filed a claim for over one million acre-feet of Colorado River water. Four years later, Los Angeles and ten other cities, also seeking expanded water supplies for urban uses, joined forces as the Metropolitan Water District of Southern California (MWD). In 1931, MWD voters approved a $220 million bond issue to finance construction of Parker Dam on the Colorado River and the Colorado River Aqueduct (Figure 3). Construction started in 1933 and partial configuration was completed in 1941. Power generated at Hoover and Parker Dams is used to pump water through the aqueduct to a maximum elevation of 1,800 feet. Water then flows by gravity through many tunnels and siphons to Lake Mathews, near Riverside. MWD's distribution network transports water to over 120 cities and water agencies, who in turn sell it to consumers or other wholesalers.

The aqueduct was built to convey the District's maximum entitlement of 1.2 million acre-feet from the Colorado River. In the 1963 Supreme Court case of *Arizona vs. California*, this amount was cut by more than half. However, that reduction will not take effect until the Central Arizona Project goes into operation in the late 1980s. To compensate for the loss of Colorado River water, MWD has contracted for an additional 500,000 acre-feet of water each year from the State Water Project. Combined with the current contract of 1.5 million acre-feet per year, MWD will thus ultimately purchase about half of the full yield of the State Water Project.

The MWD now serves nearly 11 million people—half the state's population—in an area of 4,900 square miles and comprising six counties: Los Angeles, Orange, Riverside, San Bernardino, San Diego, and Ventura.

FIGURE 4.3 Existing and authorized major features of the SWP and CVP that supply water to and divert from the Delta, 1978. *Source: California Department of Water Resources, "Delta Water Facilities," Bulletin 76, July 1978.*

FIGURE 4.4 Major aqueducts in California. *Adapted from California Department of Water Resources, Map Library.*

The Hetch Hetchy Project

San Francisco's water supply system extends nearly 150 miles from the upper Tuolumne River in Yosemite National Park to Crystal Springs Reservoir near San Francisco. The city of San Francisco sells water to many communities in the eastern half of the San Francisco Peninsula, part of northern Santa Clara County, and southwestern Alameda County.

Hydroelectric power is a very important facet of the Hetch Hetchy system. The early sale of electricity helped underwrite construction of the water supply system and income from power sales exceeds that from water. Because of its origin in the high Sierra, the Hetch Hetchy system generates more electricity per acre-foot of water than any other water supply system in the state.

East Bay Municipal Utility District

Almost all the water of the East Bay Municipal Utility District comes from the watershed of the Mokelumne River some 90 miles distant in the western Sierra Nevada. EBMUD currently serves 16 cities and large unincorporated areas in a 296-square-mile area extending from Crockett and Richmond southward to San Lorenzo, Castro Valley, and San Ramon Valley, and from San Francisco Bay eastward to Pleasant Hill—a population of over 1.1 million.

Central Valley Project (CVP)

The idea for transporting water from the Sacramento River to the San Joaquin Valley was first proposed in 1873 by B. S. Alexander. However, it was not until the 1920s that the state government was brought directly into the business of water development and an ambitious program of water resource planning was undertaken which formed the basis for what is now the Central Valley Project. Although the state originally planned to build the various dams and canals, it chose not to sell the $170 million in water-bonds approved by the voters. Instead, the state persuaded the federal government to assume responsibility for building the CVP during the 1930s. It is now operated by the U.S. Bureau of Reclamation (USBR).

The Central Valley Project is a complex system of dams and canals designed primarily for storing the highly seasonal flood waters of the Sacramento, American, Trinity, and San Joaquin rivers and conveying them to water-deficient areas of the Sacramento and San Joaquin valleys.

Lake Shasta, the principal storage of the CVP located on the upper Sacramento River, has a drainage area of 6,600 square miles and a storage capacity of 4.5 million acre-feet of water. Water released from Lake Shasta is first used to generate power but serves many purposes—irrigation, navigation, flood control, recreation, water quality control, fisheries and wildlife. Since power requirements seldom coincide with water demands, the flow from Shasta Dam is regulated by Keswick Dam, nine miles below, where more power is generated and water is moved downstream to the Sacramento and San Joaquin valleys. Major diversions are also made from the Trinity River in the north coastal area and stored in Clair Engle Lake behind Trinity Dam (2.4 million acre-feet). Water released generates electricity at Trinity Power Plant and combines with the Sacramento River above Keswick Dam to flow toward the Delta. The Trinity and Shasta Divisions have a combined storage capacity exceeding 7 million acre-feet and a power capacity of almost 900,000 kilowatts.

From Keswick, water flows south via the Sacramento River where it is joined by water from the American River. Folsom Dam, on the American River east of Sacramento, was built by the U.S. Army Corps of Engineers as a flood control project and is operated by USBR as an integral unit of the CVP. Nimbus Dam, seven miles below Folsom, creates Lake Natoma and serves as a diversion dam for the Folsom South Canal. Water not diverted here joins water from Trinity and Shasta and descends to the Sacramento-San Joaquin Delta for export to users in the south. A portion, however, is used to repel salt water from the Delta.

The Delta Cross Channel, about 29 miles downstream from Sacramento, augments the natural transfer of water from the Sacramento River into channels of the Sacramento-San Joaquin Delta. Water flows through the Delta to the Tracy Pumping Plant, 50 miles away, where electric pumps raise it 197 feet into the Delta-Mendota Canal. Water from the Delta is also diverted into the 48-mile Contra Costa Canal that transports water to homes, farms, and industries in the Antioch-Martinez area.

From the Tracy Pumping Plant, CVP water moves down the Delta-Mendota Canal for 117 miles to the Mendota Pool on the San Joaquin River to replace the natural flows of this river that are diverted at the Friant Dam. Irrigation turnouts throughout the length of the Delta-Mendota Canal now number over 250. Meanwhile, Friant Dam, in the foothills of the Sierra Nevada northeast of Fresno, stores 500,000 acre-feet of the natural flow of the San Joaquin River in Millerton Lake. It provides releases for stream requirements above Mendota Pool and diverts streamflow into two canals. The Friant-Kern Canal diverts water south from Friant to the Kern River near Bakersfield. The Madera Canal diverts water north for 36 miles to irrigate lands in the Madera Irrigation District and terminates at the Chowchilla River. However, diversion of the San Joaquin River at Friant is made possible only by resupplying the San Joaquin River at Mendota with water brought in from the Delta via the Delta-Mendota Canal. This "exchange" of Sacramento for San Joaquin River water satisfies the rights of prior water users along the San Joaquin.

The San Luis unit is a combined effort of the federal and state governments. Located about 60 miles south of the Delta, between the Delta and the Mendota Pool, the joint use San Luis Dam and Reservoir serves as an off-stream facility for both the CVP and the State Water Project. The CVP is entitled to one million acre-feet, or about 45 percent of the storage capacity at San Luis. The CVP's San Luis service area is served by the San Luis Canal, also a joint use facility flowing down the west side of the Central Valley for 101 miles to Kettleman City. An additional feature of the San Luis unit is the San Luis Drain, which will remove agricultural wastewater from the San Joaquin Valley and transport it to the Delta—but, as of 1980, only half is completed.

In 1973, CVP water deliveries totaled over 6 million acre-feet.[1] Over 83 percent of this water was used by agriculture, on nearly 2 million acres of land. The annual value of crops grown on land receiving CVP water exceeded $1.9 billion. The CVP's nine hydroelectric plants, with an estimated capacity of 1,334,080 kw, produced over 6.3 billion kwh of electricity (about 609 kwh for every acre-foot of water delivered).

The United States government, through the USBR, has and will continue to pay all costs of the CVP. USBR, in turn, decides what portion of the project costs shall be paid for by project beneficiaries. The government is reimbursed by

1. In 1981, DWR reported total developed-water capacity of the CVP at 7.9 MAF.

agricultural, municipal, and industrial consumers of water and power. It does not charge for fish and wildlife enhancement, recreation, navigation, flood control, highway relocation, and water quality. Water and power users are expected to pay a share of capital and operation costs over a 40-year period following project completion. The agricultural water user is exempted from interest payments and the rates are determined mainly by the land quality and crops that can be grown.

Numerous complicated questions have been raised and much controversy exists concerning such issues as irrigation subsidies, water pricing policies, acreage limitation definition and application, etc. In addition, construction on the controversial Auburn Dam on the American River is currently under suspension. There is much uncertainty about possible future changes in operating practices of the CVP. All of these will have important long-term implications for California agriculture. Some of these, as well as other policy issues, are briefly considered in the last part of this chapter.

California State Water Project (SWP)

The Feather River Project, basis of the SWP, was athorized by the State Legislature in 1951 and construction of the Oroville facilities began in 1957. The physical facilities of the SWP are designed to make more full use of the potential for moving water from areas of natural supply to areas of need. Thus it complements the CVP but is, for the most part, managed independently of the CVP by the California Department of Water Resources.

The key water conservation facility of the project is Oroville Dam, a 770-foot high earth-fill dam on the Feather River five miles upstream from the city of Oroville. Below Oroville Dam is the Edward Hyatt Power Plant, the project's power facility. Lake Oroville is one of the largest pumped-storage facilities in the United States. Water released from the dam passes through an underground hydroelectric power plant, through the Thermalito Afterbay, and down the Feather and Sacramento Rivers until it reaches the Sacramento-San Joaquin Delta. In the northwest corner of the Delta the North Bay Aqueduct, scheduled for full completion by 1990, will transfer water to a terminal point near the city of Napa.

Most water in the SWP continues through the Delta's lacy patterns of channels and sloughs to Clifton Court Forebay, the water regulating reservoir for the Delta pumping plant. The Delta Pumping Plant lifts water 244 feet to the California Aqueduct at a current rate of almost 7,600 cubic feet per second. The 444-mile California Aqueduct, principal water transportation facility of the SWP, beginning at the southern edge of the Delta and terminating at Castaic and Perris lakes in southern California, now includes 20 dams and reservoirs, 5 power plants, 17 pumping plants plus 100 miles of branch aqueducts. Service areas include portions of Alameda and Santa Clara counties, the San Joaquin Valley, and southern California.

The North San Joaquin Division of the California Aqueduct extends from the Delta to O'Neill Forebay at the foot of San Luis Reservoir. The San Luis Division, extending 101 miles from O'Neill Forebay to Kettleman City, includes the facilities at San Luis Dam and the San Luis Canal, operated jointly by the state and the U.S. Bureau of Reclamation. The CVP Delta-Mendota Canal is also connected to O'Neill Forebay and San Luis Reservoir. An agreement between the two operating agencies permits exchange of water when one agency must deliver water and the other wants to store excess flow from the Delta. This saves energy by eliminating the need to pump water into San Luis Reservoir.

The state-federal San Luis Canal delivers over 7,000 cubic feet per second for the state, with the remaining 6,000 cubic feet per second serving the federal San Luis service area. The Dos Amigos Pumping Plant, south of San Luis, lifts this water over 100 feet where it then resumes its downhill course until it reaches the next pumping plant at Buena Vista.

Considerable water is diverted to customers in Kings County, being lifted between 55 and 150 feet to serve areas at elevations above the main aqueduct. The South San Joaquin Division extends for 120 miles from Kettleman City to the foot of the Tehachapi Mountains. Along this route, water is lifted three times for a total of 518 feet.

At the Tehachapi Division, after a single lift of nearly 2,000 feet, water crosses the Tehachapi mountains through four concrete-lined tunnels. The Garlock and San Andreas faults are both crossed by surface pipelines, permitting easy access should the aqueduct be damaged by an earthquake. After crossing the Tehachapis, water is diverted to two branches of the California Aqueduct. The west branch goes some 22 miles to Castaic Lake while the east branch terminates 137 miles south at Perris Reservoir.

In 1974 the California State Water Project delivered over one million acre-feet of water. The Project also delivered a million acre-feet of water to the San Luis unit of the Bureau of Reclamation's Central Valley Project.[2] Of this, 865,000 acre-feet were used to irrigate 290,000 acres of land, producing crops (principally cotton) with a gross value in excess of $100 million. The population within the urban areas served by the SWP exceeds 12 million. The Edmund Hyatt-Thermalito complex produced 4.1 billion kilowatt hours of electric energy and, in addition, another 500 million kilowatt hours were produced by power recovery plants.

Construction of SWP facilities was financed primarily by the sale of general obligation and revenue bonds. To date, the state has spent $2.3 billion on the SWP. Water users will repay about 74 percent of the total costs of building the system; power users will repay 11 percent. Flood control, amounting to 2 percent of project expense, is paid by the federal government. Recreation and fish and wildlife enhancement costs (5 percent) are not assigned to any specific class of users and are paid from the state's general fund. The price of water to the user is based on a complex system of cost distribution. All consumers pay a Delta water charge, which corresponds roughly to the cost of the Oroville facilities divided by the number of acre-feet of water exported from the Delta. To this is added a transport charge which, in its simplest form, equals the cost of building and maintaining the facility (or portion thereof) from the Delta to the point of diversion, plus any pumping costs.

Administration of the SWP

The state's decisionmakers are the governor, acting through the director of the Department of Water Resources, and the legislature. The Burns-Porter Act prevents them from making significant changes in contractual commitments signed by the State Water Project and water buyers. However, the governor and legislature, the Water Resources Control Board, and the courts may

---

2. In 1981, DWR reported total developed-water capacity of the SWP at 2.5 MAF.

A pumping station in the southern San Joaquin Valley, part of the State Water Project, lifts water over the mountains which separate the valley from southern California. *Photo by Jack Clark.*

impose regulations on SWP operations. Two examples of significant changes imposed after SWP began operation are the Delta salt-water-intrusion standards established by the State Water Resources Control Board, and the Wild and Scenic Rivers Act of 1972.

In 1977, SWP entered a very important stage in its development, when many difficult choices had to be faced. While the SWP has committed itself to delivering 4.23 million acre-feet, present water development (Oroville/San Luis) yields only about 2.3 million acre-feet. Even though the demand for water is not increasing as fast as it was in the 1960s, many believe that new water must be developed by 2000 to meet projected demand.

How the SWP will meet contractual commitments and pay bond-holders, while remaining within environmental guidelines, is a hotly debated subject producing volumes of studies and plans. One problem is how to maintain water quality in the Sacramento-San Joaquin Delta while exporting more water from the Delta. Saltwater from San Francisco Bay meets freshwater from the Sacramento and San Joaquin Rivers in the Delta. High freshwater flows keep salt water from intruding the Delta area. The saltiness of the Delta water is, of course, a major concern for Delta water users.

Since freshwater enters the Delta mainly from the Sacramento River and is pumped out nearer the San Joaquin Valley by both State and Federal pumps, summer flow patterns in the southern Delta are the reverse of normal, i.e., water flows northwest to southeast, exactly opposite to the natural flow with the pumps turned off. The flow reversal confuses salmon and trout returning to spawn, lowering production of eggs and young fish.

One proposed segment to the solution of these problems involves the development of a Delta Peripheral Canal. The proposed Canal is a 43-mile channel around the eastern side of the Delta, to divert Sacramento River water to the pumps of the CVP and the SWP near the town of Tracy. The Peripheral Canal, by virtue of its 12 water-release points, also is designed to eliminate the summer flow reversal in the Delta and to maintain water-quality standards, including fish protection, with the minimum possible amount of freshwater. The Canal would have an initial capacity of 23,000 cubic feet per second and the improved efficiency of freshwater flushing could yield close to one million acre-feet for export.

The Peripheral Canal by itself would not make up the state's developed water deficiency. Although original plans had called for large dams and reservoirs on the Eel, Trinity, and Klamath Rivers, the Wild and Scenic Rivers Act of 1972 prohibited any interference with the free flowing state of those rivers. DWR is to report to the State Legislature in 1984 regarding continued inclusion of the Eel River in the SWP. During wet winters, more water flows out to sea than can be stored for later use. Present plans call for construction of additional storage facilities, but they are relatively low priority items in state planning and the timing of anticipated construction is uncertain at best.

Without the Peripheral Canal or new storage projects, the SWP will have difficulty meeting its contracted water deliveries and may find that it will also be difficult to pay off bonded indebtedness. The present high cost of energy could, of course, not be foreseen when the project was designed in the 1960s, and the state's favorable power rates expire in 1983. The new price of electricity will undoubtedly drive water prices higher, particularly in the south coastal basin. Inflationary pressure alone has more than doubled initial cost estimates for this facility. Continued controversy over the Peripheral Canal is anticipated during the decade of the 1980s.

Irrigation Systems and Irrigation Costs

The earliest diversions of water from streams for agriculture featured open ditch delivery systems and simple surface irrigation. Various modifications of the surface methods are still the dominant irrigation technology in all western states, to which have also been added newer methods involving sprinklers and trickle-drip emitters. The method used depends on highly variable physical and economic conditions such as slope, soil depth and texture, type of crop, water quantity and quality availability, water cost, and special needs such as salinity control, drainage requirements, or need for frost protection. The most important surface methods include flood, border, check and furrow application. Sprinkler methods vary from permanent set systems to center pivot, wheel line and hose drag automatic systems, to completely hand moved systems. The drip method involves relatively continuous delivery of small volumes of water under highly controlled conditions, limiting the area of soil surface wetted.

Because about 85 percent of the water used in California goes initially to agricultural uses, the various methods of irrigation application are of interest. Increasing concern over water supply uncertainty, as well as rising water and energy costs, have intensified interest in water conservation—including both optimal allocation of water between agricultural uses and non-agricultural uses, and more efficient irrigation application methods. Tables 3 and 4 present a summary of the absolute and relative importance of various irrigation methods used in California for several regions of the state and by source of water used (i.e., groundwater or surface water). Figure 5 shows the boundaries of the regions essentially as developed by the California Region Framework Committee and used as the basis for the most recent study of agricultural water use and costs in California.

It can be seen that surface irrigation methods have developed throughout the state and, in general, reflect conditions of relatively flat topography, historically low-cost water, and relatively large water deliveries. Basin and border irrigation methods are used for a wide variety of crops including rice, hay, pasture, orchards and vineyards. Furrow irrigation is used for row crops as well as for orchards and vineyards. On a state-wide basis, basin methods account for about 6 percent, border methods nearly 38 percent, and furrow methods more than 35 percent of the acreage of all irrigated crops, or just under 80 percent of the total acreage irrigated. About two-thirds of this water is supplied from surface sources and one-third from groundwater.

Over the past decade, there has been a significant increase in more efficient irrigation methods, with emphasis on sprinklers or trickle-drip emitters. These methods can be used where surface irrigation is not effective or efficient, including areas of slopes in excess of 3 percent, areas with either variable textured soils or soils with excessively high or low intake rates, areas with special problems such as frost control or leaching, and where water is either expensive or in limited supply. In the latest (1975) estimates, sprinkler methods account for slightly more than 20 percent and drip for about 0.5 percent of the irrigated acreage in California. It should be noted that a 1977 irrigation survey indicated an increase in the drip method to about 2 percent of the total and significant additional conversions are taking place steadily. It is very likely that the drought years of 1976 and 1977 plus rapid increases in

TABLE 4.3
Regional Acreage Irrigated By Surface Water, by Application Method[1]

Region	Flood	Border	Furrow	Sprinkler	Drip	Total Acreage Irr. by Surface Water
North Coast				3,579		3,579
North Bay	11,535	5,405		20,955		37,895
South Bay						
Delta		151,688	241,360[2]	29,662		422,711
Sacramento Valley	16,360	731,545	101,736	18,610		868,251
Mountain-Valley	118,925	9,600		29,449		157,974
North San Joaquin Basin		460,640	11,557	1,158		573,355
Central Coast			168,830	37,207		206,037
San Joaquin Basin		490,658	616,525	28,658		1,135,841
Westside San Joaquin	50,270		380,797	175,805	5,218	612,090
South Coast		1,246	6,465	9,747		17,458
High Desert		1,686	1,124			2,810
Imperial Valley		179,013	181,820	339,532		700,365
TOTALS	197,090	2,031,481	1,810,215	694,362	5,218	4,738,366

[1] Source: University of California, Division of Agricultural Sciences, Giannini Foundation, "Agricultural Water Use and Costs in California," Bulletin 1896, July 1980, 41 pp.
[2] Includes 133,500 acres sub-irrigated.

water costs and energy costs will continue to promote more efficient methods in the future.

Water Conveyance and Irrigation Efficiency

Irrigation efficiency refers to the percent of applied irrigation water actually used by crops, after accounting for losses from such sources as conveyance loss, weather-induced evaporation, deep percolation beyond the crop root-zone and requirements for salinity management. Water lost in conveyance may be as high as 20–40 percent if earth ditches are used or as low as zero if conveyed in tight pipeline systems. The rated efficiency of pipelines typically varies from 90-100 percent; of concrete lined ditches, from 80–90 percent; and of earth ditches, from 60–80 percent.

Irrigation efficiency is usually expressed as the ratio of the water stored in the root-zone of the soil and available for use by plants, compared to total water delivered to the field. The numerator includes both the water used by plants for growth and the much larger amount lost to the air by transpiration from plant leaves plus evaporation from soil and water surfaces. During irrigation, water may be lost as surface runoff or as deep percolation below the root zone. Some deep percolation (called leaching fraction) is needed to maintain a favorable salt balance in the root zone. In a well designed and managed sprinkler or drip system, the rate of application should be less than the intake rate of the soil so that no runoff will occur. Water application efficiency of surface methods can be increased by catching runoff and using a tailwater recovery system to recycle it. Common water-application efficiencies vary from 40–90 percent depending upon the type of irrigation system used, soil type, length of irrigation run, slope, climate, and management practices including irrigation scheduling. Surface irrigation efficiencies in California are estimated to average 65 percent and to range between 40–85 percent, reflecting a wide variation in methods and systems in use in the state. Similar estimates for sprinkler systems may range from 60–85 percent. Improved efficiency achieved with those systems is accounted for primarily by reduced runoff and elimination of seepage losses from open ditches and deep percolation. Efficiencies of even greater magnitude may be possible using drip methods, but operating problems such as clogging of emitters and salt accumulation may increase water and soil management problems and lead to increased maintenance requirements. The cost of initial installment of drip systems is also high.

Irrigation Costs

Total irrigation costs—basic water cost plus operating costs to apply the water—are highly variable in California agriculture. Basic water costs may range from less than $2 to more than $150

TABLE 4.4

Regional Acreage Irrigated by Groundwater, by Application Method

Region	Flood	Border	Furrow	Sprinkler	Drip	Total Acreage Irr. by Groundwater
North Coast				20,281		20,281
North Bay	1,104			38,504		39,608
South Bay				8,503		8,503
Delta		267,555	174,647	30,315		472,517
Sacramento Valley	39,736	279,692	99,253	76,558		495,239
Mountain-Valley	129,118	2,400		228,654		360,172
North San Joaquin Basin		32,846	7,497	14,095		54,438
Central Coast				51,067		51,067
San Joaquin Basin		300,490	358,177	150,689	21,571	830,927
Westside San Joaquin	126,280		271,361	186,038	6,147	589,826
South Coast		7,474	45,246	34,378	6,465	93,563
High Desert		6,745		41,208		47,953
Imperial Valley						
TOTALS	296,238	897,202	956,181	880,290	34,183	3,064,094

SOURCE: University of California, Division of Agricultural Sciences, Giannini Foundation, "Agricultural Water Use and Costs in California," Bulletin 1896, July 1980.

FIGURE 4.5 Regions of California by water use. *Source: University of California Agricultural Sciences, Giannini Foundation, "Agricultural Water Use and Costs in California," Bulletin 1896, July 1980.*

Of the developed water supply in California, 85 percent goes to agriculture. This farmer is building checks to control flood irrigation in his pear orchard. Many orchardists are installing trickle-drip systems of hoses with emitters placed at each tree, but they are expensive.

per acre-foot depending upon the source of the water. Total irrigation costs may range from as little as $9 to more than $375 per acre depending upon location, source of water, crop grown, and method of irrigation used.

Increasing irrigation costs may force farmers to reduce water use on low-valued crops. Recent reports have suggested that the demand for water for low-value forage crops, for example, may be reduced significantly in response to increasing water costs. Another type of response—and one that emphasizes water conservation possibilities —is to convert to more efficient irrigation methods. However, impact on total energy use will also need to be considered, especially as power

costs increase. Under comparable conditions, shifts to high efficiency technologies may add as much as $75 to $100 per acre in energy costs. This increased cost will be less of a barrier to conversion where irrigation costs are a very small part of total crop production costs than it will be if irrigation costs exceed, say, 40 percent.

Under most foreseeable circumstances, pressure to improve water conservation and water use efficiency in irrigated agricure will continue to promote shifts in irrigation methods used in California. Although irrigation efficiencies are generally higher in California than in other western states, various water management methods still offer potential for further on-farm water savings. These include basic changes in irrigation methods used, introduction of irrigation scheduling, provision for better drainage and salt management, better weed and phreatophyte control, improved seepage control and system automation.

Methods for improved water management— including the impact on irrigation costs as well as on water and energy conservation—need to be evaluated on a case by case basis. Conditions vary widely depending upon location, source of water, and the particular practice or method being evaluated. There is also need to more precisely identify the relation of on-farm water management and conservation to basin-wide water conservation and management. Increasing on-farm efficiency may not affect basin-wide efficiency significantly if deep percolating and surface runoff water is already being effectively captured for reuse within the basin.

Water Law—An Overview

This overview is designed to present only a brief description of a few of the general principles and characteristics of California water law. Specific and particular application of the principles can best be made with the advice and assistance of an attorney knowledgeable in water law.

Water rights are property rights and describe the extent and the limits on the use of water by those who hold them. Ownership of water rights permits diversion of water from a natural source of supply for reasonable and beneficial use. Water rights have value and are protected by California law. Water rights in good standing may be lost or impaired depending upon the type of right and the use, or lack of use, by the possessor of the right.

The major types of irrigation water rights affecting surface water are the *riparian rights* to divert water to land adjacent to a stream or lake and *appropriative rights* to direct water for use at sites some distance from the supply source. *Correlative rights* apply to groundwater and the right of the overlying landowner to make use of that water. Riparian rights entered California law when the first California Legislature adopted English common law as the basis of the state's legal system. However, the appropriative doctrine was introduced in fact by the practice of miners diverting water from streams for use at distant sites. Prior appropriations were recognized as valid rights and the principle led to many diversions according to the appropriation doctrine. The net result is that two sets of frequently conflicting principles governing irrigation water rights have operated since 1850 to produce a most complex body of water law in California. The struggles of riparian *versus* appropriation doctrines led to an amendment of the State Constitution in 1928 recognizing the interest of all the people of California in the state's water resources. Since then, court adjudication has essentially shaped the evolution of this dual system of rights by establishing specific rules governing the two rights.

Riparian Rights

Riparian rights attach to the natural flow of water in streams and rivers and apply to land with access bordering the stream. The right is limited to the smallest tract of land so held and the tract must lie within the watershed of the stream. Riparian water is for use only on such riparian land and cannot be used elsewhere. Diversion is normally at some point on the riparian land but may be upstream under certain conditions. Unused water must be returned to the stream and all water used must be for some beneficial purpose. Irrigation, domestic use, watering of livestock, manufacturing, development of power, and recreational uses are all recognized as beneficial uses. Riparian rights are relative and usually not absolute in that "reasonable use" includes recognition of the needs and rights of others. Riparian rights are not ordinarily lost by lack of use but a number of exceptions to this principle exist.

Because California climate makes irrigation necessary for most intensively cultivated crops, the courts have been called upon to decide whether irrigated agriculture, as well as industrial and urban uses, would be permitted riparian rights. It has been suggested that had irrigation not been decided to be a proper riparian use, California agriculture would have had to turn to appropriation as a source of water—leading to a possibly different outcome in the battle between the appropriative and riparian doctrines.

The courts have held that in situations in which there is not enough riparian water for all recognized riparian users, the courts may make an equitable apportionment of the supply. Another modification generally recognized is that upstream riparian use for personal domestic purposes may take as much water as necessary, even if it deprives downstream riparian owners of some of the supply. Storage for use at a time when riparian water would not naturally be in the river is usually permitted. These distinctions are important in governing riparian use of water in irrigated agriculture.

One difficulty with riparian rights is that, being attached to the land, they are not subject to transfer. The nature of the riparian right makes it legally impossible for a landowner to transfer water to use on other land because he has the right to use it only on riparian land. Under some particular and specially defined conditions, an individual desiring to divert water from riparian land may purchase deeds purporting to convey the right from riparian owners. If the remaining water will satisfy downstream riparian rights, then the possessor of the deeds may acquire the right to appropriate the water represented in the deeds. Such variations are possible but do not constitute an easy way around the question of lack of transferability inherent in riparian rights.

Appropriative Rights

An appropriative right is granted by the State of California, after due procedure, to divert a given quantity of water from a specified source point beginning at a specified date for an identified use to take place at a particular place for a defined period of time. The appropriative right thus carries a priority that is based on the date when the diversion is authorized, following the general rule that the first appropriator in time is the first in right. The location of the diversion does not affect the priority of the right, as several appropriators may divert from at or near the same point with the priority of right depending only upon the date at which the rights were granted. In times of short water supplies, apportionment is generally based on priority of rights. The permit to appropriate identifies a specified volume of water that may be diverted and the time during the year when the diversion may take place. The place and nature of use is identified to establish a reasonable beneficial use. Any changes in either location of use or nature of use can be made only after making application for revision. In general, appropriations may be made only of water that is (1) identified as a public source of supply, (2) naturally occurring, and (3) in excess of identified riparian and appropriative rights previously existing. The appropriative right continues so long as the appropriator continues to make use of the water in accordance with the prescribed terms of the permit. The right may be lost by failing to prevent adverse use resulting from upstream diversion of water, or by court orders resulting from specific fraudulent action on the part of the appropriator.

Although it is generally recognized in California that riparian rights are superior to appropriative rights, it is also true that the majority of California's surface water is used according to permits confirmed and granted by the State Water Resources Control Board. Water use under the appropriation doctrine is closely monitored to assure adherence to the specified terms of the permit. Because appropriated water may be used at some distance from the point of diversion and the timing of use may vary, an appropriator may store water for use at a later time. As with riparian rights, the appropriator does not own the water itself—only a right to use it. Unlike the riparian right which generally cannot be transferred independently of land ownership, the appropriative right may be sold or otherwise transferred. Thus, the appropriative right is more like other property rights in that it permits transfer of the right in accordance with economic principles of the market place. However, any transfer of the right is subject to the rule that other existing rights are not affected.

Groundwater Law—Correlative Rights
Rights to the use of groundwater have been the subject of a large number of court cases beginning in the late 1800s and continuing to the present day. Application of the doctrine of correlative rights—that is, the rights of the over-

lying landowner—have been shaped and modified by these court decisions as well as by the increasing interest of public agencies concerned with management of water resources. Prior to 1902, California law, based on English common law, permitted anyone with land lying over groundwater to extract the water and use it on any other land. In overturning this principle, the California Supreme Court substituted the correlative rights doctrine, i.e., that all overlying landowners possess coequal rights to the use of the groundwater on or in connection with their overlying land. This right extends to the quantity of water required for recognized reasonable beneficial uses and is analogous to the riparian right for surface water in that it is conferred by virtue of land ownership. Groundwater is subject to appropriation, but only if the groundwater can be identified as flowing through known and definite subterranean channels. In situations where the groundwater supply is insufficient to provide for the needs of all overlying owners, the courts have power to make equitable appointments to the overlying owners and enforce them by judgment and decree.

There are many problems in implementing groundwater law, most of which arise because of lack of full knowledge of the characteristics of the groundwater basins underlying the surface property. Thickness and extent of aquifers vary within subbasins and interconnections between subbasins are difficult to identify, let alone quantify in terms of movement of groundwater. Merely by pumping, one overlying landowner may change the character of the groundwater gradient as well as the pumping lift for neighboring pumpers.

By careful hydrologic study, average annual additions to groundwater volume in a basin from rain and runoff can be estimated fairly accurately. This is the "safe yield" of the basin. If the annual draft on the basin is in excess of the identified safe yield, one criterion for overdraft exists. Sustained pumping at overdraft rates may produce one or more of the environmental, social or economic impacts described earlier as characteristic of critical overdraft. These conditions may provide the basis for court procedures to adjudicate the basin and apportion reduced pumping rights. So far, this apportionment has occurred only in circumstances where suitable quantities of imported water could be obtained to provide for all identified water needs. Rapidly increasing costs of litigation have, over the years, led to successful negotiation of management plans for groundwater basins outside of the courtroom. Although the settlements obtained under such negotiated settlements may not conform strictly to the history of adjudicated water rights, nonetheless a new dimension and era have been reached in the field of resource planning that does emphasize equity as well as economic and social acceptability.

The Need for an Improved Legal Framework

Development of a satisfactory legal basis to conjunctively use groundwater and surface water and to manage groundwater basins is one of the most critical needs of the future for California. Many of the principles of conjunctive use are currently being applied by water management agencies within limited and restricted localized conditions. But there is need to develop an appropriate framework to permit effective transfer not only of water and water rights between subbasins in a localized area but between major basins that may not even be contiguous. Some economic and physical studies have produced evidence suggesting that more efficient and effective use of water resources could be obtained if water transfers were effected expeditiously. The major barriers to such transfer, legal and institutional, affect both groundwater and surface water. Implementation of many of the recommendations of the Governor's Commission to Review California Water Rights Law has been suggested as the basis for achieving significant improvements in this area.

Water, Agriculture and The Future

During most of California's history, water has been considered an abundant resource. Its development and use has contributed to the growth of an agricultural economy unexcelled in the nation or world. Yet Californians are learning today that water policy issues are complex and interconnected with energy, land, people and the environment. Under conditions of resource abundance, adjustments within the body politic are relatively easy—but as resources become limited, or as their fixed limits are more clearly identified, the necessary adjustments require policy tradeoffs. Water policy issue tradeoffs are particularly difficult to identify and quantify. Thus, the fu-

ture is filled with uncertainty about long-term planning needs for water and for California agriculture. Some policy issues for the next decade, however, are relatively predictable, and a review of some of these will conclude this discussion of water and California agriculture.

Water Supplies

Past water needs for an expanding irrigated California agricultural industry have been met primarily by increasing supplies through water storage and diversion projects, at first on a local level and then by storing and moving water the length and breadth of the state. In recent history the use of groundwater has become a major source of water for irrigation. However, groundwater stocks have been overdrafted in anticipation of additional imported surface supplies in some areas, and inadvertently overdrafted in others.

Conflicting views have significantly slowed the rate of new surface water supply development and may well stop it altogether. Conflicts between agriculturalists and environmentalists, as well as between water deficit areas and regions from which water might be diverted, seem to be intensifying.

Even if agreement were readily attainable, early provision of significant new supplies of water for irrigation use is hardly possible, given the time lag between authorization and funding and its eventual completion. Furthermore, low cost sources for conventional surface water developments have all been used. Inflationary pressures and specific increases in construction and energy costs will continue to push ever higher the cost per acre-foot of new water supplies. Thus, development of new water sources becomes an increasingly doubtful alternative. General resistance to expansion of tax-supported construction, combined with expanding participation in public decision-making by a wide array of constituencies competing for limited public funds, make the future prospects for significant long-run epansion of conventional surface water supply projects cloudy at best. Therefore, continued adjustment to a more clearly defined and relatively fixed water supply is in store for California during the foreseeable future.

Even before the drought years of 1976 and 1977, many of the most important groundwater basins of California were being overdrafted. Widespread drought, continued demand for agri-

Shasta Reservoir shows the effects of the 1976–1977 drought years. Large areas of shoreline were exposed as water levels dropped far below the reservoir's 4.5 million acre-feet storage capacity. *Photo by Jack Clark.*

cultural production and ever-changing energy and cost considerations have accelerated the severity of the condition. Despite some very severe local overdraft problems, however, there has not been any significant movement toward the development of groundwater management plans or legislation that would facilitate such plans.

Evidence suggests that various subregions of the Central Valley will gradually be forced to adjust cropping patterns to reflect decreased availability and increased cost of groundwater. Eventually, the limited amount of water available for "safe yield" levels of the various groundwater basins will place absolute limits on both the size and character of long-term irrigated agriculture. However, groundwater management plans to make more effective use of storage capacity in

underground aquifers could permit attainment of a more flexible and, in all probability, a more prosperous irrigated agriculture. Estimated groundwater reserves are still sizeable and there is time to permit development of the necessary management framework that could operate within normal market conditions.

What about more surface supplies? Most indications are that few if any forms of agriculture in California could pay the full price of water from new projects and remain competitive with other regions. Thus, whether surface or groundwater supplies are considered, the future of California agriculture will be increasingly dominated by more clearly defined limits of water available for irrigation use. The extent and seriousness of the adjustment will depend largely on water conservation technology available to agriculture and the willingness or incentive to change to a more conservation-oriented technology.

Current debates reflect the disagreement about the possibilities of water conservation at either the farm level or regional level. That adjustments in California agriculture will occur, however, is not questioned. Economic pressure for improved efficiency in water use technology, adjustment in cropping patterns, and more flexible institutional arrangements for transfer of water rights among areas may provide a basis for reconciliation.

Pricing and Water Conservation

Increasing water use efficiency through use of water-saving technology is a promising method of stretching limited water supplies for agriculture. It has been suggested that market-oriented pricing of available supplies is the most simple and rational approach to increased efficiency and the equitable allocation of this vital resource. Two general methods of resource allocation by using prices are possible: (1) administration of a set of fixed prices that, in effect, would ration water among users, or (2) encouragement of a system of variable prices responding to market conditions. A somewhat imperfect combination of the two general methods now operates in California.

The mechanics of price-initiated conservation suggests that a system of freely variable market prices would be more feasible than arbitrarily administered prices. But marketing institutions and legal arrangements have to be flexible enough to permit transfer of water from one user to another in response to high bidding prices. Before freely variable market prices can be used to effect water conservation in California, the institutional barriers to transfer of water between and among water agencies and users must be relaxed or eliminated. If a price-oriented system is to function, water-short districts or individuals would have to be free to negotiate for water from districts or individuals willing to sell it.

Residential water users generally will not reduce quantity demanded unless the final user sees that changes in consumption are significantly reflected in the water bill. Industrial water users generally appear to be more responsive to changes in the price of water. Either increased price of water or of effluent treatment requirements tend to cause industrial users to recycle effluent water, thus lowering total quantity demanded. Depending on the amount of the

price increases, it may even become economically feasible to introduce new, more water-efficient industrial processes.

Agricultural water price relationships are not as clear cut. In agriculture, the prime determinant of water demand is biological; and different plants consume different amounts of water. Generally, however, the evidence suggests that over time most farmers will absorb higher costs for water up to a point, and then reduce the quantity demanded either through improved water application technology or shifts in the basic pattern of crops grown. As has been noted earlier, the fact that the price of water is a minor part of the total cost of producing many high income, capital-intensive crops complicates the picture. Acreage—and hence water—committed to such crops may be determined more by market projections and labor costs than by water prices.

Freely operating markets for water and water rights might serve as a very efficient allocator of water in times of severe water shortage as well as improve allocation generally over space and time. If one user or a group of users, such as an irrigation district, values an additional acre-foot of water more highly than another, then presumably both would be made better off if a transfer of water could be made. For example, if one district grows mostly pasture and feed grains while another produces fruits and vegetables, in a water-short year the area producing high-value crops may be willing to bid a very high price for water. If the offering price becomes sufficiently high in relation to expected net income per acre-foot in the low-valued operation, it will be economical for water to be transferred and let a portion of the low-valued acreage go out of production for the season. This could lead to the establishment of a system of annual water right rentals. It should be noted that an annual water-right market is a new idea for California, but not for the West—water-rental markets have existed in Utah and Colorado for a number of years. To make such a simple solution possible in California, many issues and problems would have to be resolved—including basic changes in the California Water Code as suggested by recommendations of the Governor's Commission to Review California Water Rights Law.

In addition to legal factors, many questions are being asked concerning the probable economic effect on the agricultural industry of increased reliance on market-oriented pricing of available water supplies. What will be the nature and magnitude of the effect of periodic or long-run shifts away from relatively low-value, high water consuming crops such as rice, irrigated pasture, alfalfa, and feed grains? How might these shifts affect the livestock, dairy, and cattle-feeding industries of California? What are the likely effects of increased energy costs and increased surface water costs on groundwater pumping? What will be the degree and extent of competition between agricultural, industrial and urban users for groundwater supplies? What might the impacts of increased groundwater pumping be on groundwater storage capacity and conjunctive use water management? What are the relevant facts regarding "subsidies" for agricultural water and acreage limitation? Continued research into the physical, legal, economic and overall institutional aspects of these questions must be accelerated to provide timely answers to water planners during the coming years.

At this stage, it appears that administered pricing of water to reduce use, regulate the variety and extent of crops grown, and, in essence, control the water market is not a likely prospect for the future.

Water Quality and Quantity Issues

Quantity and quality of water are closely interrelated. With future water quantity for agriculture being increasingly viewed as limited, it becomes even more important to consider possibilities for direct use of water of impaired quality in California agriculture.

The most serious water quality problem for irrigated agriculture is the need to maintain salt balance both at on-farm and regional levels. Plants vary in their tolerance to accumulated salts in soil and water. Improved water application as well as drainage technology may provide opportunities for use of saline and brackish water.

Not all water users require the same level of water quality. Water with high nitrate levels, for example, can be used beneficially in agriculture, but would cause public health concerns if used for domestic consumption. Nevertheless, agriculture cannot always make use of the increasing amounts of saline and brackish water associated with irrigated agriculture. Harmful physical and economic effects may occur which cannot long be tolerated. Improvements in plant breeding and irrigation management may ease short-run transi-

tion problems, but the extent of their efficacy over the long run cannot be certain.

Significant relationships exist between water supply decisions and water quality decisions. Yet the current water allocation process does not adequately take water quality into account. Thus, an important concern for the future will be to determine just how the protection of water quality should be incorporated into water allocation and pricing decisions.

Competition for Water

Perhaps no issue will be more characteristic of the California water scene between now and the year 2000 than the inevitable competition between agricultural and nonagricultural uses for a strictly limited and fixed water supply.

The dominant role of the agricultural sector in the water milieu of California suggests that allocations of existing water supply away from agricultural uses will be cause for major controversy. Yet the indications are that significant expanded demands for urban-industrial and recreational-environmental uses are to be expected over the next 15–20 years. Continued population expansion with a constant or even slightly decreased per capita level of water consumption will bring forth increased demand for domestic water allocation and delivery. Anticipated industrial and manufacturing increases in California will also be accompanied by expanded water use for this sector. These water uses can—and do—pay far higher prices for water than agriculture. Economic as well as political pressures thus appear as likely forces to intensify the pressure for shifts of water away from agriculture. Statistics suggest that even though urban-industrial users may be able to introduce water conserving technology more easily and effectively than many agricultural users, nonetheless, the total level of likely urban-industrial demand for water before the year 2000 will make it very unlikely that agriculture can retain its present level of water use if the total supply remains relatively constant. Urban areas' greater ability to pay for water may also encourage use of such expensive technologies as desalinization as well as reclamation of wastewater. Such technologies, however, may be somewhat limited by increasingly expensive energy sources.

Recent controversy over water storage vs. white water rivers has highlighted the general increase in conflict between recreational and

environmental water use supporters and traditional agriculture. Greater public participation in
debates prior to decisionmaking has highlighted
and intensified the conflicts stemming from different social viewpoints or value systems of the
two groups. The concern of debate participants
for the needs of the future—food supply vs. the
esthetics of water-based environments—is
reflected in the controversy.

Legal and Institutional Issues

Because water rights are a form of property
rights and because the legal structure dominates
actual allocation and use of water, growing competition for a limited supply can be expected to
bring forth continued concern and conflict over
legal and institutional issues. The current inflex-
ibilities in both legal and institutional frameworks focuses attention upon the need to provide for effective transfer of water rights to meet
changing water demands and needs over time.

The entire system of water allocation in the
state is currently coming under legislative and
public review. Water rights long held sacred are
being questioned. The extended controversy over
the Peripheral Canal illustrates the deep concern
of various groups. How can water rights be both
protected and made more flexible to achieve
transfers among users over time? What is the relationship of individual water rights to group interests in the allocation of and pricing of water to
achieve different social objectives? The complexity of these issues, as well as the uncertainty of
jurisdiction in conflicts over legitimate claims on
limited water supply, creates a high degree of
uncertainty among all participants. Thus the increasing frequency and intensity of expressions
of extreme viewpoints, disbelief and mistrust,
and lack of cooperation in joint ventures. There
is, nevertheless, inevitably even more necessity
to obtain laws and institutions that will permit
allocation and use of limited water supplies to
the highest beneficial use.

Continued conflicts will be associated with any
effort to change existing legal and institutional
frameworks. Yet the need for a framework to
permit achievement of more free movement and
allocation of water towards optimum uses assures that pressures for modification of the current
system will continue. Public water management
and modification of individual water management practices may well characterize the closing
decades of the twentieth century.

Acknowledgments

This chapter depends heavily upon valuable advice and comment from known and unknown re-
viewers following initial assistance and
encouragement from colleagues in the Water Resources Center and the Department of Agricultural Economics. Knowledgeable readers will recognize liberal use of the many special studies
and publications listed in the bibliography—
without the work of many, this brief generalized
review would have been impossible. Readers interested in details and specifics regarding the
complexities of water in California must consult
sources that contain the detailed and specialized
descriptions and analyses.

References

California Department of Water Resources. *California Flood Management: An Evaluation of Flood Damage Prevention Programs*. Bulletin 199. Sacramento, September 1980.

California Department of Water Resources. *California's Ground Water*. Bulletin 118. Sacramento, September 1975.

California Department of Water Resources. *Delta Water Facilities*. Bulletin 76. Sacramento, July 1978.

California Department of Water Resources. *Groundwater Basins in California*. Bulletin 118–80. Sacramento, January 1980.

California Department of Water Resources. *The California State Water Project—1978 Activities and Future Management Plans*. Bulletin 132–79. Sacramento, November 1979.

California Department of Water Resources. *The California Water Plan, Outlook in 1974*. Bulletin 160–74. Sacramento, November 1974.

California Department of Water Resources. *Water Conservation in California*, Bulletin 198. Sacramento, May 1976.

California Department of Water Resources, Southern District. *Stretching California's Water Supplies: Increased Use of Colorado River Water in California*. Los Angeles. August 1980.

California Region Framework Study Committee. *Comprehensive Framework Study California Region. Appendix V: Water Resources*. Prepared for the Pacific Southwest Inter-Agency Committee, U. S. Water Resources Council, June 1971.

Fereres, E. et al. *Irrigation Costs*. University of California, Division of Agricultural Sciences, Leaflet 2875, August 1978.

Highstreet, A., Nuckton, C.F., and Horner, G.L. *Agricultural Water Use and Costs in California*. University of California, Division of Agricultural Sciences, Giannini Foundation of Agricultural Economics, Information Series 80–2 (Bulletin 1896), July 1980.

Marsh, A.W. et al. *Drip Irrigation*. University of California, Division of Agricultural Sciences, Leaflet 2740, August 1977.

Nuckton, C.F. and McCorkle, C.O. *A Statistical Picture of California's Agriculture*. University of California, Division of Agricultural Sciences, Leaflet 2292, April 1980.

State of California. *The California Water Atlas*. Prepared by the Governor's Office of Planning and Research in cooperation with the California Department of Water Resources, 1979.

U.S. Bureau of Reclamation. *California Water, A Report on Critical Problems & Proposed Actions*. California Westwide Study Team, USDI, USBR, April 1974.

U.S. Department of the Interior. *Critical Water Problems Facing the Eleven Western States*. The Westwide Study—Executive Summary, April 1975.

U.S. Water Resources Council. *Essentials of Ground-Water Hydrology Pertinent to Water-Resources Planning*. Bulletin 16 (revised), 1980.

U.S. Water Resources Council. *The Nation's Water Resources, 1975–2000. Vol. 1: Summary of the Second National Water Assessment*. December 1978.

U.S. Water Resources Council. *The Nation's Water Resources, 1975–2000. Vol. 4: California Region*. Second National Water Assessment. December 1978.

University of California, Division of Agricultural Sciences. *Agricultural Policy Challenges for California in the 1980s*. The Report of the University of California Agricultural Issues Task Force. Special Publication 3250, October 1978.

University of California, Division of Agricultural Sciences. *California Agriculture*. 31(May 1977). Special Issue: Water.

University of California, Water Resources Center. *California Water Problems and Future Research Needs*. Report No. 49, October 1980.

University of California, Water Resources Center. *Chronicle of Research, 1957–1980*. Report No. 48, June 1980.

University of California, Water Resources Center. *Directory of Water Resources Expertise*. Revised June 1979.

University of California, Water Resources Center. *Proceedings of the Twelfth Biennial Conference on Ground Water*. Report No. 45, November 1979.

Water Foundation. *Water: Will There Be Enough?* Santa Barbara, 1979.

A windmill and an electrical transmission tower symbolize forms of energy used in agriculture. *Photo by Jack Clark.*

5

Energy and Agriculture

Vashek Cervinka

The high productivity of California agriculture is based upon the optimum interaction of soils, climate, water and energy. Energy is used in agriculture in many ways: mechanized field operations, irrigation, fertilizers, handling and transportation of farm products, as well as in their processing and storage.

California is a model of energy-intensive agriculture; its development has been supported by the availability of fossil fuels and electricity. Until recent years, these energy sources were relatively inexpensive and their use was encouraged by market conditions and by the orientation of most research and development efforts. However, since 1973/74, agriculture, like other economic sectors, has been facing the twin problems of continuously increasing costs of energy and of uncertainties of supply. This situation has created new challenges as well as opportunities for California agriculture.

Overall Energy Consumption in Agriculture

The production and processing of agricultural commodities presently requires about 5 percent of the total state demand for energy. This is not a very significant amount, especially when we consider the importance of agriculture in the California economy. The value of agricultural production here was $12.1 billion in 1979. California is the leading farm state in value of production, with over 9 percent of the nation's cash receipts

from only 3 percent of the nation's farms. In comparison, energy consumption in other sectors of the state's economy is significantly higher: 48 percent of the demand is used in transportation, 16 percent in residential use, 21 percent in the industrial sector and 10 percent in the commercial sector (Figure 1).

Energy sources used in food production and processing are natural gas, electricity, aviation fuel, gasoline, diesel, fuel heating oil, and LP gas. During the last several years (1977–1980) other energy sources have also begun to be utilized in California agriculture; these include biomass, geothermal power, solar and wind energy. These new sources are expected to play an increasingly important role on farms and in the food or fiber processing industry; however, their present (1980) contribution is not yet very significant.

Although aggregate energy use in agriculture is only about a twentieth of all energy consumed in the state, the relative use of specific energy sources varies. The agricultural use of natural gas represents about 5.1 percent of total state consumption (Figure 2). The relative use of electricity is 7.8 percent; 11 percent for diesel fuel or heating oil; and about 11.5 percent for LP gas. The estimated usage of aviation fuel for agricultural aircraft (airplanes are used in California mainly for the aerial application of chemicals or fertilizer, and for the planting of rice) is relatively high at 21 percent of the total supplies used for small airplanes, both commercial and recreational. On the other hand, it is estimated that the

FIGURE 5.1 Relative use of energy in California, 1980. Agricultural production and processing uses relatively less total energy than other sectors. *Figure by author.*

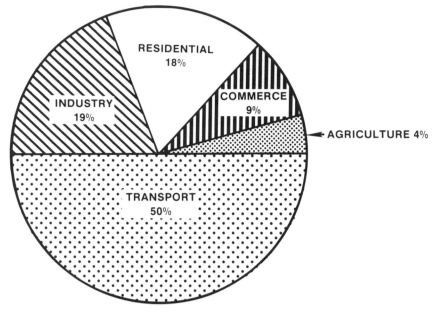

FIGURE 5.2 Proportion of California energy supplies consumed by the agricultural industry, 1980. *Figure by author.*

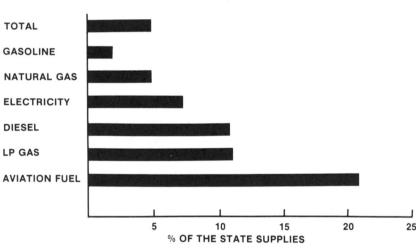

equivalent of the amount of gasoline used in California agriculture during a one-year period, is consumed by private and commercial vehicles on the state's roads in only seven days (1.9 percent of total state consumption).

The largest usage of natural gas by agriculture is in the production of commercial fertilizers and in food processing and product drying. Irrigation and milk production consume a high proportion of agricultural electricity. New tractors, trucks and other farm engines mostly operate on diesel fuel, while gasoline is generally used in farm business related to transportation. Heating oil and LP gas are used in food processing, product drying and space heating.

Over 36 percent of all requirements for agricul-

tural energy are met by natural gas. The proportion of diesel and oil is 18 percent, electricity 35 percent, and gasoline about 9 percent (Figure 3).

Trends in Agricultural Energy Use
California enjoys excellent geographical and climatological conditions, and there are many well trained men and women working on farms or in agriculture-related industries. These essential factors, combined with the availability of abundant supplies of inexpensive energy, have contributed to very high productivity in food and fiber. No other state of the union or indeed any other country in the world has achieved a comparable degree of yield and diversity in agricultural production.

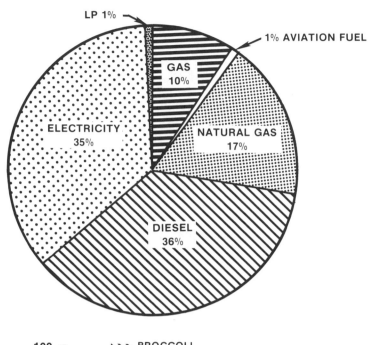

FIGURE 5.3 Total energy consumption in California agriculture, by type of supply, 1980. *Figure by author.*

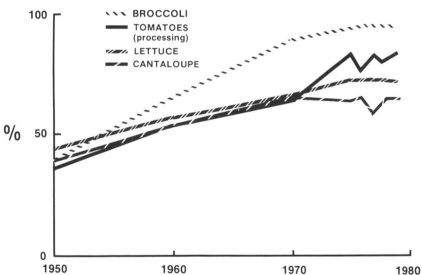

FIGURE 5.4 California share of national production of selected vegetable commodities, 1950–1980. *Figure by author.*

The ready availability of inexpensive fossil fuels and electricity in the years after World War II made it possible, under the conditions of an extremely competitive market, for California farmers to produce fruits and vegetables on irrigated land and to transport these commodities to distant markets in eastern states, Canada, and various foreign countries. California farmers and food processors have been able to supply continuously increasing proportions of various food commodities to the national market. This can be illustrated by charting trends in California's participation in the national market for several different fruits and vegetables. It can be seen from

Figure 4 that this share was increasing during the period from 1940 to 1975; these are the years characterized by the steady growth of energy-intensive agriculture. The leveling of national market shares since 1975 can be explained not only by the achievement of market potential, but also by the changing price structure of energy supplies.

The price of energy sources was relatively low until the 1970s. Using 1970 prices as a base, costs of fuels by 1980 had climbed, respectively, nearly 300 percent for electricity and nearly 500 percent for diesel. Price trends for agricultural energy are presented for electricity and diesel fuel (Figure 5).

Higher costs of energy have an impact on food prices and new technological developments on farms.

Energy trends in California agriculture have both quantitative and qualitative characteristics. While the number of units in tractors, combines or other farm equipment is constant or even declining, their size and power is increasing. This trend can be illustrated by comparing the average horsepower of tractors sold annually in California between 1940 and 1960. The change towards larger energy units can also be observed from additional data on tractor marketing (Figure 6).

Energy use in the Production and Processing of Agricultural Commodities

Commercially available energy is applied to production and processing in several diversified forms. Motor fuel energy and the energy consumed in the manufacturing of tractors and equipment is used for the field operations of soil cultivation, planting and crop harvesting. The handling and transportation of farm products and farm-related materials require significant inputs of energy. Electric motors or combustion engines are required for pumping and applying water on irrigated farmland. The production of meat on farms or feedlots, milk on dairy farms, or eggs on poultry farms requires large amounts of energy. High yields for many crops can only be achieved if sufficient levels of plant nutrients are provided. Energy is used both as a feedstock and as process energy in the production of chemical fertilizers; additional energy is needed for their transportation and application. Farm products are perishable commodities; thus, to extend the time of their availability food must be preserved by canning, freezing or dehydration. All these operations require significant energy inputs.

Energy Allocations for Specific Crops or Commodities

A 1978 survey examined the use of energy in California agriculture according to specific commodity categories, revealing some significant differences. Some crops use most energy in field

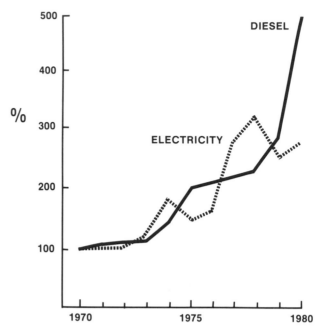

FIGURE 5.5 Price trend of diesel fuel and electricity by percent of increase, 1970–1980. *Figure by author.*

operations such as irrigation; others consume most energy in first stage processing. Sources of energy also vary according to operation. A few examples follow.

FIELD CROPS. Energy for the production and processing of field crops is typically allocated as indicated in Table 1.

In the series of steps required to produce field crops up through necessary first stage processing, irrigated grains consume most energy in irrigation and fertilizer. Sugarbeets, on the other hand, consume proportionately far more energy in the process of sugar refining. Cotton, another important California field crop, uses most energy in field operations and irrigation.

TABLE 5.1
Proportional Energy Use (%) in Field Crop Operations

Operations	Grains (Irrigated)	Sugar Beets	Cotton
Field operations	14.8	7.5	26.5
Irrigation	51.3	11.1	45.9
Fertilizers & chemicals	30.8	8.3	18.2
Aerial spraying	1.0	.1	1.4
Transport	1.6	1.9	0.7
Processing	.5	70.8	7.3

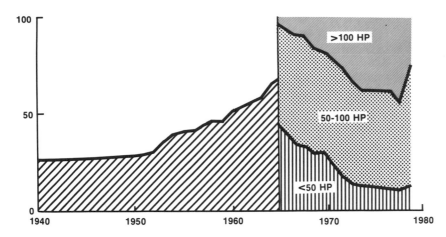

FIGURE 5.6 Average horsepower of farm tractors sold in California between 1940 and 1965. The left side of the figure shows that the average HP has approximately doubled; the right side indicates shares of tractor sales by HP between 1965 and 1979. Fewer smaller tractors have been sold as large units have increasingly dominated sales. *Figure by author.*

FRUITS AND NUTS. The production of fruit and nut crops requires about 17.2 percent of the total energy input for California agriculture. In 1979 fruits and nuts were grown on 1.7 million acres, with production of 9.5 million tons and value of $2.9 billion, requiring an estimated 43.3 trillion Btu of energy. Due to favorable climatic conditions and availability of water for irrigation, the fruit industry has reached a high degree of intensity in the production of a wide range of over 26 commodities. California is the major supplier of fruit and nut products both to the national and export markets. The production of fruit and nut crops is highly specialized, and it requires significant amounts of energy in the form of mechanized operations, fertilizers, irrigation, and frost protection. Energy consumption in fruit and nut crops is affected by production technology and final utilization of products. A significant proportion of the production of deciduous fruit and grapes is processed. Lemons are mostly supplied to the fresh produce market, while over 50 percent of orange and grapefruit production is processed.

Energy is used in the production and processing of fruits and nuts generally as shown in Table 2. Natural gas and electricity are mainly used in fruit and nut processing. Diesel or gasoline are consumed in field operations and transportation.

If fruits and nuts are ranked according to energy inputs, it is found that the highest energy requirements are for nuts, while citrus fruits require relatively the lowest inputs of energy. The percentage of energy usage for mechanized operations is lowest for lemons and highest for walnuts. Lemons also require the lowest energy inputs in the form of fertilizers and irrigation. The percentage of energy applied as fertilizers fluctuates between 11.7 percent and 18.5 percent for deciduous fruits.

Citrus fruits require significant quantities of energy for frost protection, as do some other orchard fruits. Frost protection is provided by three methods: (1) heaters, (2) wind machines, and (3) irrigation. The total energy input for frost protection is potentially equivalent to about one-third of the total energy required for the production and processing of fruits and nuts. This quantity

TABLE 5.2
Proportional Energy Use in Fruit and Nut Operations

Source of Energy	% of Total	Operations	Energy Use (%)
Electricity	39.1	Field operations	9.5
Natural gas	36.0	Irrigation	31.8
Diesel	16.7	Fertilizers & chemicals	17.0
Gasoline	7.1	Aerial spraying	0.2
LP Gas	0.9	Frost protection	11.2
Aviation Fuel	0.2	Transport	3.2
		Processing	27.1
TOTAL	100.0	TOTAL	100.0

of energy is not utilized every year, but must be instantly available when needed, since any delay in frost protection could cause significant damage to fruit production.

VEGETABLE CROPS. Vegetables are produced both for the fresh produce market and the processing industry. The major markets for fresh vegetables are in the large metropolitan areas of the United States and Canada. Relatively low energy costs for the production, irrigation and transport of fresh produce have encouraged the competitiveness of California vegetable farming and consequently California has a high share of the national market. This trend is well illustrated in Figure 4; it can also be seen that the California share of national markets has stabilized during the past five years. Increasing costs of energy may well be one factor in this trend.

The total amount of energy required for vegetable production is significantly influenced by the energy demands for the production and processing of tomatoes; processing tomatoes requires 79 percent of the natural gas, 28.6 percent of the electricity, 56 percent of the diesel and 30 percent of the gasoline used in the production and processing of all vegetables. Energy inputs for processing tomatoes are given in Table 3. Vegetable processing is an energy-intensive operation; due to its timeliness, consumption of energy is concentrated during a short season.

TABLE 5.3
Proportional Energy Use in Processing Tomato Operations

Operations	Energy Use (%)
Field operations	9.4
Irrigation	14.3
Fertilizers & chemicals	17.9
Aerial spraying	0.3
Transport	5.8
Processing	52.2
TOTAL	100.0

MILK. With more than 2,200 dairy farms, California is the second largest milk producing state. It is estimated that energy requirements for milk production are 15.1 trillion Btu (equivalent to 5.5 percent of the state's total agricultural energy requirements).

Milk production and processing is completely mechanized. Contemporary milking parlors on large dairy farms are economically sized for 16 to 20 hours of operation per day. Milk must be cooled at the same rate at which it is produced. Washup and sanitation require pressurized, hot water. The operations of milk production and processing require a constant, reliable supply of energy. Milk preservation by dehydration in dairy processing plants also requires significant quantities of energy. Nearly half of the energy used in the milk industry comes from electricity, and about 30 percent from natural gas. Transport of milk for processing and distribution also requires significant amounts of motor fuel, nearly one fourth of all energy used.

GREENHOUSES. In production value, flowers and other greenhouse plants rank in eleventh place among California crops. The principal growing areas are along the central and southern coasts of California. Plant production in greenhouses is energy-intensive because of the demands for precise control of the growing environment. Greenhouses consume about 6.6 percent of the total energy requirements in California agriculture. Natural gas represents a major portion of this energy—more than 97 percent of greenhouse energy requirements. Much of greenhouse heat energy is lost through radiation and conduction at night. Methods are being developed to prevent nightly heat loss. Insulation systems involve use of thermal curtains or blankets and liquid foam, while double plastic roofs use trapped air as an insulation blanket, saving energy. Two major energy alternatives to natural gas in the future may be solar heating and rejected heat from large industrial units.

Energy use for Systemwide Needs

Although specific commodities may use more or less energy in various stages of production and processing, certain energy-consuming activities in agriculture are common to nearly all crops grown. Conditions in California make some of these systemwide needs for energy especially significant. Although fertilizer usage is common throughout the nation, for example, irrigated agriculture is more highly developed in California than in any other state, and geographic position makes California product transportation to distant markets especially crucial.

Fertilizer Production and Application

High yields of many California crops can only be achieved with additional applications of plant nutrients. Due to the past availability of relatively inexpensive fossil fuels, the use of chemical fertilizers reached high levels in California agriculture at a rapidly increasing rate over a thirty-year period. In 1950 fertilizer usage averaged less than 40 pounds per cropland acre across the state; by 1980 average use had increased three-fold, to more than 120 pounds.

From extraction of sulfur and potash thousands of feet below the earth's surface to ammonia synthesis, fertilizer production is very dependent upon energy inputs. Synthesis of ammonia consumes the most energy in the manufacture of nitrogen fertilizers. Along with energy used in downstream conversion of products such as urea and ammonia nitrate, about 28,000 Btu are used per pound of fertilizer nitrogen. In comparison, substantially less energy is needed for the production of phosphate fertilizers (5,000 Btu/pound) and potash fertilizers (4,000 Btu/pound). Over 90 percent of the total energy represented in fertilizer usage is consumed in the manufacturing process, while transportation and application together account for only 6 percent.

High energy requirements for the production of nitrogen fertilizers, combined with continuously increasing costs of natural gas, have reduced the production of nitrogen fertilizers in California. While in the 1960s there were eight ammonia plants in California, in 1980 only two plants remained in the manufacture of nitrogen fertilizers. Consequently an increasing percentage of fertilizer needs is being met by imports; this situation results in a certain degree of vulnerability for crop production on California farms.

In the production of farm commodities, chemical fertilizers account for a high portion of total energy inputs. Data developed by the USDA identify some of the differing needs of various crops. The percentage of total energy inputs accounted for by the application of fertilizer was 52 percent for potatoes; 48 percent for corn; 45 percent for wheat; 40 percent for sugar beets; and 13 percent for soybeans.

Natural nitrogen fixation by the root system of soybeans explains the low demand of this crop for fertilizer energy. Furthermore, the energy and economic advantages of nitrogen fixation in legumes is illustrated as follows:

	Wheat	Corn	Soybeans
Crop yield/energy—1b/MBtu	606.7	641.6	732.2
Crop yield/energy—$/Mbtu	27.40	24.20	75.90

Recent research efforts in biological nitrogen fixation are intended to extend agricultural alternatives to chemical fertilizers. (See Chapter 14 for more discussion of fertilizers.)

Irrigation

The energy used for on-farm irrigation in California represents over 13 percent of total agricultural energy. This amount is equivalent to the total energy value of all fuels (diesel and gasoline) used in the production, transport, and processing of California crops. It is estimated that approximately 7 billion kwh are utilized for farm irrigation, which is 71 percent of all electricity used in California agriculture. Electricity provides more than 81 percent of energy for water pumping on California farms, the other power source being mostly diesel fuel. Most of the energy associated with irrigation is to bring water to the field, either by pumping ground water (43 percent) or moving surface water from reservoirs and rivers to fields (41 percent). The remaining 16 percent is used to pressurize sprinkler systems—17 percent of all irrigated acreage in California is serviced by the sprinkler method.

Factors influencing energy usage for irrigation on farms in general are: crop type, number of irrigated acres, application rates, type of irrigation system, climate and soil conditions, well depth or pumping needs to deliver surface water to the plants. The method of water application significantly affects energy demands. Sprinklers in general reduce the amount of water used, but require more energy per unit of water applied. It is estimated that in California the typical energy consumption is 216 kwh per acre foot for a sprinkler system. Sprinklers use, however, on the average, 20 percent less water than surface irrigation methods.

In addition to surface and sprinkler irrigation systems, drip irrigation is used in California. Because of the high cost of installed systems, this method is limited to high-value crops or to situations where special problems exist that only drip irrigation can solve. Drip irrigation generally uses lower water pressure than sprinkler irrigation, with consequent lower energy requirements.

Agricultural Transport

The diversity of California agriculture results in high demands for the transportation of farm commodities. Agriculture in other regions of the United States is characterized either by the marketing of only a few major farm commodities or by mixed crop and livestock farming systems. When only a few farm commodities are marketed, the transportation system is relatively simple. Under the conditions of a mixed farming system, most commodities are transported within the boundaries of a given farm, and only final products are shipped to a distant market In contrast, more than one hundred commodities produced on California farms must be shipped to storage, packing sheds, dehydrators, processing plants and to established markets in large metropolitan areas, both in North America and overseas.

Agricultural products are transported by trucks on farms and through the intrastate marketing channels. For the interstate market, a complex system of trucks, railcars and airplanes is used for the shipment of farm products, and ships are loaded with farm commodities in California ports for overseas markets.

Even intrastate transport is typically for fairly long distances, as observed in Table 4.

Distances between California and out-of-state markets in the large metropolitan areas are, in most cases, between 2,000 and 3,000 miles. The average fuel consumption of diesel trucks is about 4.5 MPG; consequently the transport of farm products is very sensitive to increasing costs of fuel. The case can be illustrated by the hauling of California farm products to markets in the eastern United States. The two-way distance is about 5,750 miles, which is equivalent to 1,278

A startling sight in the Delta is the view of an ocean freighter moving up the deep water channel through farm fields to the port of Sacramento. Energy is important for both production and transportation of agricultural commodities. *Photo by Sirlin Studios, courtesy of the Port of Sacramento.*

TABLE 5.4
Average Round Trip Haul Distance for Selected California Agricultural Commodities

Product	Miles from Farm to Point of Sale
Apples	160
Apricots	206
Barley	108
Broccoli	241
Cauliflower	186
Celery	228
Corn	144
Cotton	200
Oranges	200
Peaches	206
Pears	174
Rice	104
Tomatoes	200

gallons of diesel fuel used per truck. Fuel costs per a typical long distance haul have been increasing as follows:

1973	$ 367.40
1975	$ 578.80
1977	$ 646.80
1979	$ 956.20
1980	$1,252.40

This trend has profound implications for California fresh fruit and vegetable shipments, which have typically been shipped 79 percent by truck, 20 percent by rail, and 1 percent by air.

Energy Management: Options for Conservation

Continuously increasing costs of fuels and electricity impel close attention to the efficient management of energy. Major methods of achieving optimum efficiency in agricultural operations fall into several categories:

1. *Reduction of farm product losses.* Farm products which are not economically utilized in a food system can represent significant losses of energy applied during production and processing operations.

2. *Optimum application of water, fertilizers and chemicals.* Inefficient use of production inputs results in both economic and energy losses.

3. *Use of by-products and waste materials.* When crops are harvested, the main products (grains, beets, etc.) are marketed; crop residues and/or by-products have additional economic value, which can be realized either in the maintenance of soil fertility and organic structure or in their use as energy or industrial feedstocks.

4. *Well-designed tractor and equipment systems.* Optimum combinations of tractors and other equipment can result in the reduction of up to 25 percent energy needs for given operations.

5. *Cogeneration in food processing.* Improved economy of operations is often possible when high pressure steam is used both for generation of electricity and as a source of process heat.

6. *Electricity monitoring.* On dairy farms or in irrigation systems the efficient use of electricity includes peak demand control, efficient components, and good maintenance of equipment.

7. *Technological Developments.* Typical examples illustrating the use of technological developments for energy conservation are (1) heat exchangers used on dairy farms in the operations of milk cooling and water heating, (2) thermal curtains in greenhouses, or (3) reuse of heated air in the dehydration process.

Agriculture as an Energy Resource

Historically, agriculture has been not only a user but also a producer of energy. Agricultural and forestry products have traditionally been basic sources of energy. Biomass is a renewable energy source; its supply will continue as long as solar energy supports life on earth. Biomass can be converted to any of four basic types of power: direct combustion of solid fuels, or gasification, with gases used to supply heat, to run engines or to generate electricity.

Biomass includes farm crops, residues, animal manure and food processing waste. Biomass contains energy stored from the photosynthetic process in the form of sugars, starches and cellulose. Dry biomass has an energy value of 6,000 to 8,000 Btu/pound. Energy conversion of biomass does not have adverse environmental effects, and it need not create any waste problems because unused portions of biomass can be returned to the soil.

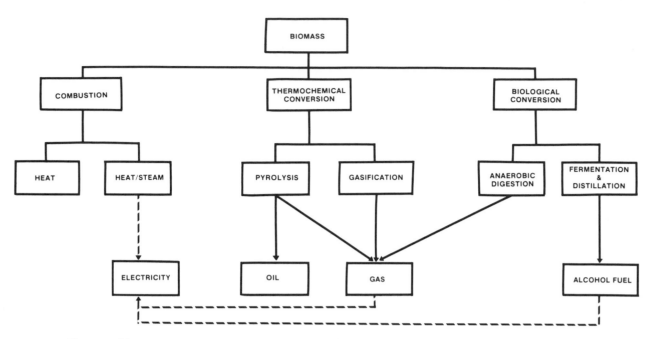

FIGURE 5.7 Process of biomass conversion into energy. *Figure by author.*

Many technologies exist for converting biomass to different forms of energy. These processes can be classified as combustion, thermochemical conversion and biological conversion (Figure 7). The technology of combustion is highly developed; it is an efficient process used for direct heating or for the production of steam and electricity. Thermochemical conversion includes energy conversion of biomass by pyrolysis or hydrocarbonization; the final products are low or medium Btu gases. Biological processes include methane production by anaerobic digestion and the production of liquid alcohol fuels using the technologies of fermentation and distillation.

Considering the importance of liquid fuels for farm operations, product transportation and the use of cars in the general transportation systems, the alcohol fuel industry is rapidly developing. The major crops presently used for the production of alcohol fuels are corn, grains, potatoes, sugarbeets and fruit waste products. Both research and practical on-farm experiments are oriented towards the use of other sugar or starch crops, such as fodder beets, sweet sorghum, corn silage, Jerusalem artichokes and additional crops with a high energy potential. The conversion of cellulosic materials into alcohol fuels is a very promising energy technology.

The conversion of biomass into different types of energy is well suited to farming systems, and various energy conversion processes are becoming a part of some on-farm operations. An overall spectrum of potential biomass-energy conversion plants ranges from on-farm plants to larger industrial operations. Other sources of renewable energy such as solar, wind energy, hydropower or geothermal power are also becoming part of revised farm technology. In the combination of biomass with other new sources of energy, farms may become not only energy independent, but also more diversified as producers of energy as well as food and fiber.

Outlook for Agricultural Energy
Rather than as a crisis, the energy problem can be viewed as a challenge and opportunity.

From a historical perspective, the twentieth century can be characterized by its energy uniqueness. Some previous civilizations also reached remarkable degrees of technological, material, and cultural wealth, but it is important to realize that their achievements were based upon renewable energy—sun, wind, water, and crops. It is unique to our civilization that for some decades we have been using a special type of energy which is not renewable. Fossil fuels will be exhausted at a given time. This energy, not created by man, has been produced through natural processes over a period of millions of years. Technologically, the use of fossil fuels has had far-reaching results; the high productivity of agriculture is a remarkable example. On the other

A co-generation system for converting walnut shells into energy was built by Diamond Walnut Growers to supply power for its Stockton plant. The cooperative also markets energy to local utility companies.

hand, dependency upon fossil fuels has led to the extreme economic and political vulnerability of our nation. Because fossil fuels are finite, we must examine the question of future energy sources.

It is generally realized that future energy systems must and will again be diversified, and that agricultural resources will have an essential role in these emerging energy systems. The ready availability of fossil fuels eclipsed agriculture's former role as a source of energy. A return to renewable sources, however, does not mean technological retrogression. Highly advanced scientific and technical knowledge exists. To utilize it for the future development of renewable energy is a challenge.

In past decades California farmers have sometimes been at an economic disadvantage because of their capacity to produce more products than the market could absorb. The diversification of farming into the production of food, fiber, and energy offers not only a technological challenge, but possibly great economic opportunity for California agriculture.

References

Cervinka, Vashek et al. *Energy Requirements for Agriculture in California.* A joint study of the California Department of Food and Agriculture and the University of California, Davis. January 1974.

Cervinka, Vashek et al. *Energy Requirements for Agriculture in California.* A joint study of the California Department of Food and Agriculture and the University of California, Davis. May 1981.

PART III 🌿 Crops:
The Pattern of
Production

An aerial view shows the patterns of the grain harvest. Some areas of the state suitable for extensive field crop farming are not adaptable to other types of agriculture because of climate, soils, or topography. *Photo by Jack Clark.*

6

Field Crops

*Members of the Department of Agronomy
and Range Science, U. C. Davis*

Chapter coordinator, *D.S. Mikkelsen*. Contributors:
R.B. Ball, seed industry; *M. George*, annual forages and
irrigated pastures; *F.J. Hills*, sugarbeets; *K.H.
Ingebretsen*, field corn and grain sorghum; *P.F.
Knowles*, oilseed crops; *D.S. Mikkelsen*, rice and cotton;
C.W. Schaller and *Linda Prato*, small grains; *L.R. Teuber*,
alfalfa; *C.L. Tucker*, beans and peas.

Mankind depends on field crops for much of its food, fiber and livestock feed, and increasingly for products that can be transformed into a variety of consumer goods. Recent developments in the area of energy, for example, suggest increasing reliance upon field crops to help meet fuel needs of the future. To meet ever changing human needs, the art and science of crop production are also changing, often in a very dynamic fashion. New crops, new varieties, new methods of crop production, protection, harvesting, storing, marketing and new uses—all are directed towards improving our living standards and providing for an ever-expanding world population.

Farmers and research workers have been adept in adapting and improving field crops originating from all parts of the world to fit the climatic and cultural conditions found in California. More than 50 different field crops are grown in the state. In 1980, 6,826,000 acres of field crops were harvested in California, representing about 20 percent of the total cropped area and producing a total value of farm products valued at $3.9 billion. The important areas of production of some major field crops, represented in the seven geographical areas of California, are shown in Figure 1. About 52 percent of the field crop acreage and 52 percent of the hay and silage acreage of California come from the San Joaquin Valley. Field crop acreage in the Sacramento Valley amounts to 29 percent and hay and silage about 10 percent. The North Coast and Northeastern Interior

accounts for about 17 percent of the hay and silage production.

Table 1 presents the acreage, yield, production, price and value of California field crops in 1980, listed in order of their value of farm production. Cotton lint and seed were responsible for the largest portion of field crop value. Alfalfa hay, rice, winter wheat and sugarbeets represent the other four leading field crops produced in California. Because farm income exerts a major influence on the selection of crops to be grown annually in the state, the rank in value may fluctuate widely from year to year.

Small Grains

The small grains, barley, wheat and oats, have played an important role in California agriculture since the days of the Spanish Missions. They are the most widely grown crops in California, with some commercial production in nearly every county of the state. They are grown under widely diverse environmental conditions, at elevations below sea level, up to 4–5,000 feet in the mountain valleys, in areas with less than 8 inches of rain (without irrigation) and in areas of 30 inches or more. They are essentially grown throughout the calendar year, with early fall planting in the south concurrent with harvest in the north and spring planting in the north just

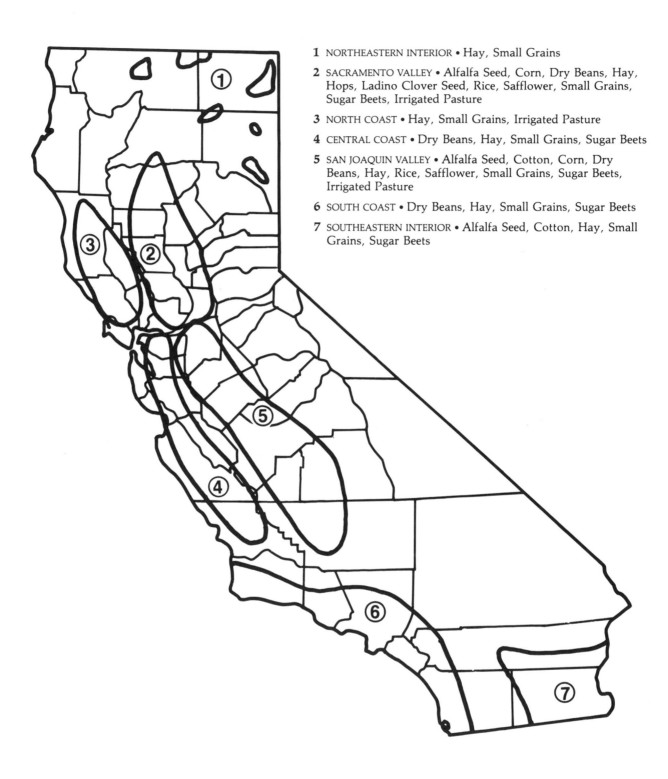

1 NORTHEASTERN INTERIOR • Hay, Small Grains

2 SACRAMENTO VALLEY • Alfalfa Seed, Corn, Dry Beans, Hay, Hops, Ladino Clover Seed, Rice, Safflower, Small Grains, Sugar Beets, Irrigated Pasture

3 NORTH COAST • Hay, Small Grains, Irrigated Pasture

4 CENTRAL COAST • Dry Beans, Hay, Small Grains, Sugar Beets

5 SAN JOAQUIN VALLEY • Alfalfa Seed, Cotton, Corn, Dry Beans, Hay, Rice, Safflower, Small Grains, Sugar Beets, Irrigated Pasture

6 SOUTH COAST • Dry Beans, Hay, Small Grains, Sugar Beets

7 SOUTHEASTERN INTERIOR • Alfalfa Seed, Cotton, Hay, Small Grains, Sugar Beets

FIGURE 6.1 California's major field crop areas and important field crops of each area. Areas are approximate only; field crops are also raised in several small areas not included within these boundaries. *Source: California Crop and Livestock Reporting Service.*

prior to harvest in the south. As might be expected, yields are equally variable, ranging from 1,000 to 8,000 pounds per acre. Collectively, they represent the largest agronomic crop acreage in California, with approximately 2.5 million acres grown annually.

The small grains—wheat, barley, oats and triticale—are cool season crops, and, except in the high mountain valleys of northern California, planting is timed to ensure the maximum use of winter rainfall and to avoid excessive summer temperatures and warm weather diseases. Varieties of the true spring growth habit have sufficient cold tolerance to permit fall planting in most areas. At higher elevations, these spring or nonhardy types will not withstand the winter temperatures and must be spring sown. In those areas, varieties with winter growth habit can be planted in the fall and are presently grown on limited acreage.

Barley is less adapted to production on heavy, poorly drained soils than either wheat or oats, but is more tolerant of alkali conditions. Consequently, wheat is the most satisfactory cereal to use in rotation with rice, whereas barley is the preferred crop on saline soils.

Grain production follows two general cultural patterns—dryland and irrigation farming. Common under dryland conditions are continuous grain; alternate crop and fallow (areas with less than 14–16 inches of rainfall); and short rotations of grain, pasture and fallow. On irrigated land, grain may be grown in rotation, either with or without irrigation. Irrigation practices vary with location, with a pre-irrigation followed by one to three crop irrigations common in areas with minimal winter rainfall. A combined total of 24–30 inches of water is normally sufficient for maximum productions. The supplemental water can be applied using large checks, or furrows, in combination with narrow or wide beds and with sprinklers.

Except in the northern valleys at elevations of 3,000 feet or more, the best planting months are November, December and early January. In the northern valleys, planting commences as soon as a satisfactory seedbed can be prepared in the spring and when the danger of severe frost is past. Grain is planted either by broadcasting the seed on top of prepared soil and covering with a harrow or disk, or placing it directly in the soil with a drill. Nitrogen fertilizer is widely used, with rates varying from 30–120 pounds of actual nitrogen per acre, usually applied just prior to

TABLE 6.1
California Field Crops: Rank, Acreage, Production and Value, 1980.

Crop	Rank in Value	Acreage Harvested	Yield Per Acre	Total Production/Unit	Season Av Price	Value of Production
		000		*000*		*$1,000*
Cotton lint	1	1,500	995 Lbs	3,109 Bales	$ 81.10[a]	1,210,272
Cotton seed	—	—	—	1,270 Tons	141.00	179,070
Hay, alfalfa	2 (est.)	1,030	6.4 T	6,592 Tons	n.a.	n.a.
Rice	3	548	6,440 Lbs	35,301 Cwt	9.70[b]	326,589[b]
Wheat, winter	4	1,050	74 Bu	77,700 Bu	4.15	322,455
Sugarbeets	5	228	25.8 T	5,882 Tons	30.80[b]	175,326[b]
Corn (grain)	6	270	135 Bu	36,450 Bu	4.15	151,268
Barley	7	712	62 Bu	44,144 Bu	3.30	145,675
Beans	8	220	1,777 Lbs	3,909 Cwt	30.70	120,006
Alfalfa seed	9	78	520 Lbs	40,560 Cwt	125.00	50,700
Sorghum (grain)	10	152	73 Bu	11,096 Bu	3.56	39,502
Safflower	11	105	2,400 Lbs	126 Tons	280.00	35,280
Oats	12	70	62 Bu	4,340 Bu	2.60	11,284
Wheat (Durum)	13	100	78 Bu	7,800 Bu	4.55	35,490
Ladino clover seed	14	8.4	345 Lbs	2,898 Cwt	136.00	3,941
Hay (ex. alfalfa)	15	520	2.2 T	1,144 Tons	n.a.	n.a.

[a] Price per lb
[b] 1979 figure

SOURCE: *Field Crop Statistics*, California Crop and Livestock Reporting Service, 1980.

planting, by broadcasting followed by harrowing. An additional spring application is occasionally made if plants show deficiency symptoms. Certain soils are phosphorus-deficient and production is increased with the application of phosphorus, which can be successfully predicted by use of soil tests. When only limited quantities are required, the fertilizer can be drilled with the seed. University field trials have shown some responses to sulfur and zinc, but little response to potassium.

The troublesome weeds in grain fields include: grassy weeds such as wild oats, annual rye grass and ripgut brome; and broad-leaved weeds, such as wild radish, mustard, star thistle and fiddleneck. Control of all types is abetted by good farming practices, including the use of weed-free seed. Chemical herbicides provide satisfactory control for most serious weeds, along with good cultural practices.

Insect pests causing economic losses to the growing crop include wireworms (immature stages of click beetles), aphids and Hessian fly, the latter formerly causing considerable damage to wheat, but of little importance in recent years. Seed treatment with approved chemicals has been fairly effective in reducing damage by wireworms. Several species of aphids transmit the virus causing the Barley Yellow Dwarf disease, one of the major diseases of these crops in California. In addition to serving as vectors for the virus, high populations of aphids can cause direct injury to the growing plants. While chemical control measures will minimize the mechanical injury, they have little effect on reducing the transmission and spread of the disease.

With the exception of oats, the small grains are normally combine-harvested. Fall planted grain is normally harvested in late May, June and July. Harvesting of spring sown grain in northern Cal-

Wheat is part of California history, and is still harvested on over a million acres. Large combines like this one are a familiar sight in wheat growing areas.

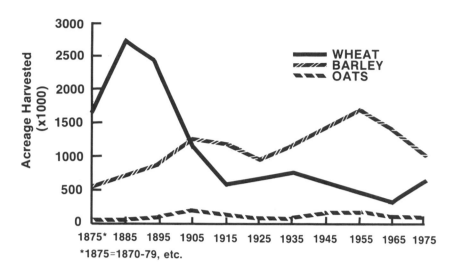

FIGURE 6.2 Average barley, oats and wheat acreage in California by decade, 1870–1979. *Data from California Crop and Livestock Reporting Service.*

ifornia usually commences about the middle of August. Grain is safely stored at 14 percent moisture. Harvesting is usually delayed until that moisture content has been reached.

Insects are a constant threat to grain in storage, particularly in warm environments, such as California. Standing grain in the field is usually free from storage insects; contamination occurs during handling and storage. Insecticides and fumigants are available to assist in controlling storage insect pests.

Wheat

Wheat was first sown in California about 1770, in the lower valley of the San Diego River near the original San Diego Mission settlement at Old Town. Until the discovery of gold, enough was planted around the expanding mission system to feed the military garrisons and some of the natives. Varieties brought by the padres from Spain were well adapted and a few were still grown on limited acreage as late as 1940. Increased demand during the Gold Rush, together with large areas suited for wheat production and the development of farm machinery, stimulated rapid expansion in acreage and, by 1888, California ranked second in the nation with more than 3,000,000 acres (Fig. 2). Beginning about 1890, California wheat acreage began to decline. Loss of foreign markets, competition with other crops following the development of irrigation systems, and the expansion of barley into the drier areas, together with the inroads of bunt, stem rust and other diseases, all contributed to this reduction. Reaching a low point in 1966, with approximately 270,000 acres, wheat acreage since has steadily

increased, with 1,200,000 acres harvested in 1980. This was the first year since 1937 when more than one million acres were harvested and, when combined with an all-time yield record of 4,400 pounds per acre, total wheat production in California was the highest in history. Wheat was California's second largest agricultural export commodity in 1979.

Major factors contributing to renewed interest in wheat include high yields per acre, together with higher prices for the grain, the latter resulting from expanding export markets, as well as increased domestic demand. Yield per acre increased three-fold during the last four decades and more than doubled in the 20 year span from 1960 to 1980 (Fig. 3). Recent development of short-statured, stiff-strawed, disease resistant varieties with a high yield potential, capable of responding to intense production practices, contributes substantially to the high yields. Increased fertilization and irrigation, judicial use of herbicide and overall effective management practices are also contributing factors.

Diseases have been a major hazard to wheat production since its introduction by the Spanish missionaries. Periodic epiphytotics of stem rust and, more recently of stripe rust, caused widespread damage and severe economic losses. These diseases, capable of totally destroying a crop, can be controlled only by using genetically resistant varieties. Other diseases of economic importance for which resistant varieties offer the only means of control include barley yellow dwarf and Septoria. Prior to the development of resistant varieties and effective fungicides, bunt (stinking smut), a seed borne disease, ranked

FIGURE 6.3 Average yields of barley, oats and wheat in California by decade, 1870–1979. *Data from California Crop and Livestock Reporting Service.*

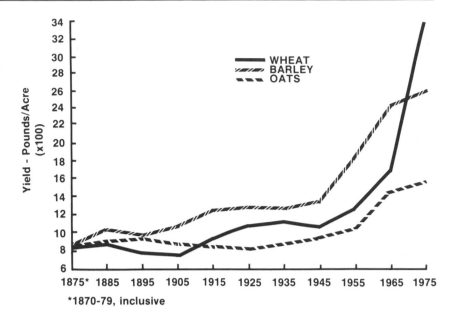

with stem rust as the two major wheat diseases in California.

In 1904, the University, cooperating with the United States Department of Agriculture, began experiments on wheat improvement. These have led to the introduction and breeding of higher yielding and disease resistant varieties. The first stem rust- and bunt-resistant varieties were released by the University in 1939. All of the varieties presently grown are resistant to stem and stripe rust, a few having tolerance to barley yellow dwarf and Septoria. The present varieties, which combine disease resistance with other desirable agronomic characteristics as short-stature and stiff-straw have been developed by cooperative breeding programs between the University and the Centro Internacional De Majoramiento De Maiz Y Trigo (CIMMYT), the latter with headquarters in Mexico City.

Although wheat is grown throughout California, the heaviest concentrations occur in the Sacramento, San Joaquin, and Imperial Valleys. Individual counties with large acreage include Yolo, Colusa, Sacramento, Butte, Kings, San Joaquin, San Luis Obispo, and Imperial. Originally grown as a dryland crop, approximately 75 percent of the present acreage receives one or more supplemental irrigations. Prior to 1965, most of the varieties had white kernels and were utilized by the milling industry in the manufacture of pastry, cake, cracker and all purpose flours. Present varieties are red, several with flour quality approaching the best bread wheats. Recent export demand for "macaroni" wheat has stimu-

lated interest and production of durum wheat in California, particularly in the southern area. Although wheat is considered less desirable than barley as a feed grain, varying quantities are utilized for feed, depending on local milling and export demands.

Barley

Barley has also been an important crop since the mission period. Seeds of the Coast-type barley, similar to that grown today, have been found in adobe bricks used in the construction of the first missions in the southwest. The first record in California is in 1771, at Jolon in the mission of San Antonio de Padua in Monterey County. The Coast type, which evolved in the drier areas of North Africa, proved to be well-adapted to the similar climatic conditions in California and, within a short time, barley became established as the principal feed grain for the livestock industry.

Barley acreage continued to increase following the increasing demand for feed grains and by 1899 more than a million acres were harvested. After reaching the 2 million acre mark in the 1950–55 period, acreage gradually declined to its present level of approximately one million acres (Fig. 2). Yield per acre continues to increase, doubling during the past four decades (Fig. 3) despite the shift of the barley acreage to more marginal land. Improved varieties and cultural practices are major contributing factors to the increase in production.

Barley is widely grown throughout California,

with the major concentration in the San Joaquin Valley. Similar to wheat in most soil and climatic requirements, barley is more productive than wheat under severe temperature and moisture stresses and on saline soils. Because of its earlier maturity, it is preferred over wheat where double cropping is practiced. Productivity can be severely reduced by several diseases, i.e. scald, net blotch, powdery mildew, barley yellow dwarf and barley stripe. These diseases can reduce yield by 50–60 per cent. Except for barley stripe which is controlled primarily by seed treatment, resistant varieties offer the only means of control. Fortunately, commercial varieties are available which provide satisfactory resistance or tolerance to all of these diseases.

Although barley is the principal raw material used in the brewing of beer, most of the California grown barley is presently used as a feed grain, either rolled into a flake or ground. During the 1935–70 period, large quantities of California grown barley were used by the local brewing industry. However, the present brewing industry consists largely of multi-location companies, with breweries located throughout the United States. In striving for uniformity between breweries, utilization of raw materials with similar quality characteristics is desirable. Since California grown varieties differ significantly in malting characteristics from those grown in the major malting barley producing areas of North Dakota and Minnesota, they are not now used by the industry. However, a limited acreage of malting barley is grown in the Tulelake basin of northern California.

Oats

Oats have never gained the prominence in California agriculture of either barley or wheat. Although its early history in California has not been well documented, early crop statistics indicate that 50,000 acres were harvested in 1867. This compares with 303,000 acres of barley and 785,000 acres of wheat for the same year. Recent acreage has ranged from 300,000 to 500,000 acres.

Oats in California are grown principally as a forage rather than as a grain crop. Grain acreage ranges from 100,000–200,000 acres, whereas acreage for oat hay varies from 200,000–400,000 acres. Limited acreage is harvested as green chop or ensiled. Oat hay is very palatable and highly nutritious and is preferred over either barley or wheat hay. Oat grain is considered a non-fattening feed and is used extensively as feed for horses, poultry, young animals and breeding stock.

Although general climatic and soil requirements are similar to wheat and barley, oats will not withstand as high temperatures, especially at flowering time. Consequently, its distribution is more restricted. Climatically, oats are best adapted to the cooler coastal regions and high mountain valleys. Production in the Central Valley is restricted mainly to the foothills and peripheral areas.

California varieties are classified as red oats, because the husk of the grain varies from light to dark red. This is in contrast to the oats grown in the humid areas of the United States which are white or yellow in color. The red oat types evolved in the warmer, drier climate of the Mediterranean area and are more adapted to the climatic conditions of California and the southwest than the white-yellow types. Grain yields have increased significantly during the past 30 years. Improved varieties, which combine disease resistance with other desirable characteristics such as shatter-resistance and improved straw quality, have contributed substantially to the increased productivity.

Stem rust, leaf rust and barley yellow dwarf are major oat diseases in California and can be controlled only by growing resistant varieties. Commercial varieties are available with resistance to the rusts, but all are susceptible to barley yellow dwarf. Other foliar diseases of limited importance include powdery mildew, leaf blotch and halo blight. Loose and covered smut, which destroy the developing kernels, are seed-borne diseases which can be of considerable importance. However, they can be effectively controlled with fungicides.

Harvesting oats for grain differs from barley or wheat in that the bulk of the acreage is cut and swathed before it is combined. This is to avoid loss of the grain before it reaches harvest moisture. Oat hay is usually cut when the developing kernel is in the soft-dough stage, although some studies have shown that cutting at an earlier stage may be more desirable.

Triticale

Triticale is a new man-made cereal grain developed from crossing wheat and rye. Although hybrids between these species date back to 1875, concerted efforts directed toward the development of high-yielding triticales as a field crop began about 1954. The variety 'Rosner' released by the University of Manitoba in 1968, was the first variety available for commercial production. A number of varieties have been released in subsequent years, including the variety 'Siskiyou' released by the University of California-Davis, for production in northern California.

Early hybrids and selections were characterized by low fertility, low yield, shriveled and shrunken kernels, tall, weak straw and late maturity. However, the grain was high in protein, with a good balance of amino acids, and a high lysine content. Considerable improvement has been made in the newer selections and varieties and, in certain environments, grain yield is equal to or approaches that of the better wheat varieties. Kernel type has also been improved.

Triticale competes for acreage with wheat and barley. It has similar environmental requirements and is grown with similar cultural methods. Flour made from the grain has a flavor distinct from wheat flour and differs in baking properties. Triticale flour and baked products are available in the markets. It appears to be satisfactory as a feed grain for most classes of livestock.

Since triticale is a new crop, acreage is limited and no definitive area of concentration has been established nor has a strong demand developed. With continued improvement in productivity and quality, triticale undoubtedly will become an important addition to the cereal grains of the world.

Rice

Rice is one of the major field crops grown in California and 3,000 farmers are engaged in its production. Acreage varies considerably according to changes in government programs, and annual yields vary with seasonal conditions. During the period of 1975–79, the average acreage harvested was 450,600 acres with an average yield of 5,760 pounds of grain per acre, with the highest state yield average of 6,450 pounds per acre attained on 527,000 acres in 1979. Two market classes of rice are produced—pearl or short grain, and medium grain rice.

The possibilities of rice production in California were explored as early as 1856 when rice imports were necessary to meet the needs of a large Chinese population. Farming ventures were es-

tablished on the river overflow land of the San Joaquin River in 1856 and the California State Agricultural Society (1858) and later the California State Legislature (1862) offered premiums to stimulate rice production. Rice did not immediately become a successful crop and experimental plantings were continued until 1908 when W. W. Mackie grew a successful crop at Biggs, California. Beginning in 1909 many rice varieties were evaluated at Biggs and two Japanese short-grain varieties produced exceptionally high yields. Success of the trials from 1909–1911 marked the beginning of the rice industry in California and rice culture spread from the Sacramento Valley into the San Joaquin Valley. The Rice Experiment Station near Biggs was established in 1912 and has been a focal point of rice research to the present.

Rice is grown in California under flooded soil conditions. Therefore production is limited to clay soils and those that conserve water, on land that has been leveled so that water depth is uniform. Climatic requirements are a long growing season (April–October), abundant sunshine, and average temperatures in the range of 68–91°F. Temperatures below 60°F retard seedling development, slow plant growth and reduce grain yields. During the growing season, 5 to 7 acre-feet of good quality water are applied per acre, although the crop only consumes about 3.2 acre-feet. Excess water either percolates into the subsoil or is drained from the lower end of the field for reuse. Fields are graded to provide a flat surface with a slope; the fields are flooded just before seeding and remain so until 2 to 4 weeks before harvest. Under the conventional water-management system, water depth is maintained at 4 to 6 inches. A newer system maintains depths of 2 to 4 inches and promotes greater yields if weeds are controlled.

Historically about 88 percent of California's rice production comes from the Sacramento Valley and the remainder from the San Joaquin Valley. California acreage distribution in the counties by percentage in 1979 was Colusa 25.9, Butte 19.1, Sutter 16.8, Glenn 13.8, Yolo 6.7, Yuba 5.7, Sacramento 2.3, and all nine San Joaquin Valley counties 9.7. The rice acreage harvested and average yields in 5 year intervals are shown in Figure 4.

Rice production in California is highly mechanized. Large tractors and heavy-duty implements are used to prepare fields for planting. Large capacity, strongly-powered grain combines,

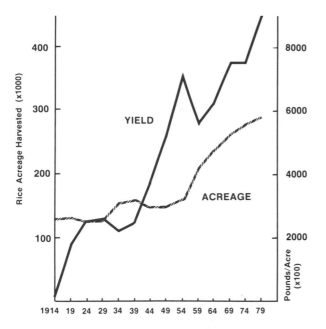

FIGURE 6.4. Average rice acreage and yield of rough rice in California by five-year intervals, 1914–1979. *California Crop and Livestock Reporting Service.*

equipped with tracks for operating in muddy fields, are used for harvesting. Seeding, fertilizing, and insect and weed control are commonly done by commercial aircraft operators. Because of mechanization, production of rice in California annually requires an average of only 7.2 man-hours of labor per acre, a sharp contrast to many less-mechanized countries where 350 to 800 man-hours per acre are needed.

To date, long-grain rice has not been commercially feasible in California, since varieties grown in the southern states have not proved well adapted to California production conditions. Adapted long grain varieties are now available, however, and California could become a major producer.

Rice varieties selected for planting are chosen in respect to the intended market, time to maturity of the crop and adaptation to the several micro-climatic areas of California. Virtually all planting seed is certified in its trueness to type and is commercially treated with fungicides to prevent seedling diseases. Seed is pregerminated in water before planting to speed seedling development and root anchorage. Weed and pest control are provided by a combination of cultural methods, including water depth and crop rotations, and by chemicals. In California, major economic pests of rice include the rice leaf-miner, rice water weevil, and tadpole shrimp. Other in-

sects, rodents, fish, and birds occasionally cause injury to rice plants in small areas.

Stem rot, a persistent, soilborne fungus disease, is becoming more of a problem in areas having long histories of rice production, and research is underway to find effective, economical control methods. In fields heavily infested with this rot, straw residues are commonly burned after harvest and the soil tilled.

Mosquito populations in rice fields can be reduced by good water management, rapid, complete surface drainage, use of mosquito-eating fish, and by chemicals.

Harvesting is done when the grain moisture is in the range of 18 to 26 percent. The paddy or rough rice is further dried to about 14 percent moisture content by fuel-heated air or forced-air drying prior to storage and subsequent milling. Most California rice is milled and marketed through grower cooperatives, although a significant amount is sold to independent millers. California rice enters the markets as quality table rice and rice products, including some for sale to brewers. About 55 percent is sold within the United States and its territories. Large quantities go to Puerto Rico and Hawaii, where special markets have been developed. The remaining supply enters the world market. California rice is principally sold to countries that demand high-quality short or medium-grain rice. Historically, California exports in cooperation with U.S. government aid programs have helped meet needs of countries faced with food shortages.

All rice production in the U.S. since 1938 has been regulated by federal laws administered by the U.S. Department of Agriculture. It has been necessary for a rice producer to own a rice allotment (acreage units) to grow and sell rice. Until January 1, 1975, rice harvested and sold without an allotment resulted in a heavy cash penalty against the producer.

Rice usually is grown year after year on the same land, primarily because its soil types are not well-suited to other crops. However, by improving soils through modern management, farmers have been able to grow more alternate crops. Rice fields also provide an excellent habitat for pheasants and waterfowl. As public pressure for recreation increases, the use of rice land for fee hunting can provide additional income. Catfish farming on soils adapted to rice farming can also provide fee recreation as well as a cash crop.

Rice fields in California are seeded by airplane in April. Here flooded rice paddies in Sutter County show the typical curved levee system controlling water flow. *Photo by Jack Clark.*

Beans and Peas

Crops identified as beans and peas grown in California fit into five species, common bean *Phaseolus vulgaris*, lima bean *p. lunatus*, blackeye or cowpea *Vigna unguiculata*, garbanzo or chickpea *Cicer arietinum*, garden or fieldpea *Pisum sativum*. Market types within each species are further distinguished by variety differences. Table 2 lists species with general production information. Two other species occupy a small acreage: bell bean, horse bean or windsor bean *Vicia faba* and mung bean *Vigna radiata*.

The most diverse species is common bean. Within the two market groups dry-bean and snap-bean, there are varieties developed for specific utilization; for instance, dark red kidney, light red kidney, white kidney, pink, pinto, small red, small black, small white, great northern, flat white, and pea beans. Most of the production is in light red kidney, dark red kidney, pink, and small white. Of the commercial pack yield, 60 percent or more are canned and the remainder sold in dry packages. Planting seeds are for local production or shipped to states in the eastern growing area such as Michigan or New York. Snap-beans are also grown locally for fresh-frozen pack.

California produces essentially all of the dry commercial limas in the U.S. and most of the crop world-wide. Large seeded fresh-frozen types are produced in small quantity in eastern states.

Peas are produced for commercial pack or for planting stock used locally or in other states. Although most U.S. peas are grown in the northwestern or northern mid-western states, California produces two specialty peas, the blackeye (cowpea) and the garbanzo (chickpea). California blackeyes account for over 90 percent of the dry-seeded cowpea production in the U.S., and are mainly marketed in the southwestern and southeastern states. Garbanzos, grown only in California, are mainly marketed in the western states.

TABLE 6.2
California Beans and Peas: Species, Utilization, Yield, and Location

Species	Utilization	Pounds/ Acre	Planting Date	Harvest Date	Location of Production*
PHASEOLUS VULGARIS					
Common dry beans	Commercial pack	2000	May 15–July 15	Aug–Sept	Upper SJV; SC; CC
	Planting stock	2000	May 15–July 15	Aug–Sept	Upper SJV; SC; CC
Garden or snap-bean	Fresh-frozen pack	3500	May 15–July 15	Aug–Sept	Upper SJV; CC
	Planting Stock	1400	June 1–July 15	Aug–Sept	Upper SJV; SC; CC
PHASEOLUS LUNATUS					
Large lima dry beans	Commercial pack	2000	May 1–June 1	Sept–Oct	Upper SJV; SC
Large lima	Planting stock	2000	May 1–June 1	Sept–Oct	Upper SJV; SC
Garden or immature seeds	Fresh-frozen pack	4000	May 1–June 15	Aug–Sept	CC; SC
	Planting stock	2000	May 1–June 1	Sept–Oct	CC; SC
Baby limas dry beans	Commercial pack	2300	May 1–June 15	Aug–Sept	Upper SJV; SV
	Planting stock	2300	May 1–June 15	Aug–Sept	Upper SJV
Baby garden or immature seeds	Fresh-frozen pack	3500	May 1–June 15	Aug–Oct	Upper SJV
	Planting stock	2000	May 1–June 15	Aug–Sept	Upper SJV
VIGNE UNGUICULATA					
Blackeye or cowpea	Commercial pack	1800	April 15–July 15	Aug–Oct	SJV
CICER ARIETINUM					
Garbanzo or chickpea	Commercial pack	800	May 1–June 1	Sept	CC
PISUM SATIVUM					
Dry peas	Planting stock	1800	Nov 1–Jan 1	May–June	Upper SJV
Garden or immature seeds	Fresh-frozen pack	3000	Nov 1–Jan 1	April–May	

*Abbreviations: San Joaquin Valley (SJV); Sacramento Valley (SV); Central Coast (CC); South Coast (SC).

Optimum yields of beans and peas are achieved only on good agricultural soil that is well drained. Most often beans are grown in rotation with tomato, sugarbeet, cereals, melons, and cotton. Garbanzo is usually not in a rotational cropping scheme. Common bean, lima bean, and blackeye are usually furrow irrigated. Peas are frequently flood irrigated and garbanzo is dryland farmed, relying on winter stored moisture. Yields vary considerably within and between species, over years of production and within a given area of production for a given variety; however, yields of large lima and baby lima of 3,000 pounds per acre are not unusual, and 2,500–3,000 pound per acre of common bean and cowpea are attainable.

Farming of beans and peas is highly mechanized from land preparation through harvest and seed processing. The only hand labor is in an occasional field where herbicides did not control certain weeds or in roguing seed fields for off-types.

Seed rot, seedling rot and root rotting soil-borne fungi can affect all of the bean species at some stage of their life cycle. Peas escape the effects of these diseases and most insects to a great extent because of cool season. A severe infection can cause stunting of mature plants and premature death. Chemical seed treatment offers partial control in early stages of growth. Tolerant varieties have also been developed. Fusarium wilt is a serious soil-borne disease of blackeye and garbanzo that causes stunting of young plants and premature death. The only economical control is plant resistance. Bean common mosaic is a virus that causes slight to severe stunting of common bean as well as leaf distortion and premature death. Most recently developed varieties are resistant. The soil-borne fungi in an environment with free moisture and moderate to high temperatures, will cause death to plants of all five species. The only control is an open canopy of plants and an environment that excludes free water over extended periods during moderate to high temperatures.

Nematodes, sub-microscopic round worms in soils, develop galls or enlargement of roots, depending on the environment and cultivar, and can cause severe stunting followed by premature death. Chemical treatment of the soil and resistant varieties reduce the effect of the organism. Lygus bugs are an insect most damaging to large lima in the San Joaquin Valley; they are also im-

portant on baby lima, common bean and black-eye. Seed yield in large lima can be reduced to less than the amount planted as the result of small lygus populations. Cowpea weevil is mainly a storage insect that can cause considerable damage to blackeyes. Eggs on the seed coats followed by larvae that bore a hole in seeds are the visible damage. Control by sanitation and fumigants are effective. Other insects and pests of less importance on beans and peas are mites, aphids, and corn earworms. Chemicals are usually an effective control.

Alfalfa

Alfalfa *Medicago sativa* was introduced to California during the 1850s. There were two primary introductions. The first, from Chile, spread throughout much of California and became the primary germplasm constituent of California common alfalfa. The second introduction was 'Smooth' and 'Hairy Peruvian,' introduced first to Peru from Spain, and more non-dormant than the Chilean germplasm. Hairy Peruvian was an important variety in southern California and southern Arizona until the early 1950s. Today California is a leading state in production of both alfalfa hay and alfalfa seed. Statewide production figures for both hay and seed from 1920–1979 are presented in Figure 5.

Alfalfa hay is grown in nearly every county in the state and ranks high in both acreage planted and in cash value. Over one million acres of alfalfa hay are harvested annually and it ranks about fifth in cash value among agricultural commodities in California. The 1979 state average yield was six tons per acre, with California second among the 50 states in yield per acre and sixth nationally in total production. Though California produces only five percent of the total U.S. crop, alfalfa is primarily a cash crop here. Nationally 20 percent of alfalfa hay is marketed, the remainder being consumed on the farm of origin. In contrast, in California 72 percent of the crop is marketed and the remaining 28 percent is consumed on the farm.

Alfalfa production is found in six climatic regions: the San Joaquin Valley, the low desert of southern California, the Sacramento Valley, the high mountain valleys, the Mojave Desert, and the coastal valleys. More than 50 percent of the total acreage is in the San Joaquin Valley with an

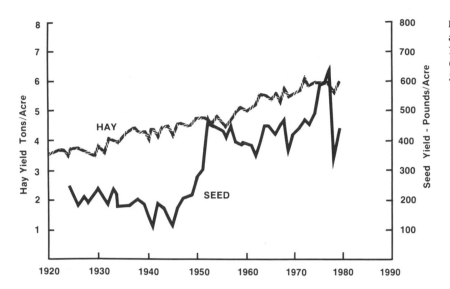

FIGURE 6.5 Average alfalfa hay and alfalfa seed production in California, 1920–1980. *Data from California Crop and Livestock Reporting Service.*

additional 15 percent in southern California. The leading alfalfa counties are Imperial, Kern, Fresno, Tulare, and Merced.

Best alfalfa yields come from well drained, deep medium textured soils. A porous subsoil is essential for deep water penetration and root development. Alfalfa can be grown on other soils with proper land preparation and crop management, however. Land preparation is especially important, since as a perennial, the crop will be harvested and irrigated for 3 to 5 years.

Alfalfa is planted in both the spring and fall, although fall planting is usually recommended for higher yields. Depending on location, fall plantings should be made in August–October (August in the mountain valleys and October in the Imperial Valley). Spring plantings are made in the San Joaquin and Imperial Valleys from February to mid-March. Late planting may result in severe stand loss during the summer because of poor plant development.

Inoculation of alfalfa seed with *Rhizobium* is generally thought to be unnecessary if the field has been planted to alfalfa previously, but seed inoculation is recommended if the field has no history of alfalfa. Most commonly, seed is broadcast into a dry loose seedbed and covered slightly by using a corregated ring roller.

The principal methods of irrigation are the border method, overhead sprinklers, and a combination method using beds and borders. The border method is most widely used but requires leveled land and a relatively large head of water. Sprinklers are used where costs of grading for flood irrigation are prohibitive, and are also helpful in

applying water evenly in small amounts during stand establishment and/or in extremely sandy soils. Bed plantings can be helpful in clay or salty soils where surface moisture control is difficult.

Alfalfa seed was produced in large quantities, mostly noncertified, in California beginning shortly after the introduction of alfalfa into the state. However, the seed produced was almost completely for consumption within the state. In the early 1940s it was shown that seed of midwestern varieties could be produced for one generation in California without any loss in genetic integrity. In 1945 the first certified field of a midwestern variety ('Ranger') was produced in the state. By the early 50s large amounts of certified seed of midwestern and eastern varieties were being produced. In 1979 37.6 million pounds of alfalfa seed were produced in California, 33.1 million pounds of which were certified "hardy" and "non-hardy" cultivars. California ranks first among the 50 states in alfalfa seed production, with about 35 percent of the U.S. total. Primary areas of production are Fresno, Kings, Imperial, Kern, and Madera counties.

Correct use of water is one of the most difficult problems of alfalfa seed production. Highest seed yields are usually obtained when irrigation practices have prevented severe stress and promoted slow, continuous growth through the entire production period without excessively stimulating forage production. Established seed fields should receive about 4 acre feet of water either as irrigation or effective rainfall.

Pollination by bees is customary with three or

four colonies per acre usually used in seed production. The first colony is placed in the field at 24 to 40 percent bloom, the second colony 10 days to two weeks later, and a third colony is added as the bloom increases to 100 percent. Bees placed in the field too soon may leave the alfalfa for other crops, especially if the adjacent crops are species more attractive than alfalfa. Groups of hives are placed 250 to 300 feet apart in the bee row with 500 to 600 feet between bee rows.

Most growers in the Central Valley spray cure to prepare the seed crop for direct combining. Windrow curing is used where fields mature late.

Numerous insects and diseases are a problem in the production of both alfalfa hay and seed. Currently a large Integrated Pest Management

(IPM) program is being conducted, involving many scientists from several campuses of the University of California. Knowledge developed in this program is being used to improve existing cultural practices and will be used in the development of improved varieties.

Coming years will probably see many changes in alfalfa. This crop is no longer viewed only as a forage crop. Alfalfa sprouts have become popular as salad and sandwich ingredients and now form a major market for seed producers. Technology now also exists to extract alfalfa leaf protein for use in human food; alfalfa can produce more protein per acre than any other crop. Alfalfa may also be developed for short term rotation so that its ability to fix nitrogen in the soil can be utilized to enhance production of other crops.

Imperial County, widely known for winter vegetable production, also raises large quantities of alfalfa hay, an important component in cattle feeding. This view shows one of the many large feedlots of the area, with alfalfa fields in the background. *Photo by Max Clover.*

Sugarbeets

Sugar purchased from the grocery store, whether from cane or beet, is about 99.96 percent pure sucrose. The average person in the United States consumes from 90 to 100 pounds of this high energy product per year. California is an abundant producer of sugar, producing about 25 percent of the nation's beet sugar supply. The sugarbeet crop is grown in most of the major agricultural counties from Tehama in the north to Imperial on the Mexican border. Usually from 200,000 to 300,000 acres are grown annually. In the five year period of 1974–78, 43 percent of the acreage was grown in the Central Valley from Stanislaus County north; 25 percent in the Central Valley from Merced County south to Kern; 25 percent in the irrigated desert area of southern California, primarily Imperial County; and 7 percent in the coastal counties.

The sugarbeet *Beta vulgaris* belongs to the same plant species as red table beet, swiss chard, and the mangle or fodder beet. Normally it grows vegetatively the first year, producing a large storage root in which sucrose is stored. The second year, after a cool winter period, the plant produces seed and the life cycle is completed. At the end of its vegetative cycle the storage roots are harvested and the sucrose extracted by a highly technical refining process.

There are presently nine sugarbeet refineries in California operated by four companies. Sugarbeets produced for sugar are grown only under contract between the individual grower and a processing company. The contract is of a participating type, guaranteeing the grower a share of the net return the producer realizes from the sale of sugar.

Since about 1969 there has been a marked increase in the state's average annual root production to a high of 28.6 tons per acre in 1976. Much of this increase is attributable to improved control of sugarbeet diseases, introduction of more productive hybrid varieties, and more intensive use of technology by farmers to provide better stand establishment, weed control, and irrigation. Since 1970, annual yield has ranged from 24 to 28 tons per acre and the sugar content of roots from 15.0 to 15.6 percent.

The sugarbeet crop is very productive in terms of food for man and animal. Based on the 1976 average state yield, the average acre grown in the state produced 8,590 pounds of sugar, 3,510 pounds of dry beet tops, 3,140 pounds of dry beet pulp, and 100 pounds of monosodium glutamate, a naturally occurring salt of an amino acid used in food flavoring. Beet tops and pulp are in demand for cattle feed. The sucrose produced amounted to about 15 million kilocalories of food energy per acre. Potentially, the crop can produce more. Many growers now produce 40 tons of roots per acre.

Sugarbeets require about five months of good growing weather to produce a profitable crop. Generally, the earlier the planting the higher the yield, provided temperatures at planting are conducive to rapid growth and plants are not retarded in growth by serious diseases or other problems. When sugarbeets are growing rapidly, the sugar content of roots is low (from 10 to 13 percent). When plant growth is slowed and light intensity is still high, roots can increase in sugar content to as high as 18 or 19 percent. Nitrogen deficiency prior to harvest and cold nights slow vegetative growth and increase sugar content.

When the growing sugarbeet undergoes a prolonged period of cold temperatures followed by a period of warmer temperatures and longer day lengths, seed stalk production ("bolting") takes place. Beet fields in certain areas of California are overwintered, that is they are held in the ground for harvest the following spring. Varieties overwintered in northern California produce seed stalks the following spring. In the Imperial Valley, sugarbeets are planted in September and grow through the winter months for harvest the following April through June. Winters in this area are not cold enough to cause much bolting where varieties have been selected for bolting-resistance.

Sugarbeet cultivars (varieties) have been developed by the United States Department of Agriculture and the major processing companies that are well adapted to California and are resistant to diseases that once caused widespread losses, such as curly top and virus yellows. All varieties grown in the state are hybrid varieties which out-yield the old open-pollinated types by from 10 to 20 percent. All varieties now have monogerm seeds, that is the corky seed pieces contain a single true seed rather than the two to four seeds of older varieties. Such seed helps eliminate doubles and multiples in beet stands, improves the operation of mechanical thinners and makes it easier to plant directly to a stand. Seed is usually purchased from a sugar company where it is processed and graded to permit precision planting and treated to protect germinating seedlings from soil fungi and insects.

Sugarbeets are usually planted on raised planting beds, 4 to 5 inches in height and 30 inches apart. If field emergence is good (50 percent or better) seeds can be planted 4 to 6 inches apart with the expectation that the resulting stand will not need to be thinned. Many beet fields are planted at closer seed spacings and require the use of a mechanical thinner or long handled hoes to space the plants from 6 to 12 inches in the row.

Nitrogen fertilizer is required for profitable beet production on many fields. Sugarbeets, however, are a nitrogen-sensitive crop. They require ample amounts of this nutrient for maximum vegetative growth but also require a nitrogen deficiency prior to harvest for proper sugar accumulation in the storage roots. Thus, applying the correct amount of nitrogen for maximum sugar production is extremely important for the individual farmer.

Irrigation is commonly by the furrow method, but sprinklers are used where topography, high water table or other special conditions make the furrow system difficult to use. Sprinkler irrigation, though more costly, has the advantage of improving seedling emergence and of using less water in the early stages of plant growth. Water requirements range from as little as 18 acre inches of water per acre per season in a cool climate where the soil is filled by plentiful winter rain, to as much as 48 acre inches per acre in a hot, dry climate with limited winter rain.

Weeds are controlled by crop rotation and a combination of chemical and mechanical means. Important dieases are curly top, a virus disease transmitted by the sugarbeet leafhopper; beet yellows, a virus complex transmitted by the green peach aphid; and powdery mildew, a leaf fungus disease. Strategies have been developed for the control of these diseases that involve attention to time of planting, isolating new plantings from old sugarbeet fields which can serve as a source of virus inoculum, and by the selective use of fungicides. In some areas of the state the sugarbeet cyst nematode is an important pest and must be controlled by careful crop rotation.

Sugarbeets are machine harvested and delivered to a beet loading station at a railroad siding or directly to a processing plant. The delivery of beets to a processing plant must be kept to the amount that can be processed in a fairly short time and thus growers are usually faced with tonnage quotas at harvest time. Growers form harvest pools to allow for continuous operation of a harvest crew and decide within a pool as to the order in which fields should be harvested. Fields that are overwintered for spring harvest will bolt (go to seed) as the spring season progresses. Bolting causes some small, but no drastic loss, in sugar yield. In fields that are overly delayed in spring harvest, seed stalks are usually removed 6 to 12 inches above the crown to prevent the seeds from maturing to become weeds in subsequent crops.

Field Corn

Field corn *Zea mays*, also known as maize (Indian corn), yellow corn, grain corn, ear corn, tall corn, and silage corn, is a warm season crop planted in late spring or early summer. It is grown primarily in the Sacramento and San Joaquin Valleys of California, with the greatest concentration in or adjacent to the Sacramento-San Joaquin Delta. Field corn has been grown continuously on the same ground with good results, especially in the Delta and river bypass flood zones. However, in California it is mainly considered a rotation crop.

California grows around 450,000 acres of field corn annually, including approximately 150,000 acres for silage. By comparison, in 1945 before the introduction of hybrids, this state produced only 50,000 acres of corn for both grain and silage. Nationally, corn has been the leading field crop for many years. United States production is around 80 million acres, compared to 75 million acres of wheat and 60 million acres of soybeans (1978 figures). California's total production of corn for grain is very small compared to the total U.S. production or even the two leading corn states, Iowa and Illinois. However, this state's per acre production of 7000 pounds in 1978, and an average production of 6400 pounds of grain per acre in the years 1974 through 1978 is currently the highest in the U.S. (Table 3).

Since California is feed-grain deficient, nearly all of the grain corn grown here, and much of what is imported, is used for poultry and livestock feed within the state. Historically, because of transportation costs, California production has commanded a premium price over imports.

TABLE 6.3
Corn Production in California, 1945–1980

Year	For Grain		For Silage	
	ACREAGE HARVESTED	YIELD LBS/A	ACREAGE HARVESTED	YIELD TONS/A
1945	29,000	1,904	25,000	10
1950	52,000	2,408	32,000	11
1955	184,000	3,696	75,000	13
1960	130,000	4,032	75,000	14
1965	144,000	4,984	79,000	19
1970	203,000	5,488	106,000	19
1975	254,000	6,104	162,000	19
1978	281,000	7,056	137,000	20
1979	260,000	6,560	168,000	22
1980	270,000	7,560	178,000	21

SOURCE: *Field Crop Statistics*, Annual Summaries, California Crop and Livestock Reporting Services.

Minor amounts of the total California supply go into human consumption, mainly through meal, flour, and specialty items like corn chips and other corn snacks. Corn grain also is used to produce the liquid corn sweetener widely used in the canning industry. In 1979 the first California plant to produce this syrup went into production near Tracy.

California growers have a wide choice of field corn varieties. In general, varieties adapted to California reach harvest maturity in 120 to 175 days. Good variety tolerance is available for major diseases, such as Fusarium ear rot, head smut, and maize dwarf mosaic.

Corn is grown on many different soil types— from the peats of the Delta to the clay adobe soils of the rice areas. Rooting is moderately deep, from 5 to 8 feet in aerated, well-drained soils. Shallow soils will give fair production if properly managed, but poorly-drained soils will not.

Corn is not as tolerant to salinity as many other California crops; yields are reduced when the electrical conductivity (EC) of the soil is 4.0 mmhos/cm, or above. Increase in production can be obtained on saline soils, however, by leaching.

With the wide range of soil conditions in California production areas, many different fertility levels exist and fertilizer rates and ratios necessarily vary. Proper soil sampling is essential for determining nutritional needs. Most growers routinely apply a nitrogen-phosphorus starter fertilizer in a band beside and below the seed row to insure early seedling vigor and earlier crop maturity. Additional nitrogen to supply the total seasonal requirements is usually applied in a second application.

Corn is planted flat or on beds, usually in rows 30 to 36 inches apart. Where drainage is a problem, 60-inch beds may be used, with two rows planted 22 or 24 inches apart. Most varieties are planted to obtain a final population of 25,000 to 30,000 plants per acre.

Corn requires 3 to 4 acre feet of water per season, with application intervals as close as 7 to 10 days on some soils. Common methods of irrigation include subirrigation, primarily on the peat soils of the Delta region, and furrow or border check irrigation on mineral soils.

Annual broadleaf weeds commonly causing problems in corn include pigweed, lambsquarter, purslane, knotweed, and nightshade. Grass weeds causing problems include barnyardgrass or watergrass, Johnsongrass, crabgrass, and wild cane. Nutsedge or nutgrass is another perennial commonly found in corn. Johnsongrass can also be a source of maize dwarf mosaic virus disease. Cultivation methods and timing can be the key to successful weed control in corn. Preemergence herbicides may also be applied to the soil surface and mechanically incorporated before the crop is planted, or postemergence herbicides aplied to the weeds after they have become established.

Corn is subject to a number of diseases and pests tht adversely affect production. Ear and kernel rots decrease yield, quality, and feeding value of the grain. Stalk diseases can lower yields and may make harvesting difficult. Differences in susceptibility among hybrids appear to exist. Spider mites, cutworms, wireworms, and garden symphlan frequently are serious problems in corn in California.

Corn is harvested with a combine equipped with a special corn header built to handle four, six or eight rows. Grain is transferred to trucks at field side or via bankout wagons for transport to drying and storage facilities. Harvesting and storage operations must take into consideration the moisture content of the corn crop. Some artificial drying may be required. For harvesting, the most desirable moisture level is 18 to 22 percent. As the crop dries below 18 percent, the risk of snapped ears, lodging, broken stalks, and grain shattering increases. Above 25 percent, corn does not shell well and kernel damage increases excessively.

For storage, corn grain can be safely held at or below 14 percent moisture (required for U.S. No. 1 grade). Grain elevators will generally accept corn up to 15.5 percent moisture (required for U.S. No. 2 grade) without additional drying charge.

Most corn is delivered to commercial elevators with drying and storage facilities, although in recent years more growers have been acquiring on-farm drying and storage equipment. On-farm facilities increase harvesting efficiency, avoid truck delays at the elevators during peak or closed periods, and provide greater marketing flexibility.

In California, a significant acreage is devoted to corn silage production. This crop is chopped in the field and then packed in pit or elevated silos near a livestock feeding site.

Grain Sorghum

Sorghum is a member of the grass family, which includes grain sorghum, forage sorghum, sweet sorgo, broomcorn, and sudangrass. Also closely related is Johnsongrass, a major weed pest. Grain sorghum *Sorghum bicolor* (L. Moench) is commonly referred to in California as milo, corn, and "gyp corn."

The history of grain sorghum in California dates back to 1874 when a small consignment of white and brown Durra was brought from Egypt. Planted in the interior valleys, these introductions demonstrated their ability to withstand the hot, dry summers and soon became established as a crop. The vast majority of California's grain sorghum is grown under irrigation. According to USDA Agricultural Statistics for 1965 to 1975, California grain sorghum acreage averaged close to 250,000 acres, although it has dropped considerably since then. Average production is near 4,000 pounds per acre, with yields ranging up to 9,000 pounds. Although there are exceptions, double-crop grain sorghum normally yields less than "full season" single-crop plantings.

Grain sorghums are of tropical origin and reach maximum development in regions having high temperatures and relatively low humidities during the growing season. Experiments and observations indicate that they can be grown successfully in all interior valleys, but production has not been good in coastal regions subjected to cool ocean breezes or at elevations above 5,000 feet. Grain sorghum may be grown on a wide range of soil types. With good management even sandy and very heavy soils can produce high yields. Sorghum is more tolerant to sodic (alkali) and saline soils than are most field crops.

Varieties are available with a wide range of maturities. Nearly all are hybrids with a high production potential, generally reaching harvest maturity in 120 to 160 days under California conditions.

Grain sorghums are usually planted in rows 20 to 40 inches apart; however, a considerable amount of grain sorghum is planted with grain drills in spaces varying from 7 to 20 inches between rows. Drill plantings are most common with grain sorghum planted in a double crop system, following such crops as wheat or barley. Since grain sorghums are planted both as a single crop and as part of a double-crop rotation, the date of seeding varies over a wide range—from March until late July. Emergence and growth will be best at soil temperatures above 65 degrees at depth of planting. Double-crop sorghum should be planted before the 10th to 15th of July in all areas of the state.

Nitrogen is generally needed, although peat soils of the Delta may need little or none. Rates of nitrogen vary from 50 to 200 pounds per acre. Phosphorus and potassium requirements as well as most minor elements are best determined by soil analysis and previous crop and fertilization history.

Sorghum, although considered a drought tolerant crop, needs 30 to 40 inches of water per acre to grow crops with high yields. Frequency of irri-

TABLE 6.4

California Grain Sorghum Acreage and Yield, 1945–1980

Year	Acreage Harvested	Yield
		Lbs/A
1945	102,000	2,185
1950	114,000	2,295
1955	162,000	3,360
1960	233,000	3,750
1965	316,000	4,200
1970	290,000	4,145
1975	207,000	4,030
1978	135,000	3,975
1980	152,000	4,080

SOURCE: *Field Crop Statistics*, Annual Summaries, California Crop and Livestock Reporting Service.

gation depends upon soil type and temperature with an interval of three weeks about average. It is especially important to have moisture near the surface soon after plant emergence to permit development of the secondary root system.

Lambsquarter, pigweed, ground cherry, and morning glory are major broadleaved weed problems. Watergrass and Johnsongrass are two of the most serious grassy-type weed problems. Grain sorghum should not be planted in fields where Johnsongrass infestations are significant. Either mechanical or chemical methods of weed control are used. Unfortunately there are few chemicals that can be used on sorghum. Specific phenoxy-type herbicides are effective in controlling annual broadleaved weeds.

Insects attacking sorghum include greenbug, sorghum midge, stink bug, corn earworm, cutworm and others. The most damaging can be controlled chemically and through the use of resistant varieties.

Diseases common in California sorghum are seedling diseases, maize dwarf mosaic, head smut, charcoal rot, Fusarium stalk rot and periconia root rot. Most of these cannot be adequately or economically controlled by chemicals. Several disease resistant varieties are available, however, and good cultural practices and management are important aids.

Grain sorghum can be successfully harvested over a reasonably wide range of grain moisture content. The upper range of moisture content for good harvestability appears to be 18 to 20 percent. If the grain moisture content is above 14 percent at harvest it is above the moisture limits for safe storage and must be dried.

Annual Forages

Annual forages are most important as a source of silage for dairies. Corn is harvested in late summer and ensiled at about 70 percent moisture when the grain is mature. At this stage total digestible nutrients per acre and dry matter ensiled per acre are maximized while field and storage losses are minimized. Because of the high water content of silage it is produced on or near the dairy to reduce transportation costs. It is possible to produce 25–30 tons of corn silage per acre.

Cereal grains are frequently double cropped with silage corn to produce an additional 15–17 tons of silage per acre during the winter. Winter forages are planted in the fall and harvested in the spring when the grain is in the soft dough stage. Oats are commonly used alone or in mixtures with barley, wheat, rye, triticale, and vetch. Winter forages, especially oats, are also harvested as hay. Oat hay is a popular horse feed. The addition of vetch or other legumes increases the protein content of the hay.

Sudangrass and sudan X sorghum hybrids are usually grown as summer hay or pasture crops but may be ensiled. They provide highly productive pasture that can provide supplemental feed when permanent pasture productivity teporarily declines in the summer. Harvesting is generally delayed until the sudangrass is 18 inches tall to reduce the risk of prussic acid poisoning.

Annual ryegrass is a versatile winter forage used in several ways. It can be seeded into permanent pasture in late summer and irrigated to produce forage into the fall and winter. In Imperial Valley it is interseeded into bermudagrass pastures as the bermudagrass goes dormant and provides feed through the winter season until bermudagrass production resumes in the spring. It is also used on rangeland where late summer irrigation is possible. Because of its rapid germination it is used extensively to reseed after brush fires to stabilize bare soil.

Irrigated Pasture

For many years irrigated pasture was the forage base for the dairy industry. Irrigated pasture acreage has decreased for several reasons: urban encroachment, dairies converted to silage and hay feed lot operations, new crop varieties that could be grown on pasture land, and land reclaimed with irrigated pasture subsequently placed in crop production.

Irrigated pasture remains the forage source for some dairies. It is used as a source of summer feed or hay for some beef and sheep operations and it continues to be used for family farms where small flocks of sheep or a few head of cattle or horses are kept. Urban fringe subdivisions have contributed to the total irrigated pasture acreage in the 1970s.

An estimate in 1977 indicated that there were approximately 1.2 million acres of irrigated pasture in the state. In 1977 the top five irrigated pasture-producing counties were Merced, Siskiyou, Imperial, Stanislaus, and San Joaquin counties. These five counties represented approximately 42.5 percent of the total irrigated pasture in the state.

Irrigated pasture in Imperial Valley and other desert areas is dominated by various varieties of bermudagrass. Bermudagrass, a warm season grass of tropical origin, is an outstanding forage in subtropical regions when properly fertilized and well managed. Bermudagrass pastures are productive in the desert areas from May through September.

In the nondesert regions of the state irrigated pastures are dominated by cool season species such as tall fescue, perennial ryegrass, orchardgrass, Ladino clover (giant white clover), strawberry clover, and trefoil. Wheatgrass and smooth brome are two other cool season grasses that are used to a lesser extent in isolated situations primarily in the northeastern part of the state. Alsike clover, red clover and other white clovers are also used to a lesser extent in various parts of the state. Dallisgrass, a warm season grass, is used to a limited extent in some irrigated pasture in the Sacramento and San Joaquin Valleys. In many locations it occurs voluntarily because the seed is distributed in the irrigation water.

Irrigated pasture is probably the least managed agricultural crop in the state. As a result, most irrigated pasture problems can be corrected by improved management. Allowing animals to graze pastures when they are muddy compacts the soil. This can lead to poor drainage, mosquito infestations, and invasion of water-loving weeds. Overgrazing reduces the vigor of the desired species and frequently allows weedy species to invade the pasture. Proper irrigation, animal management, and fertilization are the key to maintaining vigorous productive irrigated pastures.

The value of irrigated pasture is difficult to assess because it is one of the few agricultural crops not freely marketed. Because it is not a transportable commodity its value is subject to local supply and demand. As a result the rental value for irrigated pasture and other grazing lands can fluctuate widely throughout the state. Irrigated pasture is rented on a per acre basis and on an animal number basis. Few operators rent on the basis of animal gain.

Irrigated pasture will probably continue to decline in areas throughout the state. It is found on land that cannot be cropped because of some soil problem such as shallow hardpans, poor drainage, and saline and/or alkaline conditions. Furthermore, irrigated pasture has one of the lowest dollar returns per acre. As irrigated pasture inputs, such as water and fertilizer, become more expensive, our ability to manage these lands for maximum production will decrease. In some counties where water is already expensive, commercial production of irrigated pasture has ceased and pastures can only be found on noncommercial family farms.

Cotton

Cotton *Gossypium hirsutum* was among the earliest crops introduced into California by the Spanish Padres although they left little recorded information. As early as 1808 cotton was being grown on an experimental basis. In 1862 the California Legislature offered $3,000 for the first 100 bales (300 pounds each) of cotton produced in the state. Interest in cotton became active in the Palo Verde Valley around 1895. By 1909, 1,500 acres were planted in the Imperial Valley and the planted area increased ten-fold the following year. Early cotton investigations were conducted by the USDA about 1902 near Calexico in the Imperial Valley and this was later extended to the Coachella Valley. In 1916 the first commercial plantings of "Acala" cotton were made in the San Joaquin Valley after having been tested by W.B. Camp, a USDA cotton investigator. Acala cotton rapidly increased in production. The U.S. Cotton Field Station was established at Shafter in 1922. By 1925 agricultural leaders recognized the virtues of Acala cotton and convinced the State Legislature to pass the "one-variety law" prohibiting planting of any variety or species of cotton other than Acala, except in the Imperial Valley and, beginning in 1941, Riverside County. Acala cotton has undergone continuous improvement since its first introduction with significant changes in earliness, picking efficiency, yield, improved fiber quality and resistance to verticulum wilt, a soil-borne fungus disease.

Commercial plantings of cotton occur in two major areas of California, the southern countries of the San Joaquin Valley and the lower desert valleys of Imperial and Coachella. Harvested acreage increased steadily from 1910 until the mid-50s, when problems of injurious insects and disease together with lower farm profits and a declining market caused a reduction in planted

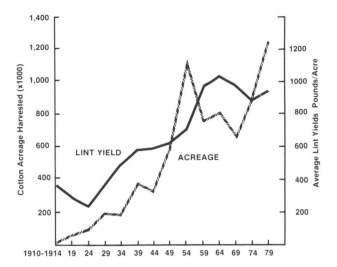

FIGURE 6.6 Average cotton acreage and lint yield in California by five-year intervals, 1914–1979. *Data from California Crop and Livestock Reporting Service.*

area. Plantings have again increased since the late 60s, and 1979 cotton acreage was at an all-time high of 1,635,000 acres. Cotton yields have also increased steadily since the mid-20s, reaching an all-time high 5-year average yield of 1,072 pounds per acre. In 1979, cotton lint was valued at about $1.1 billion and was the most important field crop grown in California.

The San Joaquin Valley and lower desert valleys of Riverside and Imperial Counties are well suited to produce high yields of quality cotton where irrigation is available. The long growing season from planting in mid-March to mid-April to harvest in October and November allows adequate time without frost for high cotton yields. Warm, dry spring weather provides for good stand establishment with a minimum of seedling diseases. High average summer temperatures and a long dry fall allow for a long picking season with little loss due to poor weather. Cotton grows well under a wide variety of soil conditions found in California but produces best on fertile, well-drained and aerated soils having a good water-holding capacity. Cotton is not a severe soil depleting crop and grows well in rotation with other field and vegetable crops.

Plowing to a depth of 10 to 12 inches is preferred. After subsequent disking, a preplant fertilizer application is usually made. Usually nitrogen and sometimes phosphorus is applied by broadcasting, banding in the bed, or field injection. Preirrigation is usually practiced during the winter season to wet the soil to a depth of at least 6 feet on deep soils, or to be the restricting layer on shallow soils. This allows the cotton seedlings to establish deep root systems during their early growth stages and provides a reserve of soil moisture.

A soil temperature of at least 60°F at a depth of 8 inches for several consecutive days is desirable before planting. Poor stands result if cotton is planted too early, and weakened seedlings may be attacked by disease organisms. Seed treatment is usual, to provide protection from such seedling diseases as *Rhizoctonia* and *Pythium*. Where normal plant growth is expected, a plant population of 40,000 to 60,000 plants per acre (approximately 2–3 inch spacing) performs best for mechanically harvested cotton.

Nitrogen fertilizer is essential for economic cotton production in all areas of California. The need for phosphorus is not so general and depends upon the soil and past cropping practices. Potassium and zinc may sometimes be deficient.

Adequate water is also critical to maximum cotton production. Where preirrigation has been practiced, subsequent irrigations are usually from early June until the end of August, their interval ranging from 7 to 30 days. Furrows are usually preferred by cotton growers, but basins are often used on relatively flat land where water penetration is slow or impaired. On land that is not level, sprinkler irrigation is used.

Cotton is attacked by a wide variety of diseases, their severity being dependent upon weather conditions and farming practices. Losses from seedling diseases, verticulum wilt, root and ball rots can be major. The use of resistant cotton varieties, seed treatment, and good farming practices help alleviate disease problems. Cultivation is the primary means of weed control, but preemergence and post emergence herbicide application, flame cultivation, and the use of geese are alternative practices. Major weeds are nutsedge, bermudagrass, Johnsongrass, pigweed and lambsquarter.

Many insects attack cotton because of its attractive foliage, flowers and fruit. The most serious pests are lygus bugs, cotton bollworms, cutworms, aphids, mites, stink bugs, and in the southern desert region the pink bollworm. Insecticides and acaracides are valuable aids to successful cotton production. Biological control, inducing suppression of pests by beneficial organisms, and integrated pest control, making use of both chemical and biological means, provide

Cotton harvesting machines march in progression across the fields of San Joaquin Valley. California produces nearly 30 percent of U.S. cotton lint and seed on over a million acres. *Photo by Mel Gagnon.*

the best all-around protection.

Boll development from bloom to open boll requires from 40 to 55 days. When the bolls have adequately matured, chemical defoliants are applied to cause leaf drop that facilitates harvesting and helps maintain the quality of the harvested cotton. Spindle-type harvesters are conventionally used for picking cotton. They remove only the seed cotton (cotton lint fibers attached to the seed) from open bolls without damage to the plant. These tractor-mounted units consist of revolving verticle drums mounted with many finger-like spindles which, synchronized with the forward movement, engage the fibers, remove them from the plants and guide the plant through the picking head. The harvested seed cotton is conveyed by air stream to the collecting basket. After harvest, ginning completes the separation of the cotton fiber from the seed.

The lint fibers (long fibers) are used extensively in textile manufacturing and the linters (short fibers) are used for felt, stuffing and a variety of chemical uses. Although cotton is produced primarily for fiber, two by-products are important commodities. Cottonseed oil is used extensively as a salad and cooking oil and the seed residue remaining after the oil is extracted contains as much as 40 percent protein. Cottonseed meal is presently fed to livestock as a protein supplement, but flour processed from cotton seed may in future become a significant food commodity, when gossypol, a toxic phenolic pigment detrimental to humans, is removed.

Oilseed Crops

The major source of vegetable oil in California is cottonseed. Other oilseeds that have been, or are being, grown in California are flax, safflower, soybeans, sunflower, castorbeans, and sesame. Except for castor and linseed oils, most of the vegetable oil that is produced in California is used in edible products. With declining supplies and higher prices of fuels from petroleum, there has been interest in vegetable oils as a substitute fuel for diesel oil. Research indicates that on a volume basis vegetable oils are equivalent to diesel fuels, though the vegetable oils have higher viscosity and do cause carbon accumulation on fuel injectors.

Oilseed meals, with the exception of castorseed meal, is a valuable feed for livestock and poultry, the value being related to the protein content. Castorseed meal is poisonous, but is used as a fertilizer.

Safflower

Safflower has been grown in the Middle East for centuries. After its introduction into California around 1900, experimentation proceeded intermittently. When cotton acreage was greatly reduced in 1950, interest turned to safflower as an alternate crop which had developed in areas of eastern Colorado and western Nebraska. Most safflower production was then in the Imperial Valley and the southern San Joaquin, but since that time the acreage has shifted to the San Joaquin and Sacramento Valleys where safflower has proven to be well adapted. In the cotton growing areas of the San Joaquin Valley, where it is preirrigated and often given some crop irrigations, safflower is sown in February and March and harvested in July and August. In the Sacramento Valley and Delta area it is usually grown dryland, but often benefits from water available in the soil after a previous crop such as rice.

Safflower has done well on deep and heavier soils with good water holding capacity. Yields have been low on shallow soils, presumably because of inadequate room for root development. Careful irrigation is essential because excess water leads to severe damage from *Phytophthora* root rot.

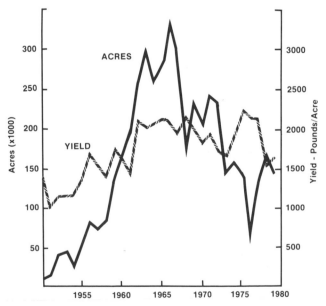

FIGURE 6.7 Average safflower acreage and yield in California, 1950–1979. *Data from California Crop and Livestock Reporting Service.*

Safflower is produced on a contract basis with oilseed processors. The contracting organization usually provides the planting seed, advisory services, and a guaranteed market and price. Most of the companies contracting acreages have their own breeding programs.

Two types of safflower are grown. The high-linoleic type has high levels (about 78 percent) of the polyunsaturated linoleic acid in the seed oil, the mono-unsaturated oleic acid making up about 15 percent of the oil. The remaining fatty acids are stearic and palmitic, both saturated. High-linoleic safflower oil is used in edible products such as soft margarines, salad oils and to a lesser extent as a frying oil. It is also used in the manufacture of surface coatings such as varnishes and paints. A high-oleic type has about 78 percent oleic acid, and 15 percent linoleic acid in the oil, which makes it chemically very similar to olive oil. The oleic-type oil is a premium frying oil. Oil contents of the seed vary between 38 and 44 percent.

Wheat competes very strongly with safflower, and when wheat prices are high relative to safflower prices, safflower acreages are reduced. Safflower has never been grown in large enough quantities for the oil to become a major ingredient of vegetable oil products marketed by major suppliers.

Soybeans

Soybeans have been under test in California for a great many years, but sustained commercial production has never been achieved. Interest has been stimulated primarily by the oilseed processing industry, principally that part of it involved with cottonseed. When cotton processing facilities ran at less than full capacity, efforts were made both to develop soybean commercial acreages and to expand research. On the other hand, when cottonseed production exceeded the amount that could be handled by local processing facilities, interest in soybeans declined. Factors discouraging interest were susceptibility to mites and low net returns compared to competing crops. Soybeans have been most promising as a summer crop in a double cropping system after winter crops such as wheat and barley.

Sunflowers

Sunflower production in the northern Great Plains of the U. S. has expanded greatly in recent years with the acreage and production in 1979 and 1980 being 5,463,000 and 3,393,000

acres and 3,506,000 and 1,738,000 metric tons respectively. Essentially no sunflower is grown for oil in California because it has not been competitive with other summer crops. However, in the Sacramento Valley a successful sunflower planting seed industry has developed which has occupied between 10,000 and 15,000 acres. Seed of hybrid cultivars is produced by several companies for sale to the main sunflower producing areas of the U.S. and abroad. For 50 or more years large-seeded, tall, and late-maturing cultivars of sunflower have been grown for confectionary purposes. The area of production has ranged between 3,000 and 10,000 acres, and much of the production has been in the Delta area. The seed is roasted and salted with or without the hulls.

Sesame

Considerable research and development has been applied to sesame in an effort to make it a viable commercial crop. The tendency of the seedpods to shatter makes mechanical harvesting difficult. Interest is primarily in the seed which is used in the bakery trade. However, if the crop could be grown successfully, there would be a strong market for the oil which is of very high quality.

Castorbean

Castor oil is an industrial oil used in the production of fast drying paints and varnishes, in plastics, all purpose greases, hydraulic fluids and other uses. Prior to 1950 there was practically no castorbean production in California since sufficient quantities were available from imports. In 1950, castor oil was classified as a strategic oil and was stockpiled by the federal government during 1951–54. Under a guaranteed minimum price, supplies increased from domestic production during 1951–54, then decreased sharply in 1954–56 when support prices were removed. Stimulated by favorable prices, research improved methods of production, improved dwarf-varieties, and developed complete mechanization of production. California produced about 6,600 acres of castorbeans in 1957 under irrigated conditions, but since 1962 the crop has not been grown commercially in California.

Flax

Research at the Imperial Valley Field Station demonstrated that the 'Punjab' cultivar was well adapted as a winter crop in the desert environment of that area. Commercial production in Imperial Valley began in 1934, and acreage ex-

TABLE 6.5
Approved Acreages for Certified Seed in California, 1960–80

Crop	1960	1965	1970	1975	1980
Alfalfa	102,284	51,962	53,402	32,035	53,643
Barley	8,692	13,267	9,482	11,985	14,078
Beans	13,327	17,377	11,993	9,960	15,757
Clover	15,242	13,076	16,609	11,609	8,032
Oats	571	1,847	783	1,001	639
Rice	7,005	9,000	11,453	20,979	23,608
Sorghum	6,940	6,432	3,311	1,247	1,591
Wheat	937	16,940	722	17,957	29,029

panded rapidly, particularly during and following World War II. Production spread to the Palo Verde Valley and to the west side of the San Joaquin Valley as the latter area was developed agriculturally. Finally flax moved to San Mateo County as a summer crop. Flax is the source of linseed oil which is used primarily in paints, varnishes and linoleums, and not in edible products. Much of the flax straw in Imperial Valley was processed to remove the fiber which was shipped to South Carolina for manufacture into high grade papers. None of the fiber of flax stems was used in the manufacture of linens.

Maximum production was achieved in 1948, when 198,000 acres were harvested and the value of production was $29,979,000. Production ceased after 1970, when 2,000 acres were grown. Several factors contributed to the disappearance of flax in California. Price was a major factor, that in 1948 being $6.18/bushel, and that in 1970 being $3.10/bushel. Barley was always a strong competing crop, and wheat became an even stronger competitor with the availability of high-yielding cultivars developed in Mexico. One problem with flax is that demand for linseed oil declined during post-World War II years because latex-base paints became much more popular than oil-base paints.

Seed Industry

The production of high quality planting seed is an important industry in California where a mild climate, controlled irrigation and rain-free periods during the growing and harvest seasons permit excellent yields and quality. Alfalfa seed, ladino clover, sunflower, beans, cereal crops, rice, and forage grasses of both public and private varieties are all important seed crops produced for local consumption and use in other states and nations. Among recent new crops, hybrid sunflower seed production has increased rapidly because F_1 hybrids are superior in crop yields.

Field crop seed production requires special technical competence as well as careful attention to planting, genetic purity, isolation, controlled pollination and customized harvesting and processing techniques.

The Foundation Seed Service and the California Crop Improvement Association distribute and certify field seed of high quality and genetic purity. Foundation seed of public varieties is distributed by the University's Foundation Seed Service in the College of Agriculture and Environmental Sciences at the University of California, Davis. The California Crop Improvement Association is the official seed certifying agency in California. It has the responsibility to see that all certified seed of both public and proprietary varieties has met the standards of genetic purity and quality by inspecting fields, testing seed samples and maintaining records.

Reference Material on Field Crops

Information on the production, utilization and processing of field crops in California is available in a series of leaflets, circulars and bulletins published by the Division of Agricultural Sciences, University of California, and distributed through the offices of the Cooperative Extension Service throughout the state.

Several publications from the California Crop and Livestock Reporting Service annually provide information on various field crops and field crop seed production.

There is no one reference on field crop production that is strictly applicable to California. There are, however, several text and reference books available that can be helpful. Examples of recent books which have some relevance to California conditions are:

Principles of Field Crop Production by J.H. Martin, W.H. Leonard and D.L. Stump. Macmillan Publishing Co, 1976.

Crop Production—Principles and Practices by S.R. Chapman and L.P. Carter. W.H. Freeman and Company, 1976.

Production of Field Crops by M.S. Kipps. McGraw-Hill Book Co., 1970.

7

Fruit and Nut Crops

Members of the Pomology and Viticulture Departments, UC Davis, and Fruit and Nut Specialists and Farm Advisors, Cooperative Extension

Chapter coordinator, *Dillon S. Brown.* Contributors: *Royce S. Bringhurst,* strawberries and caneberries; *Julian C. Crane,* figs and pistachios; *Hudson T. Hartmann,* olives; *A. Dinsmoor Webb,* grapes; *James A. Beutel,* canning peaches, pears, kiwi; *James H. LaRue,* shipping peaches, nectarines, plums, persimmons, pomegranates; *Warren C. Micke,* almonds, apples, sweet cherries; *Gordon F. Mitchell,* shipping peaches, nectarines, plums; *Joseph Osgood,* prunes; *Robert G. Platt,* avocados, citrus, dates; *David E. Ramos,* walnuts; *Donald Rough,* apricots.

California's fruit industry was born with the orchards and vineyards planted during the mission era. It grew with the state, following the Gold Rush and completion of the transcontinental railroad. Over the past fifty years it has attained a dynamic maturity which is best understood by considering each of the fruit species individually. There are, however, some aspects of fruit culture and the California industry's growth and development that are common to all.

To begin, California's varied climate continues to be the primary natural asset responsible for the state's position of leadershp in the production of a wide range of fruit crops. It is a predominantly Mediterranean climate with relatively mild winters and a long growing season, suited to a broad range of crops from sub-tropical to temperate-zone species. Except at the higher elevations, the winters are mild enough that winter injury or killing of buds and other parts of the plants is rare. Indeed, for many kinds or varieties of deciduous fruit trees, the winters are frequently too mild, especially in the southern third of the state, to provide enough winter chilling for the trees to emerge from dormancy properly. (Following mild winters, the bloom period of some fruit trees is prolonged, some of the

flower buds may drop off unopened or not grow normally, and leafing and the start of shoot growth is delayed. The crop may be seriously reduced.)

The adequacy of winter chilling may be evaluated in terms of the hours of temperatures at or below 45°F (7.2°C) experienced during the fall and winter months, with December and January being the most critical. The chilling requirement is considered as high if trees need 1000 or more hours of such temperatures, as moderate if 500 to 1,000 and as low if less than 500 hours are needed. From the mid-1920s to the early 1950s, one out of three winters were, on the average, low or only moderate in winter chilling—with some of the mildest winters of record occurring around 1940. Deciduous fruit production was adversely affected in many districts, but particularly in southern California. This contributed to the decline in deciduous fruit and nut acreage in that area and a shift of the industry to more climatically suitable parts of northern California.

While California provides a diversity of growing season climates, spring comes early throughout much of the state, so fruit and nut crops bloom earlier than in most other parts of the country. Spring frosts can, therefore, become a problem. However, with earlier bloom, some fruit varieties also ripen earlier than elsewhere, providing growers with the economic edge of an early season market. While most fruit crops benefit from the warm, dry summer weather of the interior, temperatures of 95 to 105°F (35 to 40.5°C), which are not uncommon there, may adversely affect some fruit and nut crops. Some do better in coastal districts where summer temperatures are more moderate, although for others summer temperatures near the coast are too cool for fruit to mature well. Among all the fruits, the grape, citrus, and strawberry industries have probably best adapted to and utilized a variety of climatic areas throughout the state.

In earlier years, some orchards were grown without irrigation, but today only a few remain. Changes in water availability and irrigation technology, especially in the last two decades, have also affected the fruit industry. Completion of parts of the Central Valley Water Project which made water available to western reaches of the southern San Joaquin Valley, opened up new lands for orchards. Large acreages of almonds, walnuts, and pistachio have been planted. In recent years, the use of sprinkler irrigation has increased, in many instances replacing flood and furrow systems. Drip irrigation, an even more recent innovation, is finding a place in some orchard situations. Depending on the crop, the soil, season, irrgation system and water available, annual water applications in irrigated orchards range from 1 or 2 to 6 or 7 acre feet. Water is also used in some orchards and vineyards for frost protection instead of orchard heaters or wind machines which, though still used in some places, are apt to be air-polluting.

Some California soils are better suited to fruit crops than others, either in their physical characteristics or fertility. Nitrogen remains the most generally needed fertilizer material and applications of zinc or other minor elements are a continuing necessity in special situations. Potassium deficiencies have become more prevalent in some orchard areas within the past decade or so. There has been a trend toward the elimination of cultivation, as evidenced in non-tillage systems developed for some crops—citrus and nuts in particular. These systems utilize chemicals to control weeds, mowers to clip grass or other cover crops, or a combination.

Deep tillage is sometimes used in preparing land for planting to orchard. On replanting sites

in particular, a back–hoe is used to dig extra large holes at tree sites when it is necessary to break up impervious layers in the soil profile and otherwise to make conditions more favorable for good root growth of the new trees. Soil fumigation prior to planting for control of nematodes and/or soil-borne diseases, such as oak root fungus, is sometimes used.

Pest and disease problems have not changed much in the past thirty years, at least in the kinds of organisms involved. With some crops, new insect or disease problems have arisen and older ones have been accurately diagnosed. Regulations and controls on the use of pesticides, while enhancing worker and environmental safety, have complicated pest control procedures in some instances and led to modification of most of them. In particular, there is increased interest in integrated pest management, a system which attempts to keep pests in check by using a mix of biological, cultural, and chemical controls.

Cooperatives and other organizations of growers and/or processors continue to play a significant role in the handling and marketing of California fruit crops. A number of commodities come under grower- and/or processor-approved state and federal marketing orders which provide for trade promotion, consumer advertising, production and marketing research, or in some cases, set minimum standards for grades and sizes and for orderly marketing.

Urbanization in the major metropolitan areas of southern and northern California has been responsible for many changes in the state's fruit industry over the years. Along with the development of new water resources, urbanization of former fruit-growing areas has contributed to shifts in centers of production for citrus, as well as for several deciduous species, from southern California to the San Joaquin Valley and in some cases to the Sacramento; and from the Santa Clara and adjacent Bay Area counties to the Sacramento and the northern San Joaquin areas.

Economic considerations, together with changing consumer preferences and marketing habits, have also had an impact on most of the fruit industry. Shifts in consumer tastes and development of export markets have encouraged such crops as almonds, walnuts, and avocados, while discouraging the canning of peaches and other stone fruit. Nut crops, which are amenable to mechanization of production, processing and handling operations, have gained at the expense of fruits such as apricots and peaches, which are more perishable and still have high hand-labor requirements in growing, packing or processing. Meanwhile, growers in areas of high water cost have turned from agronomic crops of low or moderate value to high value fruits and nuts.

Deciduous Fruits

Apples

Historically, apples have been grown in California primarily in coastal and foothill areas with the largest concentration of plantings in the northern coastal area around Sebastopol and the central-coastal area near Watsonville. More recently, however, sizable plantings have been made in the Central Valley, including over 1000 acres east of Merced planted for processing to wine. Even so, the statewide total acreage in apple trees of bearing age has remained fairly stable over the past 30 years, varying from 21,000 to 26,000 with an additional 4000 acres on the average in nonbearing trees annually. Presently, there are around 23,000 or 24,000 acres of apples in bearing, nearly equal to the acreage in 1950 and up some from that in the late 1960s. Usually a quarter or less of California's apple crop is sold as fresh fruit and that mostly within the state in recent years. The other three-quarters is processed into applesauce, juice, frozen slices, and dried apples. Nationwide, California usually ranks fourth in production among the major apple-growing states.

Climatically, apple trees require more winter chilling than most other deciduous fruit species in order to overcome the rest period of their buds and condition them for normal growth in the spring. In most districts in the state where apples are grown, winter chilling is barely adequate so that the period of bloom and leafing in the spring is prolonged and straggly. Winter chilling is so low in southern California that only the most mild-winter tolerant varieties can be grown at low elevations and even then not always with success. Apples bloom late enough in the spring so that the risk of frost damge is less severe than with other species, but at higher elevations or in particularly cold areas, spring frost can be limiting. Summer heat in the Central Valley can impair the edible quality of the fruit and interfere with good color development on red varieties.

Recent plantings in the state have been at densities of 150 to 400 trees per acre as compared to older plantings of 50 to 100 trees. To make these higher density plantings successful, growers are using trees on semi-dwarfing and dwarfing rootstocks and some of the compact-growing, spur-type strains of the varieties.

The varieties 'Red Delicious' and its sports, 'Yellow Newtown,' 'Gravenstein,' 'Golden Delicious,' 'Rome Beauty' and its sports, and 'Jonathan' account for 90 percent of California's apple acreage. Recently, a substantial acreage of 'Granny Smith' has been planted, mostly in the Central Valley. Since most apple varieties are not self-fruitful, though some are partially so, at least two or more varieties are planted to provide cross-pollination. Certain varieties, including 'Gravenstein,' 'Winesap' and 'Mutsu,' produce sterile pollen and are useless as pollinizers.

Trees in newer high-density plantings are trained by the central leader method rather than the vase-shape or open-center system used in older orchards. The trees are usually pruned moderately each year. The crop is thinned in most years, either chemically, by hand or a combination of both, primarily to increase fruit size and promote return bloom (reduce alternate-bearing). Apples presently are the only deciduous fruit crop in California for which chemical thinners are available. Even so, the normally prolonged bloom period, conditioned by the mild winters in the apple districts, makes the timing of chemical-thinning applications difficult and the treatments less satisfactory than in states in which the apple bloom period is short.

Although the codling moth continues to be a primary pest, the orange tortrix has in recent years also become a major pest of apples. While not as serious as in the humid East, apple scab is one of the more troublesome diseases. European canker has been increasingly a problem, particularly in the Sebastopol district, over the past decade or two.

Apricots

The apricot continues to be an important commodity in California where 97 percent of the nation's crop is produced. During the past 30 years the San Joaquin Valley has become the major producing region, centered principally in Stanislaus, San Joaquin, and Merced counties.

Other major producing counties include San Benito, Santa Clara, Solano, Yolo, and Contra Costa. Total bearing acreage during the 14-year period 1964 to 1977 dropped from 37,500 to 27,000 acres. Most of the reduction was in the coastal counties and was made up of older, poorly-producing trees. During the same period only a very few new plantings were made. Yields per acre, while showing yearly ups and downs, average about 5 tons, though some of the orchards on the west side of the San Joaquin Valley average better than 10 tons per acre. Stanislaus County orchards normally top the state in yields per acre.

In recent years there has been a gradual change in the proportions of the different varieties grown. 'Blenheim' ('Royal') for many years was the dominant variety. Today the total acreage is divided evenly between 'Tilton' and 'Blenheim'. 'Tilton' yields are higher than those of 'Blenheim,' but its dessert quality is poorer. Several new varieties have been introduced in recent years which show promise. These include 'Modesto,' 'Patterson,' 'Tracy' and 'Westley,' which were introduced by Fred Anderson, a private plant breeder. 'Patterson' may well be a good replacement for 'Tilton'. The variety tends to mature large, good-quality crops on a regular basis and appears to be well adaptable to mechanized harvesting. The new 'Castleton' variety, introduced from the U.S. Department of Agriculture's fruit breeding program, appears to have a reasonable place in the fresh-fruit market.

Apricot varieties have a moderate ('Blenheim') to high ('Tilton') winter-chilling requirement to condition the trees for growth in the spring. Time of bloom varies from year to year from late February to mid-March, so spring frosts can pose a hazard in some seasons. Most of the varieties are self-fruitful. Harvesting for fresh market begins in early to mid-May, mostly of the 'Derby Royal' and 'Royal' ('Blenheim') varieties coming from the Yolo-Solano counties area. Harvesting of 'Blenheim' for canning follows in mid-June. The 'Tilton' variety matures in early July; 'Modesto' matures a few days after 'Blenheim'; and 'Patterson' a few days after 'Tilton'. Harvesting is generally still done by hand, although equipment and technology for mechanical harvesting are available. For good fruit size at harvest the crop must be thinned, which like pruning is done by hand or sometimes mechanically.

The crop is utilized as fresh (5 percent), frozen (7 percent), dried (28 percent) and canned (60 percent) products. There has been little change in the crop utilization during the past two decades, but the number of firms processing apricots has shown a marked decline from 50 in 1960 to 11 in 1980. Prices received by growers have changed considerably from year to year, depending on crop size and processor carryover of previous year's crop. In recent years they have ranged from a low of $61 per ton in 1971 to a high of $215 per ton in 1974.

The industry appears to be stabilizing at about 27,000 acres. Expansion in the next decade will be modest, if any. In fact, the industry may be hard pressed to maintain its present markets, both fresh and processing, as it becomes more difficult for the products to compete advantageously with other fruit and nut commodities when the increasing production and processing costs for apricots are passed on to the consumer.

Canning Clingstone Peaches

Nearly all peaches raised for canning in the United States are grown in California and of these 95 percent are clingstone varieties, those with fruit in which the flesh does not separate readily from the pit. The firmness of cling peaches makes them an attractive canned product. With a total production of about 600,000 tons of grade number one peaches, they are the most important fruit canned in the state. Approximately 50,000 acres are grown by some 1,000 growers in Stanislaus, Sutter, Yuba, Merced, Butte, and San Joaquin counties. Good soils and irrigation water plus usually adequate winter chilling and moderate spring and summer temperatures make these areas desirable for cling peaches. The present acreage is approximately what it was 30 years ago, but only two-thirds of that of the mid-1960s. Competition with other fruit products, both canned and fresh, and increasing costs of production and of processing have had an increasing economic impact on the industry. Consequently, some growers have turned to other fruit or nut crops that presently seem to have more potential than canning peaches.

Cling peach fruit are more sensitive than most freestone varieties to brown rot. This disease is particularly disastrous if rains occur just prior to harvest, although the general absence of summer rains in California normally assures ideal condi-

Harvest season for clingstone peaches lasts from mid-July to September. Over-production has periodically plagued the industry.

tions for growing cling peaches. Occasionally, as in 1965, 1967 and 1973, summer rains occurring during harvest cause heavy losses in fruit harvested after the rain. Recently, new fungicides developed for brown rot control, if applied at bloom and 10 days before harvest, have given good protection against the fungus and have helped growers control a disease that, with an increasing incidence of summer rains, has severely threatened the industry. Other diseases are mostly controlled by winter applications of fungicides.

The fruit is harvested both by hand from ladders and by machine. Machines that shake the trees and catch the fruit on padded frames moved under the trees account for about 40 percent of the crop harvested. The development of harvesters has evolved slowly since 1965.

A long harvest season, from mid-July to early September, is made possible by a succession of varieties that ripen at different times with each variety accounting for the tonnage harvested over about one week of the season. The major varieties in order of ripening are 'Loadel,' 'Carson,' 'Andross,' 'Palora,' 'Klamt,' 'Carolyn,' 'Halford,' 'Starn' and 'Corona'. Two of these, 'Andross' and 'Klamt,' are from a group of higher-flavored and higher-yielding varieties developed over the last three decades in a fruit-breeding program at the University of California that was supported largely by research funds generated under an industry-wide marketing order.

Cling peach growers have made progress in bargaining for higher prices from canners (processors) through the California Canning Peach Association. The industry has learned over the last 20 years to match production with market demand. It was the first processing-fruit industry to include in its annually-bargained price incentives for fruit quality and adjustment of the base price to the tonnage processed seasonally.

Pears

Just one variety, 'Bartlett,' acounts for at least 95 percent of California's production of pears. It is a fruit suitable for both canning and marketing fresh. The trees are self-fruitful, setting fruit parthenocarpically, so cross-pollination is not essential and orchards may be planted solely to 'Bartlett'. The California pear crop, which varies from 300 to 375 thousand tons annually, is grown on 40,000 or so acres. The most prod-

uctive district, located mostly in Sacramento County along the Sacramento River, provides one-third of the state's production on just over 7,000 acres. With nearly an equal acreage in pears, Lake County produces only one-fifth of the state's production, but ships nearly half of that as fresh fruit. Other important pear areas are in Mendocino and Solano counties and the central Sacramento Valley, which produce pears mostly for canning.

The pear industry has had more ups and downs than most other California fruit industries over the last 40 years. In the 1930s, besides the economic problems typical of that time, major production problems of pear growers included control of worm damage to the fruit and control of fireblight in the trees. One of these mjor pest problems was solved during the late 1940s when first DDT and later organo-phosphate sprays were used to control worms (codling moth larvae), replacing lead arsenate which was more hazardous and gave poor control.

Fireblight, the second problem, is a bacterial disease that infects pear blossoms. The disease will move from the blossoms to the shoots and branches and ultimately to the trunks of pear trees if infected limbs are not cut off below the point to which the infection has spread. The extensive pruning practiced in the past to remove fireblight infections was not only costly in labor but also reduced tree size and yield of fruit.

The normal pear bloom occurs from mid-March to mid-April, depending upon the district and the season. Unfortunately, 'Bartlett' also produces some flowers later in April and in May after the main bloom period; these late, nuisance flowers are easily infected by fireblight, especially during periods of light, warm spring rain. In the early 1950s, copper sprays and dusts were shown to prevent fireblight infections, if applied during bloom and before rains in the month thereafter. Later, sprays with the antibiotic, streptomycin, were found to give fireblight control. After most pear growers began using copper dusts or streptomycin sprays, fireblight losses were substantially reduced.

Following widespread use of better control sprays for codling moth and fireblight, pear production increased and prices declined as the supply of pears caught up with market demand. However, the problem of declining prices was short-lived because a new disease, pear decline, began killing pear trees in Washington and

Oregon in the mid-1950s, thus reducing the supply of pears nationally. Along with and usually ahead of pear decline, a new and serious pear insect, pear psylla, moved south from Washington State through Oregon, reaching northern California in 1957. By 1958 pear trees in California began to collapse and die. Over the next 15 years nearly 2 million pear trees died, aproximately 50 percent of all the pear trees growing in California in 1957. No cure for pear decline was found until 1972, when it was determined that annual injections of about one gram per tree of terramycin kills most of the organisms (mycoplasma) that cause pear decline. Pear psylla spreads the mycoplasma from infected to healthy trees.

The pear trees killed by decline after it moved into the state were those growing on the rootstock, *Pyrus serotina*, which is highly sensitive to the disease. Research showed that certain rootstocks, includig some seedlings of the 'Winter Nelis' variety of pear and trees of *Pyrus betulaefolia*, were resistant to pear decline. Many growers in California began planting pear trees on these rootstocks with the result that by the mid-1970s the entire acreage of pears lost to decline was replaced. However, replanting was not financially possible for all growers. Indeed, most pear growers in the foothill areas of Placer, Nevada, and El Dorado counties lost their entire orchards to decline and could not afford to plant new trees.

Again, after the replanted pear orchards came into production, the pear industry had overproduction in 1977. In that year, some California pears were used for winemaking for the first time and within three years about as many tons of second-grade pears went into winemaking as there were of first-grade pears sold as fresh fruit. Today, 70 percent or more of California 'Bartlett' pears are canned as pear halves or in fruit cocktail, about 15 percent are sold as fresh fruit, nearly 15 percent is used in winemaking, and a few are dried.

Sweet Cherries

California ranks third on the average among the states in sweet cherry production, behind Oregon and Washington. The fresh market is the most important outlet for California sweet cherries because the crop, ripening earlier than in any other area in the United States, brings the premium prices accruing in an early-season market. However, even though over half of California's sweet cherries are normally sold fresh, the processing market (mainly brining) is still important. Brined cherries are made into Maraschino cherries which are packed as such or used in fruit cocktail.

Sweet cherries have fairly exacting climatic requirements. The trees need a considerable amount of winter cold in order to break dormancy normally. Though usually adequate in Central Valley, coastal valley, and foothill areas, winter chilling is generally inadequate in southern California unless the trees are grown above the 2,000 foot elevation. Spring frost also can be a limiting factor at higher elevations in the foothills. Unless modified by almost daily intrusions of marine air, such as occurs in the Lodi-Stockton district, the high summer temperatures commonly found in many parts of the interior valleys can be detrimental to sweet cherries. High temperatures may lead to misshapen, off-grade fruit, called doubles (two fruit attached at the stem). Trees may also suffer water stress with high summer temperatures and be more susceptible to sunburning of limbs. Rain occurring during bloom or harvest increases disease susceptibility and when rain occurs near harvest fruit may crack and become worthless.

'Bing' is the main variety, accounting for about three-fourths of the approximately 13,000 acres of sweet cherries grown in California. There are lesser acreages of 'Royal Ann' ('Napoleon') and 'Early Burlat'. Other varieties, used mainly as pollinizers for those three, include 'Black Tartarian,' 'Van,' and 'Larian'. All commercial sweet cherry varieties grown here are self-unfruitful and therefore, need cross-pollination in order to set a crop. A good pollinizer must bloom at the same time as the main variety and have compatible pollen. Certain varieties such as 'Bing,' 'Lambert' and 'Royal Ann' are cross-unfruitful because of incompatible pollen, and will not serve as pollinizers for one another. Pollination is done by insects, usually honeybees, so growers normally rent hives of bees and place them in the orchards during bloom.

The major rootstocks for cherries in California are 'Mahaleb' and 'Mazzard'. Each stock has its advantages and disadvantages and neither is totally satisfactory. 'Mahaleb' is severely sensitive to various root rots and gopher damage, but produces a smaller tree that usually comes into bearing at a fairly early age. 'Mazzard,' on the other hand, is less sensitive to root rots and gophers, but more sensitive to other diseases. It produces a larger tree which sometimes needs more time before it begins to bear.

Cherry trees require only limited amounts of pruning. It is best done annually, although some growers prune more severely and less frequently. Compared to other fruit crops, cherries are relatively free of insect pests, although at times they are subject to infestations of the same kinds that attack other fruit crops. Diseases that affect the tree and/or fruit are more apt to be chronic. During the 1970s, buckskin, *Phytophthora* crown and root rot, and the stem-pitting disease have caused serious losses.Birds often are serious pests in cherry orchards. Control is difficult, but some noise or scare devices and chemical repellant treatments released in the last decade have helped. Pocket gophers have also been a serious pest.

Cherries are still harvested by hand picking.

Freestone Peaches and Nectarines

FOR FRESH MARKET (SHIPPING). California has long been one of the major states growing peaches for fresh consumption. In addition, it produces nearly all of the nectarines grown in the United States. Trees of peach and nectarine are similar in appearance and growth habit, but nectarine fruits lack the pubescence (fuzz) characteristic of peaches. Actually, nectarines are a genetic variant of peach, the lack of pubescence being heritable through a recessive gene. Therefore, nectarines appear normally and predictably in the seedling populations planted by breeders of peaches and nectarines. Some have also been found as bud sports or mutations on peach trees, though the reciprocals of peach sports on nectarine trees have not—as would be expected given the recessive nature of the nectarine characteristic.

Although horticulturists describe the nectarine as a fuzzless peach, the commercial trade treats it as a separate and distinct fruit for merchandising purposes. Nectarines have increased dramatically in importance in recent years with the volume now shipped to market annually from California exceeding that of fresh-market peaches by more than a million packages. Furthermore, nectarines are marketed throughout the country, whereas the outlets for shipping peaches grown here are principally within California and the proximal western states. Producing areas, totaling about 30,000 acres of freestone peaches and 25,000 of

nectarines, are located throughout the San Joaquin Valley, centered between Fresno and Visalia, with minor production in the Sacramento Valley and in southern California.

Nearly 200 varieties of fresh peaches and nectarines are grown commercially in California. They are known as 'freestones' but many varieties are actually clingstone, semi-free or semi-cling, and may be yellow-, red- or white-fleshed. Some are patented varieties developed by private plant breeders with trees propagated and supplied exclusively through certain nurseries. Generally, varieties grown here are seldom adapted to other U.S. production areas and conversely, eastern varieties seldom perform well in California. Currently, the leading varieties of peaches for fresh market grown here are 'Fay Elberta,' 'Springcrest,' 'Redtop,' 'O'Henry,' 'Suncrest,' and 'June Lady,' and of nectarines, 'Flamekist,' 'Fantasia,' 'May Grand,' 'Autumn Grand,' 'Early Sungrand', and 'Flavortop'.

Peach and nectarine varieties vary from low to moderately high in the annual amount of winter chilling required to break dormancy and condition the trees for normal growth in the spring. In the San Joaquin Valley trees bloom in early to mid-March, so spring frost is a hazard in some years. Fruit thinning, generally by hand, is practiced to increase fruit size and saleable tonnage. Together with pruning and harvest, the necessity for thinning creates a major labor need. Mechanization has not yet proven as satisfactory as hand labor for these orchard operations. The earliest varieties are picked in late April (southern California desert), the latest in October. Heaviest production occurs in June, July and August.

Fruit is picked by hand when it has reached a stage of development that permits the ripening process to be completed successfully off the tree. The fruit is picked in buckets and transported to the packinghouse in bins containing 1,000 to 1,200 pounds, where it is packed and placed immediately in cold storage. The market containers may be packed by hand or volume (loose)-filled. The most modern packinghouses are highly automated and often pack up to 1,200 or more containers per hour. Once the fruit is packed and cooled it may be held in cold storage for later shipment or loaded immediately into refrigerated trucks, rail cars, or vans for marketing throughout the United States or for export.

FOR PROCESSING. Some freestone peach varieties serve a dual purpose with part of the crop being shipped fresh and part being processed (canned, frozen or dried). Since the mid-1970s about 40 percent of the 225,000 or so tons of freestone peaches produced annually in California have been processed, with 33 to 40,000 tons canned, 30 to 50,000 tons frozen and 14 to 20,000 tons dried. A decade earlier, with a somewhat higher total tonnage, approximately 60 percent of the freestone peaches were being processed, 90 to 100,000 tons being canned in some years. 'Elberta' and 'Fay Elberta' are the freestone peach varieties most often canned, but the acreage of both is declining as that of newer shipping varieties increases. Almost any of the freestone varieties may be used for freezing, those utilized in any one year being predominantly ones in over-supply on the fresh market.

Unlike peaches, nectarines are almost solely a fresh market fruit with only one percent or less of the crop being canned.

Plums

Plums grown for fresh consumption are mainly varieties of the Japanese plum, *Prunus salicina*. The first commercial varieties were introduced over 100 years ago from Japan, most of them improved selections by Luther Burbank. Today, over 150 varieties are grown, with new and improved ones frequently being added through introductions by private, University of California and U.S. Department of Agriculture plant breeders. While 'Santa Rosa,' an old variety, is still the leader, it is now followed in production by 'Casselman,' 'Laroda,' 'El Dorado,' 'Red Beauty,' and 'Friar,' which have taken the place of some older varieties over the last decade or so. Out of a total of about 36,000 acres of plums in the state, over 90 percent of today's production, which accounts for nearly all of that in the United States, is located in the southern San Joaquin Valley (Fresno, Tulare, and Kern counties).

Plum trees require a moderate amount of winter chilling to condition them for normal flowering and leafing in the spring. They bloom in late February and early March. Some plum varieties are self-fruitful, but many others require cross-pollination in order to bear fruit. Bees are the usual pollinating agents.

Proper pruning and fruit thinning by hand are important for the production of good crops of large-sized fruit. The earliest-maturing plum varieties ripen about mid-May and the latest in early Dctober, with the bulk of the crop being har-

vested in June, July and August. Most varieties are picked two to three times. Mechanization of pruning, thinning, and harvest has not yet reached the perfection whereby it will replace highly-skilled hand labor.

Fruit is picked in buckets, transported to the packinghouse in bins, packed and cooled immediately. A part of the crop is packed in four-basket crates, but the majority of it is packed in tight-fill or loose-fill containers holding 28 pounds of plums. After pre-cooling the fruit may be held for a short period in cold storage or shipped directly to market.

Prunes

Prune production is an intensive, specialized industry found only in California and a few select areas worldwide. All prunes are varieties of the European plum, *Prunus domestica*, that can be dried whole without fermenting at the pit. For successful drying, a high sugar content is necessary, which also makes them good as plums for eating fresh, if a very sweet fruit is desired. The European plum, grown in parts of Europe for many centuries, is believed to have originated in the Near East. Through its culture in France, the 'Prune d'Agen,' now known as the 'French Prune,' or simply as 'Agen' or as 'French,' became the leading commercial dried fruit of the world long before California was important agriculturally.

'Prune d'Agen' was introduced to California from France by Louis Pellier in 1856 in a collection of fruit scions and cuttings and many kinds of seeds. By 1869–70 there were 19,000 prune trees in California. The completion of the transcontinental railroad opened a tremendous new market, and the prune industry was established. The central coast counties of Santa Clara, Napa and Sonoma accounted for 80 percent of California's bearing acreage until about 1950. Then, with urbanization of the Santa Clara Valley and disease problems in the Napa-Sonoma district, the acreage began decreasing and production shifted to the Central Valley.

As a result of heavy planting in the 1960s, the Sacramento Valley has become the dominant prune-growing area with over 60,000 of the state's nearly 85,000 acres of prunes. It now produces 74 percent of the state's dried prunes, the San Joaquin percent, the Napa-Sonoma district 9 percent, and the Santa Clara district 7 percent. Prune trees in the interior valleys tend to have a greater producing capacity than the coastal valley plantings. Average yields have varied between 1.5 and 2.5 tons per acre with a steady upward trend over the years. 'French' is still the leading variety, accounting for over 95 percent of the state's prune acreage. Unlike most of the Japanese plums, it has a relatively high winter-chilling requirement. It blooms usually in mid– to late March. It is self-fruitful. Fruit thinning is practiced infrequently, only when the crop is very heavy. In such years thinning is essential to reduce the incidence of shoot die-back, a symptom of potassium deficiency induced by heavy cropping. When practiced, thinning is done mechanically, using the same shakers also used at harvest. The prune orchards most susceptible to the die-back induced by heavy cropping are those on heavy soils in the Sacramento Valley.

'Myrobalan' plum is the most commonly used rootstock for prunes. Although it can survive on heavy soils that tend to be wet, it is intolerant of soils that are extremely heavy. 'Myrobalan 29-C' is a rootstock selected for its resistance to root-knot nematodes. 'Marianna 2624' is moderately resistant to oak root fungus, and so far is the best prune rootstock where that fungus is a problem. Peach, almond and apricot are also occasionally used as rootstocks for prunes.

Prune fruits mature in August and early September. The crop is harvested mechanically into bins and trucked to a dehydrator where the fruit is dried to about 18 percent moisture. After drying the prunes are sized and inspected for quality, then stored in bins to await processing and packing. The packer prepares dried prunes by passing the fruits through a hot water bath to cleanse and partially rehydrate them to a moisture content of 26 to 32 percent. Fruits become softer in the process and more acceptable to the consumer. Some are rehydrated to about 38 percent moisture and packaged as moist-pack prunes. Others are processed into prune juice and juice concentrate by cooking the dried prunes in water. Dried prunes, pitted or unpitted, are also sometimes canned, packed with liquid for ready serving. Other products include purees, baby food, and diced prunes.

Over the last 30 years, the portion of the crop sold as unpitted prunes has declined to something less than half of its earlier volume, while the sales of pitted prunes has increased dramatically, nearly six-fold. Over the same period, the quantity of prunes used for juice and juice concentrate has increased one and one-half times. The amount used for baby foods, purees and other products has declined to less than 5 percent of the total tonnage of dried prunes. Unpitted, pitted, and juice and concentrate now account for about 31, 26 and 38 percent of the total tonnage, respectively. In recent years, one-third to more than half of the U.S. production of dried prunes has been exported annually.

Nut Crops

Almonds

In the United States, almonds are grown commercially only in California. With nearly 360,000 acres, they are the foremost nut crop grown in the state and second only to grapes among all fruit and nut crops. The industry has gone from a time of surplus production, few if any exports and even some imports in the 1940s and early 1950s, to a situation by the mid-1970s of no surpluses, essentially no imports and a major part of the crop being exported. Today's almond acreage is over three times that of 1950, located almost entirely (97 percent) within the Central Valley. In the 1950s the acreage of almonds in the Sacramento Valley remained nearly stable, increasing only slightly, while that in the northern San Joaquin showed a very modest gain. Moderate increases continued in the 1960s in the Sacramento Valley, principally in Butte, Glenn and Yolo counties, while in the same period the acreage in almonds rose sharply in the northern San Joaquin, more than doubling in Merced, San Joaquin and Stanislaus counties, and going from less than 2000 to about 15,000 bearing acres altogether in the counties southward in the Valley. The bearing acreage in the state more than doubled in the 1970s, with a relatively modest increase of 20,000 acres in Butte, Glenn and Tehama counties, a very substantial 56,000 acres in Merced, San Joaquin and Stanislaus counties, and a phenomenal 102,000 acres in the Valley counties further south, over half of it in Kern. The increase in bearing acres in the latter counties alone is nearly equal to the total acreage, bearing and nonbearing, in the entire state in 1950.

Almonds usually begin blooming in February, so orchard locations must be relatively free from frosts during and after bloom. Frost protection is

Most California almonds are harvested with mechanical tree shakers which vibrate the nuts to the ground, where they are swept up by other machines.

used, however, for almonds in some areas of the state. The crop requires considerable heat during the growing season, so production is usually best in the warm, dry interior valleys. Rain occurring at or shortly after bloom can interfere with pollination and encourage disease infections. Late in the season, harvest operations can also be upset by rain.

Varieties most commonly grown in California are 'Nonpareil,' which accounts for slightly over half of the acreage, 'Mission' ('Texas'), 'Merced,' 'Ne Plus Ultra,' 'Thompson,' 'Carmel' and 'Peerless,' plus small acreage of several others. The major rootstock for almond in California is peach, both 'Nemaguard' and 'Lovell,' with some plantings on almond, peach-almond hybrids and 'Marianna 2624.'

All major almond varieties require cross-pollination. Since fruit size is not especially important, no fruit thinning is done and it is important that all blossoms be pollinated for maximum production. Usually two or three varieties have been planted in an orchard with no more than two rows of one variety alternated with at least one row of a pollinator. To ensure

even better pollination, many recent plantings have been made with the varieties arranged alternately in single rows. Pollination is accomplished by insects, usually honeybees, so growers usually place rented hives of bees in their orchards at bloom-time.

Almond orchards require more nitrogen fertilizer than most fruit crops because of the relatively high nitrogen content of the almond kernel or meat. A high yielding orchard can remove 100 pounds of actual nitrogen per acre in the crop alone. Weed control is accomplished by cultivation or non-tillage methods using herbicides, mowing or combination of the two. Non-tillage when combined with sprinkler irrigation is a convenient way of handling many soils and preparing the orchard for harvest, which is highly mechanized. Most almond orchards are pruned each year, but generally fewer cuts per tree are made than with most other deciduous fruit species.

The navel orangeworm is presently the major insect pest of almond. Infectious diseases generally remain more of a problem to the north in the Sacramento Valley than to the south in the San

Joaquin where rainfall amounts are lighter. Non-infectious bud failure (crazy top), a genetic disorder affecting some almond varieties, most notably 'Non-pareil,' has become a more prevalent problem in all districts over the past two decades as the acreage of almonds has expanded. The problem can be met in part by planting varieties free of the disorder and by exercising care in selecting scion wood for propagating varieties prone to it.

Walnuts

The 'Persian' walnut (the preferred name because it indicates the area in which the species originated), commonly known as the 'English' walnut, is the second most extensively grown deciduous tree-fruit species in California with about 220,000 acres bearing in 1980.

Originally, walnuts were planted mainly in the southern coastal counties of the state with the earliest introductions by Spanish missionaries around 1770 and the first importation resulting in commercial planting coming from Spain by way of Chile in 1867. Improved selections from this initial introduction, which produced well following the mild winters of southern California, became known as 'Santa Barbara' soft shells. Of these, 'Placentia' was one widely planted in the early development of the walnut industry in southern California. That area was the principal walnut-growing region of the state until the 1920s when a progressive shift of commercial walnut acreage began to the northern half of the state.

The development of the industry in northern California was led by Felix Gillet, a nurseryman who came from France in 1852. He introduced varieties, such as 'Franquette,' which were more adaptable to areas where spring frosts could be a problem and where summer temperatures were higher than in the southern California coastal region. Today, virtually all of the commercial walnut acreage lies in the interior valleys with some small plantings also found in the coastal and Sierra foothill regions in the central and northern parts of the stte. Reasons for this shift in location include urbanization of walnut-growing areas in southern California, and higher yields and improved quality obtainable with newer varieties adaptable to northern California.

Along with 'Franquette' and other older varieties adapted to northern California, the newer varieties have a relatively high winter-chilling requirement, well above that of 'Placentia' and older varieties suited to the southern part of the state. Most of them leaf out in the period of mid-March to mid-April with the bloom period ranging about two weeks later and the varieties with the highest chilling requirement blooming latest. Low winter temperatures are not limiting to walnut production in areas where the crop is presently grown in the state, but fall and spring frosts can be a serious problem in some situations. Later-leafing varieties are generally grown in locations where spring frosts are common. Very high summer temperatures, above about 95°F (35°C), can cause darkened and shrivelled kernels of exposed nuts, particularly when the trees are under water stress. There is a marked difference, however, among varieties in heat tolerance. Rains during the spring pose a serious threat to the crop in early-leafing varieties because of damage from the bacterial disease, walnut blight.

The northern California black walnut has been the standard rootstock and is still commonly used. However, Paradox hybrid (seedlings of northern California black x English) is now generally preferred because it has greater vigor, particularly under adverse soil conditions and is more tolerant of root lesion nematode and most of the *Phytophthora* root and crown rot diseases. Northern California black walnut is still the preferred stock where salinity or Armillaria root rot (oak root fungus) problems exist.

Blackline is a serious disease affecting the longevity of walnut orchards in the central coast and northern San Joaquin Valley areas. It is caused by a virus which infects the English top, resulting in death of cells at the graft union with the rootstock and subsequent girdling of the tree. It is a graft-transmissable virus but recent evidence suggests that natural spread is mostly likely due to dispersal of infected pollen from diseased trees (walnuts are wind-pollinated).

The trend in recent years has been to plant orchards with trees closer together and with new, highly-fruitful varieties, such as 'Ashley,' 'Serr,' 'Vina,' and 'Chico.' In contrast to older varieties, the newer ones are capable of early production because of a high degree of lateral-bud fruitfulness. They also produce larger crops at tree maturity with yields commonly in excess of two tons per acre and possess a higher kernel yield (more than 50 percent of in-shell weight), which is desired because a large proportion of the crop is now sold as a shelled product. 'Hartley' continues to be the leading variety grown for in-shell sales. About one-half of the crop is marketed by a grower cooperative with the remainder sold through a number of independent handlers.

Pistachios

Pistachio production is one of California's newest industries. From less than 300 acres in 1968, rapid expansion took place during the 1970s to about 30,000 acres, over half being planted in Kern and Kings counties with Madera, Merced, and Tulare counties having about 41 percent. Much of the expansion has been speculative, with several plantings of 1,000 acres or more, prompted by a desire for potentially high-return crops to plant on new lands being opened up because of increased availability of water from the Central Valley Project. The first commercial crop of any size, 5 million pounds, was harvested in 1978, followed by one of over 17 million pounds in 1979. By the mid-1980s, production is expected to surpass the annual U.S. consumption of 30 million pounds, and there will be no need for importing the product from Iran and Turkey.

Several varieties of Mediterranean pistachio were introduced into California in the early 1900s by the United States Department of Agriculture and evaluated at the Plant Introduction Station, Chico. Relatively little interest was generated in producing the nut commercially because each of the varieties available had one or more faults that precluded its production under existing conditions. Also, mechanical harvesting and hulling equipment and techniques were not to appear until years later.

As a result of a program initiated in 1929 by the USDA to develop better varieties, a seedling produced from seeds imported from Iran was selected and named 'Kerman,' a center of pistachio production in that country. 'Kerman' is the only variety grown currently in California. It is propagated on *Pistacia atlantica* and *P. terebinthus* rootstocks because they are more resistant to nematodes and other soil organisms than is *P. vera*. The tree grows vigorously and produces relatively high yields of large nuts with excellent kernel quality. The pistachio is dioecious and wind-pollinated. The main pollinator variety is 'Peters'; one tree of it is generally planted for every eight 'Kerman' trees.

Verticillium wilt, a soil-borne fungus disease, has threatened the young industry, particularly in Kern County where some orchards have lost as much as 9 percent of the trees annually. The rootstock, *P. integerrima*, has shown resistance to the disease and is being used in replanting.

In research at the University of California the 'Kerman' pistachio has been found to be unique in several aspects in comparison with other fruit and nut trees. It is a severe alternate-bearing variety, producing a heavy crop one year but little or no crop the next. In contrast to other alternate-bearing fruit and nut trees in which there is limited flower-bud production in the year of a heavy crop, the number of inflorescence buds formed in the pistachio is actually greater in the bearing than in the nonbearing year. Most of them abscise, however, so only a few remain, producing little or no crop the following year.

The production of blank (empty) nuts is another important problem wherever pistachios are grown. It is primarily the result of seed abortion. 'Kerman' generally averages about 25 percent blanks. In other fruits and nuts, seed abortion results in fruit abscission, but aborted pistachios remain on the tree until harvest when they must be separated from filled nuts. Since most of the lateral buds produced on bearing trees are flower buds, a pruning procedure must be developed to stimulate the growth of potential fruit-bearing wood year after year. Research at the University of California is currently being conducted on such a procedure as well as on solutions to alternate bearing, blank production, and related problems.

Grapes

Grape growing has been an important part of California agriculture since vinifera grapes were introduced from Europe when the Franciscans established Mission San Diego in 1769. After a slow increase to only a few hundred acres by the time of the Gold Rush and statehood in 1849, the area in grapes grew fairly rapidly, though erratically, until 1880 when it was realized that California vineyards were being decimated by phylloxera, a plant louse which kills vinifera grapevines by feeding on the roots. Research and teaching on grape growing and wine making, stimulated by a mandate from the state legislature in 1880, helped growers meet the phylloxera and other vineyard problems.

There were about 500,000 acres of vineyard by

1919, when national prohibition caused serious dislocations within the industry. Plantings of raisin and table grapes increased, but most dramatically, good wine-grape vineyards were grafted to varieties which yielded large crops of grapes with skins tough enough to stand the rail trip to the East Coast where they were used for home wine making. Prohibition lasted long enough to nearly completely destroy California's fine wine industry. Attempts to reconstitute the wine industry following repeal of the prohibition amendment were hampered by the absence of good wine-grape varieties, the need to build new wineries, and the lack of educated and experienced enologists. Progress in solving these problems was disrupted by World War II with the result that the modern California grape and wine industry effectively dates from 1945.

Today the California grape industry consists of three segments based on the major use of the different varieties of grapes planted: raisins, wine, or table (fresh) consumption. True wine grapes constitute somewhat more than half of the total production, though in some years significant quantities of raisin and table grapes are also processed to wines and brandies. Grapes for table consumption, fresh or from cold storage, constitute a small but slowly growing segment of the total.

Raisins

The drying of grapes as a method of preservation is a very ancient practical art. Until the wide dissemination of the 'Thompson Seedless' variety in the latter part of the nineteenth century, wide acceptance of dried grapes was limited by the inconvenience and unpalatability caused by the presence of the large and numerous seeds present in most grape varieties.

As of 1980, varieties designated as raisin grapes are planted on 269,000 acres in California, concentrated in the southern San Joaquin Valley with the greatest plantings in Madera, Fresno, Kings, Tulare, and Kern counties. Of this acreage some 25,000 are non-bearing, foreshadowing a significant increase in the production of this grape type within the next few years. Of the approximately 600,000 tons average annual world production of raisins, somewhat more than half usually comes from California. Over 90 percent of California's production is from the 'Thompson Seedless' variety with only minor amounts from other seeded and nonseeded raisin varieties and

from some experimental plantings of several new types.

Approximately 90 percent of California's 'Thompson Seedless' raisins are processed by the natural sun-drying procedure. This technique, used worldwide, is suitable in areas where early fall rains are not likely and where there is sufficient room between the vineyard rows to spread the harvested fruit for drying. These raisins are normally dark, grayish-black or grayish-brown in color, with the natural bloom left largely intact. Sun-drying produces a tough skin, but with properly matured fruit the flesh is meaty and has a characteristic raisiny flavor. These raisins tend to be dry on the surface and do not stick together when pressed into a cake for packing. Usually less than 10 percent of California's raisins are produced by other procedures. One of these, also used worldwide, involves dipping the freshly-harvested fruit into an alkaline aqueous solution formulated from potassium or sodium carbonates or bicarbonates mixed with small amounts of their hydroxides and with or without some oil. The traditional oil used in the Middle East is olive oil, but in many regions other edible vegetable oils are used. The purpose of the dip is to remove the bloom or waxy coating of the skins to hasten the dehydration process, producing raisins of a light golden color. Fruit which has been treated with the alkaline-oil mixture is usually dried on trays in the shade or in a heated dehydrator. The very lightest color is obtained by bleaching the fruit by exposure to sulfur dioxide after the dipping operation and before drying in a dehydrator.

Raisins are traditionally dried to 15 percent moisture content. The more mature the fruit, the higher its percentage of dissolved solids and the greater and higher the quality of raisins upon drying. On the other hand, the total crop from the vineyard is, within limits, inversely related to the soluble solids percentage in the fruit. For maximum economic return the raisin grower thus must properly integrate in his vineyard management a number of interrelated factors including crop level, soluble solids, date of harvest, and raisin quality.

The raisin industry in California, as with all agricultural industry, is undergoing much innovation. Many of these developments are directed toward mechanizing routine hand-labor operations. For example, both the pruning and harvesting operations are being mechanized. In the production of natural sun-dried raisins, machines are available, which will spread the harvested fruit on a continuous paper tray that is unrolled in the vineyard row after the harvester has passed over the vines. A procedure is being studied intensively for severing 'Thompson Seedless' canes from the vine before harvesting the fruit and allowing it to partially or completely raisin on the cane. Drying is speeded in some cases by spraying one of the alkaline-oil mixtures on the fruit on the vine. The partially dried raisins can be harvested from the severed vines by machine. However, the majority of fruit for raisins is still harvested by hand. Studies of a more fundamental nature are directed toward identifying the compounds produced by the different raisin drying procedures. Knowledge of the biochemistry involved in these changes should help in achieving optimum raisin production and maintaining high quality in the fruit.

Table Grapes

Grapes for table use are planted in California on nearly 70,000 acres, of which a bit more than 7,000 is currently non-bearing. Just as with the raisin-grape varieties, table-grape planting is concentrated in the southern San Joaquin Valley, principally in Madera, Fresno, Kings, Tulare and Kern counties. Although a considerable quantity of 'Thompson Seedless' is used in the fresh-grape market, the varieties technically designated as table-grape varieties planted in California are principally 'Muscat of Alexandria,' 'Tokay,' 'Red Malaga,' 'Ribier,' 'Almeria,' and 'Emperor.' Smaller quantities of a few other varieties and experimental plantings of a number of new varieties are also grown.

After harvest, table grapes produced in California may go immediately to local markets or be carefully packed, cooled, fumigated and held in storage for some time before marketing. Fruit for immediate marketing does not require the elaborate and careful preparation required of fruit for long-term storage. Early-ripening varieties usually are not stored, because there is litle economic advantage in marketing them later in competition with fresh fruit of the late-ripening types. Market prices are usually high for the very earliest fruit and for the very late-season fresh or storage fruit.

Palatability of grapes for fresh or table use depends primarily on the concentrations of sugar and natural acids present in the berries. Secondary factors in taste are the characteristic varietal flavor (particularly notable in the Muscats), texture of pulp and skin, and astringency. Perhaps of more importance than taste in the marketability of grapes is the appearance. Color should be uniform and appropriate for the variety and region where it is grown. The berries should be free of defects and have the natural bloom (the white or grayish, powder-like, waxy coating on the skin of the berries) undisturbed. The stems should be the natural green or light brown characteristic of the variety at maturity. Dried, black stems and limp, dull-colored berries are symptomatic of fruit that was abused during storage and handling.

Fruit for lengthy storage must be very carefully prepared. At the time of harvest, defective berries or straggly parts of the clusters are removed. Packing into storage lugs can be done in the field but is more frequently done today in an air-conditioned, well-lighted packing-house. The clusters are examined very carefully to be sure that all defective berries have been removed. Clusters are usually packed into lugs with the stems up, the space between the rows of clusters being filled with parts of clusters or small clusters laid on their sides. Great effort is made to see that all of the fruit in the box is of comparable maturity and appearance, with very special attention to a minimum handling of the fruit in order that the natural bloom on the berries is disturbed as little as possible. Quick cooling of the fruit is very important. While grapes are much slower than many other fruits in maturing and in their deterioration following harvest, they still undergo metabolic changes mediated by the enzymes naturally present. Cooling the fruit quickly after harvest slows down these changes and keeps the fruit in good condition. Relative humidities during cooling and storage should be nearly 100 percent to prevent dehydration of the fruit. The clusters in the package and the packages in the cooling room should be properly arranged to obtain adequate circulation of cold air around and through the packages.

For long-term storage the packaged fruit must be carefully fumigated with sulfur dioxide every two to four weeks to prevent spoilage by fungus diseases. Under California conditions, the initial fumigation is usually accomplished by mixing 0.5 percent of sulfur dioxide into the storage air. After an exposure of 20 to 30 minutes the gas is vented or absorbed into an alkaline medium.

Refumigations are normally accomplished with 0.1 percent of sulfur dioxide in the circulating cold air.

Great interest has developed recently in the use of in-package, slow-release packets of sulfur-dioxide-emitting salts. In this procedure, free access of air to and through the pack of grapes is restricted rather than encouraged as it is in the older style of packing. Restriction of air circulation not only maintains a low concentration of sulfur dioxide gas within the package but also has the advantage of delaying fruit dehydration. Considerable success has been achieved with these packs in experimental and commercial trials.

Wine Grape Varieties

In 1980 the California wine grape area comprised nearly 351,000 acres, of which 36,000 was non-bearing. Although acreages of both California raisin and table grape varieties increased during the decade of the seventies, the most spectacular planting increase was in the category of true winegrape varieties. The mid-seventies, in particular, witnessed a dramatic planting of well over 100,000 acres of wine grapes. Most of these plantings were of the varieties yielding fine red table wines. The same decade also saw a dramatic increase in demand for white table wines. This resulted in the grafting over of some 8,500 acres of red varieties to white wine grape varieties in 1979 and 1980. Most of this changeover occurred in the Salinas Valley and Kern County, the regions where the greatest amount of the mid-seventies planting occurred. These are also the two regions where new large wineries have been built.

Very dramatic changes occurred in the nature of the California wine industry, as well as in its size, during the thirty-five years following World War II. At war's end about 80 percent of the wine produced was in the appetizer or dessert category—that is, with brandy added to preserve the natural grape sugar and raise the alcohol concentration to about 20 percent. In general, these wines were made from heavy-yielding varieties of grapes grown in the fertile vineyards of the warm San Joaquin Valley of central California. Little attention was paid to quality factors, the main interest being economy of production and distribution. A very small acreage of fine grape varieties in the valleys near San Francisco Bay did exist, and from these a small amount of

really fine table wine was produced. Wineries were either pre-prohibition wineries that had been refurbished or new structures emphasizing large concrete fermenters and redwood storage tanks and efficient distillation units. The American people, in general, did not know much about wines, and, based on their experience during the prohibition period, tended to evaluate wines solely on the basis of their alcoholic content. A further factor mitigating against quality in the product was the serious lack of trained enologists. Indeed, if it had not been for the inherent stability of the wine types produced, they likely would have been much poorer than they were.

Changes in the wine-grape vineyard and wine-production industries since then have been profound. In 1978, for instance, over 83 percent of the total wine produced was in the table-wine classification. Of more significance, however, is the fact that this dramatic increase in table-wine production has been in the class of fine varietal wines—those having the odor and taste characteristics of the grape varieties from which they are produced and whose names they carry. This would not have been possible, of course, without the planting of significant acreages of fine wine varietals. In 1978, for instance, 80,500 tons of 'Cabernet Sauvignon' was produced in California in contrast with less than 17,000 tons in 1973. Plantings of the other fine varietals have been on an equal scale. In 1979 and 1980, especially, fine white varietals have been planted at a very high rate.

Of a total of 724 wineries in the United States, 406 were in California, according to 1979 figures. The largest is a family corporation, the second in size a cooperative as are many others in the medium size range. Most are small, many with hardly more than a few thousand gallon capacity. Most of the wineries own and operate vineyards although in nearly all cases they must also purchase grapes to meet their crushing requirements. There are thousands of vineyard properties in California and they range in size from about 10,000 acres down to a few acres. Some have long-term contracts with single wineries while others sell each year to whoever offers the best price. There are several cooperatives of vineyard owners who operate their own wineries, as well.

Of great importance for production of good quality standard table wines has been the intro-

A University enologist checks bottles in the wine cellar of the Department of Viticulture and Enology on the Davis campus. *Photo by Ansel Adams.*

duction of heat-tolerant grape varieties that yield moderate to high crops in the warmer growing regions of California. These varieties, such as 'Ruby Cabernet,' 'Carnelian,' 'Centurion,' and a new Muscat-flavored variety (so far unnamed), yield wines in which the flavors are clean and distinct. There is also enough acid maintained under the warm growing conditions so that the final wine is pleasantly fruity in balance. When harvested carefully and fermented under cool, controlled temperatures, these new varieties yield wines of very fine quality. They are a significant factor in the ever increasing consumption of California table wines. Most of the new plantings have been made with material certified for trueness to type and free of known vine diseases. The vineyards have been established so that mechanical pruning and harvesting are possible, even if not practiced at the present.

The winery of the eighties bears little resemblance to the pre-prohibition or the immediate post-repeal winery. Today wineries are constructed of stainless steel, glass, plastic, and oak. The

old problems of wine instability because of dissolved iron or copper from the equipment are no longer troublesome since stainless steel is inert. Dramatic improvements have been made in the equipment used for crushing, for pressing, for separating the juice from the mixture of skins, seeds, and pulp, and for the fermentation. Most of this new equipment provides for rapid processing of the grapes, for maintaining a blanket of inert gas (nitrogen or carbon dioxide) over the crushed grapes to prevent oxidation, and temperature control throughout all of the processing steps. Centrifuges, which can be used either before or after fermentation, permit the enologist to control the complexity of flavors in the wine. Finally, the development of membrane filters— filters with pore size small enough to strain out yeast cells and bacteria—permits preservation of low alcohol, slightly sweet wines in the bottle without the necessity of pasturizing to prevent refermentation.

While there can be no question that California wines are significantly better today than they were 35 years ago, the advances in technology making these improvements possible have been so rapid that it is equally certain that maximum qualities have not yet been achieved. Not only can we look forward to significant improvement in the current types of table wines but it seems highly likely that several new styles will be developed. For instance, the low-alcohol or soft wines is one group that seems likely to find a niche in the spectrum of standard types. These wines depend for palatability on a skillful balance of low levels of sugar, low concentrations of alcohol, and fruity flavors to make them attractive.

The grape and wine industry in California is a relatively young one when compared with that of Europe. It has benefited in many ways by not being bound by tradition but being receptive to the results of new ideas and experimental findings, many from the Department of Viticulture and Enology, University of California, Davis. While progress since the end of World War II has been spectacular, all signs point to an even more dramatic future.

Citrus

California is second only to Florida in the production of all citrus fruits in the United States—with the exception of lemons, of which California is the leading producer.

Citrus growing in California began with the establishment of the first mission at San Diego in 1769. Not until 1841 was the first commercial orchard, with an eye to profit, planted near what is now the center of downtown Los Angeles. The Gold Rush of 1849 stimulated more plantings and many small orchards were established from San Diego to Red Bluff. It was not until the 1880s, however, that a full-fledged industry developed as the result of the successful fruiting of the 'Washington' navel orange at Riverside, the selection of the 'Valencia' orange and 'Eureka' lemon, and the completion of the first transcontinental rail service.

From 1880 to the late 1940s, citrus in California grew steadily in acreage to a high, in 1946, of 330,000 acres. During this period, production was centered in southern California where 85 percent of the state's total citrus was produced. Orchards in the central and southern San Joaquin Valley and the Sacramento Valley ac-

counted for the balance of the production.

Following the end of World War II, urbanization, industrialization and the virus disease *tristeza* forced the removal of thousands of acres in what, up to that time, were the major citrus producing counties (Los Angeles and Orange) in southern California. By 1956, total citrus in the state had dropped to 230,000 acres, with most of the loss resulting from the removal of orange orchards.

To fill the need for greater orange production, a large planting expansion, starting in 1955, took place in the southern and central San Joaquin Valley and to a lesser extent in the outlying districts of southern California including the low desert areas. Current (1978) total citrus acreage in California stands at 287,000 acres of all varieties, with slightly less than half located in southern California counties. Central California has become the leading area for 'Navel' orange production with 82 percent of the state's acreage of that variety compared to 34 percent in 1948.

It appears probable that with the continued pressures of population expansion and urbanization there will be a continued decline of citrus acreage and production in the metropolitan areas of southern California in the foreseeable future. These areas include Los Angeles, Orange and parts of western Riverside and San Bernardino counties. Other southern California counties, Ventura, Santa Barbara, San Diego, Imperial and the desert areas of Riverside, will be less affected. Citrus acreage in the central California counties of Kern, Tulare, Fresno and Madera is projected to remain fairly stable and even perhaps to increase, provided the economics of costs and returns to the grower are favorable.

Citrus fruit in California is produced primarily for fresh consumption. Fruit not meeting fresh fruit standards and fruit in excess of market demand is diverted to products including juice, pectins, oils, feed supplements, and others.

An interesting aspect of citrus production in California, not found in other citrus-producing areas of the U.S., is the diversity of climates in which commercial production occurs. Because of this, the maturity, harvest and availability for shipment of most kinds of citrus is extended over many months. Thus, with two varieties of oranges, for example, grown in several climatic zones, fresh oranges are shipped twelve months of the year.

Five climatic areas of production are recog-

California oranges, destined primarily for the fresh market, can be shipped from somewhere in the state nearly year-round. *Photo courtesy of Sunkist Growers.*

nized. A coastal area is comprised of the southern California coastal counties of San Diego, Orange, Ventura, Santa Barbara and San Luis Obispo. The climate is generally equable with cool summers, mild winters and relatively high atmospheric humidity. An intermediate area includes western Riverside and San Bernardino counties. The climate of this area is characterized by hot dry summers, bright days and cool nights. A desert area is found in the low elevation Colorado desert valleys of Coachella, Palo Verde and Imperial. Extremes of summer heat and dryness characterize this area. The central California area lies in southern and central San Joaquin Valley and includes the foothill and adjacent terrace lands of Kern, Tulare, Fresno and Madera counties. Warm to hot, dry summers and cold tule-fog winters prevail. Spring and fall weather is usually mild but occasional hot spells occur. A relatively small northern California area includes parts of Butte and Glenn counties in the Sacramento Valley. As in the San Joaquin Valley, warm to hot summers and cold, foggy winters prevail. In general order of maturity, fruit from the desert is earliest, followed by central California, northern California, intermediate, and coastal areas. In areas of milder summer climate fruit can be stored on the tree for longer periods without deterioration, thereby

extending the season of harvest.

Oranges, lemons and grapefuit are the most important kinds of citrus grown in California. Mandarins (also called tangerines), mandarin hybrids, tangelos and tangors are of lesser importance. A small acreage of limes is grown in the warmest locations. The relative importance and distribution of each is shown in Table 1. Total value of production for all citrus was $425 million in 1979.

The 'Washington' navel is the principal orange variety in the state, leading all varieties in acreage planted. It is seedless and as a dessert fruit eaten fresh is unexcelled when fully mature. Harvest usually begins in early November in central and northern California and by mid-December in southern California intermediate valleys. By utilizing tree storage, harvesting continues until June. Navel oranges are of relatively minor importance in the coastal areas and the low desert valleys.

'Valencia,' once the leading orange variety, is utilized both for fresh market and products and yields well over a wide range of climatic areas. 'Valencias' are usually harvested from February to June in the desert; from April through August in central California; and from April to November in southern California intermediate and coastal areas.

Two lemon varieties make up California's production. 'Eureka,' grown primarily in the coastal areas, is nearly everbearing with harvest and shipments occurring during all months of the year. 'Lisbon,' slightly more frost tolerant than 'Eureka,' is suited to all lemon districts. Harvest in the desert areas takes place from October through January and in central California from November through April.

'Marsh,' a white-fleshed variety, and 'Redblush,' a pigmented or red-fleshed variety, are the two principal grapefruit varieties grown in California. Commercial production is limited to two areas where both varieties are grown: the low desert valleys where the fruit is harvested during the winter months, and the intermediate valleys of southern California where the fruit matures later in the following summers. Quality of fruit in the desert areas is usually superior and color development of the pigmented fruit better.

A number of varieties fall into a class of citrus often called the "exotic" fruits. These are mandarins and mandarin hybrids, some of which are referred to in the markets as tangerines. Principal

TABLE 7.1
Total Acres Planted to Citrus in California[1] (December 31, 1978)

County	Navel	Valencia	Lemons	Grape-fruit	Tangelos	Mandarins	Tangors	Limes	Other[2]	Total
Butte	268	1	3	1	—	53	—	—	—	326
Glenn	1,629	136	—	—	—	31	—	—	—	1,796
Madera	2,745	675	96	71	279	53	—	—	56	3,975
Fresno	15,815	3,363	1,044	6	375	163	—	—	212	20,978
Tulare	57,618	22,639	5,408	302	1,151	702	—	—	187	88,007
Kern	17,738	4,711	3,974	1,972	793	285	—	—	38	29,511
San Luis Obispo	—	18	876	—	—	—	—	—	6	900
Santa Barbara	10	48	3,288	—	—	—	—	—	33	3,379
Ventura	1,542	16,040	26,635	2,444	1	6	—	—	67	46,735
Los Angeles	277	503	910	46	2	12	—	—	5	1,755
Orange	83	6,112	901	319	—	5	—	—	19	7,439
Riverside	9,818	8,460	8,481	16,140	678	2,575	877	75	35	47,139
San Bernardino	5,678	3,914	1,428	1,530	143	51	—	—	118	12,862
San Diego	1,306	9,194	4,010	1,611	929	152	—	400	15	17,617
Imperial	18	588	2,312	768	111	258	—	—	71	4,126
Other Counties[3]	253	7	40	10	9	98	40	120	—	577
	114,798	76,409	59,406	25,220	4,471	4,444	917	595	862	287,122

[1] Source: Adapted from California Crop and Livestock Reporting Service, Fruit-Nut Acreage Report—June 1979.
[2] Miscellaneous varieties.
[3] Total acreage of counties with less than 100 acres each.

varieties include the 'Clementine' and 'Dancy' tangerines; the 'Kinnow,' 'Kara,' 'Fairchild,' and 'Satsuma' mandarins; the 'Minneola' and 'Orlando' tangelos; and the 'Temple' tangor. The harvest season for these fruits is from November through April. Some of these are grown in all districts but the greatest acreage of mandarins is in the low desert valleys.

'Bearss' is the only lime variety grown commercially in California. More frost sensitive than the lemon, it is grown only on limited acreage in the most frost-free areas. The fruit is harvested in several picks during the year but heaviest production of the crop occurs in the late fall and winter, a time when demand and prices are generally lower.

Frost protection is a necessary practice particularly in the cold winter areas. Oil-burning heaters and wind machines are becoming economically unfeasible because of energy costs. The use of running water in furrows or through undertree sprinklers provides some protection from frosts and is gaining in use. Other low-energy methods are under study.

With the exception of lemons, citrus does not receive the annual pruning required by deciduous fruit trees to produce good crops of marketable fruit. Occasional topping to reduce height and make harvesting easier, and hedging to reduce crowding, are practiced in orange and grapefruit orchards. The vigorous upright growth of lemon trees, however, makes pruning of this variety an annual necessity to control tree height and enhance fruit quality and size.

Commercial citrus in California is hand-harvested, generally by picking crews under the direction of the packinghouse with whom the grower has affiliated. Both cooperative and independent organizations pack and sell the fruit.

Subtropicals Other Than Citrus

Avocados

California is the leading producer of avocados in the United States with production five times that of Florida, and is second only to Mexico worldwide.

The first recorded introduction of avocados to California was a tree imported from Nicaragua in 1856. It was not until about 1915, however, that a commercial industry based on subsequent introductions from Mexico and Central America was started.

From a small beginning of a few hundred acres, the industry grew to about 20,000 acres by 1950. For the next twenty years acreage remained relatively stable because of depressed markets. In the early 1970s acreage again started to expand due to better grower returns as a result of improved marketing procedures and an extensive trade promotion and advertising program for California avocados. By the end of 1978 acreage had more than doubled to 51,000 acres with a production of nearly 250 million pounds of fruit valued at $85 million. The rate of further planting and continued industry expansion will depend on maintenance of satisfactory grower returns from increasingly larger crops as newly-planted acreage reaches full production.

Because avocados are a frost-sensitive subtropical crop, production is limited to areas of low frost hazard in the coastal counties of southern California. San Diego County, with nearly half the state acreage, is the major producing county followed by, in descending order of acreage, Ventura, western Riverside, Santa Barbara, Orange, Los Angeles and San Luis Obispo. A small but expanding area is developing in the thermal belt areas of Tulare and Fresno counties where the more cold-hardy varieties, maturing in the fall months, can be produced.

Of the hundreds of named avocado varieties, only a few have stood the test of satisfactory production and consumer acceptance. Five varieties are currently recommended for commercial planting: 'Hass,' the leading variety in acreage and production, is harvested from spring to fall, but is frost-sensitive; 'Fuerte,' once the leading variety but an erratic producer, is harvested in the winter and spring and has moderate frost tolerance; 'Bacon,' and 'Zutano,' both more cold-hardy, are harvested in the fall; and 'Reed,' a summer-harvested, frost-sensitive variety. With this selection of varieties and production from different climatic areas, fruit is available for market throughout the year.

The major and long standing cultural problem facing the industry is the susceptibility of avocado trees to a soil-borne fungus disease, *Phytophthora cinnamomi*, which can be fatal to the tree. No certain cure is yet known. Research has developed rootstocks which have some tolerance but following practices to avoid the disease is the best method.

Dates

California is virtually the sole producer of dates in the United States. Within the state, production is confined entirely to the low-elevation desert valleys where the requirement is met for prolonged summer heat with little probability of rain or high humidity during the ripening period. Coachella Valley is the leading production area with 90 percent of the acreage.

The 'Deglet Noor,' a semi-dry date from Algeria, is the leading commercial variety in California, accounting for 76 percent of the acreage in 1978. The 'Medjhool,' a soft date prized for its confection-like consistency, accounts for about 13 percent. The balance is made up of other varieties currently of lesser importance. Total crop value in 1980 was $13.6 million.

The industry began with the importation of offshoots of selected varieties from Algeria, Iraq and Egypt in the early 1900s. By 1950 a little over 5,200 acres were planted to date gardens. Over the last thirty years acreage has declined by about 800 acres, due in part to low returns in the earlier portion of this period and to a lack of available labor willing to perform the difficult tasks of pollination, bagging and harvesting in the extremely high trees.

In recent years, market conditions have improved because of a reduction of the world supply resulting from date palm disease in Arab countries. Man-positioning equipment has been developed to ease the arduous cultural tasks. It is expected that these factors will lead to some expansion of California gardens.

Rapid expansion of date gardens has not been possible because of a shortage of sufficient offshoots used to establish new gardens of specific varieties. Recently developed methods of cloning specific varieties through tissue culture techniques, however, now provide an ample supply of planting material.

Figs

From 1950 to 1980 the total acreage of figs in California decreased gradually about 40 percent to 16,000, over 4,000 of which are currently non-bearing. Decreases of 62, 74 and 80 percent occurred for the 'Mission,' 'Adriatic,' and 'Kadota' varieties, respectively, while the most popular 'Calimyrna' ('Smyrna') variety decreased only 20 percent. With decreased acreage, production dropped from 30 to 12 thousand tons (dry basis), which still accounts for almost all that is produced in the United States. Although per capita consumption of figs in this country also decreased during the past 30 years from .35 to .18 pounds, imports, particularly of fig paste, have increased from less than one to almost seven thousand tons annually. A five-fold increase in price per ton within the past ten years has renewed grower interest and encouraged new plantings. Practically all of these have been of the 'Calimyrna' variety. The 'Conadria' variety, developed and released by the University of California in 1956, constitutes less than 4 percent of the total fig acreage.

The dried-fruit beetle, along with several other species of nitidulids, and several species of *Drosophila* (vinegar fly) are of major concern to the fig grower. They not only infest the fruit themselves, but also spread various spoilage organisms (yeasts, bacteria, fungi) to the fruits, causing enormous crop losses. Considerable research has been conducted on these insects, but no single control measure has eliminated or even greatly reduced the problem. Most new fig plantings have been made in Madera and Kern counties in somewhat isolated areas away from peaches, tomatoes, melons and oranges, fruits which ferment during deterioration and provide conditions conducive to build-up of large insect populations.

Probably the greatest boon to the fig industry during the past 30 years was the discovery in research at the University of California that ethephon applications, particularly to 'Calimyrna' figs, speed maturation of the fruit. The use of this ethylene-generating compound markedly increases the percentage of marketable fruits and reduces the usual number of pickings from three to one or two.

Olives

Olive growing started in California about 1770 with the planting of olive seeds at the early California missions by the padres who brought seeds with them from Peru and Mexico where olives were introduced earlier from Spain.

Currently, California produces almost 100 percent of the olives grown in the U.S. but accounts for less than 1 percent of the total world production. Spain, Italy, and Greece are the principal sources of olive oil and table olives, together producing about 65 percent of the world's total. Traditionally the California olive industry has been unable to compete economically with the Mediterranean countries in the production of olive oil and Spanish-green fermented olives, but this situation may change as labor costs in the Mediterranean area increase.

Tulare County leads in olive production in California, followed by Kern, Tehama, Butte, Glenn, Fresno, Madera, and Kings Counties. By 1977, California had almost 43,000 acres planted to olives, an increase of almost 13,000 acres from ten years before. Most of this increase was in large agricultural developments on the west side of the San Joaquin Valley. 'Manzanillo' is the principal olive variety grown in California, followed by 'Sevillano,' 'Mission,' and 'Ascolano.' All these produce good-sized fruits usable for table olives but 'Mission' also has a high enough oil content so that it is suitable for oil extraction.

About 75 percent of the olives produced in California are utilized for black-ripe and green-ripe table olives. The remainder is used in the production of olive oil, Spanish-green table olives, and minor products, such as chopped and sliced olives.

The climate of California's great Central Valley is uniquely suited to olive production. The winters are not cold enough to kill the trees, yet there is enough winter chilling to induce flowering. The long, hot and dry summer season allows the fruit to mature properly. The flowers, which are wind-pollinated, open about mid-May. Table olive harvest occurs mostly in October with harvest for olive oil extraction in mid-winter when the fruits become black and reach their maximum oil content.

For table olives the fruits are picked by hand when they have changed from a green to straw color. Raw olive fruits are inedible due to a bitter glucoside. In the California black-ripe and green-ripe process the bitterness is neutralized by treatments with sodium hydroxide (lye) which is subsequently leached out with water. If the fruits remain submersed under the solution throughout the lye and washing treatments they remain green and become the green-ripe olives. If they are exposed to air during these treatments they are oxidized, turning black, and become black-ripe olives. The fruits are subsequently sealed in brine in cans, then retorted at 240°F (115°C) for one hour to kill any harmful micro-organisms.

Olives are relatively free of insect and disease pests. Black scale is the worst insect problem. A bacterial disease, olive knot, and two fungus diseases, peacock spot and verticillium wilt, are the chief disease problems.

Berry Crops

Strawberries

California leads the nation in strawberry production, a position resulting from a climate fortuitously suitable for strawberries and varieties that are especially well adapted to it. The area in berries has now stabilized at about 12,000 acres, more than double that of the late 1940s and somewhat less than its peak a decade or so ago. The average yield over the last thirty years has increased from about 6 to 20 or more tons per acre. Such yields, which are four to eight times larger than in most other states where strawberries are grown, result from a fruiting season that lasts from four to six months for most varieties and up to eight or nine months for a few. The most productive plantings are near the coast, principally in Monterey, Santa Cruz and Santa Clara counties along the central coast, in Ventura, Orange, San Diego and Los Angeles counties on the south coast, and in between in Santa Barbara County near Santa Maria.

Even though strawberry plants are perennials, growers replant annually to obtain maximum yields and best quality of fruit. Strawberry plant nurseries are found in the northern Sacramento Valley at lower elevations in Shasta and Tehama counties, supplying mostly growers in the central coast area, and at high elevations (3,000 ft.) near McArthur in Shasta County, supplying growers in southern California. A few low elevation nursery plants are also grown in Stanislaus County.

Many factors have contributed to the growth of the California strawberry industry, but the primary one has been the development, through an industry-supported fruit breeding prgram at the University of California, of high-yielding varieties with good, dual-purpose fruit, suitable for both marketing fresh and processing. The first important California-bred varieties were 'Lassen' and 'Shasta,' released in 1945. Next came 'Fresno' in 1961 and 'Tioga' in 1964. The increased planting of high–yielding 'Tioga' accounts for much of the 4 tons-per-acre increase in average yields in the last decade. 'Shasta' dominated the central coast region after its introduction in 1945, but was replaced by 'Tioga' which has dominated strawberry production throughout California during the past decade or so, occupying an average of almost 50 percent of the state's strawberry acreage and reaching a zenith of 63 percent in 1974. 'Tioga' is now on the decline, however, down to about 26 percent of the 1980 acreage, and may be reduced to minor status within a few years.

In southern California, 'Tufts,' released in 1972, is outperforming 'Tioga' consistently in winter plantings which predominate there. Among other things, the 'Tufts' advantage was apparent particularly during two very wet harvest seasons in southern California, in 1978 after a very warm winter and in 1979 after a very cold one. New short-day varieties were introduced during 1979; 'Douglas' in particular shows promise with respect to earliness, larger fruit size, sustained production of quality fruit, flexibility in planting schedules, and much less runner production than 'Tufts' or 'Tioga' in fruiting plantings.

'Aiko,' released in 1975, has outperformed 'Tioga' over the past several years in the central coast district and is rapidly replacing it there. The chief advantage of 'Aiko' is its ability to sustain the production of quality fruit throughout the summer, with harvest continuing in most years until the heavy rains start, perhaps as late as Christmas. 'Aiko' plants are compact and the fruit is easier and cheaper to harvest. 'Pajaro,' introduced in 1979, also shows promise of competing favorably in the central coast areas. Its fruit is especially large and attractive.

The gains associated with the new varieties would not have been realized without research at the University of California which developed improved cultural systems designed specifically to exploit them. Practices included in these systems are: soil fumigation, annual planting, drip irrigation, improved fertilizer placement, alterations

Strawberries grow in coastal areas from February through November. Large numbers of hand pickers are still needed. *Photo by Mel Gagnon.*

in bed size and shape and associated changes in planting configurations to accommodate greater plant density. Such modifications, coupled with older innovations, have improved the efficiency of strawberry culture tremendously.

All strawberry varieties previously introduced by the University of California have been short-day types, which normally initiate flower buds only after exposure to short days (long nights). The expanded fruiting season under California conditions for short-day varieties, such as 'Tioga,' 'Tufts,' 'Aiko,' 'Douglas,' and 'Pajaro,' is due in part to their inherently low winter-chilling requirement, which keeps them from going dormant and in effect makes them nearly "everbearing." Equally important is the physiological conditioning of the plants that results from a combination of the cool growing temperatures in coastal California and control of the amount of chilling that the nursery plants receive prior to planting. However, the planting dates for the short-day types, because of their response to day length are, in effect, set by the calendar and may be varied to a limited extent only. In the south coast area, winter plantings, made about November 1 with fresh-dug plants from high-elevation nurseries near McArthur, begin to fruit in three months. Along the central coast, summer plantings, made about September 1 with stored plants from low-elevation nurseries, begin to bear in seven months.

Since the day-neutral varieties are free of day length constraints, they may, theoretically, be planted any time during the year. Already, it is obvious that the possibility of using them to shorten the time from planting to first harvest by as much as three months is good. The resulting economic advantage is very attractive. There is no intrinsic reason why the fruit quality of day-neutral varieties cannot be brought to complete equally with that of the best short-day types.

Caneberries

During the past decade the California 'Boysenberry' acreage has fluctuated around an average of 420 acres, most of which is located in the San Joaquin Valley in two districts, one in the counties of Fresno and Tulare and the other in San Joaquin and Stanislaus counties. The yield has fluctuated about an average of 4.5 tons per acre, with a high of 5.5 to 5.6 tons in 1973–74 and a low of 2.9 tons in 1979. Yield appears to be associated with variations in weather conditions.

The most important factor limiting the popularity of 'Boysenberry' with growers is the high cost of production, mostly hand-harvesting costs. It is unlikely that the acreage will increase substantially. Indeed it may decline, because alternatives to hand harvest are not promising. The harvest problem is with the strong tendency of the calyx (the leaf-like cap at the base of the fruit) to adhere to the receptacle (the structure to which the berry is attached) when the fruit is removed, frustrating mechanical shake-and-catch methods. The same problem complicates and adds to the cost of hand harvest as well. The troublesome trait is probably a legacy of the putative raspberry parentage of 'Boysenberry.'

In recent years the 'Olallie' blackberry acreage has declined significantly below the 10-year average of about 400 acres in Santa Cruz County where most of the variety is grown. During the decade the yield has fluctuated only moderately about an average of 4.3 tons per acre. The decline in 'Olallie' acreage appears to be due to the relentless pressure of urbanization in the Pajaro Valley and surrounding areas rather than to any problems associated with growing the crop. There may be further decline in acreage if the pressure continues. The 'Olallie' is satisfactory for the shake-and-catch type of mechanical harvesting.

Red raspberries have never been a very important crop in California. Over the last decade the acreage of these berries has fluctuated between a mere 90 to 120 acres, mostly in Santa Cruz and Sonoma counties. Raspberries are grown in California primarily for the fall crop, since there is no competitive advantage in producing the normal, early crop, so important in Oregon and Washington. Yield has increased significantly in recent years with a high of 4.7 tons per acre in 1978. The increase is probably due to favorable growing conditions, coupled with good cultural practices. The California acreage in red raspberries is likely to continue at about the present modest level, all that is warranted under the present and foreseeable market and competitive situations.

Exotic and Minor Crops

Kiwifruit

A new fruit industry began in California in 1967 with planting of 25 acres of a vine called kiwi (*Actinidia chinensis*), which though imported here from New Zealand is a native of the mountain areas of south China. Commercial production began in New Zealand in the 1950s, following selection of large-fruited types by local nurserymen two decades earlier. Most of the first kiwi plants planted in California were imported from New Zealand, except for a few hundred that were raised in Gridley and Chico as seedlings and grafted with scions from a single female plant at the USDA Plant Introduction Station at Chico.

During the 1970s, an organization known as the Kiwi Growers of Califonia did much through education to assist new growers in learning to grow and pack kiwifruit. Early in the decade, a Chico nursery promoting the growing of kiwi encouraged many people who had never farmed fruit before to plant the vines. This promotion resulted in plantings alog the coast of California from north of San Francisco south to San Diego and throughout the San Joaquin and Sacramento Valleys. The industry became more soundly established in the state when larger plants became available from other nurseries and more agriculturally-oriented people planted the vines. The first commercially-produced kiwifruit were sold in California in 1972. The area in vines rose from 300 acres in that year to nearly 3,000 acres in 1980. Production of kiwi increased rapidly from 1976 and by 1980 had become a $10–million industry, exporting fruit to Japan and Europe. To direct and promote orderly marketing of fruit, the Kiwifruit Commission was established in 1980.

Commercial plantings, ranging in size from 2 to 40 acres, are located mainly in Butte, Fresno, Tulare, San Diego, San Luis Obispo and Kern counties with some in several other counties in the Central Valley. The kiwi is dioecious, so male as well as female vines, are needed, planted usually at a ratio of 1 male to 8 females, with normally a total of 140 vines per acre. Pollination is by bees and other insects. The plants require moderate winter chilling, can tolerate cool coastal or hot valley temperatures in summer and grow

A relatively new crop gaining in popularity, kiwifruit grows in clusters on tall trellised vines.

well in most soils suitable for orchards. The main variety is 'Hayward.'

Kiwifruit vineyards are trained on six-foot high trellises called T-bars or onto arbors. Besides a strong trellis to support them, the large, vigorously-growing vines require frequent irrigation, normal fertilization, considerable pruning and training of canes, and protection from wind which easily breaks canes and scars fruit. The cost of growing kiwifruit is greater than for most tree-fruit crops, because of the high investment required for plants, trellis and irrigation systems, and hand labor to train the vines. Although kiwi is attacked by a few insects, nematodes and diseases, pest management has not yet become a major problem for the industry.

The fruit, which is high in vitamin C (higher than citrus), is usually eaten fresh, but also is sometimes frozen, canned or made into preserves.

Persimmons

The first imported trees of the Oriental (Japanese) persimmon were planted in California over 100 years ago. Today, slightly more than 500 acres of them are being grown in many areas of the state, mostly in the San Joaquin and Sacramento Valleys and mainly in Fresno and Tulare counties. The principal variety is 'Hachiya,' an astringent, large, acorn-shaped fruit that must be allowed to soften and lose astringency to be eaten. Another variety, 'Fuyu,' is flat-shaped and, since it is non-astringent, it can be eaten while still firm.

The persimmon is relatively free from pests and diseases. Its branches are brittle, however, and break easily, so proper tree training and pruning are important.

Depending on the variety, harvesting begins in late September and continues until early December. Most of California's crop is shipped to local markets, though an increasing portion is now going to the rest of the United States and to Canada. The fruit is used for decoration and for eating as fresh fruit. The pulp is used in cookies, bread, cakes, ice cream, preserves, and for drying.

Pomegranates

The Spanish brought the pomegranate to the New World from the Mediterranean area over 400 years ago and established it in California in the mission gardens nearly 200 years later. To-day, there are over 3,000 acres of pomegranates grown commercially in California's interior valleys, mostly in the San Joaquin Valley where high summer temperatures and low humidity allow the fruit to mature properly with good outside color. The fruit is marketed throughout the United States and some is exported. It is eaten fresh and widely used for decorative purposes. The juice is used for jelly, grenadine, and wine.

The principal variety is 'Wonderful,' followed in importance by 'Granada,' 'Flamingo,' and 'Early Foothill.' The trees bloom from late April into July. Only the fruit that sets during the first six weeks of the bloom period eventually grows to marketable size. Fruit thinning does not increase the harvest size of the fruit. The harvest season is August to October. Average production ranges from 200 to 600 packed boxes per acre. The fruit can be stored for a period of 30 to 60 days in cold storage prior to shipment, but its quality and appearance is best if it is packed and shipped as soon as possible after it matures.

Reference Material on Fruit and Nut Crops

Information on growing and handling fruit and nut crops in California is available in a series of leaflets published by The Division of Agricultural Sciences, University of California, and distributed through the offices of the Cooperative Extension Service throughout the state. In addition, the Division of Agricultural Sciences is also the publisher of *The Citrus Industry*, four volumes published one each in 1962, 1968, 1973 and 1978.

Two publications from the California Crop and Livestock Reporting Service, *California Fruit and Nut Statistics* and *California Fruit and Nut Acreage*, annually provide up-to-date information on the various crops.

There is no one reference on the deciduous fruit and nut crops that is strictly applicable to California. There are, however, several text and reference books on fruit growing that can be helpful. As an example, one of the more recent ones and one which also happens to be relevant to fruit growing in the states along the west coast is *Temperate-Zone Pomology* by W.N. Westwood, W.H. Freeman and Co., 1978.

1 TULELAKE-BUTTE VALLEY • Onions, potatoes

2 SACRAMENTO VALLEY • Honeydews, persians, other melons, tomatoes, watermelons

3 DELTA • Asparagus, sweet corn, onions, potatoes, tomatoes

4 BRENTWOOD-TRACY • Sweet corn, lettuce, tomatoes

5 SANTA CRUZ-SAN MATEO COAST • Artichokes, brussel sprouts, broccoli, cauliflower, peas

6 FREMONT-SAN JOSE • Broccoli, cauliflower, celery, sweet corn, garlic, lettuce, onions, peas, peppers, strawberries, tomatoes

7 PATTERSON-NEWMAN • Broccoli, cantaloupes, cauliflower, honeydews, persians, other melons, sweet corn, lettuce, tomatoes, peppers

8 MODESTO-TURLOCK • Carrots, honeydews, other melons, strawberries, sweet potatoes, tomatoes, watermelons

9 SALINAS-WATSONVILLE • Artichokes, snap beans, broccoli, cabbage, carrots, cauliflower, celery, garlic, lettuce, onions, peas, spinach, potatoes, strawberries, tomatoes

10 GILROY-HOLLISTER • Sweet corn, garlic, lettuce, onions, potatoes, peppers, peas, tomatoes

11 WEST SIDE • Cantaloupes, honeydews, lettuce, persians, other melons, onions, tomatoes

12 MERCED-ATWATER • Peppers, sweet potatoes, tomatoes, watermelons

13 KINGSBURY-DINUBA • Sweet potatoes, watermelons

14 CUTLER-OROSI • Tomatoes, other spring vegetables

15 KERN-TULARE • Sweet corn, cantaloupes, carrots, garlic, honeydews, lettuce, onions, peas, potatoes, sweet potatoes, watermelons

16 SANTA MARIA-OCEANO • Artichokes, snap beans, broccoli, cabbage, carrots, cauliflower, celery, lettuce, peas, potatoes, strawberries

17 OXNARD • Broccoli, cabbage, carrots, cauliflower, celery, cucumbers, lettuce, spinach, strawberries, tomatoes

18 ANTELOPE VALLEY • Cantaloupes, onions

19 LOS ANGELES-ORANGE COUNTY • Asparagus, snap beans, cabbage, carrots, cauliflower, celery, sweet corn, lettuce, peppers, strawberries, tomatoes

20 CHINO-ONTARIO • Sweet corn, onions, sweet potatoes

21 PERRIS-HEMET • Cantaloupes, other melons, carrots, onions, potatoes, watermelons

22 OCEANSIDE-SAN LUIS REY • Snap beans, cabbage, lettuce, peppers, strawberries, sweet potatoes, tomatoes

23 COACHELLA VALLEY • Asparagus, snap beans, carrots, sweet corn, cantaloupes, onions, peppers, tomatoes, watermelons

24 BLYTHE • Sweet corn, cantaloupes, honeydews, lettuce, other melons, onions

25 CHULA VISTA • Snap beans, cabbage, celery, cucumbers, lettuce, strawberries, peppers, tomatoes

26 IMPERIAL VALLEY • Asparagus, broccoli, cabbage, cantaloupes, carrots, cucumbers, garlic, lettuce, onions, tomatoes, watermelons

FIGURE 8.1 California's major fresh market vegetable, melon and potato producing districts. *Source: California Crop and Livestock Reporting Service.*

8

Vegetable Crops

Vegetable Specialists and Farm Advisors of Cooperative Extension

Robert Brendler, celery, cabbage and cauliflower; *Hunter Johnson*, sweet corn and onions; *Oscar Lorenz*, potatoes; *Keith Mayberry*, lettuce and carrots; *Norman McCalley*, spinach; *Robert Scheuerman*, sweet potatoes; *Henry Sciaroni*, Brussels sprouts; *W.L. Sims*, asparagus, peppers, and tomatoes; *Marvin Snyder*, artichokes and broccoli; and *Kent Tyler*, melons. Introduction by *O. Lorenz.*

California leads the nation in vegetable production because of its many favorable growing areas and climates, ranging from the hot interior valleys to the moderate or cool coastal climates. Many of the warm season crops can successfully be grown much of the year, their production moving with the seasons to varying locations. For several important crops, such as lettuce, broccoli, celery, cabbage, potatoes and others, there is no such thing as "out of season," since they are being harvested somewhere in California every month of the year.

Vegetables are grown in a number of geographic areas, extending from Tulelake at the Oregon border to Imperial Valley bordering Mexico.

The earliest production of vegetables is from the Imperial Valley, where the weather is warm enough during the winter months for many cool-season crops to be harvested from November through April, and warm-season crops from about May to July. Summer temperatures, however, are too hot for vegetable production.

Salinas Valley, the most important growing area in California, is known as the "salad bowl" of the United States. The cool coastal climate during the summer and moderate temperatures during the winter account for the importance of this area. Cool-season crops such as lettuce, broccoli, and celery mature from about April through November, although many crops are harvested throughout the year. The interior por-

tion of the Salinas Valley, somewhat warmer during the summer, produces warm-season crops such as tomatoes, cucumbers, and beans. The Santa Maria Valley, similar to the Salinas Valley, is geographically smaller. The South Coast production districts have moderate summer and winter climates and provide excellent climatic conditions for all vegetables.

The San Joaquin and Sacramento Valleys are favorable for summer production of warm-season crops such as melons, peppers, and tomatoes. Some spring and fall production of cool-season crops also occur. Tulelake at the Oregon border grows chiefly potatoes and onions for an early fall harvest.

In 1979 California accounted for 45 percent of the production of the nation's 22 fresh market vegetables and 53 percent of the nine major processing crops. In that year California produced 13.3 million tons of vegetables from 913,000 acres, the value of total production reaching an all-time high of over $2.2 billion. On a value basis, California produced 60 percent or more of all U.S. artichokes, broccoli, Brussels sprouts, cauliflower, celery, garlic, lettuce, melons, spinach, and tomatoes. California produces more than 20 different vegetables each having values greater than $1 million. Vegetables represent about two-thirds of all produce shipped in railway refrigerator cars. Table 1 lists the leading vegetable crops in overall value for 1980.

Vegetable production in California was signifi-

TABLE 8.1
Leading California Vegetable Crops, 1980

Vegetables	% of U.S. Production	Acres	Value of Production
		Thousands	*Millions*
Lettuce	74	158.8	$382
Processing tomatoes	89	208.3	327
Fresh market tomatoes	29	30.9	163
Potatoes	6	50.5	157
Broccoli	95	73.6	140
Celery	62	20.5	95

cant even at the time of the Civil War. Before the use of refrigerator railroad cars in the late 1800s, production was almost entirely for local consumption and was centered in market gardens near large cities, especially Los Angeles and San Francisco. In 1900 California grew about 80,000 acres of vegetables. This increased to about 200,000 acres in 1920, 530,000 in 1930, 550,000 in 1940, 600,000 in 1950 and 720,000 in 1960 and 1970. Production has averaged about 900,000 acres since 1975, the increase from 1970 to 1980 being due largely to processing tomatoes and broccoli.

For practically all vegetable crops, yields per acre have increased steadily since about 1940. New varieties, improved cultural practices and more timely harvests are primarily responsible for these increases. Although vegetable production has been characterized by high use of hand labor in cultivation and particularly in harvesting, there is a gradual trend toward mechanization of some labor intensive tasks. It is probable that mechanical harvesting of crops like broccoli, cauliflower, and lettuce will initially result in lower yields per acre, although production costs per unit will be decreased. Practically all of the harvesters to date have been used for single harvest, which presents difficulty with crops that do not mature uniformly. High labor costs may limit the production of some labor-intensive crops such as melons. Many of the high value fresh-market vegetable crops are not adapted to harvest mechanization, although practically all of the processing crops are now harvested mechanically.

High and increasing transportation costs may limit California production of some fresh market crops, especially during periods of marketing by competing eastern states closer to the principal consumer markets.

The U.S. per capita consumption of vegetables has shown slight increases during the past two decades. Leafy green and salad vegetables, including tomatoes, have shown the greatest increases, due to emphasis on these crops by dieticians and increased public interest in low calorie, high roughage nutrition. The proliferation of salad bars associated with many restaurants and fast-food outlets since about 1970 illustrates a notable tendency toward less cooking of fresh vegetables with a larger share being consumed raw.

Artichokes

The south-central coastal area of California, particularly the Monterey area, is one of the best in the world for artichoke production.

Artichokes were introduced into the United States as early as 1806 but did not become popular until this century. Commercial artichoke production in California began in the Half Moon Bay area at the start of the century. Generally production extends from this point south along the coast to San Luis Obispo and Santa Barbara counties. Other Pacific coast areas as far north as Oregon and Washington have attempted commercial production but their climate is too cool and the limited availability of suitable land, transportation, and farmers familiar with the crop were other factors.

Monterey County is the major growing area with other important areas being San Mateo, Santa Barbara, San Luis Obispo, and Santa Cruz counties. The total acreage in California is approximately 10,000 acres with Monterey alone producing close to 8,000 acres. The remaining acreage is about equally divided among the other counties. In these coastal areas year-round production is possible. The most favorable areas are frost free during the winter, and cool and foggy during the summer. The yields per acre of artichokes have recently averaged around 7,000 pounds.

The consumption of artichokes for the United States has been relatively stable for many years, at approximately 0.3 lbs. per person. Artichokes are marketed mainly as fresh market produce, but approximately 25 percent of the total production is processed as a marinated canned product or by quick freezing. The high demand in cities such as Los Angeles, San Francisco, New York, Boston, Philadelphia and New Orleans is partially attributed to their ethnic Italian and French populations.

Artichokes do best in deep, fertile, well-drained sandy loam to clay loam soils. These, however, are secondary to climatic conditions with regard to the location of plantings. California producers establish new plantings through vegetative propagation. All planting is done by hand with root sections or plant shoots being placed at intervals of 4 to 6 feet in rows 8 to 9 feet apart.

Artichokes are perennials, and through a cultural practice called cutting back or stumping almost continual production can be obtained. Maximum artichoke production occurs during the late spring but summer production is obtained from the cooler coastal areas. Artichoke plants are maintained for 5 to 10 years and occasionally longer.

Of all the production problems that artichoke growers face, the plume moth is the most significant. Economic losses caused by this insect are always significant and at times disastrous, with crop losses often exceeding well over 50 percent. This pest, native to the United States, appears to be limited to this area. Pesticides often do not give an immediately noticeable control. The principal artichoke disease is greymold or Botrytis rot, a fungus disease favored by periods of high humidity or rain and moderate temperature.

Intensive demand for labor occurs during harvesting and packing of artichokes. The harvest period generally begins in September and continues until mid-May; frosts can halt harvesting. Artichokes are harvested weekly, although this interval is extended during cold weather when plant growth is slowed. Suitability of buds for harvest is determined by their compactness and size. Production of large sizes is stressed because these are more readily marketable. Three to five

buds are allowed to develop sequentially on a single flower stalk and are hand-harvested with small knives. A 2 to 3 inch stem is left on each bud. The buds are hauled to packing sheds or to processing plants, where they are sorted, graded, and packed.

A marketing program established in 1960 promotes the sale of artichokes through market development and advertising and provides research support for cultural studies and for improvements in processing and distribution. The program is administered by a grower-elected advisory board.

Asparagus

Asparagus is an important vegetable crop in California and ranks 10th in value among the vegetables produced. With approximately 35 percent of the total U.S. acreage, California presently provides about 50 percent of the U.S. production and 70 percent of the fresh market asparagus. Asparagus acreage in California has continued to decline, however. In 1974 California had 44,000 acres of asparagus. Overall production was 128 million pounds of marketable asparagus valued at more than $36 million. Fresh green asparagus accounted for 61 million pounds and 67 million were processed. Approximately 90 percent of the processed asparagus was grown in the Sacramento-San Joaquin Delta area. By 1979 asparagus acreage in California had decreased to 26,400 harvested acres and a production of 100 million pounds. Sixty percent of the acreage is now grown outside the Delta. The principal commercial growing areas today are the Sacramento, San Joaquin, Salinas, Imperial, and Coachella valleys and Orange County. Yields per acre, however, have increased to an average of 4,000 pounds per acre. Fusarium disease, labor, and economic problems have all contributed to the decline in acreage over the past 15 years. During the early 1960s the average annual production of white asparagus in California was approximately 60 million pounds. White asparagus production has been low in California since 1970, however, due to competition from several foreign countries where labor is much cheaper.

The principal asparagus variety grown in California is 'UC72.' Other varieties grown in limited acreages are 'UC66,' 'UC309,' 'UC500W,' 'UC711,' and 'Brock's Special.' All UC lines are

After field harvesting, asparagus is trimmed and packed in sheds like this one. California accounts for about half of U.S. asparagus production.

Asparagus is comparatively free from insect pests and most diseases and does not need the highly specialized maintenance required by most cultivated vegetables. Establishing a commercial plantation requires considerable time, labor, and expense, however, and the grower receives no income until the first harvest two years after planting. A good supply of labor is a critical requirement in asparagus harvesting as mechanical harvesters are not yet widely used.

Asparagus is harvested as early as spears emerge in the spring. Cutters go over the field and harvest by hand any spears of sufficient length. Fresh market spears are cut 10 to 12 inches long. Marketable spears must have compact heads and good green color. Each production area in the state has an optimal harvest period, determined by the length of the growing season and by the requirement of a sufficient growth period after each annual harvest for plants to synthesize and store sugar reserves for the next season.

Fresh market asparagus is usually graded and packed by the growers and marketed under their own or a cooperative label. The packed crates are sent immediately through a cooling system, usually a water bath maintained at near freezing temperatures, and then stacked in a cold storage chamber until shipment to market. Asparagus spears delivered to processors are trimmed to a 7-inch length and must have a minimum of 3½ inches of green color at the tip.

related to the 'Mary Washington' variety, which was introduced into California in the late 1800s or early 1900s. The varieties 'UC72' and 'UC66' developed by the University of California possess a high degree of tolerance to the Fusarium diseases which had infested some old asparagus plantations. 'Brock's Special' was developed in the Imperial Valley for the growing conditions characteristics of that hot desert area. In 1975, 'UC157,' a new hybrid line was released by the University of California for commercial use. It has shown earliness in production, multispear initiation, and greater total production than present commercial lines. New plantings are being established principally with this new hybrid, by the seedling transplant method rather than by the older crown planting or direct seeding methods.

Broccoli

Broccoli is a very important segment of vegetable production in Monterey, Santa Barbara, and San Luis Obispo counties. Although long considered to be a cool season crop, it has become apparent that it can be grown in other areas of California. Hybrid varieties have been developed that adapt to much broader climatic conditions and acreage has increased in the last few years in the San Joaquin and the Imperial valleys.

In 1980 California produced approximately 96 percent of all the broccoli in the United States, although this represented only 5 percent of the total value of vegetable production in California. Broccoli production continues to increase in California as the vegetable has become more popular with U.S. consumers. In 1980 the crop value was over $140 million, ranking it in 5th place among the state's most valuable vegetable crops.

The Salinas Valley continues to be the leading production area with 36,500 acres, or over 50 percent of the total acreage. Next is the Santa Maria Valley with 30 percent (20,550 acres). Coastal areas are capable of producing broccoli year-round, while an area like Imperial County produces only in the winter.

The hybrid 'Green Duke' is the most popular variety grown. Open-pollinated varieties such as 'Medium Late 423' and 'Medium Late 145' are planted in the fall for winter production in the coastal areas.

The temperature range for growth of broccoli is from 40° to 80° F, with temperatures of 60° to 65° F producing the best quality product. Broccoli has some resistance to frost, but the severity of damage depends on the stage of growth and variety.

Broccoli is classed as an annual. Its growth pattern is to produce a rather large central head or flower bud on a thick stem. This is the edible portion of the plant, harvested before the buds begin to flower. After this portion of the plant has been removed, side shoots or secondary heads develop but it is not economical to harvest this secondary growth in a commercial vegetable operation.

Broccoli is cut by hand as the harvesting crew follows a trailer through the field. In some farming operations a conveyor system is used to assist in the harvest process. If broccoli is to be sent to a processing plant it is generally cut to the processor's specifications, usually 6 inches in length with few or no leaves on the head. Final trimming, cutting to length and grading are done in the plant as the product is prepared for freezing. If broccoli is harvested for the fresh market it is cut to a 8 to 10 inch length and most of the leaves are trimmed in the field. At the packing shed additional trimming is done and the broccoli heads are bound together in 1 1/2 pound bunches. The most popular method of cooling broccoli for transport is to fill the packing carton with liquid ice. As the water drains off, the ice remains in all the cavities of the carton. The carton is then shipped by refrigerated truck or rail car to terminal market.

Several insects can be serious problems in broccoli production. Of major importance among these are the cabbage looper, aphids, imported cabbage worm, and cutworms. Diseases that can affect broccoli production are downy mildew and damping off; these can be especially troublesome in the early stages of growth. In fields that have been in cole crops or sugarbeet production the sugarbeet nematode can be a problem. Soil fumigation prior to planting allows the plants to develop a good root system for a better chance of withstanding an infestation.

In the last 20 years broccoli has become a very popular item for processing and freezing, and previously volatile prices have stabilized. Growers consider broccoli to be a crop that can be grown quite economically with a fair assurance that a profit can be made as long as the yield is good.

Brussels Sprouts

The coastal district of California from Half Moon Bay to Castroville produces 75 percent of the nation's Brussels sprouts. Annual plantings range from 5,500 to 6,000 acres and are valued at $6 to $7 million.

Brussels sprouts grow best in well drained, fertile soils that are free from salt and alkali. Soil pH ranging from 6.5 to 7.4 is preferred by this crop. Brussels sprouts, like other members of the cabbage family, reach best quality if the crop develops and matures during a three to five month period with relatively cool temperatures before harvest. Warm weather during sprout development results in soft or loose sprouts.

Seed beds in the central coast area are planted from February to May. A complete fertilizer, usually 12–12–12 at the rate of 1,000 pounds per acre, is disked or side dressed in before planting. Many fields of sprouts receive 200 to 250 pounds nitrogen and 4 to 6 yards of manure. Plants are transplanted into fields beginning in April and harvest begins 90 to 120 days later, depending upon the variety. First harvest usually begins in late August and continues until February, with peak production from September to late November.

'Jade E' and 'Lunet' are the main varieties used for single harvest. Some acreage of 'Stabilo' and 'Sweet Coastal' are harvested for fresh market. 'Stabilo' and 'Gravendeel' can be harvested once or twice by hand, then machine stripped for late winter fresh market harvest. Sprouts grown for freezing must be round to fit freezer recutting machinery. Single harvest plants have their terminal growing point removed about 6 to 7 weeks before harvest.

Brussels sprouts are attacked by a number of diseases and insects. Clubroot, verticillium, and blackleg are serious soil-borne diseases. Aphids, cabbage maggot and a variety of worms can cause severe losses while nematodes are a problem in many soils. Fumigation reduces nematode population and improves yields.

Part of the acreage is now bought on the stalk by certain processing companies. In one harvesting system the stalks are cut by machine, then loaded on to a truck and hauled to a central stripping shed. The other harvesting system involves field stripping with rotary knives. The sprouts are stripped, starting from the butt end of the stalk, requiring a single pass through the machine. Harvested sprouts are hauled to central sheds and cleaned, then hand sorted. The cost of cleaning varies greatly, depending upon where performed, yield, incidence of decay, and other defects.

Cabbage

Cabbage is being harvested somewhere in California the year around. Production is highest in the winter when California, Florida and Texas are the only states producing cabbage, and lowest in the summer and fall when cabbage is being produced in many other states. California's $15 million cabbage crop accounts for about 10 percent of the nation's production.

The principal cabbage producing counties in the state are Santa Barbara, Ventura and Monterey. Most cabbage varieties grown in California are recently developed hybrids that produce uniform, moderate size heads with a high percentage of heads ready for harvest at the same time. Popular varieties are 'Headstart,' 'Princess,' 'Tuffy,' 'Tasty' and 'Stonehead.' A small percentage of California cabbage is red cabbage; a much smaller percentage is savoy cabbage.

On the Oxnard Plain of Ventura County where cabbage can be grown the year around, June plantings may mature in 70 days and November plantings may take twice that long. Land preparation for cabbage includes plowing, disking and rolling, land planing, and bed formation. Beds are usually 40 inches center to center, with two rows per bed, plants spaced at about 10 inches. Most cabbage in California is field seeded with precision planters dropping single seeds at 2 inch spacings. Plants are later thinned with long han-

dled hoes. Mixed fertilizer is applied before or at planting and nitrogen may be side dressed once or twice after planting. Irrigation is mostly by furrow. Some cabbage is sprinkler irrigated, especially for germination of seed.

Cabbage loopers, imported cabbage worms, and several kinds of aphids are the principal insect pests. Although cabbage is susceptible to several devastating fungus and bacterial diseases, serious losses from these diseases in California are unusual. Fusarium yellows, club root, black rot, and sugar beet nematodes have caused problems. Nematodes are controlled when necessary by soil fumigation.

Practically all of California's cabbage is field packed two dozen heads to a corrugated paper carton holding 50 to 60 pounds. Cartons are loaded on trucks in the field, taken for vacuumed cooling and held in a cold room until loaded in refrigerated trucks or railroad cars for shipment.

Cabbage, like other vegetables, may be grown by shippers or grown for them under contract by farmers. Shippers need to control planting dates and acres planted to have an even flow of produce at harvest time. Charges for harvesting, packing, selling, and other services or materials supplied by the shipper, are subtracted from money received for the crop and the remainder is paid to the farmer or credited to the crop growing operation of the shipper. Most California cabbage sales are made by telephone by salespersons hired by shippers. Prices fluctuate freely according to supply and demand.

Carrots

Some fresh vegetables such as lettuce, broccoli and sweet corn are highly perishable, but carrots are considered 'hardware' items capable of long term storage. Other advantages of carrots are that they are easy to grow, have few disease or insect problems, and are adaptable to mechanical harvest.

The per capita consumption of carrots in the U.S. is about 10 pounds per year, which consists of about 7 pounds fresh, 2 pounds canned, and 1 pound frozen. Both fresh and processing carrots are grown within California. In 1980 there were 37,000 acres harvested. Major growing districts are Kern, Riverside, Imperial, Monterey, and Santa Barbara counties.

California produces over half of the nation's fresh market carrots. The ideal fresh market carrot should be shaped like a 10 inch coreless cylinder. Unfortunately no varieties that fit this description have high yields. 'Imperator 58,' the most popular variety, has a long tapered root. Hybrid carrots are becoming popular. The most promising have an intense orange color, sweet taste, uniform shape and high yield. Some carrot experts believe that potentially carrots could yield 40 tons per acre instead of the normal 15–20 tons. 'Danvers' is a popular processing carrot. Roots are shorter and fatter than those for fresh market.

Carrots, which are members of the parsley family, prefer 60-70° F temperatures. In the desert districts, carrots are often planted in August when temperatures commonly exceed 110° F, to grow through winter and spring. Later planted fields may not be ready for harvest until June when high temperatures again prevail. Problems due to heat may include poor germination, woody flesh texture and undesirable flavor.

Planting and germination are the most critical phases of carrot production. Seeds are placed 6 rows to a 42 inch bed, in shallow grooves about ⅛ inch deep. Spacing between adjacent rows is 1½ inches, and approximately 1½ inches within the row. Carrot seed has not been very adaptable to precision planting and most growers still use the century-old Planet Jr. design seeders. After planting, most growers use sprinkler irrigation to germinate the seeds. Germination and emergence takes about two weeks. Too few carrots mean large roots and lower yields; too many carrots produce small, spindly roots that may not be marketable. It is not economical to thin carrots.

Carrots grow best on well-drained, sandy soils but acceptable crops may be produced on heavier soils provided there is sufficient aeration and low salinity. Phosphorus fertilizer is normally broadcast and disked into the soil prior to making the beds. If carrot roots should strike a concentrated zone of fertilizer then the tap root may fork, rendering the root unmarketable. Usually small amounts of nitrogen fertilizer are applied frequently. Carrots prefer an abundant supply of moisture throughout the growing season. Plants stressed for water have small, tough, woody roots. Excessive irrigation may be equally detrimental, causing lack of color in the root and increased susceptibility to diseases.

Nearly all carrots are mechanically harvested.

The machines grasp the tops while undercutting the beds to loosen the roots. The carrots are then lifted to the top of the machine where a knife shears off the tops, allowing the roots to fall into an adjacent truck. The carrots are then hauled to the packing shed where high pressure water hoses are used to wash them into large vats, where soaking helps to remove dirt and debris. Next, the carrots are hydrocooled with ice water and pass through automatic sizers. Carrots are packed by hand or automatic machines into one pound cello bags or may be sold in bulk for distribution by truck or rail to supermarkets across the U.S. and Canada.

Hand harvested carrots, called bunched carrots, are sold with the tops still intact. These carrots command premium prices and often are sold to fancy supermarkets and restaurants.

Cauliflower

Cauliflower belongs to the same species as cabbage, kale, collards, broccoli, Brussels sprouts and kohlrabi. It was of minor importance in the United States prior to 1920. In eating cauliflower we take advantage of the plant's peculiar habit of producing a mass of closely compacted growing stems free of chlorophyll crowded among stems that will eventually produce flowers and seeds. A head of cauliflower is composed of the tender fast growing tips of these stems.

California's $80 million crop accounted for 75 percent of all cauliflower produced in the nation in 1980. In the spring season all of the nation's cauliflower is produced in California; and in each of the other three seasons, California produces more cauliflower than all other states combined. Other states producing cauliflower are Arizona and Texas in winter and fall, New York in summer and fall, and Michigan and Oregon in the fall.

In 1978, almost half of California's cauliflower was produced in Monterey County. Other counties producing over 1,000 acres of cauliflower were Alameda, Santa Barbara, Santa Cruz, and Ventura. Most of California's cauliflower is of the "snowball" type. Varieties of the "pearl" type are harvested in February, March and April. Breeders are producing many new varieties that are being tested for yield, quality, and their adaptation to locations and seasons. Cauliflower is sensitive to climatic conditions, especially temperatures.

Most cauliflower is planted two rows to a bed, with bed centers 40 inches apart. Plantings are scheduled so that there will be a steady flow of cauliflower into the markets throughout the year. The crop is fertilized before planting and once or twice as it grows. Cauliflower is irrigated several times, mostly by furrow irrigation. Yield and quality depend on large plants and rapid growth.

The principal insect pests of cauliflower are aphids and caterpillars, including cabbage loopers, which are normally well controlled by insecticides. The diseases downy mildew and Alternaria leaf spot are often found in cauliflower fields, but they are not usually of economic importance, and can be partially controlled with fungicide sprays. Club root and black rot have recently caused serious losses in a few fields in the coastal cauliflower growing areas.

Field-seeded cauliflower has a preharvest labor requirement of about 20 hours an acre and harvesting requires around 35 hours of labor. Additional labor is required for trimming and packing in the packing shed. Although there is now some field packing, almost all of California's cauliflower is hand-cut in the field, thrown into field trailers, and trimmed, wrapped, and packed in sheds. Before shipping most cauliflower is vacuum cooled. Although at 32° F cauliflower may be stored for as long as a month, most cauliflower is not cooled to this temperature and it must be moved to retail markets with a minimum of delay for acceptable quality and shelf life.

Celery

Celery belongs to the same family of plants as carrot, Florence fennel, parsley, and parsnip. Celery with the large fleshy petioles of mild flavor that we know today was not developed in Europe until after 1750. By 1800 it was being grown in the United States.

In 1980 California's $95 million celery crop was produced on 21,000 acres and accounted for 63 percent of the nation's production. About half of California's celery comes from approximately 10,000 acres in Ventura County, a third from Monterey County. Other counties producing over 1,000 acres of celery are Orange, San Luis Obispo, and Santa Barbara.

Celery acreage in California has increased from 16,000 to over 20,000 in the past ten years. A small amount of celery is harvested for dehydration and for manufacturing of canned foods containing celery, but most is grown for the fresh market. For the nation's consumers, celery is not a seasonal crop. California's production begins early in November in Ventura, Orange, and San Diego Counties and continues in these areas until the middle of July. Production in San Luis Obispo and Santa Barbara Counties starts in May and continues until the middle of January. Production in Salinas starts in the middle of June and continues until the first of January. California's lowest production is in the summer, and highest in the fall.

Practically all celery varieties grown in California belong to the 'Utah' type. They are closely related to one another and all produce the cylindrical, compact green celery that has become standard. The principal varieties are 'Tall Utah 52–70R,' 'Tall Utah 52–70HK,' 'Tall Utah Florida 683,' 'Tall Utah 52–75,' and 'Calmario.' All of these varieties are somewhat susceptible to Fusarium yellows, which has caused severe losses in all celery-producing areas of California except San Diego County. Resistant varieties are being developed by plant breeders.

Celery growers schedule plantings to provide the market with a steady flow of celery, and the acreage of this crop fluctuates only slightly from year to year. Most celery growers plant a small amount of celery each week during their season so that harvesting will be continuous. Acceptable celery may be harvested over a period of ten days from a single planting, but during this time the yield is increasing at the rate of about fifty 60-pound boxes per day, so it is advantageous to transplant several days a week. Then continuous harvesting can go on close to maximum yield. Celery must be harvested before pithiness, yellow leaves and other defects set in.

Celery crops are fertilized two or three times while growing. The total amount of nitrogen applied per acre is from 200 to 400 pounds. Celery is irrigated more frequently than other vegetable crops. In the month before harvest, irrigation may be weekly, since for high yields and high quality celery must grow rapidly.

Transplanted celery has a pre-harvest labor requirement of 75 hours per acre. About half of this is required for transplanting and a third is required for irrigation. Hand cutting and field packing require 115 hours of labor per acre.

The principal insect pests of celery are aphids,

Celery is harvested and packed directly in the field by skilled crews of workers. Most celery is grown along the coast from Monterey to San Diego. *Photo by Mel Gagnon.*

cabbage worms, army worms, loopers and leaf miners. These are controlled by insecticides sprayed, usually by ground equipment, when the populations are high enough to be hazardous to the crop. Celery is also susceptible to root-knot nematodes which may be controlled by soil fumigation prior to planting. The most common and sometimes devastating fungus diseases of celery are late blight and Fusarium yellows. Late blight, also known as Septoria leaf spot, is most severe in rainy weather and may be aggravated by sprinkler irrigation. Fusarium yellows is a soil-borne fungus disease that can severely stunt plant growth to the extent that severely infested areas are a total loss.

Although one type of mechanical harvester is now being used, most celery for fresh market is cut by hand. Field packers use small platforms on wheels, pushing them through the field as they work. Celery stalks are cut to a length of 14½ inches, older petioles are pulled off and the stalks are segregated for size. Uniform sizes are packed in cartons holding 60 pounds. Stalks too small for the four-dozen size are sent to a packing shed where they are trimmed to a length of about eight inches, packed two or three to a plastic bag and sold as celery hearts.

At one time it was common to load celery in railroad cars or trucks without precoooling, covering the load with a foot or more of ice. This practice is being replaced by precooling to approximately 40° F. Vacuum cooling, especially in warm weather when plants are under moisture stress in the field, may extract an excessive amount of water from the celery. This disadvantage has been overcome by development of hydro-vac cooling in which produce in the vacuum tube is drenched with water while cooling is in progress. Although it is seldom done in California, celery can be stored for two or three months. For long storage it needs to be cooled to 32° F and held in a room with relative humidity of 95 percent.

Cucumbers, Melons and Squashes

Cucumbers, melons and squashes are members of the Cucurbitaceae family of plants which also includes pumpkins, gourds, marrow and chayote. Cucurbits, as they are commonly called, require warm, preferably dry, weather for successful production. A frost-free period of 120 to 140 days is essential. They grow well on almost any fertile, well-drained soil, but do not tolerate wet, poorly-drained soils.

Optimum soil temperature for germinating cucumber and melon seed is 95° F. To achieve sufficient warmth for early plantings in the low desert valleys, peaked beds are prepared which run east and west and have a broad slope to the south into which the seed is planted. Seed is generally drilled into the soil in rows at rates sufficient to produce an even and fairly dense stand of plants. Establishment of melon stands through transplanting does not usually justify the costs unless seed is extremely expensive, as is the case with seedless hybrid watermelons. Some early fresh market cucumber fields are transplanted rather than direct-seeded.

Pollination using honeybees is important for successful cucurbit production. Field research has shown that one active colony per acre provides the minimum effective concentration of bees.

Irrigation of cucumbers, melons and squashes is necessary because their production occurs in the drier areas of the state and during the dry season of the year. Irrigation is important in the desert valleys to provide moisture with which to germinate cucurbit seed. In the Central Valley the practice has been to plant seed in sufficient depth that soil moisture already present germinates the seed.

A number of diseases which attack cucumbers, melons and squashes include: Fusarum wilt, Verticillium wilt, powdery mildew, mosaic virus diseases, crown blight and root-rot diseases. Pests include root knot nematode, aphids, leafhoppers, leafminers, darkling ground-beetles, cutworms, wireworms and spider mites.

Cucumbers

Field-grown cucumbers are categorized as fresh market or pickling types. Fresh market or slicing cucumbers vary in length from 6 to 12 inches and are usually harvested when more mature than those grown for pickling. Pickling cucumbers are harvested when 2 to 6 inches long, with the smaller sizes bringing higher returns.

California cucumber acreage during the last decade averaged 8,100 acres per year, with 4,900 acres grown for pickles and nearly 3,200 for fresh market. Value of the combined crop ($20 million in 1980) places it within the 20 most important vegetable crops produced in California. Most of the state's pickling cucumbers are produced in

San Joaquin, Alameda and Monterey Counties. Fresh market production is mainly in Ventura, San Diego, Imperial and Orange counties. California pickling cucumber yields are the highest in the nation, averaging slightly over 15 tons per acre for the last 5 years. The state ranks fourth nationwide in total production, while in pickling cucumber acreage the state usually ranks eighth.

Pickling cucumbers in the Central Valley are planted from March to July for harvest during June to September. Planting for the central coast is from April to August for harvests during July through October. Pickling cucumbers are direct-seeded by drilling in double rows 12 to 14 inches apart on 40-inch beds. Spacing in the row should provide about 6 plants per foot. The crop is usually ready for harvest 45 to 60 days from planting. Fresh market cucumbers, grown mainly in the southern counties, may be transplanted or direct-seeded in the field. They are produced for early spring harvests when prices are high. Clear plastic tunnels are used extensively in San Diego County to cover the plants and enable early production of the crop. Protective hot caps and brushing with paper strips have been successful in other areas for very early plantings.

Muskmelons

California ranks first in muskmelons in the U.S. with more than half of the total acreage, two-thirds of the production volume, and about 80 percent of the value. Most of the state's production is shipped out-of-state by long-haul trucks and rail to eastern and midwestern cities. The cantaloupe is the most important muskmelon grown in California with the 1980 crop at nearly 48,000 acres and total crop value at $96 million. The Honeydew melon ranked second in importance, with 12,000 acres for a value of more than $26 million. Casaba, Santa Claus, Crenshaw and Persian melons were also produced, with combined acreage total of 2,800 acres and value of more than $6 million.

Spring and fall cantaloupe production, roughly 30 percent of the total California cantaloupe crop, is located in the Imperial and Palo Verde Valleys. Almost all of the summer cantaloupe production is in the San Joaquin Valley. More than 90 percent of the other melons are produced in the Central Valley, with about half of the acreage in the San Joaquin Valley and half in the Sacramento Valley.

Spring cantaloupes grown in Riverside and Imperial counties are planted in February for harvest in late May to July. Summer cantaloupes, in Kern, Fresno, Kings and Merced counties, are planted from March to mid-July and harvested during late June through early October. The fall crop grown in the Imperial and Palo Verde valleys, is planted in July or August for harvest in October and November. Spring cantaloupes usually account for about 20 percent of the total cantaloupe acreage, summer cantaloupes nearly 75 percent, and fall cantaloupes about 5 percent.

The cantaloupe variety 'PMR 45' was introduced in 1936 and remains the principal variety for all districts except Imperial Valley. 'PMR 45's susceptibility to crown blight which devastated Imperial Valley cantaloupe production during the 1940s and 1950s led to the development of the crown blight-tolerant variety, 'Topmark,' which is now the main variety for the Imperial Valley and is also grown in significant acreage in the San Joaquin and Palo Verde Valleys. Although not resistant to powdery mildew, 'Topmark' can be dusted with sulfur to control this disease. Both 'PMR 45' and 'Topmark' are extemely high quality shipping varieties which have successfully withstood challenges from new varieties over the years. The newest developments in cantaloupe breeding are the F1 hybrids which are being tested in commercial plantings.

Cantaloupes reach maturity ahead of other muskmelons and are usually harvested for the first time about 90 days from planting. Fruit maturity is judged by the development of an abscission layer where the stem attaches to the fruit. This abscission is known as the 'slip' and the fruit can be easily separated from the vine when the visible crack or separation encircles the stem where it joins the fruit. Cantaloupes are commercially harvested when they reach the ¾ to fullslip stage.

Watermelons

Production of watermelons in California is mainly for markets on the west coast and in the mountain states. Almost half the crop is marketed within the state. There were 11,400 acres grown in 1980 which was close to the annual average for the past decade. Average yield for 1980 was 13.5 tons per acre and with total value of $26.5 million. Watermelons ranked 14th among California vegetable crops.

The spring production area for watermelon consists of the low desert valleys of Imperial and

Palo Verde. Summer production is primarily in the San Joaquin Valley, with Kern, San Joaquin, Merced and Stanislaus the leading counties. Other counties producing summer watermelons include Riverside, Los Angeles, Fresno, Sutter, Tulare and Yuba. Planting dates for the spring crop are from January through March for harvests in May, June and July. The summer crop is planted from March through June and harvested from July to October.

West coast markets prefer medium-sized watermelons weighing 18 to 25 pounds. The 'Peacock' and 'Klondike' varieties produce medium-sized melons and are the predominent types for the state. Seedless hybrid watermelons have been grown in small acreages in California during the last few years. Cost of seedless watermelon production is substantially higher than open pollinated varieties, while yields remain comparable to 'Klondike' and 'Peacock.'

Watermelons are grown on wide beds to provide sufficient space for vine extension. Southern California growers use rows 8 to 9 feet apart. San Joaquin Valley growers use alternate rows 7 and 10 feet wide and turn vines into the 10-foot space as the vines grow. The 7-foot space is used for the irrigation channel. Watermelons are usually thinned by hand with long-handled hoes when plants have 2 or 3 true leaves. Spacing varies from 1 to 3½ feet between plants in the row with a single row on each bed.

Proper irrigation of watermelons is very important to avoid blossom-end rot and misshapen fruit. From 20 to 30 inches of irrigation water are required, depending on the soil and weather conditions. Nitrogen fertilization is usually required.

Harvesting the watermelon crop usually occurs 90 to 130 days after planting, depending on the season and the variety. A planting may be harvested several times, usually at 5-day intervals. All harvesting is done by hand. Melons are cut from the vines, windrowed, then hand loaded onto trucks or trailers.

Squashes
In 1976, the estimated combined squash acreage in this state was about 5,000 acres with value at more than $10 million. Leading counties were Imperial, Riverside, Fresno and San Diego which produced more than 60 percent of the acreage and value.

Summer squash types include bush scallop,

yellow crookneck and zucchini. All are rapid maturing, bush-type varieties. In the desert valleys the fall plantings are seeded in late August and September for harvesting in late October to early December. The spring crop is planted in late December and January for harvestig in March, April and May. The early spring crop is often protected from wind and frost by brown paper barriers held upright on the beds with stakes, wire and arrow weed stalks. This practice, known as 'brushing,' adds as much as $600 per acre to the cost of production. Mosaic virus is the most serious disease problem of desert-grown summer squash. Currently there are no control measures although some varieties appear to tolerate the virus better than others. Brushing tends to reduce the incidence of mosaic virus compared to the open grown squash.

All squashes are hand harvested. Zucchini fruits are cut from vines when 8 to 10 inches long and hauled to nearby sheds where they are graded, sized and packed.

Winter squash are also grown commercially in California. Growing season is much longer for these long-vined types, since their fruits are allowed to completely mature before harvesting. While summer squash varieties take 40 to 50 days from planting to harvest, the winter squash varieties usually take 85 to 105 days. Popular winter squashes for California production include banana, butternut, and Hubbard types.

Lettuce

Crisphead lettuce competes with processing tomatoes as the leading vegetable crop in California. In 1980, about 160,000 acres were grown in the state. California lettuce is shipped throughout the United States as well as to Canada and Western Europe. Unlike many seasonal vegetables, lettuce is produced in about equal quantities every month of the year. California has many climatic zones that provide near optimal conditions in one district or another throughout the year. Monterey County produces more lettuce than any other area in the world. In 1979, 61,750 harvested acres produced over 2.2 million tons of lettuce destined for salads, hamburgers and tacos. The harvest season in Salinas-Watsonville starts in early April and closes in early November. Imperial County ranks second (42,900 acres) and ships from early December

A Salinas Valley lettuce crew harvests, wraps and packs in the field behind a slowly moving machine. *Photo courtesy of Grower-Shipper Vegetable Association of Central California.*

until mid-March. Other counties producing appreciable quantities of lettuce include: Contra Costa, Fresno, Kern, Kings, Riverside, San Benito, San Luis Obispo, Santa Barbara, Santa Clara, Santa Cruz, and Ventura.

Lettuce is a cool season crop but intensive plant breeding has provided a wide selection of varieties adaptable to some very severe climatic conditions. There are varieties that are planted when air temperature is 118° F and others that mature when ground freezes occur and fields do not thaw until 10 a.m. In the Salinas area with more moderate temperatures growers may plant only one or two varieties such as 'Calmar' or 'Salinas.' In the variable climate of Imperial Valley, growers use at least five varieties and often eight or more. Some varieties are so limited in adaptability that they must be planted during a single week in late September or not at all.

In almost all areas lettuce is precision planted with pelleted seed. The standard crop is grown with two rows on a 40-inch bed. Growers either furrow irrigate or sprinkle to germinate seed and achieve a stand. Furrow irrigation is practiced through the remainder of the season. Lettuce re-

quires moderately high amounts of nitrogen and growers apply 200–250 pounds nitrogen per acre for the season, usually a third at preplant and two-thirds during active growth. Phosphorus is applied preplant at 60-200 pounds P_2O_5 per acre depending upon the soil type.

Weeds are a serious problem in lettuce culture. Good chemical herbicides are available, and additional weed control is practiced as the fields are thinned by special crews using long handled hoes. Mechanical weed control is practiced during cultivations with sled or tractor-mounted knives.

Lettuce is attacked by a myriad of insects including the cabbage looper, the beet armyworm, the tobacco budworm, aphids, fleabeetles and thrips. Good insect control is essential as insects in the heads, either dead or alive, are objectionable to the consumer. Both air and ground insect control measures are practiced. Big vein, Sclerotinia, downy mildew, aster yellows, Botrytis and bacterial spot are examples of the numerous diseases attacking lettuce; some may be controlled while others may not. Lettuce mosaic virus has been successfully controlled by seed indexing, thus eliminating seedborne virus problems. Tip-burn is a serious disorder that occurs during warm weather. Lettuce also suffers from many postharvest diseases and disorders which are either due to contamination or poor transit conditions.

Once a lettuce field is mature, the shipper decides whether to ground pack or film wrap. The ground pack system is presently the standard harvest method. A carton making machine is surrounded by field workers who cut and pack the heads by hand. The workers are divided into units called "trios," which consist of two cutters and a packer. Twenty four heads are hand packed into each cardboard carton; the cartons are then hand loaded onto widetrack trucks which haul the load to the cooler, where the lettuce is vacuum cooled to 34° F and stored for truck transit to terminal markets. A small amount of lettuce is also shipped by rail. Film-wrapped lettuce is becoming more popular since more saleable lettuce is shipped when the wrapper leaves are trimmed off and left in the field. In this system, large self-propelled harvesting units move slowly across the field, with conveyer belt wings spread 20 feet to each side of the main unit. The ground crew cuts and hand trims heads which are placed on the machine. Workers riding on the "wings" wrap and seal balled heads in plastic film. The heads are then conveyed to the center of the unit for packing.

If all the heads in an average lettuce field could be marketed, the yield would be over 1,100 cartons per acre. In reality the average yield is only about 600 cartons per acre. More lettuce is often planted than can be marketed. Lettuce is truly a gambler's crop since an active market can make a grower wealthy almost overnight. On the other hand, a string of poor markets can cause him to lose everything. The lure of "Green Gold" is strong and new investors enter the business yearly. Oversupply can occur not only due to excess acreage but also due to weather. Favorable weather in the growing district may cause many fields to ripen at one time. Snowy weather at terminal markets may delay unloading and distribution. Sales of lettuce diminish when consumers are unable to drive to the supermarket and the housewife changes to stew or roast instead of chef's salad. Success in the lettuce industry is for those shippers who are able to survive the poor seasons and capitalize on the high markets.

Onions

California ranks first in the U.S. for dry onion production, contributing about 30 percent of the nation's annual supply. The harvested acreage has increased by nearly 50 percent during the past ten years, and in 1980, 33,000 acres of onions were grown in the state with a market value of nearly $80 million. Dry onions are produced for the fresh market and also for dehydration. Onions for dehydration comprise about 60 percent of the crop volume annually, and California provides nearly all of the dried onion products consumed in the United States. Most of the fresh-market dry onions are not stored, as in many states, but are delivered directly to market channels. California also grows several thousand acres of green bunching onions which are marketed throughout the year from one or more producing districts.

Dry onions are grown throughout the state from Tulelake in the north to the Imperial Valley along the Mexican border. The largest acreage is located in the counties of Kern and Fresno. The next largest production is in the southern desert. The earliest production begins in the Imperial Valley in late April from plantings which were seeded the previous October. Harvest in that area continues into early June and overlaps the beginning of harvest in the Fresno area which lasts about six weeks. Production in the Bakersfield area, which accounts for about 30 percent of the total fresh-market crop, also begins in early June and continues until September. The Stockton area, another major producer, harvests from mid-June through July. The latest production in the state occurs in September and October from the Antelope Valley in southern California. Spring plantings of dehydrator onions in the counties of Monterey, San Benito, Modoc and Siskiyou are also harvested during the fall months.

Dry onions for the fresh market are almost all harvested by hand. Onions are harvested when the necks weaken above the bulb causing the tops to fall over. To loosen the soil and sever the roots, the bulbs are first undercut by a horizontal blade mounted behind a wheel tractor. Then the onions are hand-pulled and tops and roots are clipped off with shears. The trimmed bulbs are placed in burlap bags and cured in the field for four to ten days before transporting to the packing shed for grading and packing in market bags.

Green onions are bunched in the field by truck farmers in the Delta for delivery to local outlets.

Producers have experimented with various types of mechanical harvesters for years, but none has been entirely acceptable. Recently, however, some mechanical harvest of market onion acreage has occurred with excellent results.

The harvest of dehydrator onions has been highly mechanized for many years. When the onion tops have matured, they are cut off by a tractor-drawn rotary beater or rotating blade. After several days of curing in the ground, the bulbs are then undercut, lifted and windrowed. A large grading machine picks up the onions mechanically as it moves through the field; clods and defective bulbs are removed by hand. The graded bulbs are then transferred to a bulk truck for transport to the processing plant.

Onions for both fresh market and dehydration are grown on raised beds, usually 40 inches on center, with two to eight rows of plants per bed. The dehydrator crop, which is direct-seeded, is planted with six, sometimes eight, rows per bed.

Irrigation is required for onion crops in all districts. From 3 to 6 acre-feet of water per acre are required depending upon soil type, season and method of irrigation. Onions are very responsive to fertilizer because of their shallow root systems and dense plant populations. Nitrogen is the most important element in all areas with 200 to 300 pounds per acre applied to most crops. Onions on most soils will respond to 100 to 150 pounds of phosphorus per acre.

Weeds can be a very serious problem in onion production. Hand weeding is very costly, especially if the weeds are grasses. Experienced growers have learned to select land very carefully for their onion crops in order to minimize weed problems. Selective herbicides are widely used. Several diseases are of economic importance in California onion crops. Downy mildew can be a serious problem during periods of humid weather and moderate temperatures. Pink rot, caused by a soilborne fungus in growing areas throughout the state, is increased when onions are continuously cropped and yields can be seriously affected. Resistant varieties are the primary means of control, but are not completely effective. Black mold, neck rot, basal rot and white rot are also common disease problems. Insects are a relatively minor problem on California onions, although thrips, onion maggot and leafminers cause some damage occasionally.

The choice of onion varieties and the date of planting are of paramount importance to a suc-cessful crop because of the effect of day length on bulb development. Varieties which are successfully grown in one district will not necessarily perform well in another district. Varieties classified as "short-day" types are grown in the Imperial and Palo Verde valleys for early spring maturity. "Intermediate" day length types are grown across central California, and "long-day" types are grown in the areas north of Stockton and Sacramento. Many varieties within each of these types are grown. The short-day types, such as those grown in the low desert valleys, are very mild in flavor but have a very short storage life. Intermediate types, which are grown from Bakersfield to Stockton, are of mild to medium pungency, and some varieties can be stored for several months under proper conditions. Long-day types are more pungent and have good storage capabilities. Dehydrator types are very high in soluble solids and very pungent.

In addition to market and dehydrator onions, California is also a major producer of onion seed. Several hundred acres are grown annually in the San Joaquin and Imperial valleys and in the vicinities of Salinas, Gilroy and Hollister.

Peppers

California is a leading producer of many types of peppers, both sweet and hot. Pepper production centers in the central and south coastal areas, with Monterey, Ventura, and Santa Barbara counties leading. Smaller acreages are grown in Orange, Tulare, Kern, Riverside, Santa Clara, Fresno, and San Diego counties. Peppers are a warm-season crop and need a long growing season for maximum production. Light frosts injure or kill the plants. California's bell pepper crop is planted in early March through April and harvested from early May to late November and early December. The chili and pimiento pepper harvest begins in mid-September and ends in November. Direct seeding or transplants are used.

Several kinds of insects attack peppers and can be a serious problem. Insects that attack seedlings are flea beetles, darkling ground beetles, cutworms, aphids, spring-tails, vegetable weevils, grasshoppers, wireworms, and seed corn maggots. Later in the growing season, pepper weevils, leaf miners, aphids, corn earworms, cabbage loopers, beet armyworms, and other caterpillars can develop into damaging populations if not controlled. Peppers are also subject to a number of diseases that reduce both yield and market value of the fruit. These are the tobacco mosaic virus, western yellows or curly top, blossom-end rot, spotted wilt, Phytophthora root rot and the damping-off organisms.

Yields of fresh market and processing bell and pimiento peppers vary greatly. Yields of 400 30-pound cartons (6 tons) per acre for the bell pepper fresh market are not uncommon, while bell peppers for processing average 12 to 14 tons per acre. The chili yield is usually between 1 and 1½ tons of dried peppers per acre.

Bell peppers are usually harvested when full grown or when at a good marketable size but still green, firm and crisp. Fruits are snapped off by hand and carried from the field in sacks or buckets. Recently, however, the conveyor belt or mobile field loader has proved to be more economical on large fields, also causing less plant breakage. The use of bins and trailers in hauling is on the increase. After culling, the marketable fruit is wiped clean, graded for size, and generally packed in corrugated paper cartons for shipment.

Potatoes

Traditionally, potatoes, lettuce, and tomatoes have ranked as the $100 million vegetable crops grown in California; however, the present value of potatoes is only about one-third that of tomatoes or lettuce. Potatoes became important in California soon after the Civil War with 30,000 acres grown in 1875, 45,000 acres in 1900, and 75,000 acres by 1910. Production in 1929 decreased to about 32,000 acres, but from this low it increased rapidly again to over 70,000 acres in the early 1940s. Peak production was reached in 1950 with 122,000 acres. During the 1960s acreage decreased again, and since 1975 production has stabilized at about 60,000 acres per year.

Potatoes can be grown in most areas of California, but there have been major shifts in the areas of production over the years. Before 1900 most of the potatoes were grown on the islands of the San Joaquin-Sacramento Delta and near Half Moon Bay, north of San Francisco. By 1920, production was centered in Los Angeles, San Joaquin and Sonoma counties. Production is now concentrated in four areas: Kern County, River-side County, San Joaquin County, and the Tulelake area, including both Siskiyou and Modoc Counties. Kern County grows primarily for the spring market; in 1978 this county accounted for 27,200 acres out of a state total of 29,000 acres. Production in Kern County essentially began in the 1930s, reached a peak about 1950, and has continued to decrease since then. Fall potatoes are grown mostly in the Tulelake area—13,000 acres out of a state total of 17,300 in 1978. Potato production at Tulelake began in the 1930s, rapidly increasing to about 9,000 acres in 1940. Summer production of potatoes on about 8,000 acres in 1978 was mostly in Riverside County, with minor acreage in San Joaquin and Santa Barbara Counties.

Potato yields per acre have increased dramatically over the past fifty years. Before 1920, yields were less than 80 hundredweights per acre. These increased to about 100 hundredweight in the late 1920s, 150 in the 1930s, 250 in the 1940s, 300 in the 1960s and are now approaching 400.

Most potato acreage in California is planted to the same varieties grown 20 to 30 years ago: 'White Rose,' 'Russet Burbank,' and 'Kennebec.' Of these only the 'Kennebec' variety has been developed since 1930. 'White Rose' has been the leading variety grown in California. It is grown mostly as the spring and summer crop in Kern and Riverside Counties and is also important in the Tulelake ara. 'Russet Burbank' or 'Netted Gem' is grown primarily in the Tulelake area. 'La Soda,' a red-skinned variety, is grown in all areas, as is 'Kennebec,' which is used almost exclusively for the chipping industry.

Since the peak years of the 1950s there has been a decline in the production of California potatoes because of a shift in consumption patterns toward more processed than fresh. The cooler regions in Washington and Idaho obtain higher yields of the versatile 'Russet Burbank,' which is in greater demand than California's 'White Rose.' Most potatoes in California are grown for fresh market and are consumed soon after harvest. (The fall crop at Tulelake, however, is stored and then marketed during the winter and early spring.) Except for chips, California potatoes are not processed, in contrast with the other major producing states where about half of the potatoes are processed as frozen french fries or bakers, or dehydrated and marketed as granules, flakes or powders.

An extensive varietal improvement program

has been undertaken by the University of California in cooperation with the USDA and the Experiment Stations of Colorado and Idaho. One goal of this program is an early maturing russet-skinned variety that will produce well in the warmer growing areas such as Kern and Riverside Counties. 'Centennial' and 'Norgold Russet,' recently introduced varieties, have gained some prominence in attaining this goal. The potato improvement program is supported by the California Potato Research Advisory Board using funds provided by an assessment of all potato growers in California. More than 100,000 different lines have been evaluated in these studies for disease resistance, earliness, market appeal, high percentage of No. 1's, heat tolerance and high specific gravity.

Poor quality seed potatoes infected with a multitude of tuber-borne diseases were responsible for many of the low yields experienced during the past decades. Better production techniques and inspection programs have continually improved the quality of seed potatoes. Pathologists have isolated tissue cultures from the standard California-grown varieties to make them pathogen-free. Foundation seed from these isolates is now available to certified seed potato growers.

The method of potato harvesting has changed greatly during the past 30 to 40 years. Previous to about 1945, potatoes were dug by machines which lifted the tubers from the soil and laid them on the soil surface, after which they were picked up by hand and placed in sacks to be hauled to the packing shed. Today mechanical harvesting and bulk handling are the preferred practices for almost all growers. Twin-row harvesters capable of handling up to 5,000 hundredweights per day are common. In mechanical harvesting great care is taken to avoid excessive bruising and injury of the potatoes.

Spinach

California ranks first in the U.S. in spinach production, providing about half of the nation's supply, both fresh and processed.

Spinach acreage in the state steadily increased in the decade 1969–79. In 1969, 55,750 tons of processing spinach valued at over $2 million were grown on 6,800 acres; by 1979 this had increased to 10,400 acres yielding 102,000 tons val-

ued at $8.3 million. Fresh market spinach production meanwhile more than tripled, from 7,300 tons grown on 1,360 acres valued at $2.1 million, to 25,200 tons grown on 3,200 acres valued at $9.5 million. The leading counties for fresh market spinach are Ventura, Monterey, and Santa Barbara, which provide a year-round supply for U.S. consumption. Spinach for processing is grown principally in Stanislaus, Monterey, Yolo, and Ventura counties, which harvest for this market typically March 15 to May 10. Planting dates are from November 20 to February 15.

The varieties of spinach now being grown in California are undergoing change as a result of loss of resistance to downy mildew in the popular hybrid variety 'H–424' and similar types. At present, varieties which have the required resistance plus other desirable horticultural traits are 'Polka,' 'Mazurka,' and '501.'

Thick planting helps spinach seedlings compete with weeds, but herbicides and hand weeding are needed to supplement cultivation by tractor. Liberal quantities of fertilizer are used to obtain fast growth, high yields and dark green leaf color. On lighter sandy loam soils, about 100 pounds per acre each of nitrogen, phosporic acid and potash are applied. Well-drained soils are important, particularly when growing winter spinach. If excessive rains cause waterlogged soil, spinach becomes stunted and chlorotic.

The most important disease of spinach grown in California is downy mildew, also known as blue mold. Other diseases which can be important under certain conditions include Cercospora leaf splot, anthracnose, fusarium wilt, rust, scab, or black mold, smut, crown rot, damping off, and root rot. Seed treatments with fungicides give some protection. Spinach is also susceptible to numerous viruses which can cause stunting, distortion, and yellowing of leaves. Beet mosaic, beet yellow, and cucumber mosaic virus (spinach blight) are some of the most important. Transmission of these viruses is by various aphid species, the most important vector being the green peach aphid. Cucumber mosaic is also transmitted by the striped cucumber beetle and western spotted cucumber beetle. Destruction of weeds surrounding fields is a practice recommended to control virus diseases, to eliminate breeding sources for the insect vectors as well as host plants for viruses.

The pea leafminer and several other leafminers are other pests of spinach. During the early seedling state, damage can also occur from the spinach crown mite. In the coastal regions of California, this can approach 30 to 40 percent of the crop in years of heavy infestation. These mites feed on the young interfolded leaves damaging the tissue which develops shot holes, blackened areas, and ragged, deformed leaves. Several species of worms may attack spinach, but the alfalfa looper is the most serious. Many insecticides are presently used to control these pests.

Quality standards for processing spinach are especially rigid. Spinach for processing is harvested usually twice by mechanical harvesters which mow the plants several inches above the crown. Trailers are hauled with the loose spinach to freezer plants or canneries by trucks. Fresh spinach is cut by hand. The tap root is severed with a knife just below the basal leaves. Some loose packed spinach is sold in 20-pound cartons, but most is bunched and packed in two-dozen bunches per carton in the field. The packed spinach is vacuum cooled and shipped to market in refrigerated trucks or rail cars.

Sweet Corn

California is not noted for national importance as a sweet corn producer even though the state ranks about tenth in production volume. Almost all of California's sweet corn is grown for the fresh market, and about two-thirds of it is consumed within the state. A small amount is delivered to processors.

Sweet corn is grown to some extent in every county in the state, in backyards, market gardens or in large commercial acreages. The principal producing counties are Riverside, Los Angeles, Orange and Solano. The largest acreage is in Riverside County's Coachella Valley where both the earliest and latest crops are produced. California-grown sweet corn is available in markets from May to November.

Almost all of the hand labor has disappeared from commercially grown sweet corn plantings. It is precision seeded, herbicides and cultivation are used for weed control, and it is largely harvested by machine. Mechanical harvesting has developed during the past ten years. Prior to that, hand crews selectively snapped mature ears from the stalks, tossing them into a trailer which was pulled through the field by a tractor.

Sweet corn is a warm-season crop, preferring daytime temperatures in the 80 to 90° F range with warm nights above 60° F. Cooler temperatures slow growth and delay maturity. Very hot temperatures, especially if windy, can cause poor pollination, leaving blank kernels on the ears. Early plantings in the Coachella Valley may require as much as 120 days from seeding to market maturity because of the cool temperatures following the January to February planting dates. Plantings during April or May in the San Joaquin Valley mature as early as 75 days after seeding.

The corn plant is relatively shallow rooted and requires frequent irrigation (every four to seven days in warm weather) to maintain the vigorous rate of growth which is necessary to produce high quality ears. Poultry manure is often used as a preplant fertilizer, with supplemental applications during growth from inorganic sources of nitrogen.

Sweet corn is grown in single rows usually spaced 30 to 36 inches apart, with plants spaced at 8 to 12 inches in the row, providing about 16,000 to 20,000 plants per acre. Each plant is capable of producing one marketable ear (although two are produced in some cases), which adds up to 250 to 350 crates per acre with five dozen ears per crate.

After germination and seedling emergence, the plant goes through a rapid stage of leaf development for a period of about eight weeks, while the tassel (the male flower which eventually appears at the top of the plant) and the ear initials (the female flower plants) are developing deep inside of the plant. At about the nine- or ten-week stage, the tassel emerges and begins to shed pollen, while the silks are beginning to emerge from the young ears. Pollen grains fall in a shower over a period of days on the silks where they germinate. Each silk leads to a separate kernel which must be fertilized by the germ tube from a pollen grain in order to develop into one of the many fruits which line the sides of the cob. When environmental factors interfere with this process, blank kernels develop which reduce the ear quality.

The major pest problem in sweet corn plantings is the corn earworm, which can make the ear unfit for market. To control corn earworm, farmers apply pesticides to the crop by large tractor-mounted spray equipment of special design. Applications begin at the first appearance of the silks and continue (often 10 to 15 times under conditions of heavy infestation) through the two- to three-week period while the ear is maturing. Other pest and disease problems occur sporadically in various producing districts but all are secondary to the earworm problem.

After harvest, sweet corn must be cooled as soon as possible, usually by a cold-water process called hydrocooling, to maintain sweetness and high quality. Many growers harvest at night when the corn is coolest in order to reduce the amount of field heat which must be removed from the product. After cooling, the ears are sorted to remove those with mechanical or earworm damage and packed in wire-bound crates or wax-treated fiberboard containers containing four to six dozen ears. The shelf life of sweet corn is short, and it should be stored at 35° F to maintain good quality. Long-distance shipments are generally limited to the early season when supplies are short.

Sweet Potatoes

Nationally, California ranks third in total acreage and total production of sweet potatoes— although first in yield per acre—and produces almost 11 percent of the national crop. (North Carolina and Louisiana are first and second, respectively.) Acreage varies somewhat from year to year, but generally ranges between 8 and 10 thousand acres, producing about 160 hundred weight per acre.

Merced County produces about 78 percent of the state's sweet potatoes, with Fresno and Stanislaus Counties producing most of the rest. Most of the sweet potatoes are grown within a 15-mile radius of Livingston, California, where the soils are sandy and the growing season long.

California grows three major varieties—the 'Garnet,' which has a deep red or purple skin with an orange flesh; the 'Jewel,' which has a copper colored skin with a deep orange flesh; and the 'Jersey,' which has a creamy colored skin and a yellow flesh. A fourth variety, grown strictly for the processing industry, is called the '779,' also known as 'Golden Pride.' About 5 percent of the sweet potatoes grown in California go to the processor.

The 'Garnet' and the 'Jewel' are 'yam' types. The term 'yam' is only a market term which denotes the *moist* texture of the potato flesh after it is baked. The 'Jersey' has a dry type flesh after it is baked, and this variety is sometimes referred to as a 'sweet.' All varieties are, however, 'sweet potatoes.'

Sweet potato transplants are grown from mother roots (seed potatoes), which are regular sweet potatoes. Each grower usually grows his own plants. The seed potatoes are covered with sand in specially prepared plant beds which are sheltered by plastic tents. Temperatures in these beds are critical and must be monitored every day. The plastic must be removed when temperatures get too high and for irrigation. Plants (sometimes called sprouts) grow mainly from the stem end of the potato. When they are 6 to 8 inches high, they are pulled from the mother roots and taken to the field for transplanting after the danger of frost, usually the last week in April and during the month of May.

Sweet potatoes are harvested when they reach market size. Some fields may be harvested by late August or early September, but general harvest begins during October. After harvest, the potatoes go through a special "curing" process for 4 to 6 days, in rooms where the temperature is 80–85° F with humidity of 85 to 90 percent. During the curing process, all injuries are healed rapidly through the formation of new corky layers beneath the wounded areas. This procedure keeps decay to a minimum. After curing, the potatoes are moved to a cool room where the temperature is kept from 55° to 60° F. Relative humidity of 85 to 90 percent is needed to keep shrinkage to a minimum. Potatoes held at temperatures below 50° F for a period of time are subject to chilling injury which will cause them to rot. Sweet potatoes can be stored from 6 to 8 months or longer, depending on the type of storage facilities.

From storage, the potatoes are taken to packing sheds, where they are washed, treated with a fungicide to lengthen shelf life in the stores, and hand-packed into cartons holding 40 pounds. The main market area consists of the three western states: California, Oregon, and Washington. Some potatoes are also shipped into western Canada; late in the shipping season, shipments are made into Colorado, Texas and some midwestern states.

Growers have recently formed a "Sweet Potato Council of California." This is a volunteer organization where growers are assessed on a per acre basis, shippers on a per carton basis, for funds to be used for promotion and research.

Tomatoes

Tomatoes usually rank as the number one vegetable crop in California. They are grown for both processing and the fresh market.

Fresh Market Tomatoes

California and Florida together grow approximately 72 percent of the total U.S. supply, each state accounting for about 36 percent. Usually around 30,000 acres are planted to fresh market tomatoes in California. In 1978 the state produced 7.6 million hundredweight with a value of $144 million.

Two distinct growing methods are used in California. Some fresh market tomatoes are grown on stakes or trellises and harvested at the pink fruit stage in several picks. Other are grown as bush types with one or two picks in the mature-green stage. The major districts that produce staked or pink-stage tomatoes are in San Diego, Ventura, and Tulare counties, with smaller acreages in several other districts. The primary bush or mature-green producing districts are in Imperial, Fresno, Kern, Kings, Merced, Monterey, Stanislaus, and San Joaquin counties.

The production districts and harvesting periods are as follows: Imperial Valley (May 25–July 1), Cutler-Orosi (June 15–August 5), Merced (June 24–August 15), northern San Joaquin Valley (July 15–November 15), Gonzales-King City (August 1–October 5), and southern California (May 20–December 30).

Most fresh market tomatoes in California have been hand harvested. In 1980, however, approximately 30 percent of the bush type tomatoes were expected to be machine harvested. Over the past decade, new varieties more suitable for mechanized harvest have been developed, while tomato harvesters have been modified, cultural practices necessary for a once-over harvest have been adopted, and post-harvest handling has been changed.

New varieties of fresh market tomatoes had to be developed for machine harvest. The older varieties—'Ace,' 'Early Pak 7,' and 'Pearson A-1'—lacked wilt disease resistance and were not adaptable to machine harvest. Some bush-type varieties presently grown are 'Royal Flush,' 'Jackpot,' 'Valley Pride,' 'Castlemart,' 'Blazer' and 'Sunlight'. Staked varieties are 'Peto 6718' and 'Peto 7718.' The fresh market tomato industry has converted rather decisively to the use of

F$_1$ hybrid varieties. Several of the new varieties also have nematode, tobacco mosaic and Alternaria stem canker resistance and the jointless gene (easy stem release) is being incorporated into the bush types. The long range objective of the breeding program is to develop needed and useful characteristics including disease and pest resistance, tolerance to environmental stresses, new plant types, and qualities such as flavor, color, firmness, and high vitamin content.

Processing Tomatoes

From 1974 to 1978 approximately 6 million tons of processing tomatoes were produced per year on an average of 260,000 acres, with annual raw product value of over $380 million. The combination of favorable climate, good soils, ample water supplies and applied technology has allowed the industry in California to grow from an average production of 2 million tons annually in the 1950s to 6.4 million in 1979. In that time approximately 100,000 additional acres were planted, yields were increased 41 percent, raw-product value increased from $60 million to over $426 million and California's share of total U.S. tomato production grew from 52 to 86 percent.

With the end of the bracero program after 1965, a major change began in the growing, harvesting and processing of tomatoes, the necessary research for this shift having been conducted over the previous decade. With the shortage of harvest labor, a dramatic transition occurred in processing tomato production as harvesting became almost completely mechanized. Growers have found that exact management precision in cultural practices and much personal attention are necessary to grow and harvest processing tomatoes. Everything must be considered in planning: varieties, soil, irrigation, cultural practices, type and maintenance of machines, and postharvest handling.

In California, commercial production occurs in the Imperial and Palo Verde desert valleys, through the San Joaquin and Sacramento valleys north to Butte County, and in the south and central coastal counties. Harvesting begins in the desert valleys in mid-June and ends in the southern coastal areas in November. Five counties produce two-thirds of the state's processing tomatoes: Yolo, Fresno, San Joaquin, Sutter, and Solano.

Many varieties of tomatoes have been tried by the tomato-processing industry since the start of

A mechanical harvester for processing tomatoes uses a sorting crew to cull out defective tomatoes and dirt, then transfers the good fruit by conveyor to gondola cars for delivery to the cannery. *Photo by Jack Clark.*

machine harvesting in 1961. New varieties and new strains of existing varieties are continually being offered to the industry. A tomato variety suitable for mechanical harvesting should: (1) have adaptability and good yield potential; (2) be able to withstand mechanical harvesting; (3) set fruit during a short span of time and under a wide range of climatic conditions, so that a high proportion of fruit will ripen at the same time; (4) be resistant to Verticillium and Fusarium wilt diseases; (5) have a firm fruit with a small stem scar, capable of remaining on the vine in good color and condition for a reasonable time; (6) have fruit with desirable processing factors, such as low pH, high soluble solids, good consistency, favorable solids/acid ratio, high vitamin A and C content, and good color, flavor and peelability.

Of the various strains of 'VF145' that have dominated the processing tomato variety picture for many years, 'VF145-B-7879' has been most widely planted. Other varieties now being increasingly used include the irregular shapes, the oblongs, the square-rounds such as 'UC82,' and

'UC134,' and some hybrids. Acreages planted of lesser-used special varieties vary from year to year, but are an important part of the industry because they fill specific needs, such as for juice, paste, or whole-peeled tomato. With the advent of 10- to 12-ton size containers for hauling, emphasis has been on varieties that have firmer fruit and can withstand the rigors of bulk delivery.

All harvested processing tomatoes must be inspected, and a certificate of compliance issued, by the Canning Tomato Inspection Service at an inspection station as prescribed by regulations of the California Department of Food and Agriculture. Tomatoes may be rejected on a basis of color, worm damage, mold, green fruit, and extraneous material, such as dirt, detached stems, etc. Tolerances cannot be exceeded. A new "limited use" category sets limits on overripe, sunscald and mechanical damage.

U. S. tomato production has risen steadily (with short-term fluctuations) since 1922. Most of the increase has been in the crop destined for

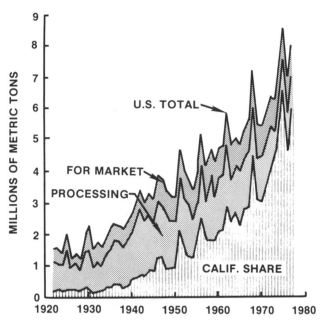

FIGURE 8.2 Tomato production in the U.S. and in California, 1922–1978. *Source: W. Sims.*

processing: canning, sauce, juice, catsup and so on. California has accounted for an increasing share of that crop, while the U.S. commercial fresh-market crop has remained about the same. Figure 2 displays production trends since 1922.

California now produces nearly 90 percent of all processing tomatoes in the U.S. This is equivalent to consumption of about 60 pounds per person. The tonnage of processed tomatoes greatly exceeds that of any other fruit or vegetable.

Reference Material on Vegetable Crops

Information on the culture, utilization and storage of vegetables is available in a series of leaflets, circulars and bulletins published by the Division of Agricultural Sciences, University of California, and distributed through the offices of the Cooperative Extension Service.

Statistical information on vegetable production and utilization is available in many publications of the California Crop and Livestock Reporting Service.

Because there are so many different vegetable crops and because they are grown under so many different conditions in California, there is no single reference available. The following recent books provide good information on a large number of crops:

Knott's Handbook for Vegetable Growers by O. A. Lorenz and D. N. Maynard. Wiley-Interscience. 1980.
Producing Vegetable Crops by G. W. Ware and J. P. McCollum. The Interstate Printers and Publishers, Inc. 1980.
Leafy Salad Vegetables by E. J. Ryder. Avi Publishing Co. 1979.

A Marin County dairy scene shows cows on pasture. Large numbers of dairy cattle are now fed in drylots under confinement. *Photo by Jack Clark.*

9

Animal Products

Animal Science, Avian Science, Aquaculture Science, and Apiculture Science Specialists of Cooperative Extension, UC Davis

Reuben Albaugh, horses; *Fred Conte*, aquaculture; *John Dunbar*, cattle feeding and swine; *Ken Ellis*, beef cattle, sheep, and Angora goats; *Ralph Ernst*, poultry; *Frank Murrill*, dairy goats; *Eric Mussen*, bees; *C.L. Pelissier*, dairy.

California has been livestock country since the early explorers and settlers. From the time of the first Spanish missions and land grants until today, livestock have had an important role in the building of California's economy and in maintaining a position as our nation's leading agricultural state.

As in other parts of the West, the opening of the California frontier in the 19th century brought many would-be farmers to a semi-arid land which proved in its early days to be best suited for grazing animals. Its vast acreages of natural forage included perennial and annual grasslands, coupled with foothill and mountain brushlands. The rainless months and scattered natural waterways meant that the California territory was not easily adaptable to traditional forms of agriculture as practiced along the eastern seaboard and in the prairie soils of the midwest. Without ruminant animals to convert forage and brush indigestible to man, it is unlikely that the valleys and mountains would have been populated so thoroughly and transformed into the world's richest agricultural production area in just a few generations. While technological progress has changed the face of California agriculture, the sheep, cow, goat and horse have provided food, clothing and transportation throughout the state's history.

Many of California's foothill, valley and mountain areas, because of climate and terrain, are still productive only with ruminant animals. Without ruminants to convert forage, millions of acres in California could not be utilized by man. Even certain wildlife populations benefit from proper utilization of forage and brush by livestock.

While irrigated agriculture and urban encroachment into agricultural areas have reduced the resources available for sheep and cattle grazing, California remains a major livestock state and is projected to continue so in future. This chapter reviews the development of the California livestock industry in the decades since World War II.[1] Sections are included on apiculture (beekeeping) and on aquaculture (culture of aquatic plants and animals)—industries which have become more significant in recent years.

Beef Cattle

Even though California produces more than 200 crops and agricultural commodities on a commercial basis, in most of the years since 1950 beef cattle have been classed as its number one agricultural commodity—if the value of cattle imported, finished and marketed within the state is included. Production of cattle and calves accounts for about 13 percent of the total gross ag-

---

1. The story of California livestock through the turn of the century and in the years preceding World War II is told in a chapter of *California Agriculture* (1946), a book produced by members of the faculty for the 75th anniversary of the University of California.

ricultural income in California.[2] If the value of land, facilities and total investment is coupled with the value of production, the beef cattle industry in California clearly outranks other commodities in its financial dimensions.

More than two-thirds of California land area is essentially nontillable because of steep slopes or poor soils. Much of this area is, however, available for grazing animals, and the forage is best utilized by ruminants. Some of this land is appropriate for grazing cattle and sheep together; however, vast areas of California are not suitable, or at least are not used for sheep because of terrain, plant species, or predators. These areas are typically utilized as rangeland for beef cattle. In addition, beef cattle are produced on irrigated pasture lands in the low foothills where irrigation water is available and economically possible to use; and in the great valleys of California (Sacramento, San Joaquin, and to a lesser extent, Imperial). The cattle feeding industry, i.e., the fattening of cattle under confinement in feedlots, is an important segment of California's beef cattle production, located primarily in the Imperial and San Joaquin valleys.

California ranked seventh in the United States in beef cattle production in 1978, surpassed only by Texas, Iowa, Kansas, Nebraska, Colorado and Oklahoma. Significant numbers of calves and yearlings are shipped from California's rangeland and pastures each year to finishing feedlots throughout the West and Midwest. This is because investment, labor and feed costs are usually lower in other states. Because California must import beef to feed a large and growing urban population, many of these California calves, finished and slaughtered in other states, find their way back to the state in the form of carcasses or wholesale boxed beef cuts, plus a multitude of by-products (hides, medicines, gelatins, etc.). In recent years, differences in freight rates and lower costs have made it possible for midwestern meat packers to ship beef into California and compete with west coast slaughterhouses for a burgeoning meat market.

Beef cattle are distributed throughout California, with commercial production taking place in every county except San Francisco. Climate, to-

pography, and overall conditions vary widely within the state and so, too, do the sizes and types of beef cattle operations. The leading counties in California for beef cows and heifers that have calves are: Modoc (northeast section); Merced (upper San Joaquin Valley); San Luis Obispo (central coast); and Siskiyou (northern). Even the counties of Los Angeles, Orange, San Bernardino, and Riverside, which have large highly urbanized areas, have considerable numbers of beef cattle grazing foothill, high desert and mountain areas. Figure 1 displays the distribution of cattle throughout California.

On January 1, 1980, beef cattle were reported on 23,300 farms in California and the total inventory of beef cows and calves plus heifer replacements was 1,078,000 head. Over 80 percent of California beef cattle operations are in herds less than 100 head. These herds, however, contain only about 10 percent of the beef cattle. Herds of 100 to 500 head comprise 11.5 percent of the ownerships, with 23 percent of the beef cows. Larger herds (500+ cows) make up only 5.5 percent of the ownerships, but contain 66.5 percent of the state's beef cows. Cattle numbers in California during 1980 were down approximately 18 percent from the all-time high in 1974. Current numbers are more than double the 1920–29 population and nearly a third more than the 1950–59 period. Calves under 500 pounds numbered 1,019,000 on January 1, 1980.

In addition to cows, calves and replacement heifers, a large number of stocker or yearling cattle are purchased by cattlemen to harvest rangeland forage growth or to consume feeds produced as by-products from other agricultural enterprises. Stocker cattle (steers and heifers) numbered just under one million head in January 1, 1980. These cattle are normally placed on rangelands in the fall of the year, utilize natural forage growth during the winter and spring months, and are then removed from the range when the annual plants mature and dry in late spring and early summer. Stockers may be purchased from other areas of the state or they may be shipped in from other states.

In most areas of California, beef cow breeding takes place in winter to early spring with calves being born in the fall to early winter. In some mountain areas, however, cows are bred for calves to be born in the spring and early summer, to avoid winter weather. The average annual calf crop or calving percentage is approxi-

2. In recent years the value of milk produced from dairy cattle has exceeded the value of production (calves) from the state's beef cow herd. Also, of the animals marketed for beef in the state, culled cows and bulls from California's dairy herd make up a significant part (nearly 20 percent).

FIGURE **9.1** California cattle population, January 1980. 1 dot = 1,000 head. *Source: California Crop and Livestock Reporting Service.*

mately 85 percent. A major goal of brood cow owners is to increase the percentage of calves born, for this is a major factor in the level of economic return for the cow-calf operation. Careful attention to nutrition, health and herd management helps increase the number of calves conceived and born. Research at the University of California and other leading institutions into improved breeding practices, handling techniques, physiology, artificial insemination, increasing ovulation rates and behavior are helping to improve reproduction and calving rates.

Research in genetics, improved selection, and electronic processing of performance data and information has also contributed to improving efficiency of production. New understanding of the endocrine system of cattle is revolutionizing the cattle industry. For example, discovery of prostaglandins (natural hormones that govern body functions) is allowing beef cattle producers to control breeding programs, facilitate embryo transfer, encourage artificial breeding in commercial herds and make multiple ovulations (twins) possible. Research at the University of California has also produced improved methods of feeding beef cattle (net energy system), computerized ration formulation programs, new vaccines and health products, and methods for producing leaner beef carcasses. Major industry groups, such as the California Cattlemen's Association, the California Cattle Feeders Association, the Western States Meat Packers Association, and the California Farm Bureau Federation have co-operated with and supported this research.

Almost all breeds of beef cattle are raised in California, either in purebred form or under crossbreeding plans. The "British" breeds such as Hereford, Polled Hereford, Angus and Shorthorn form the historical and current basis of the commercial beef cow herd, with significant infusions of other breeds newer to the United States being used in crossbreeding programs to utilize specific traits. Examples of these breeds are the Brahman (India), Charolais (France), Simmental (Switzerland), Limousin (France), Gelbvieh (Germany), Chianina (Italy). The British breeds are utilized for their adaptability and usefulness under a wide variety of conditions, while the "Continental" or exotic breeds from Europe contribute improvement in traits such as growth rate, mature size and milking ability.

Management conditions vary widely for beef cattle in California—irrigated pasturelands in the interior valleys with intensive production; arid high desert in southern California requiring vast areas to support cattle; south and central coast and north coast areas with mixtures of high rainfall and cool climate; rolling foothills adjacent to inland valleys producing annual forage in winter and spring; the intermountain region of the northern and northeastern part of the state with warm summers and cold winters; and finally the great Sierra, which provides summer grazing but cannot support year-round operations because of severe winters. All of these areas present separate and distinct challenges to beef cattle producers in terms of rainfall, temperature patterns, topography, breeding and calving conditions, transportation, marketing, urban development, and cattle rustling and vandalism.

Beef cattle enterprise returns for the owner can be quite variable, depending on production costs (land, labor, interest on borrowed and invested capital), weather conditions that affect forage production and cattle health (drought, flood, extreme cold or heat, parasites, diseases) and fluctuating prices. Beef cattle production is usually considered a "narrow margin" agricultural enterprise which requires strong and capable management as well as dedication. Investment costs for cattleland and equipment are high. Some years, especially during prolonged drought, bring poor forage growth on rangeland and pastures. Market prices are volatile. The costs of producing cattle can actually be more than the return when market prices are low.

Historically, beef cattle population and prices have been governed by what is known as the "cattle cycle." When cattle numbers are large, prices usually decline and many production units become unprofitable. This prompts the sale of cows and heifers for slaughter, with a resulting decline in calves born in subsequent years. As cattle numbers are reduced, demand for beef eventually causes an upward trend in prices. When profit margins return, beef cattle producers tend to retain cows and keep more replacement heifers and thus the cattle population increases again. In the past, cattle populations have completed a "cycle" each five to seven years—depending on prices, profits and losses sustained by producers. Unusual weather patterns, such as the drought of 1976 and 1977 in California, can also affect cattle numbers and herd sizes. If range forage is not available, feed must be supplemented or herds reduced. Often, beef herds are combined with other agricultural operations to form diversified farms, or herds are managed primarily to supplement incomes.

From Range to Feedlot

The raising of cattle on farms and ranches, whether on irrigated pasture or on open rangeland, is only part of the total picture of beef production for human consumption. Since 1950 a real revolution has taken place in the fattening of cattle for market. Commercial or custom feed yards first were established in California in the 1920s and 1930s primarily for winter feeding or for "finishing" cattle before slaughter, but increasing knowledge of scientific feeding for weight gain, and desire to utilize agricultural by-products for animal growth and finishing, have brought about great changes in cattle production and management.

Although cattle feeding had its start in the 1930s, growth was modest until after World War II. Then the industry exploded. Cattle feeding became big business, characterized by large, highly mechanized feedlot operations. On January 1, 1945, there were only 125,000 head of cattle on feed in California. Twenty years later, on January 1, 1965, the figure had climbed to over one million head. Numbers of feedlots and cattle on feed further increased in California in the 1960s and then declined. (See Table 1). Decline in numbers of feedlots has been almost fivefold from the peak in 1965 to the recent low in 1979. During the same period, cattle marketed declined 40 percent.

The rapid growth and subsequent decline in cattle feeding in California is the result of many factors. Growth can be attributed to the warm dry climate ideal for cattle feeding, the population boom following World War II, and the high per capita income in California, which is associated with higher consumption of beef. In addition, the availability of cheap roughage and by-product feeds encouraged feedlot expansion. Two other important factors in the growth of cattle feeding in California were an early lead in technology and expertise as a result of research here, and greater availability of credit in California in the 1950s. These advantages had, however, dissipated by the 1970s.

On the negative side, the cattle industry is cyclical in nature and fluctuates when certain forces are imposed on it. Aside from cattle numbers, other factors which influence the cattle feeding industry and its growth or decline are: 1) excess or shortage of feed supplies, 2) the general consumer economy, 3) government programs, 4) weather, and 5) wars. California cattle feeders are now also facing intense competitive pressure from both within and outside of the state for supplies of feed commodities and feeder cattle.

The San Joaquin, Kern, Imperial and Palo Verde valleys, along with the Central Coast area, are the most important cattle feeding areas in the state (see Table 2).

Cattle fed in California feedlots today are relatively light compared to years past. Approximately two-thirds of the cattle fed weigh less than 500 pounds and some as light as 150 pounds. About 80 percent of the cattle fed originate from outside the state (see Table 3). Texas is the largest supplier of cattle for California feedlots, leading the way with 38 percent. The majority of feeder cattle are shipped into California by truck.

With the exception of a few feedlots that are linked with farming operations, the feeding industry depends almost entirely on commercial sources for feed grains, by-product feeds, roughages, proteins, minerals, vitamins and feed additive supplements. Two-thirds of the concentrate feeds are imported from sources outside of Cali-

TABLE 9.1

Trends in California Feedlots

Year	Number of Feedlots	Capacity	Marketed
1960	559	1,271,000	No data
1965	563	1,845,000	2,282,000
1970	410	2,037,000	1,966,000
1975	156	1,657,000	1,649,000
1980	101	1,102,000	1,253,000

SOURCE: California Crop and Livestock Reporting Service, 1980.

TABLE 9.2

Number of Feedlots and Capacity by District, 1980

District	Feedlots	Capacity
Sacramento Valley and other northern counties	13	33,300
Central Coast	7	117,000
San Joaquin Valley and Kern County	27	267,000
Kern, San Luis Obispo and southern counties	20	171,900
Imperial Valley	34	512,800
STATE TOTAL	101	1,102,000

SOURCE: California Crop and Livestock Reporting Service, 1980.

TABLE 9.3

Cattle Shipped Into California, 1978

State of Origin	Stockers and Feeders 1978	Immediate Slaughter 1978
Arizona	397,000 head	184,000
Colorado	22,000	8,000
Idaho	96,000	30,000
Louisiana	25,000	n.a.
Mexico	13,000	—
Mississippi	26,000	—
Montana	33,000	4,000
Nevada	131,000	44,000
New Mexico	37,000	2,000
Oklahoma	25,000	—
Oregon	221,000	20,000
Texas	731,000	10,000
Utah	41,000	15,000
Washington	10,000	4,000
Wyoming	14,000	2,000
Other	93,000	n.a.
TOTAL	1,915,000	323,000

TABLE 9.4
Cattle Products and By-Products

MEATS	EDIBLE BY-PRODUCTS	PHARMACEUTICALS
Steaks	Oleo stock	Rennet
Roasts	Oleo oil	Epinephrine
Ground beef	Gelatin	Thrombin
	Canned meat	Insulin
VARIETY MEATS	Natural sausage casings	Heparin
Liver		Progesterone
Brains	INEDIBLE BY-PRODUCTS	Thyroid extract
Tongue	Leather	Cholesterol
Ox joints	Sports equipment	TSH
Kidneys	Surgical sutures	Estrogen
Tripe	Soap	ACTH
Sweetbreads	Cosmetics	Trypsin
Heart	Buttons	Bile salts
	China	Cortisone
	Photographic film	Liver extract
	Violin strings	
	"Camel hair" brushes	
	Explosives	
	Blood	

California cattle, reared on rangeland or pasture, are usually finished in feedlots. Nearly every county reports cattle as a commodity, but most feedlots are in the southern San Joaquin Valley or the southern desert areas.

fornia, but California feedlots purchase 95 percent of their roughage from sources within California.

Most feedlots are currently using computer programs to develop least-cost rations. Computer programs are also being developed to supply detailed information of feed inventory, cattle moving onto feed, cattle on feed, performance and cost projections of the cattle, and fed cattle sold.

The majority of fed cattle in California are sold directly to California packers. Cattle are sold on a direct liveweight basis with a standard shrinkage assessment. Nearly all beef carcasses are inspected and graded with government grades, packers' brand or chain store specifications before being sold for distribution in retail markets. Packinghouses feel the economic pressure to make every conceivable item useful and by-products from beef carcasses are continually being utilized in new and better ways. Table 4 identifies just some of about 80 by-product materials produced from cattle.

The California Dairy Industry

The modern history of California's dairy industry does not begin until the Gold Rush that followed the discovery at Coloma in 1849. Cows tramped overland behind covered wagons and supplied milk to the infants and children of the pioneer families. By 1859 there were 264,000 people in California and 104,000 cows. The 1860 census also indicated that there were two principal dairy regions in California, the San Francisco area and the Sacramento Valley. Importation of good dairy stock was increasing; herds of cattle selected specifically for milk production ringed the population centers. As cow numbers increased, milk in excess of fluid needs was used for butter and cheese production. Table 5, developed primarily from census data, provides an indication of the early growth of the dairy industry in California.

As roads improved, the production of butter and cheese moved farther afield where seasonal pasture was more abundant. But transportation limitations kept these operations within horse and wagon reach of the metropolitan areas until the railroad era.

The Humboldt Bay area was unique in its rise to prominence as a dairy area. Good cows and excellent pasture combined to produce butter and cheese economically for shipment by sea to San Francisco and other ports. As the population of the Bay Area increased, the counties to the north and south became important milk supply areas for San Francisco; East Bay distributors reached eastward to Contra Costa County and later into the interior valleys.

Dairying development in Southern California followed the same evolutionary process. Los Angeles required a milk supply to match its phenomenal growth and the dairy industry responded. In 1925, Los Angeles County became the state's top milk producer. As population growth accelerated, all of the adjacent counties became absorbed into the Los Angeles milkshed. Because of greater input costs, Southern California producers concentrated their efforts on meeting the demand for fluid milk primarily, leaving the production of milk for manufacturing to distant producers. With this specialization came a departure from the traditional methods and new management concepts emerged. Hay was more expensive than in other areas, so more concentrates (feed grains and by-product feeds) were fed. Cows responded with higher milk production, an important profit factor. Work organization, specialized labor, and mechanization increased labor output. Higher hay costs were offset by greater production efficiency and a shorter milk haul as compared to San Joaquin Valley producers. These innovations were eventually to have considerable impact worldwide.

The vast dairy potential of the San Joaquin Valley had to await a series of successive developments before it could be realized. As the luster of the Mother Lode wore off and farming became the conventional way of life, the population growth of the valley lagged, and so did the market for fluid milk. Butter and cheese had a limited market until the railroads provided access to

TABLE 9.5

Historic Dairy Data—Through 1910

Year	Milk Cow Numbers	Butter Manufactured	Cheese Manufactured
	Thousands	*Million Pounds*	*Million Pounds*
1860	205	3.1	1.3
1870	164	8.0	3.4
1880	210	16.2	3.7
1890	317	27.0	4.9
1900	307	34.0	6.9
1910	382	52.6	4.3

the major markets. The centrifugal cream separator and power churn both gave butter production a substantial boost. Equipment and technology for the production of condensed and evaporated milk was improving rapidly at the turn of the century and provided another attractive alternative for milk. The production of non-fat dried milk soon followed and provided an outlet for that fraction of whole milk not used for butter. With such versatility, the San Joaquin Valley became firmly established as California's principal dairy region by 1910. But for many years to follow, Valley milk was used predominantly for manufacturing puposes, which netted producers considerably less than milk used for fluid sales. As milk supplies surrounding the metropolitan areas grew tighter during the 1930s, sweet cream began moving from the Valley to Los Angeles and the Bay Area. Subsequently, the Bay Area began to draw fluid milk supplies from San Joaquin and Stanislaus Counties and later Merced.

The rapid increase in need for milk in Southern California during World War II provided some Valley producers with a temporary market for fluid milk. However, the postwar population explosion increased demand and milk began moving southward routinely from Kern, Tulare, and Kings Counties. The Central Valley area, primarily Fresno and Madera Counties, became the pivotal area; fluid milk beyond local needs moved both north and south, depending on which metropolitan area was in greater need.

Ultimately, the San Joaquin Valley emerged as a vast reservoir of milk from which both major population centers drew to supplement nearby sources. The milk tanker, a large truck specifically designed to transport milk in refrigerated stainless steel tanks, played a key role in providing the dairy industry mobility and flexibility. Availability of low cost fuel and improvement of the highway system allowed development of these large trucks in the late fifties. The tanker fleet is now sizable, capable of balancing milk supplies from border to border. Milk not needed for fluid use (about 42 percent statewide) remains in the area of origin and is utilized for manufacturing purposes to minimize hauling costs.

Table 6 displays the dairy industry's growth through the years. Table 7 shows trends in the production of the major manufactured dairy products. Table 8 shows the distribution of California's dairy cow population and milk production by region.

The road to progress for the dairy industry was not smooth. The early years were marked with instability; dairying became very dependent on feed production. The U.S. butter market alternated between an import and export market from year to year, and this caused sharp price changes.

Determined to expand and improve its College of Agriculture, the University of California in 1913 had created a new division, Agricultural Extension. After study, this fledgling organization concluded that the dairy industry was an "in and out business" and that the average dairyman only made money when hay was cheap. The economic environment was not conducive to herd improvement; the 1920 census showed that the average production per cow in California was

TABLE 9.6
Growth of Dairy Industry in California

Year	Average Number Cows Milked	Average Production Per Cow		Total Production		Cash Receipts
	Thousand	Pounds		Million Pounds		Thousand Dollars
		MILK	BUTTERFAT	MILK	BUTTERFAT	
1924*	569	5,870	223	3,340	127	74,546
1930	611	6,550	249	4,002	152	90,121
1940	705	6,940	267	4,893	188	91,539
1950	777	7,710	301	5,991	234	237,042
1960	824	9,780	362	8,059	298	380,624
1970	755	12,526	450	9,457	340	534,049
1978	846	14,027	506	11,867	428	1,228,958

*Earliest comprehensive statistics available.

TABLE 9.7
Manufacture of Dairy Products in California

Year	Butter	Cheese	Cottage Cheese	Nonfat Dry Milk	Condensed & Evaporated Milk	Ice Cream
	Million Pounds	*Million Pounds*	*Million Pounds*	*Million Pounds*	*Million Pounds*	*Thousand Gallons*
1900	34.0	6.9	—	—	—	4.3
1910	52.6	4.3	—	—	—	16.9
1920	68.1	13.0	1.3	5.3	83.5	6.5
1930	73.9	9.8	18.0	32.6	200.2	14.2
1940	69.8	16.0	25.1	46.0	271.0	20.4
1950	34.6	10.1	77.4	47.1	405.0	38.8
1960	34.3	18.3	129.7	72.6	246.9	56.4
1970	86.0	17.5	165.3	118.5	214.9	72.0
1978	138.9	137.0	146.4	158.03	109.5*	88.2

*Condensed skim milk only; evaporated milk not reported

TABLE 9.8
Distribution of Milk Cows and Milk Production, 1978

Region	Percentage of Milk Cows	Percentage of Milk Produced
North Coast	2.4	1.6
Central Coast	8.7	7.4
Sacramento Valley	5.5	4.2
San Joaquin Valley	53.0	54.4
Southern California	29.5	31.9
Other Regions	0.9	0.5

only 182 pounds of butterfat annually. At the Pacific Slope Dairy Show in Stockton on December 5, 1921, B. H. Crocheron, Director of the Agricultural Extension Service, announced that a dairy campaign was being launched to raise the average production per cow in California to 265 pounds of butterfat by the end of 1930. This goal was selected because the lower half of the cows in the Ferndale Cow Testing Association averaged 265 pounds of butterfat the previous year. (Established in 1909, this cow testing association—now known as the Ferndale Dairy Herd Improvement Association (DHIA), established in 1909—was the first in California and is the oldest in continuous service in the U.S.) The goal seemed impossible because during a nine year period an increase in production per cow 15 times as great as achieved the previous 20 years would be required.

However, preliminary data for 1930 indicated that the average butterfat production per cow totalled 265.6 pounds; the campaign seemed to have achieved its goal. Subsequent data (Table 6) show that the goal was not actually achieved in that year, but the alliance between education and application remained a permanent fixture. Greater participation in production recording programs, intensified genetic iprovement through sire selection and cullig, improved nutrition, herd health, and management became permanent goals. In terms of production per cow, California dairymen became the nation's pacesetters and remained unchallenged until the 1970s.

Important though it may be, high production per cow does not assure prosperity. Prices of dairy products dropped sharply following the stock market crash in 1929 and lagged badly below parity for several years. Not until milk production was stimulated by federal subsidy in 1943, as food supply insurance under the threat of war, did milk prices rise to parity level. Once the pump was primed, milk production increased continuously for almost two decades. The post-World War II population explosion created a strong demand for fluid milk. European demand for dairy products remained strong during the post-war recovery period. As Europe recovered, however, storage stocks of dairy products piled up. Prices broke sharply in 1953, touching off a long period of market turbulence. The price gap between milk for fluid use and manufacturing milk widened, and "shipping rights" for Class I milk became a valuable asset. Competition for a Class I market became fierce.

Market abuses became so prevalent that sufficient unanimity was finally achieved among producers to pass the referendum which put the Milk Pooling Act of 1967 into effect. Producer morale soared to new heights in 1969 when pooling actually began, even though producer prices remained relatively unchanged. Producers now were able to sharply reduce their output of manufacturing milk if its production was not profitable.

Feed grain exports increased substantially in 1972 and feed prices increased sharply. The feed market became very sensitive to export fluctuations for several years, but producers were saddled with a price determination system that was slow to respond. Periodically, producers were left in a serious deficit position for several months before milk prices could be adjusted. Formula pricing became effective in 1978 and has allowed milk prices to be more responsive to prevailing economic circumstances. This formula gives consideration to production costs, manufacturing milk prices, and the consumers' "real net spendable earnings."

The Dairy Revolution

Milk producers in Southern California established themselves as renegades from traditional dairying. Dairymen elsewhere scoffed at the level of concentrates they fed and wondered how a dairy farm that produced little or no forage could survive. But as milk-per-cow output in Los Angeles and Orange Counties increased, many critics went south for a second look. Levels of concentrate feeding increased elsewhere; the number of cows milked per milker increased as the Los Angeles system spread northward. After shaking off the war-oriented economics of the 1940s, the dairy revolution spread statewide. California already had the highest percentage of cows on test in DHIA in the U.S., and this contributed significantly to the data needed to manage dairy herds efficiently. Dairying became an independent enterprise, not just another segment of the farm enterprise—even in feed producing areas. Specialization had arrived on the dairy farm.

Mechanization took giant steps forward following World War II. Improved field choppers and feed wagons made it feasible to bring freshly chopped alfalfa to the manger instead of allowing the cows to do their own harvesting. Field chopping allowed better alfalfa management and provided greater yields than with grazing. Irri-gated pasture gave way to alfalfa in most areas because higher land values demanded greater forage yields; irrigated pasture became more difficult to manage properly as herds became larger. These factors combined to hasten the conversion to drylot dairying where cows are maintained without pastureland. In such arrangements the cows are concentrated in lots or corrals, and all feed and water brought to them.

Mechanization and specialization also demanded drastic changes in milking and milk handling systems. Pipeline milkers eliminated the need of carrying milk from cow-side to the dump tank. Milk tankers stimulated the conversion of farm storage to refrigerated bulk tanks; the milk can rapidly became a collector's item. Surface (aerated) milk coolers gave way to plate coolers; now milk could be transported through a closed stainless steel system from the cow to the bulk tank. Refrigeration systems with improved capacity and efficiency facilitated this transition. Herd growth hastened the transition to improved milking equipment, because systems with insufficient capacity frequently caused milk quality and herd health problems.

Full benefit of milking equipment improvement could not be realized, however, until milking barns were modified to complement the equipment. Few barns met their life expectancy without extensive modernization or replacement; many have undergone this transition within the last ten years. Most large milking barns that accommodated a large portion of the herd at one time have been replaced with milking parlors of several designs. In the parlors and most large barns remaining, the cows are channeled to the milker and his stationary milking units instead of the milker moving his machines from cow to cow as in years past.

Mechanization, specialization and building modernization achieved during the dairy revolution required enormous capitalization. Such improvements were feasible only if the cost could be shared by a large enough number of cows. As a result, herds grew rapidly in size and at an increasing rate, while many small dairy operations opted out.

Neither specialization nor mechanization could happen spontaneously; both required an entrepreneur. A new breed of dairyman was emerging, combining financial skills, knowledge, ability, and initiative to integrate innovations into a productive and profitable system. Capitalization required extensive financing and allowed little tolerance for management error.

By the 1970s, electronic data processing of individual cow records provided better management information. Participation in DHIA increased significantly. Submarginal cows were detected and culled more quickly. Artificial insemination moved ahead rapidly and elevated the genetic base of the cow population significantly. DHIA records provided a means of evaluating the genetic performance of sires to speed genetic improvement. "Least cost" formulation of concentrate mixtures was quickly adopted by feed manufacturers. Drylot dairying allowed producers to feed each cow more precisely to her need to sustain higher production. The use of computerized "complete rations" is increasing rapidly and computers are being harnessed to an increasing number of management decisions. Table 9 gives some indication of the net impact of the dairy revolution on the make-up and performance of dairy herds. There is ample indication that this revolution is sweeping eastward.

In 1940, California ranked thirteenth in cow numbers but the high output per cow raised her to seventh in total milk production and fifth in cash receipts for milk and cream. In 1978, California ranked third in cow numbers but second in both total milk production and cash receipts. Milk production alone (excluding the value of dairy beef and veal) contributed approximately one-eighth of the state's total value of agricultural production.

It was common knowledge years ago that cows in large herds could not produce as much milk as those in small herds because they could not receive the individual attention they needed. This "fact" became another casualty of the dairy revolution. Since 1960 cows in large herds have produced more than those in small herds by a significant margin. DHIA data for 1979 show that cows in herds of 500 or more produced roughly 15 percent more milk and 7 percent more fat than cows in herds of less than 100 cows.

Born under economic stress, the dairy revolution had its share of casualties as Table 9 reveals. The sharp break in milk prices during 1953 led to almost two decades of economic stress. Milk production efficiency became a necessity for economic survival and many producers could not comply. Almost 20,000 dairy farms were operating in 1950 while fewer than 3,000 remain in 1980. Dairymen were culled more heavily than their herds; those that survived are a select group.

DHIA data for 1979 show that 40 percent of the herds on test had fewer than 200 cows—the approximate maximum of family operation. Sixty-five percent had fewer than 300 cows. Small herds are even more prevalent among herds not in DHIA. Though the various input costs vary widely, labor costs usually approximate 15 percent of the total cost of milk production; this accrues to the operator of the family dairy. This advantage can provide the family farm a headstart to survival, providing it grows and improves as economics demand.

Who Reaps the Benefit of Milk Production Efficiency?

Milk production efficiency improvements first benefit the efficient dairyman, but only until those improvements become standard practice.

TABLE 9.9
California Dairy Trends

Year	Cows Milked	Dairy Herds	Average Herd Size	Average Milk Production Per Cow
	Thousands		*Cows*	*Pounds*
1945	800	25,265	32	7,150
1950	777	19,428	40	7,710
1955	840	17,224	49	8,620
1960	824	9,764	84	9,780
1965	783	6,183	127	10,830
1970	755	4,265	177	12,526
1975	800	3,402	235	13,566
1979*	871	2,864	304	14,408

*Based on preliminary data.

Consumers eventually reap the benefits in the form of lower milk prices—and these gains are permanent. Table 10 illustrates this point.

One hour's labor in 1970 purchased more than twice as much milk as in 1940—even more in 1979. If milk prices had increased as much as wages since 1950, a half-gallon of milk would have been priced at $1.55 at the supermarket in December 1979 in Los Angeles, and at $1.71 in San Francisco.

California's leadership in milk production efficiency becomes apparent when producer and consumer milk prices are compared with those of other states. California prices are among the lowest in the U.S., even though input costs in California are among the highest of the major dairy states. Feed grains, grain by-product feeds, and hay prices are much lower in the Central States. California is deficient in feed grains, and freight rates add substantially to grain purchase price. Feed is the largest cost factor in milk production, usually approximately 60 percent of the total cost. Land values, wages, and most equipment costs also are higher in California than in other major dairy states. These higher input costs, however, usually are more than offset in California by greater production per cow. During 1979, for example, the average California cow produced 37 percent more milk than cows in Iowa, 22 percent more milk than New York cows, and 86 percent more than cows in Alabama—25 percent more than the national average.

Edwin C. Voorhies, Professor of Agricultural Economics, University of California, stated in 1931, "If the state is to continue to produce large amounts of milk fat for other than market milk purposes, it must do so on the basis of cows of higher producing ability." The genetic advantage

in cow output is even more critical today. But genetics is not likely to sustain the current production advantage because the almost exclusive use of frozen semen in artificial insemination allows every dairyman in the U.S. an equal opportunity to secure semen from superior sires. Management proficiency is the California dairyman's only secret weapon.

Sheep and Wool

Early in California's history huge flocks of sheep grazed the lush mountain meadows and brushy mountain areas as well as the foothills. Sheep were an integral part of the extensive grain farming operations in the San Joaquin and Sacramento valley. Grazing of grain stubble to glean fields following horse and mule drawn harvestors was a common and useful practice. Many of these large flocks were trailed or driven to summer pastures in the Sierras and coastal mountain ranges during the summer and early fall months. The sheep population in California at the turn of the century was many thousands higher than today. The 1930s recorded California's highest sheep population, with an yearly average for the decade of 3,020,000 head. Early chronicles listed more than 300,000 sheep and goats in Tehama County alone in the early 1900s. During that time, while steamboats and barges still navigated the Sacramento River, Red Bluff rivaled San Francisco as a wool shipping point.

Restrictions on controlled burning of brush fields and reduction of grazing permits on public lands plus an increase in losses from predators drastically reduced the area available for sheep grazing and contributed to a significant reduction

TABLE 9.10
Milk Purchased by Hourly Earnings in Manufacturing Industries

Year	Average Store Price Milk		Average Hourly Earnings		Quarts Milk Purchased By Hourly Earnings	
	Cents/Half-Gallon		Dollars			
	LOS ANGELES	SAN FRANCISCO	LOS ANGELES	SAN FRANCISCO	LOS ANGELES	SAN FRANCISCO
1940	21.2	22.4	.74	.82	7.0	7.3
1950	36.0	36.0	1.62	1.71	9.0	9.5
1960	44.5	47.6	2.59	2.79	11.6	11.7
1970	51.2	52.7	3.66	4.25	14.3	16.1
Dec. 1979*	90.0	85.0	6.96	8.12	15.5	19.1

*Data no longer available to calculate annual averages.

FIGURE 9.2 California sheep population, January 1980. 1 dot = 1,000 head. Source: California Crop and Livestock Reporting Service.

in sheep numbers in California, and a resulting change in how sheep are raised and managed in the state. Sheep flocks still utilize some forage and brush areas on public lands today, but trucks have replaced the trail drive as the principal means of transporting sheep from the valleys to mountain areas.

Today sheep are raised in all parts of the state, in small as well as large commercial flocks. California ranks second only to Texas in the nation in sheep production. In 1980, over one million sheep and lambs were recorded in California on 4,000 farms and ranches (numerical average per flock, 250 head). In addition, more than 175,000 lambs were on feed in special feedlots or pastures within the state (primarily in Imperial, San Joaquin and Sacramento Valleys) as of January

1980. California is one of the few states to post a significant increase in sheep and lamb numbers in recent years (up from 900,000 in January 1977) and the trend is expected to continue at a slow but steady rate. The larger flock ownerships (1,000 ewes and more) are expected to remain fairly constant but in a strong position, and they will continue to be a major source of the state's sheep and wool production.

Many of the same topographical and climatic features that favor beef cattle production in California also apply to the raising of sheep. In some areas the availability of extensive grazing lands along with irrigated and dryland crop residues has encouraged the establishment of large flocks. Figure 2 displays the distribution of sheep throughout the state. Kern County recorded the largest sheep population in California with 165,000 sheep and lambs as of January 1980. Fresno County ranked second with 110,000. Both of these southern San Joaquin Valley counties have large areas of native foothill rangelands, plus extensive field crop areas that provide opportunities for the grazing of sheep on residues that would normally not be of value. The existence of large areas of productive rangeland plus close proximity to summer grazing in the surrounding mountain ranges has helped to maintain a viable sheep industry in northern California as well. Tehama County in the northern Sacramento Valley ranked third in the state in 1980 with 74,000 head of sheep and lambs. Solano County ranked fourth with 49,000, and Sonoma County ranked fifth with 47,000. A healthy mix of large, medium and small size flocks in these counties has been responsible for increasing sheep numbers. Los Angeles County, noted worldwide for its vast metropolitan areas, ranked seventh in 1980 with 44,000, most of these in larger sized flocks that utilize the high desert and rangelands to the east and north of the city of Los Angeles and surrounding suburbs. California is also a significant lamb feeding state, finishing a large number of feeder lambs from resident state flocks and importing additional lambs from surrounding western states.

Ewes normally have a five-month gestation period. Sheep are naturally seasonal breeders, with the normal breeding season in the fall of the year when days are shorter and temperatures cooler, and an anestrus period in spring and early summer. Many purebred flocks and smaller sized flocks are managed to produce lambs in the late winter or early spring. However, most of the medium and large flocks that utilize foothill rangeland, dryland pastures and mountain ranges, are managed for an early breeding season in late May, June and July, in order to have lambs born in the fall and early winter. Although this plan takes extra attention to the flock and increases the responsibility of the shepherd at lambing time, it helps toward better utilization of winter annual grasses and clovers. In addition, with reasonably normal rainfall and temperature patterns, many of the lambs born in the late fall and early winter can be ready for the spring market when consumer demand is strong. However, conception and twinning rates are lower in this early lambing system than in flocks managed for spring lambing.

Lambs are marketed as "fats" or finished at approximately 95 to 120 pounds, with the preferred range in weight at 100 to 110 pounds. In average seasons on good quality rangeland, a majority of the fall-born lambs can be finished entirely on mother's milk and forage within six months.

Sheep owners pride themselves in producing a species of livestock that can provide a desirable and highly palatable meat product and natural wool fiber for human use from forage alone. In some years, however, when drought strikes or temperature patterns greatly restrict the natural forage growth, hay, grain and by-product feeds required to finish "feeder" lambs to market weights. The number of feeder lambs therefore varies with environmental conditions in California.

Breeds of sheep in the state vary widely, partly because of environment and partly because of the interest and desire of the flock owner. In general, the commercial production ewe (regardless of large or small flock) is a "whiteface" crossbred. Some of the whiteface sheep used (but not a totally inclusive list) are Rambouillet (origin France), Corriedale (origin New Zealand), Columbia (origin America, developed from several breeds), Dorset (origin England), and Targhee (origin America, developed from several breeds). The whiteface breeds are noted for finer grade wool and hardiness under range conditions. Blackface breeds of sheep are also used widely in California sheep production for market lambs, and for producing rams to breed to whiteface ewes for high quality, improved conformation market lambs. Blackface breeds usually have faster growth rates and can reach heavier mature weights. The two most popular blackface breeds are Suffolk (origin England), and Hampshire (origin England). Shropshires (origin England) are also raised in the state to some extent. Some outstanding purebred flocks of both whiteface and blackface breeds have been developed and maintained in California; several have attained national reputation.

Sheep have a higher reproductive potential than cattle. The normal birthrate is an average of one lamb per ewe per year (100 percent); however, ewes in almost all breeds of sheep have a natural ability to ovulate more than one egg. Twins are common and triplets are not rare. Birth rate per flock can vary from 100 to 150 percent or even higher, depending upon breed, nutrition, season of breeding and lambing, and

A rancher in Amador County checks on a flock of sheep. Skilled sheepdogs help with herding.

flock management. Improved management (nutrition, selection, and facilities) can enhance the ability of sheep to attain a higher reproductive rate. Survival of baby lambs at birth and for the first 24 to 48 hours is critical and extra care must be taken at lambing time by the shepherd.

In recent years, new breeds of sheep have been utilized in the U.S. for specific characteristics. For example, Finnish Landrance (Finnsheep), developed centuries ago in Finland, are noted for prolific birth rates; triplets are common, with litters of four or five baby lambs from pure Finn ewes common. Finnsheep are being crossbred with California ewes to produce replacement females that will conceive and bear more lambs. A line with one-fourth Finn blood can give birth to about 20 percent more lambs than a straightbred ewe. Other breeds, such as the Barbados which have hair instead of wool, are being researched for crossbreeding because their anestrus period is shorter or absent, and thus they breed more readily out of season.

Interest and demand for wool has increased in recent years as consumers have returned to natural fibers both for comfort and for economy as the price of manmade fibers increases due to the rising cost of petroleum. Blends of wool and manmade fibers are also increasing in popularity, to utilize the best qualities of both. Wool also contains a natural oil, lanolin, which is utilized in a wide variety of products for skin care and lubricants.

Sheep are normally shorn in the spring in California. The average fleece weight for the wool breeds is ten pounds per head; for dual purpose breeds it is six pounds. More than 10 million pounds of wool were produced from California flocks in 1980, approximately 10 percent of the total wool shorn in the United States. Almost all of the wool produced in the state, however, is shipped to other areas or countries for scouring, cleaning, carding (combing), and spinning into yarn and fabric. Some fleeces are marketed in California for home spinning and weaving, an ancient art that has seen a strong revival in recent years.

Most commercial sheep ranches maintain shearing sheds or barns that can accommodate large shearing crews which travel from southern California, where shearing begins first, to the north part of the state. As many as 12 to 20 shearers in a crew will complete the process in a few days for each large flock. Usually the sheep owner contracts with a "crew boss" for shearing, the price being determined on a per head basis. Smaller flocks can be pooled in a central location or shorn by individual shearers on the ranch or farm. Shearing schools are held at several locations in the state each year to train individuals in the art of shearing sheep. Shearing is done primarily today with electric clippers that cut the fleece very close to the skin.

Increasing numbers of individuals and families are acquiring sheep in California, especially in the mid- to small-size flocks. More Californians are "re-discovering" sheep for their adaptability to small acreages, ease of handling, and compatibility with other ranch enterprises.

California should retain its national ranking in sheep and wool production in the foreseeable future. However, producers in the state face a declining slaughter and packer facility situation, along with increased pressure from predators, particularly coyotes and domestic dogs.

Swine

Because of the profitability in swine raising, the pig has become known among farmers as the "mortgage lifter." Today the hog industry in the U.S. is centered in the intensive corn-producing states from Ohio to Nebraska, primarily for two main reasons. First, hogs can be produced most economically in the midwest because corn, a major ingredient in hog feeding, is abundant in these states. Second, because of geography and transportation facilities, pork can be shipped quickly and economically to most of the major metropolitan areas in the U.S., including those of the west coast.

California is a state deficient in hog production, importing more than 80 percent of the pork it consumes. The number of hogs on farms in California on January 1, 1980, was 180,000 head, or less than 0.27 percent of the nation's total swine number. In 1979 a record of 1,440,000 live hogs were shipped into California for slaughter, and large tonnages of fresh and cured pork were also imported from out-of-state.

The value of hogs and pigs sold by California farms in 1980 was $28,412,000. The figures for 1970 ($11,064,000) and 1975 ($20,744,000) indicate a rapid increase in value of production for hogs during the 1970s, even accounting for the influence of inflation. Nevertheless, the value of hogs and pigs sold by California farms accounted for only 2 percent of the value of all meat animals sold in in the state in 1979.

California hogs are raised under many different conditions, from finishing one animal to large scale commercial operations. Large numbers of hogs are scattered widely throughout the Sacramento, San Joaquin and other valleys of California located near consumer markets in Los Angeles and San Francisco. The California hog industry is made up of generally small operations, often carried on as a side enterprise. Some large-scale, highly commercialized hog operations have developed, but growth is slow. Cost of land and facilities required for handling large numbers of hogs, pollution problems, and shortage of trained management personnel are factors which deter rapid growth of large hog operations in California, even though the state possesses an excellent climate and produces a diversity of feedstuffs suitable for pork production.

Goats

Goats have been widely raised in other areas of the world not unlike California in native environment, including neighboring Mexico, but in this state these animals have never been widely utilized commercially. They are, nevertheless, quite useful and highly adaptable animals, and recent years have seen a resurgence of interest in their potential as meat, milk, and fiber producers, as well as recognition of their value in brush control on rangeland.

There are distinct breeds of goats just as there are of cattle, and their uses and habits are quite different. Dairy goats have been bred specifically for milk production, while "Angora" goats are known for their fine fleece coats, useful in producing fiber for blankets, furniture coverings, and human apparel. "Spanish" goats are the ordinary goat frequently seen in the countryside of Mexico, where the meat of the young kid goats is frequently eaten roasted or barbecued.

Dairy Goats

Dairy goats have become very popular in California during recent years. The number of dairy goat owners continues to increase. Most seem to be family units who want dairy goats for home milk production, meat, breeder stock, or for 4-H Club projects. It is difficult for these small units to expand into commercial enterprises, since most non-commercial hobby owners seem to be emotionally attached to their animals and therefore do not operate with sound business principles for profit. State regulations governing commercial milk production are also severe, particularly for small herds, and since considerable capital is required for facilities, most goat owners do not wish to take the financial leap necessary to establish themselves in business. The number of commercial dairy goat herds in California is therefore extremely small in comparison with the number of family units milking goats.

Other economic factors inhibit the growth of commercial goat dairies. Goat milk producers have the same general problems as do cow milk producers regarding collection, processing, and distribution of milk and dairy products. Dairy goat owners are usually more widely dispersed or may be in marginal areas, making milk collection difficult and expensive. Processing in small batches also is inefficient and results in higher prices for comparable dairy products to consumers. If dairy goat milk and products were priced more competitively, perhaps there would be more demand, but because of the higher labor requirement per unit of dairy goat milk and the lower milk production per animal it is not surprising that prices for dairy goat milk must contain greater margins than for cows' milk.

Goat milk is almost pure white. The fat globules of goat milk are smaller and more easily digested and the curd is finer and more flocculent and is therefore more easily digested than cows' milk. For this reason, babies, invalids and others who cannot take cows' milk often thrive on goat milk. Goat milk, like cows' milk, varies with the breed, stage of lactation and individual animal.

The seasonal variation in goat milk production may also discourage consumer acceptance, since peak production occurs during the period of least consumer demand. The seasonal breeding of dairy goats is responsible for this paradox. Selection programs for year-round breeding should receive high priority if commercial dairy goat milk production is to become a reality in the future.

Like cows, goats are ruminants, so the principles applied to the feeding of dairy cows are appropriate to the feeding of dairy goats. Goats are browsers by nature, but dairy goats need additional grain and hay for milk production. About six to eight goats can be kept on the amount of feed required for one cow.

Hundreds of dairy goat owners are participating in the National Cooperative Dairy Herd Improvement Program. This program provides production testing and record keeping for management, breed and herd improvement programs, education and research programs. Many of the participants are serious breeders of purebred dairy goats for sale purposes.

Angora and Spanish Goats

Introduced into California in 1848, Angora goats were exhibited in the 1861 State Fair. The Angora fleece grows rapidly and produces a very fine and soft fiber called "mohair," used in apparel and fabrics where a soft finish or "hand" is desired. Although fine and soft, the mohair fiber has strength and durability and has been in increasing demand throughout the world in recent years, especially for blending with wool and other fibers. The price for the fiber has been quite strong. Mohair is processed into fiber much as fleece is converted into wool, although it does not have as high a lanolin content as wool.

Texas is the center of Angora goat and mohair production in the United States, with 95 percent of the national population. California ranks fourth among other states. The center of Angora population here is in Amador and Calaveras Counties, with additional populations in Tehama, Tulare, and several other counties ranging from Orange in the south to Trinity in the north. Angora goat kids, while not normally raised for slaughter, are a delicacy and in strong demand by several ethnic groups in California at certain times of the year.

Spanish or "hair" goats are used primarily in California for control and conversion of brush species in certain areas. They are also raised throughout California for production of kids and as pets.

More individuals and families have discovered the adaptability of goats to small acreages and the ability they have to keep pastures, lots and brushy areas free of unwanted vegetation. This is proving particularly valuable for brush areas where residential housing has been built. Several research and demonstration projects directed through the University of California at Davis have illustrated the advantage of goats in controlling unwanted vegetation for fire protection and control, watershed improvement and wildlife habitat improvement without the use of chemicals, while producing food and fiber products.

Horses

The first horse to reach California was brought here by Jose de Galvez in 1768. Under sunny California skies horses multiplied rapidly, and their numbers were often controlled by government-sanctioned slaughtering, in order to protect the golden bunchgrass of the cattle ranges.

The horse was the basis of all early California life, both work and play. Without the horse, utilization of this empire of free grass would have been impossible and the history of the conquistadores would have been a blank page. Afoot the Spaniard was just another man, but on horseback he excelled. Many of the Spaniards eventually married Indian women and in time passed their skills in horsemanship, speech, and dress to their offspring. These vaqueros became the backbone of the large cattle ranches of California. Spanish influence on our horsemanship, equipment, dress, hospitality, and ways of entertainment still exists in the Golden State.

In 1980 there were about 800,000 horses in California. Over 100,000 horses are used in managing cattle ranches and feed yards; thus the horse is still important in California agriculture as a working animal. Today, however, the horse is primarily a pleasure animal although it has other roles as well.

In 1978 over 10 million people attended horse races, and the state's share of the pari-mutuel betting was more than $117 million. Of this amount, $13 million went to support district and county fairs, where better practices in agriculture are demonstrated; $750,000 went to support research work on wildlife problems; and over $1 million was donated to charity organizations by the racing association.

The aesthetic value of the horse is an asset that is difficult to measure, but much of the color and romance of the West would be lost without the horse—both wild and domestic. Pleasure horses

also have an economic importance to the state of California. Horse shows and rodeos are popular forms of recreation for both participants and spectators. Many of these shows funnel some of their profits into charity organizations. In 1975, 595 rodeos were held in 40 states, attracting over 40 million spectators.

Horses are valuable research animals in both human medicine and in the field of animal science. Hormones from pregnant mare serum and the urine of stallions are used in human medicine. Pregnant mare serum is also used to stimulate multiple ovulation in beef cattle in connection with embryo transfer. Horses also produce the tetanus antitoxin used to control tetanus in humans. The number of horse research projects underway in the United States and Canada in 1975 totaled 268. This research, being carried out in 34 states, was funded by nearly $4 million.

Horses also contribute to the meat-packing industry in many states, where most of the horsemeat is shipped overseas for human consumption. In California, where horses are slaughtered only for use as pet food, three slaughterhouses process old and crippled horses for this industry. In 1979 3,265 head valued at $587,000 were processed.

Currently there are an estimated 1,500 head of wild (feral) horses roaming the ranges in Lassen and Modoc counties. These animals are under the supervision of the Bureau of Land Management and the U.S. Forest Service, which control the wild horse population through a horse adoption program and by disposing of the old, crippled, and diseased animals.

Poultry and Poultry Products

Chicken Eggs
Egg production in California started as a sideline on the early homesteads. By the 1920s, specialized egg farms began to appear largely in the coastal valleys close to population centers. By the 1930s the town of Petaluma in Sonoma County had developed into the state's major center of egg production, calling itself the "Egg Basket of the World." During depression years more than a thousand egg farms were located among the Petaluma hills; the sandy soils allowed for good natural drainage of chicken yards and neigh-

boring dairy farms supplied plenty of skim milk for feed. Eggs from over two million birds, raised in flocks of 100 to 3,000 birds, were transported to San Francisco for shipment throughout the state and to locations as far away as New York. During the early 1950s Petaluma production peaked at seven million laying hens, but only a few years later the crash in egg prices drove many small poultrymen out of business, particularly during the season of 1959–60 when prices were at their lowest point in 18 years.

It was during the 1950s that California became an egg-surplus state, as changing practices in poultry production encouraged growth in flock size while keeping costs and market prices down. Though the number of small operations has declined dramatically in the Petaluma area, over two million hens are still kept today on about 20 farms in Sonoma County. The typical flock is no longer 3,000 birds but ranges from 50,000 to 100,000, no longer roaming freely between chickenyard and henhouse, but confined in ranks of wire cages which expedite sanitation, feeding, and gathering of eggs.

Petaluma is a major example of the revolution that has occurred in poultry and egg production in California during the last 30 years. Concentration of production on fewer and larger farms has been encouraged by advancements in nutrition, breeding, mechanized feed handling, improved disease control, and the development of wire cages. As farms grew larger, they were forced out of coastal areas by urban pressures. Today poultry and egg production is predominantly in the San Joaquin Valley and the interior areas of southern California.

Changes in the egg industry are illustrated in Table 11. The cost of producing a dozen eggs

was about the same in 1978 (41¢) as it was in 1950, even though during the same period labor costs increased more than threefold and feed costs increased 61 percent.

Today California leads all states with the annual production of about 725 million dozen eggs, or about 13 percent of total U.S. production. About two-thirds of the state's egg production is concentrated in Riverside, San Bernardino, San Diego and Ventura Counties in the south. In northern California, Stanislaus and San Joaquin Counties lead, with significant production also in Sonoma and Merced.

Most egg production is now on large specialized ranches where hens are held in multiple bird cages. Egg collection, feeding and watering is often completely or partially mechanized. The 1960s and early 1970s saw a trend toward construction of environmentally controlled henhouses, but soaring energy costs have discouraged this in recent years. Today most new housing, though highly mechanized, is naturally ventilated to reduce energy use. The modern hen's environment is closely controlled, and her feed is carefully formulated to provide all of the nutrients needed for high egg production.

The eggs are washed, graded, weighed and cartoned by machines in modern egg processing plants at rates as high as 4,000 dozen per hour. Machines lift the eggs from the flats and carry them on specially designed conveyors until they are deposited in the cartons. During this process only the undergrade eggs (e.g., cracks, stains, blood spots, etc.) are touched by the workers who remove these undergrades as the eggs pass over a bright light. The eggs are washed during this process with warm water containing a detergent-sanitizer and brushing action. The

TABLE 9.11
Trends in California Egg Production, 1949–1978

Year	Average Flock Size	Eggs Per Hen	Feed Per Doz.	Labor Cost Per Hen	Estimated Labor in Min./Hen/Year	Cost of Production
	×1000		Lbs.			¢/Doz.
1949–50	2	215	5.5	$1.25	80	38.8
1959–60	8	242	4.5	.78	35	29.8
1969–70	46	238[1,2]	4.6	.40	10	27.0
1977–78	86	237	4.3	.46	8	41.0

[1] A change in management (molting) extended the length of a laying cycle.
[2] Four year average (1968–71).
SOURCE: Cooperative Extension, San Diego County Poultry Mangement Studies.

Chickens start life as they emerge from their shells in an incubator. The poultry industry has seen great changes since World War II.

washing is followed by drying (using blowers) and application of a light coating of mineral oil to reduce loss of moisture and preserve the firmness of the albumin. This entire process is accomplished in a few minutes and the eggs are returned to a refrigerated holding room for subsequent shipment to retail stores. Eggs are usually cartoned on the day after they are laid, and it is common for eggs to reach retail outlets during the same week. Eggs which fail to meet the Grade A standard are usually broken for use in the baking or food industry.

Today's improved housing, feeding, and management, combined with the breeding of hens for continuous laying, have resulted in consistent egg production all year round. Labor-efficient processing and improved transportation have enabled the producer to supply a much fresher egg to the consumer. As a result, "storage eggs" are a thing of the past.

Broilers and Fryers

During the last 30 years dramatic changes have also occurred in chicken meat production. In the 1950s it took 13 weeks and 10 pounds of feed to produce a three-pound fryer. Today three-pound birds are grown in six weeks with 5.4 pounds of feed. These advances in technology have made chicken the most economical meat available today. Consumption of chicken has risen steadily since 1940 to a present level of about 45 pounds per capita. The Republican political campaign slogan of 1928—"A chicken in every pot"—is now a reality taken for granted.

California produced 700 million pounds of fryers (sometimes called broilers) in 1980 with a farm value of $220 million. Production of fryers is now concentrated in the San Joaquin Valley. Five companies account for the majority of fryers processed, with the birds raised on either company-owned or contract ranches. Virtually all

A turkey grower inspects his birds. These broad-breasted white turkeys are many generations removed from their original wild forebears.

fryers produced in California are sold fresh dressed and command a premium price compared with imported fryers. California fryers are typically grown to heavier weights (average 4.4 pounds each) than imported broilers (average 3.9 pounds each). This practice results in a dressed bird with more finish, and also reduces the processing cost per pound. Despite the increased production of fryers in California, however, imports still account for about 40 percent of chickens consumed within the state.

Fryers today are usually grown in large open-type houses, using shavings or rice hulls for litter. Heat is provided until the birds are feathered. Ventilation is controlled by raising or lowering curtains which cover openings along the sides of the houses.

Turkeys

Improvements have also occurred in the turkey industry. In 1950, a tom turkey would reach a weight of 18 pounds in 26 weeks on 75 pounds of feed. Today, a tom will reach 18 pounds in 16 weeks on 45 pounds of feed. The modern turkey also has more breast meat and less waste.

Although turkeys, like chickens, were widely grown in small operations dispersed throughout the state before World War II, turkey production in California is now concentrated on large farms in the San Joaquin Valley. In 1980 over 415 million pounds of turkey were produced with a farm value of $178 million. California turkey production has ranged from 15 to 19 million birds per year over the last decade, or about 14 percent of the U.S. total. California usually ranks second or third among states in production of turkeys, its close competitors being Minnesota and North Carolina.

A young turkey called a poult begins life in a large machine called a "hatcher" where temperature and moisture are carefully controlled to assure hatching success. The entire hatchery is kept hospital clean so that the poults remain healthy. After sex separation and injection with a vitamin-antibiotic mixture, the poults are placed in clean plastic or new cardboard boxes for their journey to the turkey ranch.

A typical turkey growing operation consists of a brooding house, where the poults stay until they are feathered, and a growing house or yard. The floor of the brooder is covered with litter and poults are kept under the brooders (heaters) with special low fences until they learn to return to the heaters. They will then be given the run of the house or pen which typically holds several thousand poults. At about eight weeks of age, they are moved to a growing house and/or yard where they remain until marketing.

A new feature in turkey processing has been the development of specialized processed products. The meat is removed from the bones or the carcass is cut up to yield parts. The result has been a wide variety of new consumer products such as turkey roasts, parts, ham, hot dogs, pastrami and bologna.

Bees

Bees of various types are required to pollinate a large number of commercial crops grown in California. Honey bees (*Apis mellifera*), originally imported from Europe, are used most extensively in crop pollination. The alfalfa leafcutting bee (*Megachile rotundata*) and the alkali bee (*Nomia melanderi*) are particularly useful for pollinating dormant-type alfalfas in northern California. These three bees, with limited help from wild pollinators, make an important contribution to California's agricultural production. Other species of honey bees, adapted to different habitats, are little used here.

The honey bee industry of California began in 1853 when Christopher Shelton moved 12 colonies to a location just north of San Jose. During the 1860s and 1870s, thousands of colonies were moved across the country or sailed around the Isthmus of Panama from the East Coast, lured by the extremely good sage and buckwheat honey crops obtained in San Diego County. Currently, California has approximately 300 commercial beekeepers who are exclusively engaged in production of queens and bulk bees, production of honey and beeswax, or rental of bees for pollination. About 1,250 part-time beekeepers earn a portion of their income from beekeeping operations and nearly 20,000 beekeepers have a few hives which they operate as a hobby. Approximately 80 percent of the bees are owned by commercial beekeepers who collectively produce about 40 percent of their income from honey and wax, 40 percent from renting bees, and 20 percent by raising queens and bulk bees.

Bees must be kept in areas where nectar and pollen are available whenever production of brood (immature bees) is desired. Solitary bees,

This apiculturalist has smoked his bees into somnolence and is now examining the racks from the hive.

like *M. rotundata* and *N. melanderi,* rear a generation or two during the summer when food is available, then rest until the next year. Honey bees maintain an active population year round. The size of the population varies with the season. In spring the abundance of nectar and pollen producing flowers stimulates honey bees to increase brood production to 1,000 to 3,000 new bees per day. If a queen bee cannot perform at this level, the worker bees replace her. Hive populations reach a peak of about 45,000 individuals in early summer and maintain that level until food availability decreases. Populations decrease in the fall, reaching a low of approximately 15,000 individuals during winter when brood rearing is minimal or nonexistent. At this time beekeepers locate their bees as near to home as possible in case they require feeding to avoid starvation.

Honey production, with beeswax as an important by-product, can occur in many areas of the state but only where wild or cultivated plants are producing large amounts of nectar which can be collected by foraging bees and stored for future use. Wild plants which occasionally produce good honey crops are sages, wild buckwheats, manzanitas, yellow star thistle, and blue curls. Cultivated plants which may be good honey sources are citrus, alfalfa, cotton, lima bean and eucalyptus.

Approximately one-third of the commercial beekeepers run queen-rearing and packaged bee operations during March, April, and May in the Sacramento Valley. Packages consist of a queen and 2 or 3 pounds of bulk bees in a screened cage. Demand is great from beekeepers in the western provinces of Canada and the northern states of the United States who kill their bees to avoid feeding them honey all winter, or who lose bees which did not survive the winter. Northern California bee breeders ship approximately 350,000 packages to these out-of-state honey producers each year, and provide approximately 250,000 additional queens used by other beekeepers to requeen hives during the year.

Nearly all commercial and most part-time beekeepers are involved in pollination service to some extent. California has vast acreages of almonds which require cross-pollination in order to produce a crop. Almond growers rent up to four hives per acre during the time when bees are building populations rapidly in the spring and demand for pollen is high. Following al-

monds, bees are required for pollination of vegetable seed crops (see Table 12) in the Central and Imperial Valleys. These nuts, fruits, vegetables, and seed crops comprise about 10 percent of California's total agricultural output. However, if the value of other commodities dependent on these vegetable and forage seed crops (beef cattle, dairy cattle, etc.) are included, then bees are involved in the production of about half of California's agricultural output.

Hives of bees cannot be moved from one pollination rental to another indefinitely, because honey bee population densities which are ideal for maximum crop pollination are so dense that the bees are slowly starved in the fields. Therefore, hives periodically must be moved to areas where adequate food is available to allow bee populations to rebuild. Often these areas are where honey crops are produced, although there are not enough of these locations to meet beekeepers' needs. Bee populations can be maintained or increased on substitutes for nectar and pollen, but the expense is too high to be practical at current prices for honey and beeswax or pollination services. Frequently hives of bees must be placed adjacent to flowering commercial crops in order to obtain nectar and pollen. Pest control operations on those crops can be very detrimental to the bees.

California beekeeping is highly migratory. Most commercial beekeepers use boom loaders or fork lifts to load from 100 to 300 hives per truck at night and move them to new locations within the state or in other states, depending upon what the bees are to do next. Beekeeping is a labor intensive business, since bees must be examined at regular intervals to determine whether or not there is adequate food, whether or not there is brood being produced, and whether or not there is disease in the bees. If honey is produced, extra help is required to harvest and process the honey. Special operations, such as packaged bee production, require many extra people to visit apiary locations, collect bulk bees, catch queens, or fill packages with measured amounts of bees, a queen, and a can of food.

A bacterial disease, American foulbrood, is considered to be the most harmful infectious disease of bees, but it can be prevented by feeding antibiotics to bees during times of nectar dearth. However, American foulbrood causes very little loss in comparison with insecticides. Each year approximately 50,000 (or 10 percent) of the hives

TABLE 9.12
Crops Pollinated by Honey Bees

	Crops Dependent[1]		Crops Increased[2]

FRUIT AND NUT CROPS	Almond
	Avocado
	Apple—most
	  varieties
	Apricot—some
	  varieties
	Cherry
	Kiwi
		Pear—some
		  varieties; Bartletts
		  in unfavorable
		  weather years
		Plum
		Prune
		Tangelo
		Tangerine
			Apricot
			Bushberry
			Macadamia nut
			Olive
			Peach
			Pear
			Persimmon

VEGETABLE CROPS	Cucumber
	Melons:
	  Cantaloupe
	  Honey dew
	  Persian
	  Watermelon
		Pumpkin
		Squash	

OIL SEED CROPS			Flaxseed Safflower

FORAGE SEED CROPS	Alfalfa
	Alsike
	Berseem
	Birdsfoot
	  trefoil
	Ladino clover
		Red clover
		Sanfoin
		Crown vetch
		Vetch (purple,
		  common, and
		  hairy)
			Crimson Clover

VEGETABLE SEED CROPS	Asparagus Broccoli Brussels sprouts Cabbage Carrot Cauliflower Celery Chinese cabbage Collards Kale Kohlrabi	Leek Melons Mustard Onion Parsley Parsnip Pumpkin Radish Rutabaga Squash Sunflower Turnip	Eggplant

[1] These are unable to produce a commercial crop without cross-pollination.
[2] These generally produce a larger crop when honey bee pollinated.

in California are killed by insecticides. Specific laws have been passed to protect bees, but lack of cooperation has limited their effectiveness. Insecticides with reduced toxicity to bees are disappearing from the market for various reasons, and new formulations of old products, such as microencapsulation (insecticide wrapped in a polymer capsule the size of a pollen grain which can be carried back with the pollen), are potentially more damaging to bees than ever. Members of the California State Beekeeper's Association view pesticides as their most significant problem.

The University of California has the largest state-supported bee research laboratory in the country. Research emphasis is directed toward honey bee behavior, honey bee nutrition, and pollination biology.

Aquaculture

Aquaculture is the cultivation of freshwater and marine species. Recently interest in this industry has increased significantly as a result of a three-fold increase in public demand for aquatic products in the past 20 years and the realization that natural fisheries are limited in their ability to meet this demand. Aquaculture now accounts for approximately 10 to 12 percent of fishery products consumed worldwide. Traditional marine fisheries, once considered unlimited, are now thought capable of producing a maximum sustained level of harvest at 100 to 120 million metric tons per year, a limit that may be reached or exceeded by the turn of the century. Rapid development of the aquaculture industry is necessary to increase natural populations through enhancement programs and to fill the void between the supply and the demand for fishery products.

Aquaculture is not new to California, having originated almost 130 years ago with the establishment of oyster culture in San Francisco Bay. The first public fish hatchery was established on the University of California's Berkeley campus in 1870, while at the same time a privately owned rainbow trout hatchery operated on the Truckee River.

Because of California's diverse natural resources, a multitude of aquatic products can be cultured. The wide range of climates, the diversity of water conditions from cold mountain runoff to geothermal springs, and the 1100 mile coastline all contribute to California's potential as a major aquaculture producer. Commercial production already includes oysters, trout, catfish, baitfish, and ornamental fish, in addition to a large variety of plant and animal species undergoing commercial development.

Oysters

Early oyster culture consisted of transporting young oysters from the eastern coast of the U.S., and replanting them in San Francisco Bay. The eastern oyster, although unable to reproduce in the state's colder waters, did grow to market size. This transplanting, growout and marketing process continued until the 1930s when declining water quality made the Bay Area unsuitable for oyster culture. Because the oyster industry was not successful with the eastern oyster in other California bays, the industry remained depressed until the introduction of the Pacific oyster from Japan in the 1930s. The Pacific oyster, however, also does not naturally reproduce in California bays; therefore, the industry now receives its seed stock primarily from Canada, Washington state, and California hatcheries. The young oysters are planted on bay bottoms, grown on racks, stakes or trays, or suspended from lines attached to rafts in the bay waters. Today aquaculture provides over 99 percent of all the oysters produced in California with less than 1 percent from the natural fishery. The major growing areas are located in Humboldt, Drakes, Tomales, and Morro bays. Oyster hatcheries are expanding in California, producing seed oysters for U.S., European and Asian markets, and developing techniques for a number of new products.

Trout and Catfish

Rainbow trout and the channel catfish are produced both as stock fish for the multi-million dollar recreational fishing industry and as retail products supplied directly to the processed food market.

Trout have been grown in California for over 100 years, but in the past 15 years production has rapidly expanded. The technology of culture is well established. Trout are routinely reared in flow-through raceways in cold water (10–19° C), and fed a prepared diet. Well-managed operations using high-quality water average 15,000 pounds per year for each 450 gallons per minute of water flow. The market for this industry includes stocking of recreational fee-fishing lakes and of public waters. In addition, direct sales to food retailers have doubled since 1977. California is now recognized as a major producer of trout eggs, supplying both domestic and international markets.

Channel catfish are warm water fish, and obtain maximum growth rates at water temperatures above 25° C. They are also fed a prepared diet which contributes to their light, delicate flavor. Production systems range in degree of intensity from 2 to 40 acre ponds to newly developed controlled intensive tank systems employing geothermal water and oxygen regulation. In pond systems production averages 2,000 pounds per acre annually, although intensive tank culture shows promise of increasing this yield many times. As with trout, catfish in California are primarily supplied as stock for recreational fee-fishing lakes. The industry is expanding, however, to include fresh and processed fish, and California-grown channel catfish are appearing more frequently in neighborhood markets.

Sunfish, Carp, Bait Minnows and Feederfish

These fish are cultured exclusively for recreational fisheries, hobbyists or the aquarium industry. Sunfish, including bass, crappie, bluegill and red-ear sunfish, and Chinese carp are cultured in low densities in ponds located in warm water areas of the state. The sunfish and carp are used as stock in private and public fee-fishing operations. Ornamental koi, colorful members of the carp family, are also raised in pond systems; unique individuals command a premium price from fish hobbyists. Less valuable koi are sold in pet stores or served as feederfish in the aquarium industry.

Bait minnows are also produced for the recreational fishing industry, and millions of these fish are distributed to recreation areas throughout the state. Fathead minnows and golden shiners are primary members of this group. These fish are raised in extensive pond systems and shipped to retail distributors throughout California, the producers facing a highly competitive market. Goldfish are used as bait in the Colorado River recreation fishery, but the majority of these fish are distributed to the aquarium industry where they are used to feed other species of fish.

Table 13 reports recent aquaculture production statistics in California.

Research and Development

The state's variety of commercial aquaculture enterprises has attracted investors researching new aquaculture products. Areas undergoing rapid development toward commercialization include abalone culture in onshore facilities, in habitats suspended from offshore platforms, or on the ocean floor. Salmon ranching, the spawning and rearing of young salmon for release in the ocean and later harvest, is being explored at the pilot commercial level on the central California coast. Commercial oyster hatcheries are now investigating the cultivation of new species for both domestic and international markets. Figure 4 displays distribution of aquaculture facilities in the state.

Considerable aquaculture research is also underway in California's educational institutions, often in cooperation with state agencies and private enterprise. Broad university research programs, similar in scope to those for terrestrial agriculture, include work in nutrition, pathology and microbiology, aquatic engineering, systems analysis and economics, genetics, and toxicology. Such studies are applied to the culture of freshwater and marine species and for varying sites and scales of production. Research has led to the establishment of pilot-scale production of lobsters, rock scallops and kelp. Other projects have resulted in significant strides in sturgeon and striped bass culture. Public and private research efforts are investigating aquaculture culture of a number of species including abalone, mussels, finfish, crayfish, and freshwater and marine shrimp.

In 1980, over 300 freshwater and marine aquaculturists were licensed in California, including commercial producers and firms exploring new species development. Aquaculture represents a substantial contribution to the state economy through enhancement of our natural fisheries.

TABLE 9.13
California Aquaculture Production* 1980

Oyster (shucked meat)	150,000 gallons	Bass	50,000 lbs.
		Crappie	15,000 lbs.
Oyster Seed (hatchery)	100,000,000 spat	Bluegill	80,000 lbs.
		Chinese Carp	100,000 fish
Oyster Seed (advanced)	25,000,000 seed	Bait Minnows	20,000,000 fish
		Goldfish (feeders)	15,000,000 fish
Trout	2,500,000 lbs.		
Trout Eggs	150,000,000 eggs	Koi Carp	1,000,000 fish
Catfish	1,445,000 lbs.		
Catfish Eggs	6,500,000 eggs		

*These figures are approximations based on data collected from the state aquaculture industry. The units presented reflect how the product is sold on the market. Bait animals are usually graded to size, then sold in units of quantity. Goldfish and koi are usually treated the same, however all three are reported as numerical units. Data compiled by Fred Conte.

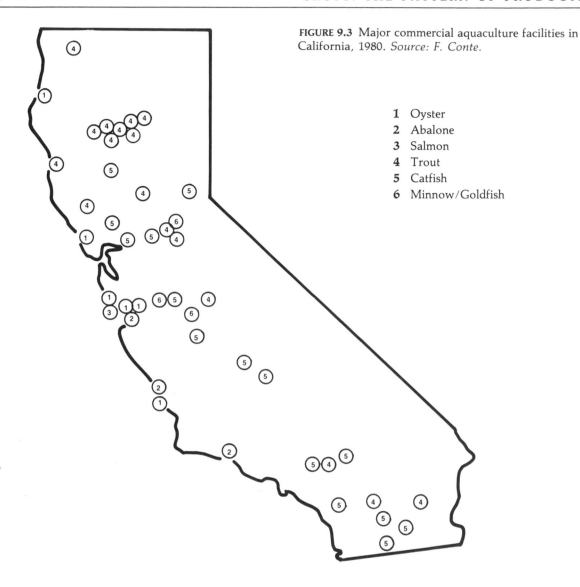

FIGURE 9.3 Major commercial aquaculture facilities in California, 1980. *Source: F. Conte.*

1 Oyster
2 Abalone
3 Salmon
4 Trout
5 Catfish
6 Minnow/Goldfish

References

The following references were selected with a view to providing further information to readers of this chapter, and do not necessarily represent a bibliography for information contained in the chapter.

Bardch, J.E., Ryteher, J.H., and McLarney, W.O. *Aquaculture: The Fishing and Husbandry of Fresh Water and Marine Organisms.* Wiley & Sons, Inc., 1972.

Cole, H.H. and Ronning, M. *Animal Agriculture.* W.H. Freeman & Co., 1974.

Cunha, T.J. *Swine Feeding and Nutrition.* Academic Press, 1977.

Eckert, J.E., and Shaw, F.R. *Beekeeping.* Macmillan Publishing Co., 1960.

Ensminger, M.E. *Sheep Husbandry.* The Interstate Printers and Publishers, 1952.

Evans, J.W., Barton, A., Hintz, H., and Van Vleck, L.D. *The Horse.* W.H. Freeman & Co., 1977.

Lebritz, E. *Trout and Salmon Culture.* Agricultural Sciences Publications, University of California, Berkeley, 1980.

University of California, Division of Agricultural Sciences. *Fundamentals of California Beekeeping.* Manual 42.

University of California, Division of Agricultural Sciences. "Goats for California Brushland." Leaflet 21044.

Wagnon, K., Albauagh R., and Hart, G. *Beef Cattle Production.* The McMillan Company, 1960.

10

Nursery and Greenhouse Products

Raymond F. Hasek

For many, the term "agriculture" means the production of food and fiber, a definition which excludes ornamental plants since they are neither. Yet although the production of ornamentals may be peripheral to mainstream farming, it is definitely part of agriculture, because of its very nature.

Compared to producers of ornamental crops in the rest of the United States, Californians produce a wider selection and greater number of most categories—thousands of plant species and cultivars. The ornamentals industry in California is comprised of a large number of loosely associated sub-industries, making production highly varied and multifaceted.

Prior to World War II the ornamentals industry in California was confined to a relatively limited market, in the states bordering the Pacific Ocean and those immediately adjacent. Shipment by air was virtually unheard of. Railway Express required such extended time in transit that only the hardiest of plants or flowers having a relatively high unit value could be shipped by this means to midwest markets. The highway system west to east or even southeast also left much to be desired. It was not until the late 1940s that many segments of the ornamentals industry began to flourish. The boom in California population had begun; servicemen stationed at the military installations within the state during the war, finding a more moderate climate and less formal way of life, returned to the West Coast later to seek their fortunes. Although the number of potential customers of ornamentals was increased, many of these young people were starting families, buying homes, and other "hard goods". Little was left to spend for ornamentals, and even a modest expansion in some lines saturated the market. By the early 1950s, however, improved aircraft service—both in equipment and numbers of flights—plus improved rail service enabled some producers of ornamentals to expand their markets. Blessed with a climate suited to virtually year-round production for many crops and a substantial pool of relatively cheap labor, California producers were able to compete with other states on quality and cost, in spite of shipping costs.

From modest beginnings after World War II, the wholesale value of ornamental crops produced in California rose to $823 million during 1980 (see Table 1). Some of the recent increases in value have been due to inflation; the percentage of increase due solely to this, however, is far below the annual inflation rate reported by the federal government. Increased labor costs and sharp increases in energy and transportation costs have made producers explore every avenue of conservation to make their operations as efficient as possible, not only to sustain a profit but in many instances to survive. Competition can be keen; cut flower producers are faced with ever increasing imports from Central and South America as well as Israel and Holland. Well managed firms with efficient production and sales methods can still do well in the marketplace, but the margin for error continues to decrease. Nevertheless, throughout California more firms and more people are now involved in the production of ornamentals than

TABLE 10.1
Trends in California Ornamental Production

Category	1965	1972	1980
	(thousand dollars)		
Deciduous and evergreen trees, shrubs and vines	$ 43,930	$ 81,560	$263,237
Cut flowers and cultivated greens	53,879	95,406	186,081
Potted plants, flowering foliage	12,809	28,736	199,056
Bedding plants	7,525	15,451	54,772
Turf seed and sod	2,308	6,600	32,405
Rose plants	5,033	7,916	26,331
Propagative materials	3,087	6,393	17,509
Christmas trees	—	4,314	13,341
Bulbs, corms, etc.	5,465	4,312	9,646
Flower seed production	3,016	2,688	7,616
Herbaceous plants	2,689	2,726	13,066
TOTALS	$139,753	$256,107	$823,064

TABLE 10.2
Categories of Ornamental Production by Major Locations

Category	Percent of Total Dollar Value	Leading Counties
Deciduous and evergreen trees, shrubs and vines	34.2	Orange, Los Angeles, San Diego, Sacramento
Cut flowers and cultivated greens	24.6	Monterey, Santa Clara, San Diego, San Mateo, Santa Barbara, Alameda, Santa Cruz
Potted plants, flowering and foliage	22.3	San Mateo, San Diego, Orange, Los Angeles, Santa Barbara
Bedding plants	5.8	Los Angeles, Contra Costa, Alameda, Santa Barbara
Turf seed and sod	4.4	Ventura, Riverside, Stanislaus, Kern
Rose plants	2.8	Kern, Glenn, San Bernardino
Propagative materials	2.1	Monterey, Alameda, San Diego
Christmas trees	1.6	San Diego, Orange, Shasta, Los Angeles
Bulbs, corms, etc.	1.0	Del Norte, San Diego, Santa Cruz
Flower seed production	.8	Santa Barbara, Ventura, Kings, Santa Clara
Herbaceous plants	.4	Sacramento, Contra Costa, San Diego, San Mateo
TOTAL	100.0	

twenty, or even five years ago. Not all segments of the industry have been equally affected. As in any other business marginal operators drop by the wayside when competition gets too tough. Often, they are replaced by new firms which perceive good opportunities in the ornamentals field.

Environmental requirements for producing the plant species classified as ornamentals are met in many areas throughout California. Table 2 shows the major categories of ornamental production and the counties most important in their production. Some counties contribute a wide variety of ornamentals to the industry (Los

TABLE 10.3

California Counties Reporting $10 Million or More Farm Gate Value of Ornamental Production in 1979

Rank	County	Dollar Value (000)	
		ORNAMENTALS	NURSERY OTHER THAN ORNAMENTALS
1	Los Angeles	121,900	—
2	Orange	121,848	10
3	San Mateo	93,144	—
4	San Diego	89,305	—
5	Monterey	52,945	—
6	Santa Barbara	46,301	229
7	Santa Clara	43,654	606
8	Ventura	34,679	2,407
9	Alameda	34,449	333
10	Kern	28,806	5,526
11	Santa Cruz	27,084	—
12	Sacramento	18,357	313
13	Contra Costa	14,669	2,310
14	San Joaquin	14,532	2,924
15	Riverside	13,112	1,870

Angeles, San Diego, Orange, San Mateo, Santa Clara, Monterey and several others), while some are "one crop counties," ornamentally speaking. Kern County is an example, since its $20 million output is virtually all the result of rose plant production. The ranking of counties having ornamental production in excess of $10 million wholesale value annually is shown in Table 3. These widely dispersed geographical locations attest to the diversity of the ornamental industry.

The following sections describe in more detail the major categories of California ornamental production.

Deciduous and Evergreen Trees, Shrubs and Vines

Plants of this nature are grown in outdoor nurseries for eventual use in the landscape around homes and industrial buildings, in cemeteries, on city streets, and in parks and other recreational areas. Although available plants number a thousand species or more, most large nurseries grow a few hundred bread-and-butter cultivars having the greatest demand. In addition, they may produce specialized lines of plants which can command relatively high prices. Specialist nurseries are also found in the industry. They may be large or small in size but the plants they produce

have a common environmental requirement. Examples of specialized nurseries would be those which primarily grow palms, rhododendrons, cacti, succulents, native plants or any other plant group of a similar nature.

The growth rate of the nursery segment of the ornamental production industry in California has been phenomenal. From 1965 to 1979 the wholesale value of nursery plants increased sixfold, even though the volume of sales was substantial at the outset of that period.

Figure 1 shows that nurseries producing ornamental plants are located in many sections of the state. The majority of large growers are located in the Los Angeles basin. Land under high voltage power transmission lines is sometimes used for this purpose. Since no permanent structures can be constructed within a designated area under power line right-of-ways, the power companies and nurserymen have found the use of land for the production of nursery stock to be of mutual benefit. Several hundred acres in Los Angeles and Orange counties are presently under lease by nursery operators. A few large growers are also located in the San Francisco Bay area, in Santa Cruz County to the south and in Sacramento County to the east.

Approximately 80 percent of the dollar value for ornamental nurseries is produced by no more than 20 percent of the total producing firms. A

FIGURE 10.1 Locations of nursery production of ornamental trees, shrubs and vines in California, 1980. Sizes of circles indicate relative volume of production. *Figure by author.*

dozen or so of the largest firms produce a substantial portion of the 80 percent figure. Several hundred small nurseries are scattered throughout the state, many of which grow part of their plant lines but purchase other plant species from large producers to supplement their inventory. Garden centers in small towns or other rural areas often operate in this fashion. At the other extreme is the large chain of garden centers which has growing grounds of its own from which it supplies its various retail outlets.

In years past, nursery plants were grown directly in the ground. At harvest time the plants were dug leaving a large ball of soil around the roots, which was wrapped tightly with burlap. This procedure, though quite effective, required a great deal of labor. In California the "ball and burlap" method began to be supplanted by containerization as early as the 1930s. Today practically all nursery stock, with the exception of some fruit trees or specialty items sold bare root, is grown in containers. Container sizes vary with the size, age and species of the plant, although the most popular size has been the one gallon metal can. Plants in containers of this size are easy for the consumer to handle while not re-

quiring a large cash outlay in order to complete a landscape planting. Progressively larger specimens of plants are grown in 2, 5, 7 1/2, and even 15 gallon containers. Under some conditions specimen trees and shrubs are grown in specially constructed square wooden boxes measuring as much as two to six feet on a side and up to four feet in depth. Even at the wholesale level, nursery stock can vary in price from a few dollars for those in small containers up to a thousand dollars or more for large specimen plants.

Many nurseries today are more diversified than their predecessors. When housing starts dropped off sharply during the mid 1960s, and with them the need for landscape plants, there was a general decline in the rate of growth in the California nursery business. Nurserymen cast about for means of increasing business, and shortly thereafter the foliage plant boom occurred. Many firms expanded into this field by building greenhouses and growing a wide assortment of potted foliage or "green plants." The expansion acted as a relief valve until housing starts and the demand for nursery stock again increased. Today an urban garden center can usually order nearly the full line of plants it needs from a single large supplier. Among the more popular types of plants in demand by the consumer are various species of juniper, euonymous, raphiolepis, pyracantha, photinia, pittosporum, ligustrum, nandina, cotoneaster, azaleas, and camellias.

There are no accurate figures on the amount of nursery stock produced in California and sold out of state. With approximately 10 percent of the United States population living in California, it is safe to assume that a substantial proportion of the stock produced in California is sold within the state.

Cut Flowers and Cut Greens

Perhaps the greatest boost for the development of the field and green-house-grown cut flower industry was the advent of the four engine aircraft used by transcontinental airlines in the early 1950s. Aircraft flying from east to west were well loaded with passengers and freight which made these flights highly profitable. However, the return flights from west to east carried little more than passengers and mail. When cut flower grower-shippers, and eventually strawberry growers, discovered they had a ready-made market in the eastern and southeastern United States which could be serviced rapidly by air, the airlines jumped at the chance to provide the service. This condition persisted until the early 1970s, when California with its burgeoning electronics and computer businesses began to ship quantities of "hard" goods to the east. Airlines became disenchanted with perishables such as cut flowers and strawberries since they had to be handled very expeditiously. With the oil shortage of 1973-74 and subsequent increases in oil prices, the cost of air shipment for all commodities rose sharply. Prices of cut flowers did not rise proportionately, and it became imperative to look to other means of conveyance. Shipment by refrigerated truck had been another method used by some progressive growers for several years before the oil shortage. As air freight rates continued to increase, cut flower shipments by refrigerated trucks also increased, until today at least half of all transcontinental cut flower shipments are made in this way. To insure the arrival of California cut flowers in the best possible condition upon arrival at eastern destinations, the packed cartons of flowers are precooled before being loaded onto refrigerated trucks.

The major production areas today are found in the counties surrounding San Francisco Bay and extending to Salinas, and in the coastal regions of San Diego County. Several growers of field flowers are found in the Santa Barbara-Ventura County region but they represent a lesser percentage of the total cut flower production. Figure 2 shows cut flower production and its relative importance at various locations.

Three major cut flower crops grown in greenhouses account for two-thirds of the annual income from cut flowers and cultivated greens—roses, chrysanthemums and carnations.

Rose plantings have increased substantially since 1975 due to the conversion of greenhouse growing areas from carnation to rose production. The 1979 USDA annual floricultural survey showed that 55 percent of all hybrid tea roses and 45 percent of all the smaller, sweetheart roses sold in the United States came from California greenhouses. Production areas for this crop are found in San Diego County in southern California and the region between Salinas in Monterey County northward through the counties surrounding San Francisco Bay to Petaluma in Sonoma County.

Cut chrysanthemums are produced in gener-

FIGURE 10.2 Locations of cut flowers and cut greens production in California, 1980. Sizes of circles indicate relative volume of production. *Figure by author.*

ally the same locations throughout the state as roses. At present, chrysanthemum growers are plagued by a severe infestation of leaf miners. Registered pesticides are relatively ineffective while those which are effective cannot be used legally within the state. This single species of insect is presently threatening the entire chrysanthemum production industry in California.

Carnation growers have had difficulty in meeting competition from imported flowers shipped to the United States from Mexico, Israel and, most important, Colombia. In general, the imported flowers are of similar quality to the domestic product while the cost of production in the exporting countries is much less than in California. In northern California, for example, energy costs for heating greenhouses during the winter months are becoming prohibitive. Many former carnation growers have switched crops and are now producing foliage plants or roses. A few growers with aging greenhouses completely surrounded by urban sprawl have sold out to developers. Carnation production has declined from 403.3 million stems in 1975 to 325.5 million stems in 1979, while imported carnations rose from 163.3 million stems in 1975 to 376.4 million stems in 1979. The latter figure represented 44 percent of the total carnations sold in the U.S. With the ever increasing costs of energy and labor, carnation growers in northern California may be hard pressed to show a profit in coming years. Growers in San Diego County have a better change for survival since carnations can be grown in cheaply constructed unheated greenhouses whose polyethylene film coverings do little more than shed rainwater during the winter months. Although labor is still high, at least energy use is minimal.

Other crops of cut flowers grown in greenhouses are orchids, snapdragons, gardenias, stephanotis and freezias, to name a few.

At least 65 types of cut flowers are grown outdoors throughout California. Major counties are San Diego, Ventura, San Mateo, Santa Barbara, and Santa Cruz. Some of the more important species are gladiolus, gypsophila, strelitzia, daisies of all types, statice, bulbous iris, column stock, heather and yarrow. In spite of extensive plantings of many of the outdoor cut flower crops, the total dollar value of these crops is relatively small compared to the "big three" of greenhouse production.

At least 25 types of cut greens are available at various times of the year on the San Francisco and Los Angeles flower markets. Cut foliage is used as a filler in flower arrangements, as backing for wreaths and in several other ways by retail florists. A popular item for use in floral arrangements is the blue leaf eucalyptus, juvenile foliage. Extensive plantings of up to 75 acres can be found in the mild coastal climates. Relatively smaller amounts of myrtle, strelitzia, ivy and several types of fern such as asparagus, brake, maidenhair, and leatherleaf are grown in various climates throughout the state.

Other miscellaneous seasonal flowering plant materials such as acacia, flowering peach, flowering quince, Queen Anne's lace and others are collected from plantation plantings or even landscape plantings and offered for sale at the two major flower markets. Still another fast growing business is supplying the dried flower market, a development of recent vintage. Several firms are now engaged in this multimillion dollar business, and a wide variety of plant materials (strawflowers, lunaria, eucalyptus and gypsophila) are grown especially for drying. Others are collected from the wild in this and many other countries of the world.

Marketing of cut flowers in California is ex-tremely intricate and complex. The classic arrangement for years has been the grower consignment to a local shipper or wholesaler. The shipper sells the flowers to a second wholesaler in some eastern or southern city, who then sells the flowers to local retailers, who finally sell them to consumers. Direct shipments from grower to eastern wholesalers or retailers have lately become more common as growers attempt to reduce handling costs. Another sales outlet developed during the past fifteen years is the mass market or supermarket stores, and large volumes of flowers have been moving through this channel. Potted plants have fared better here than cut flowers since they are better able to withstand the rigors of air conditioned buildings. One of the most economical sales methods is the "clock" auction which is popular in Holland and some parts of Canada. In this system, wholesalers bid for flowers by use of a clock which starts at a high price and gradually descends to lower prices. Buyers stop the clock at the price they wish to pay, and must pay cash for the product before taking possession. Growers own the market and are paid once each week instead of the present 60 to 90 days, an advantage for their cash flow.

Flowering and Foliage Potted Plants

Potted plant production was originally located in the San Francisco Bay area and in Los Angeles County. Very little production occurred outside of these areas except for the azaleas and rhododendrons grown along the coast in the northern part of the state, and these were grown primarily for landscape use. There was relatively slow growth in the industry until the late 1960s. Few new greenhouse establishments were built for the purpose of pot plant production in the 1950s, but the mid and late 1960s saw a great surge in the demand for "natural" plants throughout the entire United States. Greenhouse operators in San Diego County, Santa Barbara County, the Lodi and Half Moon Bay areas expanded rapidly to take advantage of the sudden demand. Several greenhouses which had been built during the grape propagation boom in Sonoma County converted to foliage plant production in the 1970s. Many nurseries which formerly had been devoted to the production of containerized plants for landscape use also expanded into production for the foliage plant market.

Growers looking for more outlets for their stock developed working relations with large mass market chains. Instead of production being funnelled only through conventional florist shops, the buying public was soon exposed to the sight of cut flower and potted plant sections in grocery supermarkets, discount stores and even an occasional large drug store. Plant "boutiques" sprang up at numerous sites in big cities and small towns alike. The bonanza continued for several years until the mid- to late-1970s, when tremendously expanded production from growers in California and Florida slowly saturated the market. Meanwhile, consumer preference shifted to larger plants in six or eight inch pots rather than those grown in small two to four inch pots. To add to the problem, increased pressures on the California cut flower industry caused a shift within the state; several large carnation growers began to produce foliage plants. As a result of all these changes, the present outlook for foliage plant production is not as bright as it once was.

During the heyday of the green plant boom some of the major genera grown were dieffenbachia, philodendron, epipremnum, ficus, maranta, cissus, crassula, pilea, dizygotheca, brassaia, zebrina, fittonia, dracaena, nephrolepsis and other ferns, chamaedorea palms and many others. In all, over 250 species and varieties of foliage plants were being offered for sale in the trade.

Flowering potted plants had been all but forgotten by many growers in the rush to cash in on the "green plant" rush. By the beginning of the 1980s, however, the pendulum had swung back toward the flowering plants which by now had seen the introduction of several new species and many improved varieties of the old.

The favorite flowering plant grown year around is the pot chrysanthemum. Other crops such as African violets, kalanchoe and the like are in lesser demand. Several major crops are relatively seasonal or are produced almost exclusively for holidays, such as poinsettias for Thanksgiving and Christmas, lilies for Easter and hydrangea for both Easter and Mother's Day. Several minor crops are produced on a more or less seasonal basis, for example, gloxinia, cyclamen and azaleas.

Some larger growers rely heavily on out of state trade, shipping to many widely scattered points from Texas to New York. Other firms are more reliant on domestic sales within California. The future for pot plant sales appears to be good.

Bedding Plants

A gradual increase in the size of the bedding plant industry in California has taken place since World War II, with an ever increasing population and the building starts which accompany it. Until recently, bedding plants were primarily annuals and a few perennials, used for mass effect in large landscapes, gardens and border plantings. They were sold in small multiunit plastic containers or "packs," and in bulk as flatted plants. In the latter instance customers could either purchase the entire flat or have the desired small amounts removed and placed in cardboard containers to take home.

More recently, apartment and condominium living has reduced the need for individual home garden plantings of bedding plants. The apartment dweller does not care to plant small seedlings and wait for bloom late in summer, but prefers annuals which have been grown to the bloom stage in medium size pots for instant color. In winter such plants as pansies, stock or primrose can follow a summer scene of marigolds, zinnias, celosia or other splashes of color.

Most bedding plant production is located in or near the heavily populated counties of southern California, while northern producers are scattered throughout the San Francisco Bay area and near the larger towns of the San Joaquin and Sacramento valleys.

Great strides have been made in the production practices of bedding plant nurseries during the past twenty years. Mechanization of soil pasteurization and mixing, seeding, watering and fertilization have improved vastly in attempts to keep down production costs. Lightweight greenhouse structures have been developed to start seedlings, which are moved outdoors to harden before shipment.

Most bedding plants produced in California are sold within the state. Estimates by knowledgeable growers indicate that between 15 percent and 20 percent are sold elsewhere, primarily the western states.

Turf Seed and Sod

The sale of sod for use in home lawns began around 1920 in the eastern United States, but in California sod production before 1958 was on a very small scale. The major use was for instant landscape sets for the movie industry. The source of sod in California was not pasture land as in the eastern United States, but plantings of hybrids and improved cultivars of species adapted to the production area. After 1958, a slight increase in sales occurred annually but about 1969 there was a real upsurge in sod planting. At first, the primary customers for sod were golf courses, sports fields and the like. However, model home landscaping required instant lawns for best showing and the concept spread to purchasers of new homes. Today the greatest demand for sod is for new home lawns and replacement or repair of existing lawns by home owners.

The development of the powered sod cutter during the 1950s has been credited with being the most important factor in the advancement of sod production during the following decade.

In the 1970s, sod sales increased nearly tenfold in eight years. The great advances which took place in sod production during this time were largely due to improved cultural techniques, better cultivars and efficient mechanical handling procedures. By far the primary production area for sod is located near Oxnard in Ventura County, where a relatively few growers produce much of the available supply.

Although dichondra is often considered to be a ground cover, it also has importance for use as turf. Seed for this crop is produced largely within the state, primarily in Sacramento and San Joaquin Counties. Other. than this, no grass seed is actually produced in California.

Production and sales of sod are dwarfed by the dollars spent to maintain and repair turf areas for all uses in California. The University of California in 1979 estimated the total money spent by Californians for this purpose as about $690 million dollars. Turf and its care are truly big business.

Rose Plants

Rose plants grown in California have two general uses. The greatest number by far are those which can be grown in backyards, in municipal park rose gardens and recreation areas. Then there are the cultivars which are grown in greenhouses for the production of commercial cut flowers. Generally, a given cultivar is satisfactory for one or the other of these uses, but not both.

In the early days of rose plant production, major growing grounds were in the San Francisco Bay area. Some growing grounds located in Riverside County were of lesser size. As land become more valuable due to urban sprawl, rose growing grounds were moved out of the Bay Area and new plantings appeared in the San Joaquin Valley around Newman and Patterson. Today Kern County is the site for at least half a dozen major rose producers. Another is located in Glenn County while several smaller growers are elsewhere in the Sacramento and San Joaquin valleys and inland in Riverside County.

The dollar value of the California rose crop more than tripled during the 1970s, from $7.1 million in 1969 to $26.3 million in 1980. This was partly due to more production, but inflation also increased the unit sale price substantially.

It has been estimated that approximately 3,500 acres of land in California are devoted to rose production annually. Not all of that is harvested each year since plants grown for the home owner require two years to produce while those destined for commercial cut flower production can be finished in a single season.

Rose breeding is being carried on by a handful of large producers. Intensive breeding programs are carried on by the Jackson and Perkins Company at Santa Ana and the Devor Nurseries, Inc., at Pleasanton.

Propagative Materials

The three major ornamental crops which are vegetatively propagated by specialist firms in California are chrysanthemums, poinsettias, and carnations. Several minor crops such as geraniums, pelargoniums and many woody plants are also propagated to some extent but the bulk of the dollar value of this category comes from the three major crops. Each of them has a unique history and possible future.

This interior view of a Half Moon Bay greenhouse shows thousands of poinsettias being prepared for Christmas holiday sales. *Photo by Ray Hasek.*

Poinsettias

Poinsettia stock plants, cuttings, finished potted plants and cut flowers have been produced primarily by a single firm in southern California which has been operated by the same family for three generations. The original growing grounds were located in what is now West Los Angeles at the beginning of the century. Subsequently, the operation was moved to Encinitas where Ecke Poinsettias is now located. For many years this firm has been a primary supplier of poinsettia cuttings and stock plants in the United States.

Rooted and unrooted cuttings are the major propagative forms of poinsettias. Shipments made in late spring and early summer are used by growers throughout the country as stock plants for their own summer and early fall cutting supply. When the shipping season is over, the stock plants at Encinitas are allowed to flower and the long stems and colorful bracts are harvested as cut flowers for the holidays. Even more popular are the potted plants which by

their prevalence in mass market outlets and florist shops attest to the popularity of this plant.

Chrysanthemums

For many years, most California chrysanthemums, both standard and pompon, were grown during summer months under tobacco cloth covered structures. Unfortunately, shortly after World War II a very serious viral disease called "chrysanthemum stunt" struck plants in the eastern states and soon spread throughout the country. Through extremely expensive methods of heat treatment, indexing, and grafting, the disease was brought under control and "clean" plants free of the virus became available.

As the chrysanthemum growing industry again flourished, plastic greenhouses were constructed and year around production began. Individual growers bought disease-free cuttings and established their own stockplant areas from which they made cuttings as needed. Two major producers of disease-free cuttings became established in northern California in the mid-1950s. They continue to supply cuttings to growers of potted plants and cut flowers alike, not only in California but the rest of the country as well. The production of cuttings for sale to growers is now a highly specialized operation requiring special laboratory techniques and extremely sanitary greenhouse conditions in order to produce high quality, predictably responsive plants.

Carnations

Carnations are also propagated asexually by cuttings from commercial cut flower production. In the late 1940s a red cultivar named 'William Sim' was introduced and immediately became the major red carnation grown in this country. Its free growth habit and quick return cropping after flowers are harvested made it a profitable item. The cultivar has "sported" frequently over the years, producing a half a hundred or more new cultivars with a rainbow of colors including pink, white, orange and an assortment of bicolored flowers. Today no less than two dozen named colors are available.

In the mid-1950s a large eastern carnation propagator established a greenhouse range south of Salinas where the majority of cuttings used by California growers of standard carnations has originated ever since. Like chrysanthemums, carnation propagative materials require heat treatment, meristem culture, aseptic growing conditions and isolation in special greenhouses devoted to the process. An added problem in carnation propagation is the free sporting characteristic of the 'William Sim' cultivars. Each stock plant must be grown to produce a flower true to type before cuttings are taken.

Recently there has been an increase in miniature or spray carnations being grown in California. Propagation of these cultivars is spread among several firms which are carrying out breeding programs. Plant patents are being sought for many of them, to restrict propagation rights to the developers.

Geraniums

Geranium and pelargonium stock plants are still being grown in southern California for out-of-state growers of potted plants. Several years ago, however, failure to maintain disease-free stock in California led to an embargo on importation of California cuttings by such eastern states as Pennsylvania, Ohio, and Michigan. Since these states were those which absorbed the bulk of west coast production, the industry was sorely hurt. Development of cultured cutting procedures by eastern firms plus the increased popularity of growing geranium plants from seed has kept the California propagating industry from regaining its former stature.

Christmas Trees

Christmas tree plantations for choose-and-cut have become big business since 1960. In previous years, woodlot regrowth had served as a source for trees harvested by the consumer. Though this continues, plantations have become more significant. Container-grown trees have been available from commercial nurseries for many years, but relatively high cost and bulky handling have limited the demand.

The industry is separated into production for the wholesale and retail markets. In general, wholesale producers are located in foothill, mountainous or less accessible areas of the state. Plantations can be located in back canyons or hilly terrain not well suited for the growing of other agricultural crops. Northern growers produce such types as Douglas fir, silver tip fir, white fir, red fir, Scotch pine and other pines. In southern California, Douglas fir, Scotch pine, and perhaps grand fir have been available.

For "choose and cut" retail sales, plantations of trees located within relatively short distances from large cities or towns are most popular. Hundreds of acres of land under high voltage power lines, some in the heart of cities, have become profitable growing grounds for Christmas trees.

Monterey pine is by far the most popular tree grown in plantation plantings, in both northern and southern California. In northern California, Douglas fir, Scotch pine and white fir are also grown in quantities. Perhaps the primary reason for the popularity of Monterey pine among Christmas tree growers is its rapid growth. Plantations only three to four years old produce saleable trees whereas other species can take two to four times that period to reach the desired height.

Plantation-grown trees usually have an excellent shape as the result of pruning. Variations in the growth habits of the various species determine the number of times shearing might be needed during the growing season. The trees

Christmas trees are grown in some areas under power transmission lines. *Photo by Ray Hasek.*

must also be protected from insects and diseases. Judicious fertilization and spraying can keep most diseases under control without too much difficulty since a healthy tree is less susceptible to problems.

Christmas tree production in California was valued at more than $13 million wholesale in 1979, but this probably is a conservative figure. As transportation costs in the family car increase, the annual trek to Christmas tree woodlots far from home may diminish.

Bulbs, Corms and Tubers

Bulbs

The most important crop in the bulbous category is the lily. Several species and cultivars are produced by growers located in the northwestern counties of Humboldt and Del Norte, particularly Del Norte which borders on the state of Oregon. The cool coastal climate with abundant rainfall during the winter months is ideal for the production of lily bulbs. Easter lily (Lilium longiflorum)

bulbs are produced for sale to greenhouse potted plant growers throughout the United States, who force them into bloom for the Easter holiday. Since the Easter lily types are highly hybridized, they do not breed true, so propagation is exclusively vegetative. Cultivars such as 'Creole,' 'Croft,' 'Ace' and others were popular but have since been supplanted by the better forcing cultivar 'Nellie White.' Scales removed from flowering bulbs are planted in beds through the first year. They are then lifted and replanted in field rows for the second year at which time they are allowed to bloom. In late summer, the bulbs are dug, graded, packed and shipped to greenhouse operators. Highly colored species are also produced in this area, known in the trade by such exotic names as 'Rubrum,' 'Erabu,' 'Mid Century Hybrids' and others. Most of the 'Rubrum' and 'Erabu' types are tall growing, making them unsuited for potted plants. They do well in the home garden or are grown for the cut flower market. Many 'Mid Century' cultivars are grown for all three purposes.

In the same general growing area as lilies,

daffodil or narcissus bulbs are grown for both cut flower purposes and bulb stock. The most popular cultivars used in the cut flower market are the 'King Alfred' types. Several other narcissus types are produced for bulb sales, but are of relatively minor importance in the florist trade.

Tubers

Along the coast from Santa Cruz to Monterey, the tuberous begonia is produced by many growers. Tuber production takes place under a variety of conditions from greenhouses and cloth houses to outdoor field plantings. These large flowered plants make excellent pot plants for use in a marine climate. A field or greenhouse full of flowering begonias is an awesome sight with a multitude of colors.

Dahlias are tuberous plants grown mainly in the northern San Joaquin Valley. Although produced primarily for the tubers, some dahlia flowers are harvested for the cut flower market.

Some plantings of agapanthus are grown as cut flowers, but container nurseries throughout the state have this plant available for use in the home garden. The same is true for hemerocallis cultivars.

Corms

Southern California, especially San Diego County, has the primary growing grounds for many types of corms, which include crocus, freesia, ixia and gladiolus. By far the most important is the gladiolus. Where once the growers of this crop for corms and cut flowers numbered in the dozens, today only six remain. One relatively large grower in the San Francisco Bay Area and a large grower in San Diego County account for three quarters or more of the estimated 900 acres remaining in cultivation. This flower has lost favor in the cut flower market during the past twenty years. A smaller version of the old large flowered type has been developed in an attempt to diversify usage of the gladiolus, but at this time, the 'Pixiola Glad' has not made substantial inroads into either the corm or cut flower market.

Flower Seed Production

Flower seed production started in California a little over one hundred years ago. The relative freedom from rainfall during the growing and harvesting season and the mild climate of California coastal valleys provided ideal conditions for early profitable ventures into this field. From 1880, when the industry was established in California, until after World War II, flower seed production shifted gradually to the drier climate of the West. By 1949 over 75 percent of the flower seed crop value in the United States was produced by only five firms, all located in California.

Today most of the production comes from the Santa Maria, Lompoc and Salinas valleys. Large producers maintain breeding and development programs for new cultivars. Skilled hand labor is an absolute necessity for the success of flower seed operations. Due to the relatively high cost of labor, some firms now use the California climate and greenhouse culture to develop and test potential cultivars, while the actual seed production fields are often located in other countries such as Guatemala and Costa Rica, where a cheap and plentiful labor supply exists. In this highly specialized business, practically all the flower seed produced in California at this time comes from five or six firms. In 1970, over 90 percent of national flower seed value was generated by California growers.

One has but to examine a few flower seed catalogues to be impressed by the hundreds or even thousands of cultivars available to the amateur as well as the professional who grows flowers for profit. Each has its own planting time, cultural practices, pest and weed control problems, as well as pollination and harvesting procedures. The number of variables involved in a full scale flower seed production schedule is virtually astronomical. Some of the fast blooming cultivars can require as little as six months from planting to seed harvest; others might need as much as eight to eleven months to complete the cycle. Maintaining seed purity to type involves further specialized conditions. Cultivars of the same species can be planted in fields relatively close together unless blooming dates coincide, which increases the possibility of cross pollination by insects. Where bloom dates are similar, a separation of a quarter of a mile or more between plantings is needed to maintain cultivar purity.

The diversity of cultivars makes culture and harvest of each flower a specialty. Gross income per acre is high, but when the risks of the market and high hand labor costs are computed, the profit picture is not always pleasing. Flower seed production is a conglomeration of small plots compared to other forms of agriculture, and the acreage devoted to individual cultivars is usually very small. Demand for many flower seeds can be satisfied from as little as a half acre. Harvesting can be a real problem under these fragmented conditions. Much of the seed is harvested by machinery similar to that which is used for grain. Other seed is harvested by cutting and placing the plants in windrows to dry. The run for any one cultivar is usually rather short, but the stringent cleaning process required for all machinery between plots can result in a time consuming operation.

Popularity of individual flowers varies with the passing of time. Those which have stood the test over the years are zinnia, marigold, sweet pea, stock, petunia, aster and snapdragon, not necessarily in that order.

Use of Resources in the Ornamentals Industry

Mechanization has reduced the costs of production for many segments of the ornamentals industry, but the need still exists for much hand labor in the production of cut flowers, potted plants, outdoor nursery stock and other crops. The labor force used in ornamentals production does not have to be highly skilled except in relatively few operations. Some greenhouse operations and nurseries provide nearly year around employment, thus furnishing stability for the worker not characteristic of most other types of agriculture.

Energy use in the production of ornamentals grown outdoors is not greatly different from that required for many row or pomological crops. Production of ornamentals under greenhouse conditions is often another matter. In the northern part of the state the energy use as measured by natural gas consumption has been in the magnitude of 1 to 1.5 therms per square foot of ground area covered per year. Variability in actual use can be attributed to environmental requirements of the crop, efficiency of heating equipment and the heat-efficient construction of the greenhouse. Many energy conservation measures have already been applied—insulated liners, double polyethene coverings, better heat distribution systems and a general elimination of hitherto wasteful practices. Natural gas costs quadrupled from 1975 to 1980, and other fuels

have risen in price at about the same rate, providing the incentive for producers to make their operations as efficient as possible.

Water conservation is also practiced by growers of ornamental plants. Drip irrigation has proven satisfactory for many field operations but not all. Overhead irrigation can be wasteful if a large portion of the applied water does not enter individual containers; however, runoff waters can be captured and held in outdoor reservoirs for recycling. Some greenhouse operators collect rainwater runoff from the tops of the greenhouses during the rainy season and store it for later use in outdoor reservoirs. A unique system of water conservation is in effect at Half Moon Bay where two greenhouse operators and a local golf course superintendent cooperate in the reuse of waters.

The ornamentals industry shares many of the same problems of resource management which characterize the rest of agriculture. The discretionary nature of ornamental product purchases, however, exerts high pressure on growers to maximize efficiency and economy in production. In the spectacular growth of the industry since 1950, expanding knowledge and horticultural techniques have enabled many new entrepreneurs to enter the business. Their continued success will depend on their ability to manage resources efficiently while meeting the desires of the marketplace.

Reference Material for Ornamental Crops

Statistics for the wholesale value of ornamental crops are reported in the Agricultural Crop and Livestock Report published annually by the Agricultural Commissioner's office in each county. Compilation of statewide figures are available from each county office of Cooperative Extension.

Some of the information contained in this chapter was obtained from University of California periodical publications, namely "Flower and Nursery Report for Commercial Growers", "California Turfgrass Culture" plus leaflets such as "California Forest-Grown Christmas Trees" and others.

For historical information on some facets of the ornamentals industry, the United States Department of Agriculture Yearbook of Agriculture 1961, entitled *Seeds*, and back copies of the bulletins of the California State Florists Association are helpful.

Additional publications such as *Bedding Plants*, edited by John W. Mastalerz of Pennsylvania State University, and the *Poinsettia Manual* published by Paul Ecke Poinsettias of Encinitas, California, provide substantial information for these crops. No one text, however, provides complete cultural and marketing procedures for all ornamental crops grown under California conditions.

11

Timber, Forage, and Wildlife

Harold F. Heady, A. Starker Leopold, Henry J. Vaux, and John Zivnuska

Forest and Rangeland Resources

California forests and rangelands produce a diverse mix of economic and social values, including timber, forage, wildlife, watershed protection, recreational environments, and landscape amenities. Forest products and livestock forage comprise the major economic values entering the market system. Unmarketed values include protection of water quality and amelioration of flows on the state's major watersheds, provision of sites for a wide range of outdoor recreation activities, habitat for much of our fish and wildlife, and visual surroundings which affect the amenity of large areas of the state.

Area

Forests and rangelands account for about 60 percent of California's land area. As of 1975, 39.5 million acres were classed as rangeland, which includes land on which the existing vegetation is suitable for grazing or browsing use and is dominated by vegetation other than trees. Some 32.6 million acres were classed as forest land, defined as land having at least a 10 percent tree canopy cover at maturity and not developed for nonforest purposes—including land that could be reforested. A total of 52.2 million acres were actually grazed by domestic animals in 1976, so there is significant overlap between the areas defined as forests and rangelands. Resource areas are summarized in Table 1. (Characterization of the soils, climate and vegetation of the major forest and range areas is provided in Chapter 3.)

Potential Productivity

Forest and rangelands exhibit wide variations in potential productivity. For example, nearly half of all California forestland does not have the biological potential to grow 20 cubic feet of wood per acre per year. As a result, it is considered too low in productivity to produce commercial timber crops. Of the 17.9 million forest acres considered "productive," about 25 percent are capable of growing at least 120 cubic feet per year, and the remainder range in productivity from 20 to 120 cubic feet per acre per year. Almost two-thirds of the potential for wood production in the state is on 7.1 million acres of forestland capable of growing 85 cubic feet or more of wood per acre per year.

Rangeland exhibits somewhat comparable variability in productivity. The Soil Conservation Service has estimated that about 17 million acres of

TABLE 11.1

Forest and Rangeland Areas in California, 1975

Type of Land	Forest	Range	Forest & Range
	Million Acres		
Forest, not usable as range	24.1	—	24.1
Forest and range	8.5	8.5	8.5
Range, not forested	—	27.5	27.5
TOTAL	32.6	36.0	60.1

SOURCE: California Department of Forestry. *California's Forest Resources: Preliminary Assessment*. Sacramento, 1979.

range are of "prime" productive capability, with the remaining 19 million acres of lesser productivity. USFS estimates that annual grasslands produce an average of 2,000 lbs/acre/year on a land area of 10 million acres. Rangelands produce more than 60 percent of the total annual forage required by cow and calf herds, by beef stockers, and by ewes and lambs in California.

Ownership

In contrast with other sectors of California agriculture, major portions of the forests and sizeable areas of rangelands are in public ownership. This is particularly true for the "productive" forest lands where 57.5 percent of the area is owned by public agencies. Public ownership is less dominant, but by no means unimportant, on "unproductive" forests and rangelands, accounting for about 35 percent of all land in each of these categories. Although 35 percent of rangeland is in public ownership, these lands produce about 7 percent of the total grazing capacity in the state. Federal lands are most important in northeastern California, where 31 percent of the forage supply is produced on federal land. Table 2 summarizes ownership patterns for the major types of land.

Privately owned forests are divided among three quite different sorts of enterprises. Forest industries and other owners who manage their lands with the primary objective of producing timber or wood products own slightly over half of the private commercial forest. Farmers and ranchers own 22 percent, and the remaining 25

percent is owned by an array of other sorts of ownerships. In contrast, forest industries own little or no unproductive forest or rangeland. In such areas farm or ranch ownership predominates but with significant acreages held by owners who have no entrepreneurial ties either to agriculture or to forest industry.

Of the 15.5 million acres of publicly owned forest land, almost 88 percent is in the national forest system, administered in the U.S. Department of Agriculture. Another 9 percent is in the national park system, with the remainder divided among a number of federal, state and local government agencies. The Bureau of Land Management in the U.S. Department of Interior is responsible for the biggest share of publicly owned rangeland, administering much of the desert lands and large blocks of chaparral which amount to one-third of the total rangeland. The U.S. Forest Service is also a major range owner with 10 percent of the total rangeland. About 65 percent of the state's rangeland is in private ownership.

These intermingled and complex ownership patterns are important as determinants both of the mix of products and services for which the forest and rangelands of the state are managed, and for the managerial and investment resources allocated to such purposes. For example, industry-owned forest lands and rancher-owned rangelands are usually managed relatively intensively for marketable timber and forage values. Much of the national forest system land, however, is managed for a variable "mix" of tim-

TABLE 11.2

Percentage of California Forests and Rangelands by Major Type of Ownership, 1975

Class of Land	Ownership		Total Area
	% PUBLIC	% PRIVATE	MM ACRES
Productive forest			
Commercial	53.2	46.8	16,299
Reserved	100.0	0.0	1,645
TOTAL	57.5	42.5	17,944
Unproductive forest	35.2	64.8	14,630
Total forest[1]	47.5	52.5	32,574
Total rangeland[1]	35.4	64.4	35,964

[1] An undetermined area of forest range is included both in "Total Forest" and in "Total Rangeland."

SOURCE: Calif. Dept. of Forestry. *California's Forest Resources: Preliminary Assessment.* Sacramento, 1970.

ber, domestic animals, wildlife, dispersed recreation, and watershed values. National and state park system lands, by contrast, are managed either for preservation of the ecosystems involved or for relatively intensive recreational uses, frequently for both purposes.

Apart from these significant differences in the management policies applied by different classes of owners, rational management of forest and rangelands for whatever purpose is often severely complicated or constrained by the intermingling of ownership and by the uneconomic size of many individual parcels. In many areas, the ownership pattern is literally checkerboarded; and the rectangular survey system which underlay the establishment of most property boundaries nowhere bears a significant relationship to the biological realities of resource management. Reliable data on the sizes of ownerships are fragmentary, but it appears that the average size of holdings of commercial forest land by non-industrial owners is of the order of 120 acres. This is generally considered to be well below the minimum size of an economic timber growing unit.

Frameworks for Resource Administration

Unlike commodity agriculture, which is regulated mainly through the market system with some governmental control of specific aspects, the institutional framework within which forest and public rangelands are administered to meet the needs of the public is both complex and varied. It results in systems of legal and market controls over forest and range use which differ substantially, depending on whether the land is publicly or privately owned, and whether one is considering timber products, livestock forage, wildlife, water, recreation or some other forest and range value.

In general, the administration of forest and rangelands involves three elements: protection of the resource against destructive agents, management of the resources for one or more beneficial uses, and distribution of the products of the land among users. Under California climatic conditions, protection of forest and range resources from natural and man-made fires has attained special importance. Adequate resource protection is a necessary prerequisite for rational management of lands for productive use. Although wildfire is usually considered the principal enemy of forest and rangeland management, in-

sects, diseases, plant pests, unrestricted development, and improper use of offroad vehicles are other very serious protection problems. Prescribed and controlled fire is a valuable tool in brushland management for livestock and wildlife.

PUBLIC LANDS. National forests (administered by the Forest Service, USDA) and public resource lands (administered by the Bureau of Land Management, USDI) are managed under federal statutes which generally prescribe policies of sustained yield use for all of the multiple values which such lands can produce. Mature timber and forage for livestock are sold to private producers of forest and livestock products either at the market price (for timber and some forage) or at an administratively determined price (for most grazing leases). User fees are charged for some recreational facilities, but in general recreational use of the public lands, whether for hunting, fishing, hiking, cross-country skiing, or wilderness enjoyment, is provided without cost to the user. In the past, use of both water and wildlife produced on national forest and the public resource lands have been assumed to be under the regulatory control of the state, although some recent legal actions have raised questions about this doctrine. Protection and management of the forest and range resources, including determination of the mix of resource values to be produced, is the responsibility of the federal agencies concerned and is limited by the appropriations Congress provides for these purposes. National parks management differs from that of the national forests and public resource lands in two respects. The parks are managed solely for preservation of their resources and for recreational use consistent with that objective. Wildlife in the national parks is regulated exclusively by the managing agency.

PRIVATE LANDS. Control of wildfires on virtually all privately owned forest and rangeland in California is provided, through the California Department of Forestry, at the expense of the state's general fund. This program was designed to minimize damage from wildfire to those private lands which are primary watershed lands of the state.

Fifty years of fire control activities have not eliminated and perhaps not reduced wildfire problems. Many fires occur every year and considerable acreage is burned. These costly programs of wildfire control need to be supplemented with fuel management programs, usually with controlled burning, and control over land use in zones of high fire hazard. Rational management of the resources for goods and services will continue to depend on minimizing risk of fire loss.

State regulation of private timber harvesting practices was initiated in 1945. Since 1974 a comprehensive system of state regulation has been applied to private timber harvesting in the interest of protecting the public interest in permanent timber productivity, water quality, and soil erosion prevention, and with some protection for wildlife habitat and other environmental values. As on public lands, the state controls use of water and wildlife produced on the forest and range areas. Except for fishing and hunting, recreational uses of these lands are of limited and local importance.

Against this rather general background, the nature and significance of timber, forage, wildlife, and related forest and range resources in the state can best be presented by discussing each of these resources separately. But the reader should bear in mind that at the level of the land base, each may be intimately linked to many of the others.

Forest Products Industries

Since the discovery of gold in the tail race of John Marshall's sawmill in 1848, the production and consumption of forest products have been important in the state's economy. Production during the nineteenth century was directed almost entirely to markets within the state, but following the development of the transcontinental railroads and the opening of the Panama Canal, markets in other regions of the United States and even export markets for certain specialty products such as redwood also became important to California mills. As of 1980, California ranked third among the states in volume of timber harvest and produced nearly 16 percent of the nation's total softwood lumber production. Despite this relatively large forest products output, the state consumes far more wood than it produces. Thus, the volume of timber cut in the state represents less than half of that required to produce the forests products consumed within the state.

The Timber Harvest

In the middle of the 1920s and again in 1940 the timber harvest in California rose to more than 2 billion board feet. Immediately following World War II the harvest jumped to more than 4 billion board feet and then continued to rise to an all time peak of more than 6 billion board feet in the late 1950s. Since then the harvest level has declined gradually, fluctuating between 4.5 and 5.0 billion board feet during the 1970s. In total, more than 260 billion board feet of timber have been harvested for the production of lumber and related products over the last 130 years, with roughly 70 percent of this having been cut since World War II. In addition, large but unrecorded volumes of timber have been removed for other purposes, including extensive conversions of land to grazing and agricultural uses as well as many other forms of land clearing.

The expansion in harvest following World War II was particularly pronounced in the north coast region, although output in other parts of the state also expanded. There have also been significant shifts in the pattern of sources of the timber harvest by ownership classes during the last 30 years. During the 1950s small and medium sized nonindustrial private holdings were the primary source, providing well over half of the harvest volume. In the subsequent years production from these lands fell off sharply, while the harvest from industrial forests and public forests rose. The volume of timber sales from the national forests in California rose to a peak level averaging 2.1 billion board feet per year in 1963–68, but then declined by 13 percent to an average of 1.8 billion board feet in 1973–78. In 1976 about 40 percent of the harvest came from national forest lands and another 2 percent from other public forests; 42 percent came from industrial forests; and the remaining 16 percent came from nonindustrial private holdings.

While timber is harvested in appreciable volume in more than 30 counties, 75 percent of the harvest comes from the ten leading counties. Humboldt County alone is currently the source of about 20 percent of the harvest. The other leading counties in declining order of output are Mendocino, Shasta, Siskiyou, Trinity, Del Norte, Plumas, El Dorado, Tehama, and Lassen. Although the ten most northern counties of the state are thus the center of its timber and forest products activities, significant forest products production continues southward for the length of

Sawmills like this one in Mendocino County are important to the economy of many northern California communities. *Photo courtesy of Louisiana-Pacific Corporation.*

the Sierra Nevada. In terms of species, reported log consumption in 1976 at mills in California was 27 percent Douglas fir, 26 percent ponderosa and sugar pine, 19 percent white and red fir, 19 percent redwood, 4 percent incense cedar, and 5 percent other species.

Lumber

The level of timber harvest and the consumption of logs in California is tied to the lumber industry to a degree which is not equalled in any other major forest products state. In 1976 some 86 percent of the roundwood volume was initially processed by sawmills. The veneer and plywood industry took 11.5 percent of the volume and another 2 percent of the roundwood volume was exported from the state as logs. The state's wood pulp, hardboard, and particleboard industries are based almost entirely on wood residues

from the lumber and plywood mills and are not direct consumers of roundwood.

Lumber production in the state has fluctuated around 5 billion board feet annually for more than 20 years, with perhaps a very slight downward trend. Thus California ranks second only to Oregon in volume of lumber production, with the output typically being about 16 percent of the nation's total softwood lumber production. Over the years the pattern has been for production to be concentrated in fewer but larger mills, with a drop in the total number of operating sawmills in the state from 263 in 1962 to 142 in 1976.

While this production level is relatively high, it is less than estimated consumption of softwood lumber in the state, which is more than 6.5 billion board feet annually. Moreover, available data show that only 63 percent of the softwood lumber produced in California is shipped to destinations within the state, with 21 percent being shipped to other western and midwestern states, 14 percent to other regions of the U.S., and 2 percent to foreign countries. This reflects the specialty nature of some of California's lumber products and the nature of the railroad and other transportation systems available to producers, particularly in the northeastern portions of the state. Thus, although softwood lumber production within the state amounts to about three-fourths of consumption in the state, only about half of the softwood lumber actually consumed in California is produced within the state. Another one-quarter is brought in from Oregon and most of the balance comes from Washington and the northern Rocky Mountain states.

Other Forest Products

The production of softwood plywood, the other major solid wood product in addition to lumber, was late in developing in California, but then rose rapidly to a peak of 1.3 billion square feet, 3⁄8" equivalent, in 1963 and again in 1964. Subsequently, however, there has been a persistent decline in output, with the volume in 1976 being less than half of the peak level. Consumption within the state is believed to be nearly 3.5 times as large as recent production levels, with more than 70 percent of the softwood plywood consumed in California being brought in primarily from Oregon, with lesser amounts from Washington, Idaho, and Montana.

The wood pulp industry has had only limited development in California, with the estimated 1976 production of more than 800,000 tons representing less than 2 percent of total U.S. production. This is notably low relative to the magnitude of the state's timber harvest and the level of consumption of paper and paperboard products within the state. Costs and other restrictions related to water development and environmental quality standards in California appear to be important in explaining this limited development, but other factors may also be involved, including the effects of the combined geographic and ownership patterns of California's forests on the possibilities of developing control of adequate long term wood supplies. With a part of the state's wood pulp production being shipped to offshore markets, it appears that California produces less than 15 percent of the wood basis for the paper and paperboard consumed within the state. The large volumes of paper and paper-

board shipped into the state come almost entirely from the major producing regions within the U.S., with the exception of newsprint which is also obtained from British Columbia.

As a limited offset to the declining trend in plywood and the limited development of wood pulp production, the state has experienced an expanded production of reconstituted wood panel products in recent years, with the growth being mainly in particleboard production. By 1979 the state had six particleboard plants, two hardboard plants, and one medium density fiberboard plant. Such panel products are alternatives to plywood in various uses, while their production is based on much the same sort of wood residues as is wood pulp production.

One substantial pole and piling operation and various specialty product operations, typically small in size, complete the roster of forest products industries in the state.

Future Prospects

Despite some gradual decline in the volume of timber harvested over the last twenty years, California continues to rank high among the leading forest products states in terms of volume of wood produced, employment generated, and value of output. However, the volume of production within the state is well below the volume of consumption in all of the important forest products, with the possible exception of particleboard. Considering softwood lumber and plywood, paper and paperboard, and reconstituted panel products together, it is estimated that California is a net importer of roundwood to the extent of one-half of its domestic consumption of such products. In addition, the state is almost entirely dependent on external sources for the roundwood volume represented by its consumption of hardwood lumber and hardwood plywood. In the years ahead, this very heavy degree of reliance on external sources for the forest products consumed in California seems certain to increase.

Various projections have been made as to future timber harvest levels in the state. There is general agreement that the cut from industrial forest lands will fall off markedly by the 1990s as inventory volumes are reduced by the current heavy cutting level. On the small and medium nonindustrial private holdings, heavily cut in the past, growth is now well above harvest, and a substantial increase in harvest levels should become sustainable in the future. While it has generally been anticipated that this increase will not be achieved until after the year 2000, some analysts believe it could occur somewhat earlier in response to the rapid increase in the real price of the cutting rights to timber which is anticipated over the decades ahead. The key element in the future harvest level appears to be the harvest from the national forests. The inventory volume there is sufficient to sustain an appreciable increase in cut for several decades without reducing currently planned cutting levels over the longer term. However, the intensity of demand for the other goods and services obtained from these lands, concern as to the environmental effects of any substantial increase in harvest levels, and the proliferating regulatory network controlling action on these lands all indicate that such an increase is not in prospect. In the absence of any pronounced change in public policies controlling the harvest of timber on the national forests, some further decrease in the cut from these lands appears more likely than any increase. Putting these various elements together, the highest projections of the future harvest show a continuation at about the current level, while projections allowing for a decline in the harvest of public timber indicate that an appreciable decline in timber harvest may be in prospect.

No increase in the production of solid wood products in California can be expected, although some limited increase in the production of by-products may be possible. At the same time, further increases in demand for forest products must be expected, as a result of continued growth in California's population and total economic activity. Under such conditions, the state's dependence on external sources of forest products will undoubtedly increase. Unhappily, the state's primary supply region for such products, the Pacific Northwest, also faces a probable decline in the level of timber harvests, along with increased demands from all of its market regions. Thus it appears likely that California will have to develop new sources of supply for forest products. The most promising possibility seems to be in the southern United States, which now seems to be the main region in North America from which substantial increases in forest products output can be expected over the balance of this century.

Thus the outlook in California is for very tight timber supplies, with a continued rapid rise in the real price of cutting rights for the timber which is made available for harvesting. In addition, a further rise in the price of forest products sufficient to check sharply the expanded consumption of such products must be expected. Under such conditions, actions to ensure protection of the current level of timber harvests, to increase timber growth and expand harvest possibilities over the longer term, to achieve increased recovery of forest products in the processing of the timber which is harvested, and to achieve more efficient use of the forest products which are consumed will all be highly important in California's forest product future.

The Range Livestock Industry

Although the average acre of rangeland in California produces the feed equivalent of only about 300 pounds of alfalfa hay, the millions of acres make rangeland an important resource. Range forage provides over 60 percent of the feed required for breeding beef cattle, about 70 percent for stocker cattle, and about 50 percent for sheep in California. Table 3 shows that the San Joaquin Valley counties and the central coast between Lake and San Luis Obispo Counties produce about half of the rangeland grazing in California —most of it on private land.

TABLE 11.3
Animal Unit Months (AUM)[1] of Grazing in California and the Proportion Furnished by Rangeland

	Total AUM Available	Percent Furnished by Rangelands
	000	
North Coast	1,081	95
North Central	1,034	43
Northeast	1,551	49
Central Coast	4,318	79
Sacramento Valley	2,992	43
San Joaquin Valley	6,516	52
Eastern Sierra	301	58
Western Sierra	1,634	80
Southern California	1,555	69
STATE TOTAL	20,982	62

[1] An animal unit month is enough feed to maintain a mature cow or equivalent for one month.

SOURCE: "The Contribution of Rangeland to the Economy of California." Cooperative Extension MA-82, 1974.

The natural rangelands also provide major habitat for wildlife as well as areas for outdoor recreation and building sites. Conflicting demands, many with higher value than forage for livestock have resulted in modification or even elimination of livestock grazing and wildlife habitat. The rangeland resource is over six million acres less than it was 20 years ago. Closure of land to livestock for the most part has occurred on public land but private rangeland taken for living purposes, roads, and recreation has also reduced the rangeland area.

Estimates for 1980 made in response to the Resources Planning Act indicate that maximum biological potential for rangeland-produced forage is about double that being produced and harvested. This assumes investments in rangeland improvements and efficient management. Predictions of greater demand for meat, decreasing land resources devoted to food and fiber production, and lessening energy supplies point to

increasing demands on both the public and private rangelands in California.

History of Rangeland Grazing

Of necessity, range grazing and ranching became the first local industry upon arrival of the Spanish colonists at San Diego in 1769. Herds of livestock were mainstays of all the Spanish missions as the animals furnished meat, milk, hides, and wool for the soldiers and others in the missions. Soldiers of the early expeditions became the earliest ranchers in California. Free range was there for the using, labor was available from the local peoples; the Spanish-type animals were hardy stock bred for rangeland grazing; and markets were developing, first for hides, tallow and wool, and later for meat during the Gold Rush days. However, the rangelands of California were not fully stocked until after the Gold Rush. U.S. census reports showed an increase from

Thousands of acres of California rangeland provide forage for beef cattle. Winter rain encourages lush spring grass in foothill areas; in some parts of the state cattle are shifted to higher locations as snowpack melts in the mountains. *Photo by Ansel Adams.*

300,000 animals (cattle and sheep) in 1850 to nearly 5 million in 1880.

The major vegetation types grazed by early livestock were the California prairie in the Central Valley and numerous coastal valleys, the oak woodland adjoining the prairie, and the chaparral—where it was burned, allowing grasses to dominate for a few years. Sagebrush-grassland occurred in northeastern California and desert shrub in the Mojave and Colorado deserts. After 1850 and reaching a peak in 1910 to 1920, much grazing occurred in the open conifer types and mountain meadows. Presently, the principal rangeland types supporting livestock are the open annual grasslands, oak woodland (with an annual grassland understory), sagebrush grass, converted chaparral, and mountain meadows.

Changes in the Native Vegetation

The grasslands west of the Sierra Nevada were greatly altered by grazing and cultivation. Most explorers plying the Pacific along California came from England and southern Europe. They brought animals and, of course, feed for them. During inland forays from the sailing vessels, seeds of many European annual grasses and broadleaved plants were accidentally introduced. These hardy annual species became abundant as the pristine grassland vegetation was reduced by overgrazing, cultivation, road building, severe droughts, urbanization and other causes. Records in diaries, early botanical collections, interviews, and vegetational studies suggest that the replacement of the largely perennial California prairie by annual grassland with few perennials occurred in the 20 to 30 years after 1850.

This replacement of vegetation by introduced species, perhaps the most striking of any such change in the world, should not be described as good or bad. Apparently, none of the original dominants have become extinct. Relic perennial grasslands are present throughout the climatic range of the prairie, from 4 inches average precipitation along the west side of the southern San Joaquin Valley to well over 50 inches in Humboldt County. Everywhere can be seen mixtures of the native plants and the so-called annual exotics. These "new-native" species are not likely to be replaced as they are well adapted to California habitats. The annual grassland present in 1980 is permanent and as "native" as the perennial grassland of 1780.

In several respects the present California annual grassland may be superior or equal to the original California prairie for grazing purposes. Except in the northwest coastal area where high rainfall encourages range management that fosters the perennials, the annuals furnish as much forage for grazing livestock as did the native grassland. While the annuals may not be green and nutritious late into the spring and early summer, they grow more abundantly in fall, winter and spring. Both perennials and annuals respond poorly to drought but the annuals germinate and grow quickly after the first rains, giving protection to the soil. The annual grassland is more resistant to overgrazing and as productive as the pristine perennial grassland.

Range Improvement and Management

Many attempts have been made to improve the California annual vegetation for grazing purposes. Fertilization increases forage productivity and can pay when practiced on certain soil types and in specified rainfall belts. Seeding of subclover varieties and other legumes improves forage quality and enhances livestock production. Livestock distribution, hence grazing pressure, is made relatively even over the rangelands through fencing to reduce pasture size, development of livestock watering places, and placement of salt away from water. Grazing can be year long on the annual grassland but only if sufficient forage remains on the ground when the hills turn golden to feed the livestock until a new crop germinates in the fall. Another common grazing system is one where young cattle are purchased by the landowner in the late summer and fall. They are allowed to graze the winter's annual vegetation until it dries out in the late spring, when all the animals are sold; thus the range is not grazed during the summer.

Grazing in the coniferous zones, the sagebrush grass types, and the southeastern deserts is largely on public lands. Numbers of animals and season of grazing on the national forest and national resource land allotments are controlled by lease-permit systems which require light to moderate grazing. This maintains vegetation and soil in stable condition. Grazing on the public lands increasingly has come into conflict with other uses of the forests and rangelands and is subject to analysis of relative damage and values in the process of preparation of environmental impact statements.

*Rangeland Grazing as a Part
of the Livestock System*

The major value of California rangelands for grazing lies in its relationship to other sources of livestock feed and to the total system of meat production. California is a deficit state in the production of meat—cattle and lambs for immediate slaughter and also for fattening before slaughter. As a result, this state produces many but not all of the stocker cattle it needs for fattening. These animals are raised first on rangeland and later with the use of irrigated pastures, harvested forage crops, grains, and agricultural by-products. Hence, cow and calf operations on rangeland begin the cycle that eventually furnishes fattened animals for slaughter. The same is true for sheep except that most of the slaughter lambs come directly from rangelands.

Wildlife Resources

Wildlife is a resource of the forest and range lands of California, but unlike trees and livestock forage, the ownership of wildlife is vested in the State and not in the proprietors of the land. This situation creates a different pattern of responsibility for husbandry and management than in the case of resources that are privately owned.

Historically, public interest in wildlife was concerned almost wholly with game or fur-bearing species that supplied meat for the table or furs for the market. Unregulated hunting and trapping in California led to local scarcity of some species in the late 1850s and the legislature passed regulations intended to protect important species from depletion. Over time these legal strictures became more formalized and integrated into state law, leading ultimately to the firm principle that wild animals were wards of the state—which was fully responsible for their protection and management. Between 1910 to 1920 all of the states, including California, adopted the plan of requiring hunting licenses for those who would take wild animals, using the funds from license sales to employ wardens to enforce the game protective regulations and to serve other functions in game husbandry. This system remains the basic scheme of financing wildlife conservation.

The federal government became involved in two ways: (1) public lands, including national forests, national parks, and public domain lands (now public resource lands) constituted a substantial part of the habitat supporting wildlife; and (2) the Migratory Bird Treaty Act of 1918 formally transferred responsibility for migratory birds from the individual states to the U.S. Department of Interior.

Under present operational rules, wildlife management is administered through a complex set of interlocking state and federal responsibilities and regulations.

The Game Animals

Of the several species of larger hoofed mammals native to California, only one—the deer—has adapted easily and successfully to the changes accompanying settlement. Larger deer in the northern and eastern parts of the state are called mule deer; the small coastal races are black-tailed deer. However, all belong to the same species and interbreed freely where races adjoin. The creation of brushfields where forests have been cleared or logged favored the welfare of deer in both the Sierra and coastal mountains. Improved forage conditions coupled with predator control and close regulation of hunting led to an enormous increase of deer in the 1940s. The ranges then became overpopulated with deer, leading to frequent starvation die-offs and severe damage to the browse plants on which the animals wintered. In 1952 the state deer population was estimated at 1,123,000. In 1980 it is probably less than one-half of that. But deer are adaptable, and despite deterioration of their range they doubtless will persist in numbers adequate to supply sport hunting for a great many Californians. The annual buck kill has been running about 30,000 in recent years, down from 75,000 during the peak years.

The bighorn sheep, which was a highly favored game animal during the Gold Rush, was quickly reduced to scarcity, mostly by overhunting but in part by disease and food competition that came with flocks of domestic sheep. Bighorns are considered an endangered species now in the Sierra, and they are faring poorly in the desert ranges where they have to compete for food with feral livestock—mostly burros.

Likewise pronghorn antelope were numerous in the San Joaquin Valley and on the grass and sagebrush ranges east and north of the Sierra Nevada. Today only a remnant population remains in Modoc and Lassen counties.

Elk were very abundant in the Central Valley

Extensive research on livestock management and foothill land use is carried on at the 5800-acre University of California Sierra Foothill Range Field Station east of Marysville. Beef cattle reproduction and feeding studies are a major part of the station's program. Research is also done on soil erosion, vegetative cover, watershed and wildlife management.

and the coast ranges until the Gold Rush when most of them were converted to meat to feed the hungry miners. A few small herds remain, none of them thrifty enough to withstand hunting.

The only new species of hoofed game to be added to the roster of California species is the wild-pig, which is thriving! The original introduction was unintended, when barnyard hogs wandered off in the woods and adopted an independent life style. Later some European wild boar were imported and released in Monterey County. They have crossed freely with the feral pigs, producing a highly adaptable hybrid. Whereas all other ungulates in the state are decreasing or barely holding their own, the wild pig is increasing steadily in number and in occupied range. A recent survey of the pig population in San Benito County indicated a local density of 12 per square kilometer, higher than the local deer density. In 1977 a post-card survey of a 3.5 percent sample of California hunters indicated a yearly kill of 32,000 pigs, almost equal to the deer kill. The wild pig may soon be the number one big game animal in California. It may also

prove to be a serious pest, rooting up the forest floor and pasture lands, and invading agricultural lands. The pig may be hunted throughout the year, yet it continues to increase.

Of the carnivorous mammals of California, only the grizzly bear was hunted to extinction, the last individual being killed in the mountains back of Los Angeles in 1922. The black bear, mountain lion, coyote, and bobcat are all thriving despite warnings to the contrary by concerned conservationists. There is no valid evidence to support the assertion that any of these common predators is threatened by existing levels of hunting or trapping. Raccoons and gray foxes are widely abundant, and the red fox—once very scarce in the state—is now common in the Sacramento Valley and is at least present in the Sierra Nevada. Even the little kit fox, which has been evicted from much of its desert habitat by extended irrigation projects, is found to occupy some secure foothill ranges. The wolverine, fisher, pine marten and river otter have all increased under legal protection.

By and large, the carnivorous animals are

adjusting to human competition much better than the native hoofed ungulates.

Upland Game Birds

Of the California species of grouse, the two that live in the forests or mountains—the ruffed grouse and the blue grouse—are doing very well. The two species of the grassland or shrub-grassland are failing. The sharp-tailed grouse of the grasslands of Modoc and Lassen counties quickly became extinct when these grazing areas were overrun with livestock. The sage grouse, or sagehen, of eastern and northern California is barely holding on in an environment made untenable by livestock grazing. It would appear that conflicting forms of land use are not favorable to the welfare of species of grouse adapted to grassland.

Mountain quail and California quail are holding their own, although at a level far below that of years past. During the early years of land clearing and exploitation in California these native quails thrived in the frontier habitat of weeds and brush. As land use became more intensive, agricultural fields were enlarged, fence rows were removed, brushfields were more heavily grazed, and the aggregate effect has been a lowering of carrying capacity for quail.

Three exotic fowls, however, have added substantially to the game bird population of the state. The ring-necked pheasant is now widespread and abundant in the Central Valley, though it is regressing in abundance as cultivation becomes more intensive. The chukar partridge is filling an unoccupied niche in the desert grasslands and rockpiles of eastern California. And last but by no means least, the wild turkey is spreading and increasing its numbers in the forests and brushlands of the coast ranges and the west slope of the Sierra Nevada. The present population justifies an open hunting season in all counties except San Diego. The wild turkey was native to California in the Pleistocene, but for some reason became extinct late in the Ice Age. Its reintroduction is singularly successful.

The band-tailed pigeon and mourning dove complete the roster of upland game birds. The dove is, in fact, the most important game bird in California, supplying an annual harvest in excess of four million birds. The species is highly adaptable both to agricultural areas and to urban development, and will doubtless be abundant indefinitely into the future.

Wildlife on Agricultural Lands

Agriculture is a prime industry in California and virtually all usable land is now in tillage. The efficiency of farm production has been greatly increased by irrigation and by mechanization. Both processes favor the consolidation of land ownership into fewer farm units, with larger individual fields and less unused space in field borders and roadsides. These trends toward higher intensity use of every fractional acre of soil have substantially reduced the carrying capacity of agricultural areas for birds and mammals of all kinds, game and nongame alike. The native California quail, once called "valley quail," is now virtually gone from the valleys with the elimination of brushy field borders and weedy corners. Marshes and poorly-drained basins are being ditched or filled, lessening the habitat for waterfowl and other marsh-dwelling species. Even the ring-necked pheasant which can exist with a minimum of cover is being crowded to the fringes of farmlands by intensive row-cropping. Grain stubbles are used by waterfowl, pheasants, doves and some smaller birds for winter food, but during breeding season there are very few vertebrates living in croplands.

Freshwater Fishes

Just as most native birds and mammals suffered adversity with the habitat changes accompanying settlement, so also the native freshwater fishes were partially displaced by the same chages. Clearing, logging, grazing, and road-building altered the hydrologic characterstcs of most rivers in California, leading to higher runoff and more frequent flooding. Streambeds were scoured, and some gravel spawning beds are smothered in silt. Populations of trout and of many native minnows and nongame fishes are substantially reduced, primarily as a result of habitat degradation. Introduction of the European carp further muddied waters. Many of the rivers are dammed or diverted from their original channels. In addition, the game species are held down by high fishing pressure.

Some native forms of trout like the Piute and Lahontan cutthroats are nearly extinct. Former strongholds such as Pyramid Lake at the mouth of the Truckee River, Lake Tahoe, and Independence Lake are essentially devoid of their native cutthroat populations. Heavy infusion of hatchery-raised rainbows has accelerated the demise of the cutthroats.

Today most of the important fisheries of fresh waters in California are composed of exotic species, introduced from elsewhere. Brown trout, brook trout, black bass, crappie, bluegill, catfish, shad, and striped bass now supply most of the sport fishing in the state. The rainbow trout, which is native, continues to be important, and there are still modest salmon runs in a few of the undammed rivers. But by and large, California fishermen now seek mostly exotic fishes for their sport.

A conscientious effort is being made to restore the productivity of some California lakes and streams. Newly adopted regulations regarding allowable land use practices on riparian strips are designed to protect stream banks and water channels from erosion. Watershed protection is the objective of new limitations imposed on logging, grazing, and urban sprawl. A few rivers are being designated as "wild rivers," to be preserved for recreation of many kinds, including fishing.

Nongame Fish and Wildlife

Wildlife conservation from about 1910 to 1970 was heavily weighted toward preservation and restoration of game birds, game and fur-bearing species of mammals, and game fishes. Protection for nongame was routinely provided by legislative edict (i.e., the Migratory Bird Treaty Act) and by the assumption, which was accurate enough, that establishment of wilderness areas and of refuges for game species also provided habitat for nongame species. But specific management programs for nongame were generally lacking. During the 1970s this casual attitude changed abruptly. Both the federal government and the California Department of Fish and Game became deeply concerned about the status of all native animals, especially those that were considered rare or endangered. New legislation was passed and new field programs were initiated to assure the security of even the most obscure snake or mouse or minnow. Serious questions were raised about the impact of various land use practices on native vertebrates. The effects of logging on forest birds like the spotted owl or on hole-nesting species that depended on snags for home sites led to new forest practice rules. The effects of grazing on wildlife of riparian strips, especially in arid zones, came under intensive scrutiny and led to new regulation, particularly on public ranges. Irrigation projects on the east side of the Sierra have been stopped to preserve water tables for desert pup-fishes in springholes. Even oceanic wildlife such as marine mammals were caught up in this new wave of conservation. The long-established pattern of financing wildlife conservation by licenses and taxes imposed on hunters and fishermen came under challenge, since nongame clearly was the responsibility of the public at large and not just the sporting public.

Future historical appraisal of the wildlife conservation movement will inevitably mark the decade of the 1970s as the turning point when public responsibility was declared for the preservation of all wild animals and not just game species. The "ecologic conscience" was substantially broadened in the process.

Other Forest Resources

California forests are also the source of a number of other kinds of values of great importance to the state. Among these, water yields and recreational environments are of particular importance.

Water Yields

The average annual surface runoff in California is estimated at 78 million acre feet per year. The geographic distribution of the precipitation which produces this yield is highly correlated with forest cover. As a result, it has been estimated that well over 90 percent of the runoff originates on the forested lands. Some 50 million acre feet of the total originates on the 21 million acres of conifer forests of west slope of the Sierra Nevada, the Siskiyou, and the Coast Range.

Forests can influence the amount, the timing, and the quality of these water yields. For example, reduction in the density of the forest cover (particularly in the riparian zone) reduces water loss resulting from transpiration and may therefore increase surface runoff. Manipulation of tree cover in the snow zone can induce local increases in snow packs and delay the occurrence of snowmelt in the spring. Maintenance of a full cover of vegetation promotes infiltration of water into the soil and minimizes soil erosion and consequent stream pollution.

The latter relationship is of the most importance. Some forested watersheds in California are extremely unstable because of their geologic character and thus have unusually high rates of erosion even under undisturbed conditions. Any acceleration of those natural rates of erosion is of concern, both because of adverse effects on the quality of the water for beneficial uses such as irrigation and because of more rapid deposition of silt in storage reservoirs.

Because of these relationships, water quality protection has become a major objective of resource management on public lands and a major goal of the state's system of forest practice regulation on private lands.

Recreational environments

Use of the forests for recreational purposes in 1977 amounted to 45 million visitor days—more than two visitor-days per year per capita of the state's population. Although checked by the drought years of 1976–1977, the growth in forest recreation use has been increasing for some time at a rate of over 3 percent per year. Eighty percent of the recreational use occurs on public lands—national forests, parks, and resource lands, and state parks.

Family and organization camping is the principal type of forest recreation activity, accounting for over half the total use. Fishing, winter sports, hiking, hunting, and picnicking are also important. Off-road vehicle use has increased rapidly in recent years, amounting to close to one million visitor-days in 1978.

Since 1970 the growth in forest recreation has exceeded the growth in available recreation facilities. Since further increases in recreation demand are expected, more intensive use of recreation facilities and consequent increased overcrowding are to be expected. Such problems have already led to such controls on forest recreation use as the institution of permits to enter certain wilderness areas and reservation systems on popular campgrounds. But unless the demand for forest recreation use is severely dampened by rising transportation costs, more positive means of balancing recreation supply with demand will be needed if critical problems of recreational quality and resource deterioration are to be avoided.

References

California Department of Forestry. *California's Forest Resources: Preliminary Assessment.* Sacramento, 1979.

Reed, A.D. "The Contribution of Rangeland to the California Economy." Publication MA-82. University of California Cooperative Extension, Davis, June 1974.

U.S.D.A. Forest Service. *An Assessment of the Forest Range Land Situation in the U.S.* (Review Draft). Washington, D.C., 1979.

PART IV 🌿 Institutions: Private and Public

12

Organizations in Agriculture

Ann Foley Scheuring and Refugio I. Rochin

Topography and climate combined with modern technology have made California agriculture unique in the world. Natural resources have been brought to a high degree of productivity, however, not only because of technological advances but because of the existence and evolution of a complex social and economic infrastructure which has encouraged large-scale development. The chapters which follow describe the private and public institutions which have supported and advanced, or attempted to control and regulate, the development of California agriculture. This chapter looks briefly at agricultural organizations in the private sector. Farm worker unions are also described since they too are agricultural organizations which have had significant influence in recent years.

Farm Organizations

California agriculture is characterized by a high degree of social and economic organization, perhaps far more than in most other regions. One reason for this is the diversity of production in the state, and the necessity for strong marketing organizations to deal with the problems of transporting 100 or more commodities to distant markets. Another impetus to organizing has been the necessity for group action in dealing with problems of resource development in a vast and semi-arid land. The number of farm-related groups has also multiplied in recent decades to defend the interests of agriculture in a world that is increasingly urbanized and regulated. While reapportionment in the state legislature has resulted in fewer lawmakers sympathetic to rural constituencies, the increasing number of regulatory agencies imposes increasing burdens on farmers. Thus, farmers have found that they

must, in self-defense, become more politically active.

Like California farmers themselves, the organizations in which they participate are diverse, ranging from local associations with only a few members to the monolithic California Farm Bureau Federation with its budget of millions of dollars and membership nearing 100,000. The California Agricultural Directory of 1980 contains 66 pages of names and addresses for farm and farm-related organizations, divided into 42 categories. A few are general interest organizations with activities spanning the state. Some are local or state commodity organizations through which growers market their crops or pursue common interests: for example, Monterey Winegrowers Council, San Joaquin Valley Hay Growers Association, and Northern California Worm Growers. Some organizations are formed for special purposes, for example, irrigation districts, labor associations, farm supply companies, and farm credit agencies. Others represent processors and handlers of agricultural products. Many farmers belong to several different organizations, depending on their interests and economic or political involvement.

General Organizations

CALIFORNIA STATE GRANGE. The oldest farm organization in the state is the California State Grange, started in 1873 when farmers nationally were organizing as "Patrons of Husbandry" to fight low grain prices, high railroad charges, and political corruption. An initial business venture by the California Grange into grain marketing was a failure. Thereafter the organization primarily served fraternal and social needs in isolated rural areas, particularly in the northern grain-growing counties. The local Grange hall often served as a community center for education and self-improvement. In the early years of the century the Grange was opposed to organized gambling, liquor, and tobacco. The Grange has consistently represented smaller farmers, many of them part-time, in seeking redress on economic grievances. It has supported youth activities and a housing project for the elderly as well as several service programs for members, including insurance and wholesale purchasing of hardware and building supplies.

In the 1930s the organization surfaced in Sacramento as a political entity, due to the leadership of George Sehlmeyer, Master of the State Grange from 1929 to 1959. Under Sehlmeyer, the Grange supported a fairly broad program of political action: public ownership of utilities, banking reform, equality of taxation, and purity in politics. The Grange often took a position to the left of most other farm organizations. In the 1960s the organization, with its small staff and modest budget, was relatively inactive, but by 1979 membership had increased to 48,000. In recent years, the Grange has tended to take its political stands more in conjunction with other farm groups, although on the issue of acreage limitations in federal water deliveries, it has stood alone in supporting the expressed intent of the 1902 Reclamation Act. The political clout of the Grange has been considerably less than that of the Farm Bureau because of the organization's continuing focus on fraternal and community affairs.

CALIFORNIA FARM BUREAU. In 1980, membership in the California Farm Bureau Federation was 96,000; only about half of these are farm families, since business firms and other interested parties may join to take advantage of insurance programs and other services. The 1979 move of Farm Bureau headquarters to Sacramento into a newly constructed building minutes away from the Capitol reflected the expanding activities and ambitions of the organization as it stretched increasingly into new roles very different from its original one.

Farm Bureau began as a kind of stepchild to the Agricultural Extension Service which was established following passage of the Smith-Lever Act in 1914. B. H. Crocheron, first director of Extension, would place farm advisors from the Uni-

versity only in counties which had first established farmer organizations to cooperate with the Service in disseminating technical information being developed through the Experiment Station. County Farm Bureaus, with their local centers in country schools or churches, were formed to host monthly meetings for farm families who came to attend lectures and demonstrations, and, equally important, to socialize. Working closely with Extension to encourage farm and home improvements, Farm Bureau was also active in community affairs.

By 1938 the activities of the state organization had become increasingly commercial with the addition of purchasing programs in petroleum, fertilizer, and other farm supplies. As the Farm Bureau started to become involved in local and state politics, its headquarters were moved off the University campus. During and after the war years the farm center programs declined in importance.

In recent decades the Farm Bureau Federation has substantially increased both its membership and its perceived roles. Its affiliated companies now offer health, life, and property insurance programs, farm supplies, a farm labor employer advisory service, and travel tours, while FARM-PAC, its political action affiliate, has proved a potent fund raiser for political campaigns. Seven registered lobbyists analyze legislation and carry the Farm Bureau position to the legislature and to regulatory agencies. Generally Farm Bureau's policy positions have been consistent with its free enterprise philosophy. Its leaders have taken strong anti-regulatory stands on such issues as water rights, the Reclamation Act acreage limitation, marketing, land use, farm labor, and international trade. While the membership reflects a spectrum, leadership has tended to represent the more conservative commercial family farmers. County Farm Bureaus have always been particularly strong in the San Joaquin Valley.

AGRICULTURAL COUNCIL OF CALIFORNIA. Another organization active in the capitol is the Agricultural Council of California, founded in 1919, the same year as the California Farm Bureau Federation. The Agricultural Council is an umbrella organization for 71 major California farmer cooperatives, and does not include individuals directly in its membership. It serves as an informational clearinghouse and advocacy group for the passage of legislation favorable to agriculture, particularly on marketing. Because of the financial strength of the cooperatives involved, the Council has been able to exert considerable influence, and works closely with other farm groups. In addition to its lobbying activities the Council sponsors educational publications and agribusiness tours for high school and college teachers of agriculture, as well as training workshops for members. As part of its public relations role the Council publishes a weekly newspaper opinion column.

OTHER GENERAL ORGANIZATIONS. In the Coachella Valley in 1975, a new group, California Women for Agriculture, was organized. Developing from the desire of farm women to have their own organization rather than to work solely as an adjunct to existing groups, CWA numbered about 3,000 members statewide in 1980. Its chief focus is to monitor legislation and agency actions relevant to agriculture, but the group also sponsors public relations projects, particularly those educating urban consumers on farm matters. Not a formal lobbying organization, CWA has used letter-writing and telephone campaigns in an effort to shape legislative and public opinion on such issues as the United Farm Workers lettuce strike of 1979.

Discussion of agricultural organizations in California must include the role of the California Chamber of Commerce. Nearly 40 percent of the Chamber's 1980 Board of Directors had ties, directly or indirectly, with agriculture. The Agricultural Committee of the State Chamber is comprised of about 100 individuals in agribusiness across the state. This group has represented the interests of agriculture from the point of view of the property holder, allying itself with other segments of the business community in its positions on tax and regulatory issues. A chief function of the Agricultural Committee has been to formulate policy recommendations on farm labor, pesticides, and land use. Since nearly all farm organizations in California belong to the Chamber, its Agriculture Department has served as coordinator for meetings on important issues affecting agribusiness interests. In the late 1960s the Chamber began coordinating the "Agricultural Roundtable," a weekly breakfast meeting for the legislative advocates of farm organizations with offices in Sacramento. Strictly an informational forum to keep members in touch with the flow of events during legislative sessions, this informal liaison has helped to avoid duplication of efforts and to coordinate activities.

Local Chambers of Commerce have also been more or less active in agricultural matters, depending on the leadership of the times, particularly in Fresno, Bakersfield, Los Angeles, and San Francisco.

Commodity Groups

Some of the larger commodity groups maintain legislative representatives in Sacramento, who attempt to secure legislation favorable to their members. The powerful League of California Milk Producers dominates the heavily regulated milk industry. The California Cattlemen's Association has consistently stood for cattle ranchers on the issues of land use and tax legislation, while Western Growers Association represents nearly all fresh produce growers in the state in the areas of labor and marketing. The California Forest Protective Association is comprised mostly of larger forest owners, while the Forest Landowners of California represents the smaller ones. The California Nurserymen's Association is another large and strong commodity group employing a lobbyist.

Trade Associations

Some trade associations represent processors of commodities. The Canners League includes nearly all fruit and vegetable canners in the state, both proprietary and cooperative. Other important processor associations are the American Food Institute, the Wine Institute, the Dairy Institute, the Pacific Coast Meat Association, and the California branch of the National Food Processors. Trade associations commonly provide technical services to their members, sponsor research, and employ legislative advocates.

Special Purpose Organizations

Some agricultural organizations were formed for special purposes and do not engage in formal lobbying. Among them is the Council of California Growers, formed in the 1960s as a general membership public relations firm which attempts to present a more favorable image of agribusiness to the consuming public via media releases and a speakers' bureau.

Associated Farmers, an association of farm employers organized during the Depression to combat union organizing activities in farm areas, gained notoriety for heavy-handed tactics during episodes of labor unrest. Now much reduced in scope, it operates out of San Francisco doing labor relations work for a dozen or so clients in southern California.

Alternative Groups

Not part of the agricultural establishment in California, but sporadically part of the farm scene, is the loose coalition of growers supporting the goals of the National Farm Organization and the American Agriculture Movement. In the late 1960s and early 1970s vigorous attempts were made by the NFO, particularly in the Sacramento Valley, to organize farmers and ranchers into "collective bargaining" blocks, to negotiate for higher commodity prices by pooling individual supplies. NFO "blocked" basics of grain, meat, rice and cotton, plus specialties of beans, potatoes, walnuts, hay, and almond hulls. NFO organizers claim that as many as 5,000 growers were organized at the peak of efforts into a block comprising a third of all California grain production, but the sudden sharp increase in grain prices following massive Russian purchases in 1973 obliterated the power of the block and dissipated the energies of the organization.

The American Agricultural Movement, a volunteer group rather than a structured organization, never attracted many followers in California, although a tractor cavalcade demonstrated in Sacramento in December 1978.

National Farmers Union, active in the Midwest for decades, at one time chartered a California affiliate, the California Farmers Union, but this charter was revoked in 1948. Since then NFU has done no reorganizing in the state, although in 1980 it claimed over 3,000 dues-paying California members.

Farm Worker Organizations

Farm workers, unlike farmers, have not until relatively recently been organized into effective group efforts to improve their lot. Disorganization has been intrinsic because of transient jobs and dispersal of workers across sometimes great distances. Also, California farm workers historically have been composed of fragmented ethnic groups lacking in general education and possessing little discretionary income or opportunity for pursuit of group goals. Often enough, farm work has not been an occupation with which workers desire permanent identification, but a more or less temporary job to move out of when possible.

Since California's agricultural workforce has been composed of such diverse groups differing from each other and from society in nationality, language, and culture (see Chapter 1), a labor system evolved that coped with the situation but kept workers from cooperating to better their lot. The inability of most workers to speak English ensured their dependence on a bilingual labor contractor who dealt with the grower, sometimes at considerable personal profit to himself. Farmer-laborer relations often remained impersonal, the workers sometimes being identified only by their Social Security card numbers. Employers did have racial-group preferences, however, and differing wage rates were once paid to different nationalities. Thus, those in one ethnic group would regard those in another as competitors and mutual suspicion was aroused. Sometimes mixed crews were used to destroy the solidarity that inevitably grew up in homogeneous groups. The system meant that effective worker bargaining power was a long time coming.

Nevertheless, various short-lived unionization efforts were aimed at organizing California's farm workers between 1900 and 1960. Such attempts were made by several unions: The Industrial Workers of the World (circa 1910); the Filipino Labor Union (FLU); the Cannery and Agricultural Workers Industrial Union of the Trade Union Unity League and the Confederation General de Obreros y Campesinos Mexicanos, the American Federation of Labor (early 1930s); the Teamsters; the United Cannery, Agricultural Packing and Allied Workers of America of the CIO (late 1930s); and the National Farm Labor Union in collaboration with the Southern Tenant Farmers Union (late 1940s and 1960).

Labor organization activity was particularly strong during the Great Depression, with 140 recorded farm labor strikes in California during the years 1930–39. Low farm prices and surplus labor conditions reduced wages to bare subsistence levels and worker discontent was widespread. Public opinion, however, usually sided with farm employers, and unionism did not achieve widespread sanction or support. Probably the most effective efforts at union organizing during the 1930s were made by Filipino field workers, who formed collective bargaining associations to win wage concessions in the Salinas and Stockton areas. The Filipino Agricultural Labor Association (FALA) in Stockton developed intense worker loyalty by providing community services to members, and at one point numbered over 6,000 laborers and contractors harvesting asparagus in the Delta. Mobilization of workers in World War II, however, ended unionizing activities and brought the bracero program, and not until the early 1960s did the farm labor movement coalesce again into unified group action.

Several reasons can be cited for the failure of early efforts. The general problems inherent in low levels of income, uncertain employment, discrimination, lack of organizing skills, and ineffective communication thwarted labor's attempts to organize effectively against employer wealth and power. The size and structure of California farms also made organizing workers difficult. Hired workers on smaller farms often developed strong ties with their employers, and even if unions succeeded in organizing small farms, they faced relatively high costs of servicing workers across many widely dispersed units. Large farms also posed barriers to unionization by denying access of union organizers to workers, and by participating in employers' associations which recruited braceros or other contract workers.

A major factor in frustrating the organization of farm workers was the absence of a legal framework for settling disputes and lending legitimacy to labor movements. Agricultural labor was specifically excluded in the National Labor Relations Act (NLRA) of 1935 and its amendments, which regulated the procedures under which labor and management might interact.

The early 1960s witnessed the rise of further unionization efforts among farm workers in California. The Teamsters signed an agreement in May 1961 with Bud Antle, Inc., the largest lettuce producer in California and the United States. The Agricultural Workers Organizing Committee (AWOC), established and underwritten by the AFL-CIO in 1959, incorporated Filipinos of the FALA days and began union activities in Delano in 1965. While the Teamsters waited for the bulk of the produce industry to follow Antle's lead, Cesar Chavez began building a union in 1962 exclusively for field workers throughout the San Joaquin Valley.

Chavez chose Delano as his National Farm Workers Association (NFWA) headquarters, concentrating on organization of resident grape workers. The early years of his efforts were spent building a strong sense of solidarity in the farm labor community. NFWA services to members included group insurance, a credit union, a cooperative gas station, and staff help with personal problems. By 1964 over one thousand Mexican-American families in seven counties had joined the association. The events surrounding the passage of civil rights legislation in 1964 heightened public awareness of ethnicity and equal rights, permitting Chavez to integrate both economic and sociopolitical concerns into his organizing drives.

Union organizing began in earnest in 1965, with work stoppages, picketing, and mass jailings. The NFWA engaged in strategies and tactics common in pre-1935 industrial labor relations, including field strikes and secondary boycotts of business selling nonunion and rival union products. Chavez was able to combine the AFL-CIO's Agricultural Workers Organizing Committee with his own National Farm Workers Association to form the United Farm Workers Organizing Committee (UFWOC) in 1967.

Public awareness of the farm labor issue was further aroused by the dramatic national boycott of California table grapes in 1968–69. During the early 1970s, as growers began to agree to collective bargaining by workers, there was intense inter-union rivalry between the (renamed) United Farm Workers Union (UFW) and the Teamsters, who secured field worker contracts to complement their cannery and transportation membership. Continued pressure finally forced legislative hearings on agricultural labor problems, and the inevitability of some type of farm labor legislation was gradually realized despite initial employer antagonism.

Before the passage of the California Agricultural Labor Relations Act (ALRA) in 1975, Fuller noted that:

...farmworkers were being forced into one union or the other without the exercise of self-determination; farm employers were being forced to choose one of three options: to resist unionization, to capitulate to coercion, to engage in collusion. None of these choices brought tranquillity to the parties involved or to the community at large. To the contrary, the numerous frictions involved were regularly disruptive and occasionally bloody.

The ALRA, modeled after the NLRA, created an Agricultural Labor Relations Board (ALRB) to supervise and certify farm worker elections, creating a framework for orderly settlement of disputes.

In its first months, the ALRB conducted 429 elections involving 32,239 ballots, but when funds were exhausted, opponents prevented refunding from February to July 1976. After a

November 1976 ballot proposition making ALRA funding a constitutional mandate was defeated, the ALRB settled down to "normal" operations, opening several field offices in December.

In February 1977 the Teamsters halted their field worker organization drives, leaving harvest labor's representation to the UFW. In fiscal year 1977–78, 133 elections were held in which nearly 10,000 farm workers voted. Although the UFW dominated these elections, other unions appeared on some ballots: the Christian Labor Association, the Fresh Fruit and Vegetable Workers, the International Union of Agricultural Workers, the California Independent Union, the Independent Union of Agricultural Workers, and the Amalgamated Meatcutters and Butcher Workmen of North America. Of total votes cast, 60 percent were for the UFW, 8 percent for other unions, 25 percent for no union representation, and about 6 percent were challenged as having been cast by ineligible voters.

ALRB procedures are still being worked out in court hearings. Determining eligibility to vote for unionization has been a major problem, given the composition of the agricultural labor force. Since a majority of hired workers do less than 25 days of farm work each year, an equal weighting of votes would give those with only a casual attachment to farm work relatively more voice than that of permanent employees. The ALRA declares that "employees...whose names appear on the payroll . . . immediately preceding the filing of the petition . . . for an election should be eligible to vote." The administration of this seemingly simple ALRA instruction, however, is quite complex since payment time periods differ from employer to employer (e.g., daily, weekly, biweekly) and the eligibility of workers under various conditions is not always clear. The eligibility of strikers, for example, is not certain, nor the eligibility of workers in runoff elections. Opinions differ about whether those on temporary layoff because of bad weather or delayed crop ripening should be eligible. These and other matters have yet to be resolved as the ALRA struggles to administer the law fairly.

Some strike activity occurred in the lettuce fields in 1979, continuing the labor-management conflict in spite of marked improvement in labor relations in many areas of the state. Also in 1979, the UFW filed charges of "bad-faith" bargaining against 28 lettuce and vegetable growers. Union concern has been centered recently on clarifying established contracts and on contesting elections.

From its headquarters near Keene in the Tehachapi Mountains, the UFW in 1980 claimed over 150 contracts. As we enter the decade of the 1980s, however, less than 10 percent of California's table grapes are harvested by members of the UFW. The union's active membership stands at about 30,000, down from a peak of about 50,000 at the end of the seventies. No one fact can explain why the union did not grow to encompass a larger fraction of the potential membership of more than 200,000 farm workers in California. Although collective bargaining for seasonal farm workers, now clearly legitimized, is a reality, we should not expect future successes of farm worker unions to come easily in California.

References

Chambers, Clarke. *California Farm Organizations.* Berkeley: University of California Press, 1952.

Daniel, C. E. "Agricultural Unionism and the Early New Deal: The California Experience." *Southern California Quarterly*, Summer 1977.

Friedland, W. H. "Seasonal Farm Labor and Worker Consciousness." *Research in the Sociology of Work* 1(1981):351–380.

Fuller, V. "The Struggle for Public Policy on Farm Labor-Management Relations." Unpublished manuscript. University of California, Davis, Department of Agricultural Economics, 1976.

Geyer, W. H. "State Agricultural Policy Formulation in the 60's: The California Experience." Unpublished paper delivered at the symposium, "Agriculture in the Development of the Far West." University of California, Davis, June 1974.

Rochin, R. I. "New Perspectives on Agricultural Labor Relations in California." *Labor Law Journal*, July 1977, pp. 395–402.

Segur, W. H., Jr. *Representative Elections for Farm Workers: Voting Power Under Alternates Rules of Eligibility.* Ph.D. dissertation, University of California, Davis, Department of Agricultural Economics, 1980.

Sosnick, S. H. *Hired Hands: Seasonal Farm Workers in the United States.* Santa Barbara: McNally and Loftin, 1978.

Sufrin, S. C. "Labor Organization in Agricultural America, 1930–35." *American Journal of Sociology* 43(January 1930):544–559.

13

Agricultural Credit

Walter W. Minger

Dramatic changes in California agriculture in the period since World War II have been accompanied by an unprecedented increase in productivity. The development of technology both at the farm level and in public systems such as irrigation projects has been responsible for much of this increase. Concurrent with technological development have been the consolidation and expansion of farming operations, as marginal units have been absorbed into more economically viable ones. A necessary component in the adoption of new technology, however, as well as in the expansion of operations, is the availability of support money. An understanding of the credit system is essential to an understanding of the state's agriculture.

The purpose of this chapter is to give readers some perception of the scope, the complexity, and the challenges of agricultural financing and the efforts of lenders to meet the need for money and related services in the farm and ranch sector of California's economy. The following pages describe the structure of the credit system, discuss the evolution of current practices in financing, and point out some implications for the future.

Agriculture and Credit in California Since 1950

Since 1950, the use of credit by California's farmers and ranchers has steadily increased, as shown in Table 1.

Up to the spring of 1980, there was no lack of capital available from the state's institutional lenders to California's farmers and ranchers. The ability of the farmer and rancher to expand operations, to purchase equipment to substitute for labor, and to invest as needed to intensify agricultural operations has been supported to a great extent by the increase in land values in California.

Annual farm income over the 10 years 1970–1979 outpaced the inflation rates. At the same time the increase in assets was financed by increasing amounts of debt. Over the last 10 years the amount of assets and the amount of debt grew at an average annual rate of 9 percent and 18 percent respectively. The asset-to-debt ratio in 1970 was over 5; at the end of 1979 it was down to a little over 4.

Asset values, standing at $22 billion in 1970, increased sharply to $42 billion by the end of 1979. Debt grew 2.95 times during this ten-year period, however, while asset values increased only about 1.9 times. Farmers' equities increased 1.72 times over the same ten-year period.

Although the 1970s were generally profitable years for agriculture, the profit on gross sales declines and debt remains high, so there is some pressure on net income after taxes to provide enough debt repayment capacity. While lenders were relatively comfortable with the farm financial picture in the seventies, they will be less so if the squeeze on profits and the high level of debt continues for long into the eighties.

The Agricultural Credit Delivery System
The agricultural credit delivery system in California today can be described by reviewing a number of its characteristics:

(a) Many institutional lenders, headquartered both within the state and elsewhere, serve agriculture in California. For example, the Federal Reserve Bank reported that as of December 31, 1979, there were approximately 516 corporations organized to do general banking. Of those 516, 150 national and state chartered branch banking systems own and operate almost 4,000 banking offices scattered throughout California. Aside from the chain banks, there are 96 unit banks. The latter are less likely to have significant agricultural loan portfolios although there are some noteworthy exceptions in the Sacramento and San Joaquin Valleys, where a number of small unit banks have commercial loan portfolios made up principally of loans to agriculture.

In addition to the commercial banking system, the Farm Credit System has 76 field offices out of which the Production Credit Associations operate, and the local Federal Land Bank Associations have another 37 offices. The latter two institutions are also well dispersed over the state of California in all agricultural areas.

(b) California has a highly competitive lending environment. The agricultural loan market is eagerly pursued by a number of institutional lenders. Farming has been a reasonably profitable business for a number of years earning, on average, approximately 20 to 25 percent of pretax net income on gross sales of farm products at the farm gate. Agriculture is the state's major commercial activity, and the industry is a major user of credit within the state. For example, total funds utilized in 1979 amounted to approximately $10 billion, of which about $9.5 billion was borrowed. Particularly since the credit crunch of the mid-70s and the subsequent lacklustre demand for funds from the nonagricultural commercial sector, the lending community

has recognized the ongoing need for credit by the agricultural sector, and the wooing of agricultural borrowers has gone on apace. As one would suspect, varying degrees of commitment are made by lenders to agriculture. These commitments range from a willingness to make an occasional agricultural loan, to a spasmodic venture into agricultural lending on a "hot and cold" basis, to the ongoing serious commitment of resources, both of people and money. Enough major lenders have made serious commitments, however, so that the lending environment is competitive. This has been a major factor aiding in the development of agriculture to its present level.

(c) Financial services are available for almost every conceivable need of the farmer, the rancher, or the agribusiness firm. The following list identifies financial services available to the agricultural community:

Seasonal loans for crop production
Seasonal loans for livestock production
Medium term loans for herd/flock expansion
Seasonal hay loans for dairymen
Equipment loans; equipment leases
Storage and drying facility loans
Farm real estate
Working capital loans for agribusiness
Medium and long term loans to cooperatives and agribusiness
Loans to purchase milk quota and various growing rights
Hedging loans
Feeder cattle loans
Livestock fattening loans
Heifer grow-out program loans
Dairy cattle loans
Dairy operating loans

TABLE 13.1
Trends in Credit Usage for California Agriculture

Year	Real Estate	Non-Real Estate*	Total Debt
1950	$ 461,000,000	$ 185,000,000	$ 646,000,000
1960	1,241,000,000	534,000,000	1,775,000,000
1970	2,353,000,000	1,073,000,000	3,426,000,000
1979	6,428,000,000	3,686,000,000	10,114,000,000

*These are loans outstanding at year-end at commercial banks, units of the cooperative Farm Credit System, and Farmers Home Administration. Loans to farmers held by individuals or credit provided by vendors, trade suppliers and merchants are not included.

SOURCE: USDA Economics, Statistics, and Cooperative Service, Washington, D.C.

Home/farmstead improvement loans
Land development loans
Farm housing loans
Water system/irrigation system loans
Poultry loans—flock replacement
Poultry loans—broilers
Commodity loans
Export letters of credit
4-H, FFA, FHA project loans
Orchard, vineyard development loans
Loans for conservation practices
Consumer loans—appliances, autos
Loans for investment in cooperatives or
   proprietary companies
Foreign exchange
Deposit-savings accounts
Investment accounts, i.e. Treasury Bills,
   Certificates of Deposit
Pre-authorized lines of credit
Safe deposit boxes
Investment counseling
Farm management
Retirement accounts
Credit cards
Escrow services
Collection services
Travelers cheques
Payroll service
Farm enterprise planning
Trust services
Farm estate planning
Agri-credit life insurance

(d) Generally, a fairly sophisticated, experienced and well-trained group of lending or credit officers is found in the institutions that grant credit to farmers and ranchers. Trade groups such as the California Bankers Association and some of the major lenders themselves regularly conduct training courses in commercial and agricultural credit and in other subjects such as financial statement analysis. Seminars on various subjects of interest to agricultural lending officers are held throughout the state and are well attended.

(e) There is no shortage of capital to meet agricultural financial needs, at least not up to the time of this writing. It is anticipated that agriculture's financial needs, even if they grow very rapidly, can be met by the lenders committed to agriculture. For years farmers enjoyed favorable interest rates, principally because of the competitive presence of the Farm Credit System, as well as the inherent reluctance of farmers to pay market rates of interest. Interest rates for agricultural borrowers have recently, however, tended to rise toward the level of those charged in other commercial activities such as manufacturing and consumer services. In the future, if there should be some shortage of capital to meet all the needs of our economy, farmers and ranchers will have to compete with other borrowers.

(f) The state of California has a legal system and a legal documentation procedure that fairly apportions equity to both farmer and lender. Both have some assurance that their interests are protected not only in periods of ordinary creditor/debtor relations but also in adversity when, for one reason or another, the farmer's or rancher's financial position declines to the point where operations are liquidated to satisfy outstanding debt.

(g) Lenders in the state of California, particularly large institutions which have more formalized lending programs, have ready access to the nation's money markets to obtain the funds needed by their agricultural customers. In that manner, lenders can supplement local sources of funds which may be inadequate, at times, to the borrowing requirements of agriculture.

(h) The state has a good system of ancillary services that facilitate and encourage lending to agriculture. These include legal, accounting, computer, and insurance services, plus all the infrastructure of a modern and sophisticated food system to accommodate the tremendous annual outpouring of both specialty and staple crops.

(i) Enhancing competition in the lending environment is the amount of information about the industry that is available to competitive lenders. At least twice a year banks issue "call reports," which show their financial positions. The Farm Credit System releases quarterly statistics on its loan portfolio, while its Production Credit Associations and Land Bank Associations issue annual reports. The University of California and the United States Department of Agriculture issue statistics that detail the various types of loans utilized by California's farmers and ranchers. The county agricultural commissioners report crop acreages and the cash values of annual sales of the commodities. A number of sources issue market share reports indicating how the seasonal agricultural loan market is divided among commercial banks and Production Credit Associations. The formal statistics issued by these institutions are read avidly by their competitors. The sheer size of the agricultural credit market, which includes loans to agribusiness, creates intense interest among lenders. In addition, the perception that farm lending can be quite profitable is backed up by the earnings reports of various lenders, the public utterances of financial officers, and Farm Credit System statements.

A summary comparison of commercial and Farm Credit System loans in 1979 follows.

TABLE 13.2
Non-Real Estate Loans to Farmers By Selected Institutions in California

Lender	Dollars Outstanding 1/1/80	Percent of Total*
	Thousands	
Commercial banks	$2,654,543	57.8
Production Credit Associations	1,424,193	31.0
Federal Intermediate Credit Bank	37,041	.8
Farmers Home Administration	472,440	10.2
	$4,588,217	

*Figures do not add to 100 due to rounding.
SOURCE: USDA-ESCS

TABLE 13.3
Real Estate Loans to Farmers By Selected Institutions in California

Lender	Dollars Outstanding 1/1/80	Percent of Total*
	Thousands	
Federal Land Banks	$1,955,525	26.7
Farmers Home Administration	87,980	1.2
Life insurance companies	1,731,456	23.6
All banks	398,232	5.4
Individuals and others	3,142,561	42.9
	$7,315,754	

*Figures do not add to 100 due to rounding.
SOURCE: USDA-ESCS

Trends in Financing Since World War II

The war years of 1941–45 were difficult for California's agriculture. When the U.S. war machine was increasing in production, the industrial complex recruited labor from every source imaginable, including farm labor. The latter were attractive employees inasmuch as they generally possessed good manual skills and were experienced in handling equipment and machinery. Wartime wage and price controls and rationing programs affected many of the items used in agriculture, and scarcities (and occasional gluts) of resources made life difficult for those who were left on the farm. Little or no new equipment was available, and consequently prices of used equipment were sky high. Generally, there were input shortages. However, whatever was produced generated reasonably good dollar profits although at great expense if measured by the frustrations and tribulations experienced by farm operators.

Returning veterans came home to worn-out equipment. The economy was slow in converting to peacetime production, and it was a number of years before equipment and machinery could be replaced. Along with the numerous changes in California farming which have been detailed in other chapters of this book, the period after the war saw a significant expansion in use of credit by farmers.

Changes in Credit Practices

If one goes back far enough, perhaps 75 to 100 years, one would generally find farmers operating in a cash or barter economy. If a farmer did have some short-term debt, he usually pledged the farm. Collateralized lending—still practiced in many parts of the country—was then and is now basically secured lending. If repayment was not made, the mortgaged assets were either delivered to the lender by the farmer, or were possessed by the sheriff and would then be sold to satisfy the requirements of the lender's defaulted note. Financial leverage played no part in a cash/barter economy or when collateralized lending was the rule.

Since World War II, farmers have increasingly been borrowing on a "line of credit" basis. This requires a cash flow analysis in which an important part is the demand for payment of debt. The degree of financial leverage and the terms of repayment of long-term debt sharply influence the annual cash needs of the farmer.

It was about 1950 before this method of financing became general in the major banking institutions and the Farm Credit System. Agricultural lending was considered to be highly risky; consequently, loan commitments were often hedged by restrictive conditions. One such restriction on advances of credit for seasonal operations was that no money was to be advanced until after danger of frost had passed. That generally meant that farmers had a good deal more of their own money up front in a crop for land preparation, seeding, fertilizing and pest controls. They were forced to carry the crop until the lender was assured that his security would not be killed by a late frost. Some crops under this restriction were tree fruits, such as plums, nectarines, peaches; various seed crops; table grapes; almonds and walnuts; tomatoes, and some fresh vegetable crops.

Vendor or trade credit has been used traditionally by farmers to augment their credit. In the credit crunches of 1967 and 1975, the usual trade sources of credit began to dry up as suppliers found farmers increasingly utilizing this previously "free" credit. Dealers began to have large dollar amounts of accounts receivable on their books. Trade terms were such that farmers were generally not required to pay interest on their accounts payable; suppliers began to find themselves short of money and paying a very high price for cash that they needed to borrow to carry on their businesses. Trade credit has continued to be offered, but more recently farmers are paying interest on these bills.

Most institutional lenders at one time had policies that discouraged the use of credit from more than one source by their borrowers. This policy was difficult to implement, due to the tremendous amount of competition for the business of agriculture. As a result, farmers have increasingly been able to utilize a number of sources of credit, as they offer apparent cost advantages in borrowing money.

What have been some of the other changes in agricultural finance over this post-World War II period?

(1) Credit-takers (borrowers) are fewer in number, but they borrow much more money. Better records are now required to support the credit requests of farm borrowers.

(2) More financial services have been developed to meet the needs of farmers and ranchers.

(3) Almost all seasonal credit is extended on the basis of cash flow projections. There are some exceptions to this. Financially affluent farmers continue to borrow on an open note basis as they always have.

(4) More financial institutions are making credit available, and average loans are much more substantial, averaging somewhere between $180,000 and $240,000 per borrower for a seasonal credit commitment.

(5) Farmers are involved in more complex business arrangements. While the sole proprietor family farmer and the family partnership is still the most common, family farming corporations—where control of the corporation is in the hands of a few related people—are growing in number. Large conglomerate type corporations and non-agricultural companies that own some farmland have not grown much in number. Only about 2.25 percent of California's farms and ranches are owned by the latter type of business organizations. These do, however, produce a far greater percentage of total farm gate sales in California than their percentage of ownershp of farms would imply.

The complexity of operations arises because of the trend for combining various different businesses under one management. For example, sole proprietorships, partnerships, corporations, Sub-Chapter S corporations, cooperatives, and joint ventures are often part of a large and financially complex farming venture. These combinations are further complicated by having some business arrangements on a basis of cash accounting, others on an accrual basis, while still others use a hybrid type of accounting combining some features of both. Some ventures even have different business years; for example, one will close its year on December 31, another on April 30, and yet another on June 30. Trying to make sense out of financial transactions that take place between entities such as these is difficult. This is particularly so in cases where there are intercompany transactions such as growing feed and selling it to a related company, loaning money to one another within the business family, and so on.

(6) A growing percentage of farm products now go through cooperative processors and handlers. The system commonly used by cooperatives in California is a procedure whereby capital for the cooperative's operations is furnished by the grower in the form of deferred payments for his crops. Full payment to the grower for the raw product is not made until sales of canned, frozen, or manufactured products have been made. These deferred payments have become another factor in credit administration, inasmuch as these growers must, of necessity, carry their loans outstanding with lenders until the processor can generate enough cash flow to pay off his raw product supplier. Hence, the farmer is delayed in paying off his seasonal credit, while the cooperative in effect has the use of the farmer's credit to carry on operations. This is in contrast to proprietary processors, who generally pay soon after delivery or within 30 days, unless such terms are waived. In light of the recent lack of profitability in their businesses, some proprietary processors have now begun to contract for new products on a profit-sharing basis—essentially the same system the cooperatives are using to acquire raw product.

Differences in California

Other changes have been evident in the financial structure of California's farming ventures over the last 30 years. In the late 1940s as the nation recovered from the war years, the financial position of farmers in California was on a par with farmers in other areas, such as the Middle West. Changes in California's farm economy, however, came swiftly after the conversion to peacetime production. Huge acreages of land were converted as water sources for irrigation were developed. A number of dams, water storage facilities and water diversion projects were completed in the years 1950 to 1960. Plantings of specialty crops were made in regions that had formerly been in dryland crops or used for grazing livestock. Along with these increased plantings came increased farm size and the processing and marketing infrastructure needed to accompany the growth in production. Farmers relied more and more on borrowed money.

Today in looking at the individual state and consolidated national balance sheets of agriculture, one is struck by the contrasts. The average California farmer's balance sheet now differs substantially from his peer in Iowa or Illinois.

The midwestern farmer has a net worth that has grown over the years at a rate slightly higher than inflation. While more money has been borrowed, profits from operations and increases in the value of his assets—particularly his land—have combined to give him a plus growth in his net worth, over and above inflation.

What of the California grower? His operation has grown larger and the majority of his crops tend to be specialty crops returning high gross sales volume. His land is generally all irrigated, and he has a full complement of tools and equipment to meet the critical timing needs of specialty crops. A superficial look reveals a huge, well-equipped, well-maintained, irrigated row and permanent crop type farm operated by a 50-year-old confident, well-educated, broadly-experienced farmer-businessman.

Deeper research would unearth several other facts. The farmer has come to rely more and more on borrowed money to keep the operation going, to improve the efficiency of the operation, and to maintain a lifestyle that is commensurate with that enjoyed by his city cousins. His intensive specialized cropping pattern results not only in higher per acre production costs, but also does not permit his own labor, home grown products, and traded work to make nearly as large a proportional contribution as under a less complex and demanding kind of farming—say, a barley, wheat and sheep operation. The sharply increasing scale of operations has also reduced the impact of family contributions and has required more outside sources of input.

The number of the farmer's memberships in processing and marketing cooperatives is up. His investments in the processing and marketing infrastructure are substantial and these appear on his balance sheet as additional assets. He also has marketing arrangements with proprietary companies who buy the farm's products that do not go to the cooperative.

The assets side of his balance sheet has experienced dramatic change, as has the assets side of the farmer's balance sheet in Illinois. Equipment values, land values, commodity values, are all up sharply. Along the liabilities side, both short- and long-term debts are up sharply too, and, in fact, have increased faster than assets. Consequently, the farmer's net worth has suffered, and his equity in his business has grown less than have the claims of creditors against the farm. In

California, the farm family's net worth has grown at a rate *below* the rate of inflation.

The Role of Leverage

What does this portend? California's farmers are more financially leveraged than are those in other important farm areas. The USDA's 1979 report of farmers' financial condition showed $820.2 billion in total assets, $137.5 billion in liabilities, and $682.7 billion in proprietors' equity. Therefore, financial leverage of farmers nationally, as a ratio of liabilities to assets, was 16.8 percent. Put another way, farmers across the nation owed $16.80 for every $100 of farm resources owned. By contrast, California's farmers' assets in 1979 were $42 billion, liabilities $10.1 billion, and proprietors' equity $31.9 billion. Financial leverage was 24 percent. For every $100 of resources owned by California farmers, they owed $24 of debt.

California's farmers have enjoyed generally good income over the last decade but have also expanded operations, by increasing debt; hence, their financial leverage has gone up over time. In 1970, California farmers' financial leverage ratios compared favorably with all U.S. farmers', at which time every dollar of debt was covered by over $5.00 of assets. In 1979, each dollar of debt was covered by only $4.00 of assets—an increase of about 20 percent in financial leverage over the last ten years.

This ratio implies that California's farmers have fewer assets to cover debt in the event of liquidation. There is also a greater demand on annual cash flow to service debt, which leaves less available for capital investment and growth. Constant pressure is maintained on the farmer to lower unit costs by improving efficiency or by increasing yields, and to convert low value crops to high value crops such as tree fruits, vines and nuts. The latter, however, requires additional investment, not only for the physical development of the orchard or vineyard, but also to cover costs until the plantings bear a crop adequate to provide reimbursement.

All these pressures put a premium on superior management, and also give the farmer who is financially well positioned great advantage during periods of natural disaster, disease, or economic adversity. Unless the after-tax profit picture improves, the farmer in California in the 80s will rely more and more on borrowed capital to continue the job of farming.

Types of Credit

When farm balance sheets are presented to lenders, the extent to which the modern commercial farmer/rancher in California operates on credit becomes clear. The average balance sheet shows numerous uses of short-term financing for immediate needs, as well as longer term financing for the acquisition of farm equipment or farm equipment or farm real estate. Many sources of credit—dealer credit, insurance company financing, individuals, Farm Credit System and bank financing—may be utilized.

Short-Term Sources of Credit

Short-term funding is frequently used by farmers and ranchers to provide some of the inputs that go into the growing of seasonal crops and livestock. Trade and supplier credit is a very common form of short-term credit. It is often supplied by companies selling fertilizers, pesticides, packaging materials, fuel, spare parts, seeds, veterinary supplies and services, and the like. As examples, growers who order seasonal supplies from vendors, water from water districts, or power from public utilities to run pumps and irrigation equipment, commonly use trade credit for varying periods of time.

Processors and handlers of the commodities that farmers and ranchers produce often provide either in-kind or cash advances to producers to assist them in paying for the inputs required by seasonal operations. In many cases, so-called grower-shipper arrangements, which essentially are joint ventures, provide that handlers or processors make advances to the producers for some of the cash needs required in the production process. Processors and shippers very often are responsible for providing seedstock and for harvesting the crops.

In our complex society, many agricultural ventures are often structured as sole proprietorships, partnerships, joint ventures, corporations, and/or Sub-Chapter S corporations, or combinations of two or more in more complicated organizations. Very often inter-company loans facilitate the production of crops. These loans are usually made when it is more advantageous for a company to provide funding to an affiliate or subsidiary or to another company which is under common control, than to borrow from an external source. These are reflected on financial statements as inter-company loans or advances.

Private financing and family financing is common among sole proprietorships and partnerships. For a young or beginning farmer, the contributions toward seasonal financing needs made by family members or private individuals are an important source of credit inasmuch as the young or entry-level farmer may not have the financial strength or secondary financial support that would permit the use of institutional credit to the extent required.

Institutional lenders, however, are the largest source of seasonal financing. This includes banks and the Farm Credit System. Additional institutional sources of credit for seasonal needs of farmers can include credit unions, some of the newer nonagricultural lenders such as General Electric Credit, Ford Motor Company Credit Corporation, Commercial Credit Corporation, and similar types of lending institutions. The Commodity Credit Corporation (known as the CCC) has again become a major short-term lender to farmers who grow staple crops that can be put into warehousing to collateralize loans made under the Farm Credit Act of 1979. Banks are important lenders, and a number of banks are active in lending to California agriculture. These include banks domiciled within the state, usually either unit banks or branches of large branch banking systems, but they can also include out-of-state or foreign-based banks. An example of the latter would be Credit Agricole, a large cooperative farmer-owned bank in France with an office in Chicago, Illinois.

A number of large nonfarm publicly-held corporations operate farms within the state of California. Financial affairs of many of these are centralized in the company headquarters. More often than not, the financing office of the large corporation, wherever domiciled, will "downstream" funds into the farming subsidiaries of the corporation. Hence, these farming operations have no need to borrow from the same sources that most proprietary farmers utilize in financing seasonal crop and livestock operations.

The Farmers Home Administration and the Small Business Administration are government entities that provide seasonal financing to farmers and ranchers. This has been particularly true in the last few years when a number of natural disasters—sandstorms, floods, hurricanes, drought, windstorms, disease—have affected farmers. Both the Farmers Home Administration and the Small Business Administration have programs to provide disaster loans and short-term operating loans to farmers who have difficulty in continuing their borrowing relationships with their conventional lenders.

The Farmers Home Administration also has an on-going program called "Farm Operating Loans," designed to assist financially marginal farmers with seasonal operating loans up to the limit of $100,000 if made directly. Private lenders can also make these loans. In the latter event, a private lender can commit a loan of $150,000 and 90 percent of this, including interest, will be guaranteed by the Farmers Home Administration under its Guaranteed Loan Program.

As a group, Production Credit Associations (PCAs) are the most important institutions providing seasonal credit to California's farmers and ranchers. PCAs not only utilize their own equity funds to make loans, but they can discount the farmer's notes with the Federal Intermediate Credit Bank located in Sacramento.

Banks and other institutional lenders can also organize what is called an "Other Financial Institution." These are financing institutions with the privilege of discounting agricultural loan paper with the Federal Intermediate Credit Bank in a manner similar to that used by the Production Credit Associations. "Tri-State Livestock Corporation" is a cooperative credit association owned by its members which finances the needs of livestock producer members in several states. It is an "OFI" and discounts its farm paper with the Federal Intermediate Credit Bank in Sacramento.

Medium-Term Credit Sources

This type of credit is utilized whenever the farmer or rancher is investing in resources that cost more money than can be reasonably repaid out of cash flow in a given year. Examples of some of these investments are large items of farm equipment, breeding cattle, land development such as leveling and orchard planting, drilling of wells and installation of irrigation pumps, interim financial needs for land purchases, and surface structures such as grain dryers and storage facilities.

Dealer credit is among the more important of these sources. All the automobile, truck, and farm equipment companies have dealer credit programs available to purchasers of their products. Companies such as GE, Ford Motor Company, Chrysler Motor Company, International Harvester, John Deere, Caterpillar Tractor Company, Harvestore, and similar companies are involved in providing credit to purchasers of their products. Companies that manufacture irrigation systems, such as Rain-for-Rent and Valmont, also provide credit programs available to farmers and ranchers.

The Farm Credit System and banks also have credit programs providing terms from one to ten years. Credit unions and insurance companies, to a modest extent, are also involved in financing machinery purchases.

In the last decade, leasing companies have become a significant provider of credit to farmers. This method of financing permits a farmer to conserve his working capital. However, the equipment owned by a leasing company is usually leased on a contract written for a term of years no longer than the economic life of the equipment. At the conclusion of the financial lease, the farmer has the option of renewing the lease for a short period, or of paying the cash value of the equipment at that time and acquiring ownership of the depreciated equipment.

Investment funds and private placements of various kinds have begun to appear as providers of intermediate-term credit to farmers and ranchers. Investment funds and the limited-partnership type of funds have a checkered history, particularly in the livestock business. In the late 60s to the mid-70s, a substantial amount of credit was provided to cattle feeding operations by these funds, and critics have suggested that this practice led to the unwarranted increase in livestock inventories that resulted in a sharp break in livestock prices in 1973 and an exacerbation of the problems usually associated with the cattle cycle. Investment funds have again, however, begun to be providers of money for both intermediate-term needs and for longer term uses such as farmland investment.

Long-Term Sources of Funds

Long-term credit is usually used for the purposes of acquiring additional real estate. Growers may also occasionally find it necessary to improve their liquidity; in these cases, they will borrow on their real estate. These funds are then utilized in current operations.

Important providers of long-term funds are: (1) individuals, (2) Federal Land Bank Associations, (3) insurance companies, (4) banks, and savings

and loan companies to a modest extent, (5) Farmers Home Administration under their Farm Ownership Programs, (6) corporations, both foreign and domestic, and (7) pension funds.

Some foreign capital has also come into the state, though to a lesser extent than was originally thought by critics of foreign investment in U.S. farmlands. Subsequent research has indicated a relatively constant percentage of foreign ownership. Recent legislative attempts have been made in both Sacramento and Washington, D.C. to prohibit farmland ownership by foreign investors and nonfarm corporations. To date, foreign investors are required to report their farmland holdings to the U. S. Secretary of Agriculture.

Financial Institutions and Programs

Commercial Banks

Commercial banks in California are the leading source of non-real estate farm loans but rank a poor fourth in providing real estate loans.

Banks are prominent in providing operating credit to the farm sector for several important reasons. They are readily accessible to farmers, since they are located in nearly every town in all of the major farming regions. Banks are able to give prompt credit service at competitive interest rates, and they can provide a full range of financial services, as detailed in an earlier section.

Farm Lending by Commercial Banks

Banks make all types of farm loans, but the most numerous are those for operating expenses. These usually mature annually or at the end of the production cycle. Because a local bank usually has a personal working knowledge of the borrower's business and financial ability through its other services, it is often capable of assessing the borrower's credit rating quickly and accurately and may provide substantial amounts of short-term funds on an unsecured basis.

Smaller loan amounts are for intermediate-term needs such as machinery, breeding livestock, and buildings. These carry maturities of from one to ten years. Loans for the purchase of real estate are typically for periods exceeding ten years.

This emphasis on short maturities reflects the

source of a large portion of loan funds: bank deposits which can be withdrawn quickly, thereby causing wide swings in funds available. Partially to accommodate this fluctuation, some loans for intermediate-term purposes are written with short maturities but with the understanding that any unpaid principal can be renewed. Mortgage loans are very helpful in financing add-on acreage, the dominant type of farmland transfer.

Meeting the credit needs of farmers may require a different kind of expertise than that needed to service other kinds of borrowing. Farm loans typically are for relatively long periods of time, procedures for assessing risk and the repayment ability of the firm are different, and competition may keep rates relatively low. Assessment of risk and repayment ability can be difficult because factors such as weather, insects, disease, and prices may cause great change in a short period of time. These characteristics of farm loans frequently require loan analysis by specialists. However, banks with agricultural expertise find this segment of their loan portfolios to be profitable, contributing significantly to the bank's overall performance and to the growth of the community. Many banks have loan officers with technical training in agriculture, who understand the needs of the farm sector and can evaluate credit needs. If funds are short, these specialists, working closely with senior management, can spearhead efforts for channeling outside sources of funds through the bank into the farm community.

Although banks are a major institutional source of credit to the California farm sector, sharply rising capital and credit needs coupled with increasing costs of money have made it difficult for some banks to accommodate these needs adequately. Farm debts have grown much faster than farm net worth. As this trend continues, loan-deposit ratios have in many cases reached a maximum level considered prudent by bank management. When this happens, banks have difficulty meeting further increases in farm credit requests, particularly if the bank's own leverage is high.

Banks have competition, mainly from Production Credit Associations (PCAs). Outstanding non-real estate farm loans of banks are, however, almost double the amount held by the Sacramento Federal Intermediate Credit Bank (FICB) and the PCAs in California.

The Cooperative Farm Credit System

The Cooperative Farm Credit System is made up of member-owned institutions operating under the supervision of the Farm Credit Administration, an independent agency of the federal government. The member associations were organized and put into operation by the U.S. government as a means of helping farmers help themselves. The System is designed to provide credit and closely related services to farmers, their cooperatives, and selected farm-related businesses. The System also provides credit for rural homes, to producers and harvesters of aquatic products, and to associations of such producers.

Credit is distributed through the three component parts of the System: (1) Federal Land Bank (FLBs) and Federal Land Bank Associations (FLBAs); (2) Federal Intermediate Credit Banks, Production Credit Associations, and Other Financial Institutions (OFIs); and (3) the Banks for Cooperatives. Some related business services are provided to the banks by a fourth component of the system, Farmbank Services, in Denver, Colorado.

The System divides the country into 12 Farm Credit districts. The California district Federal Land Bank, Federal Intermediate Credit Bank, and Bank for Cooperatives are located in Sacramento. These serve Federal Land Bank Associations, Production Credit Associations, Other Financial Institutions, and farmer cooperative associations in local areas.

Agricultural borrowers control local FLBAs and PCAs by directly electing their boards of directors. Beyond that, the combined boards of the PCAs and FLBAs as well as the cooperatives owning stock in the Sacramento Bank for Cooperatives each elect two members to the district Farm Credit Board. A seventh member of the district board (the director-at-large) is appointed by the governor of the Farm Credit Administration. The district board serves as the policy making body for all three banks in their respective districts, and for joint functions of the banks.

The Federal Land Bank System

Federal Land Banks were established by the Federal Farm Loan Act of 1916. Also authorized by the act were local National Farm Loan Associations, now called Federal Land Bank Associations (FLBAs). Loans in California are made through more than 55 local FLBAs. The banks were initially capitalized by the federal government, but became completely owned by their borrowers in 1947 when all federal "seed money" was repaid.

LOAN TERMS AND PURPOSES. The Federal Land Banks make loans secured by first mortgages on real estate. Loans may be made to farmers and ranchers for any agricultural purposes or other credit needs of the applicant. Loans may range from 5 to 40 years, and repayment plans are designed to accommodate borrowers' cash flows. The Sacramento Federal Land Bank has a variable interest rate plan for its loans. The plan provides a means by which interest rates charged borrowers may be lowered or raised depending on the average cost of the money to the bank. Loans to persons or businesses furnishing farm-related services may be made for necessary capital structures, equipment, and initial working capital.

OWNERSHIP AND CONTROL. An individual applies for a loan at his local area FLBA. A representative of the association appraises the property offered as security and checks the credit-worthiness of the applicant. When the loan application is approved and accepted, the loan is closed and the farmer or rancher becomes a member of the association with the purchase of voting stock. The amount of stock purchase is between 5 and 10 percent of the amount of the loan; this amount may be included in the funds borrowed. Funds paid for stock help capitalize the association.

Participation certificates also are used to help raise capital. These are issued to rural home borrowers and those obtaining loans to finance services related to farming. Holders of certificates do not have voting rights. This helps keep control of the association in the hands of bona fide farmers.

The Federal Land Bank System functions with private capital loan funds that are obtained mainly through the sale of Federal Land Bank bonds backed by first mortgages on farm real estate. The FLB, after providing for reserves as required by law and for net worth objectives, may distribute any net earnings among the associations in the form of dividends. The associations, in turn, may pass dividends on to their members.

The Production Credit System

The Production Credit System in California is composed of the Sacramento district Federal Intermediate Credit Bank (FICB) and 72 Production Credit Association offices serving 13,000 borrowers.

The Federal Intermediate Credit Banks were authorized by the Farm Credit Act of 1923 for the purpose of discounting the short- and intermediate-term notes of farmers held by commercial banks and other financial institutions. In 1933, Congress authorized the establishment of local PCAs which could discount with the FICBs. In effect, the PCAs became the major retail outlets for credit available at wholesale from the FICBs. The FICBs continue to discount eligible paper from commercial banks and discount loans made by other financial institutions. These uses of funds, however, are a minor portion of the total activity of the FICBs.

The Federal Intermediate Credit Banks and Production Credit Associations were initially capitalized by the government. In 1956 changes in the laws governing these institutions permitted them to repay the government's investment, and since 1968 they have been completely owned by their farmer borrowers.

PCA loan terms may range up to seven years in length. Repayment plans are designed to accommodate the borrowers' cash flows. While some loans are written with a fixed number of annual, semi-annual, or monthly installments, many PCAs have instituted budget or line-of-credit financing plans. Under these plans, a borrower arrranges in advance for a loan to cover all his financial requirements for an entire season or agricultural cycle. He draws the money as he needs it and repays it according to a prearranged schedule. Loan service fees and stock ownership requirements affect the net cost of borrowed funds. Loans may also be made to farm-related business for working capital, equipment, or operating needs related to custom-type services performed on the farm. In addition to credit and financial advice, many PCAs provide borrowers with other services, including credit life insurance, and crop hail insurance.

The Farmers Home Administration

The Farmers Home Administration (FmHA) is a government lending agency operating within the U.S. Department of Agriculture. It has two main objectives. The first is to provide supervised credit to farmers unable to obtain adequate credit from commercial lenders at reasonable rates and terms. This is done primarily through loans for operating, farm ownership, and emergency purposes in disaster areas. These loan programs are intended to maintain and strengthen the family farm structure by helping farmers who could not get credit elsewhere, and by providing credit to beginning farmers. Most lenders do not view the FmHA as a competitive institution. In fact, they may be pleased to be able to refer prospective customers who do not meet their credit standards to FmHA. In these farmer-oriented loan programs, a borrower is required to "graduate" to commercial credit when able to do so.

The second objective of FmHA is to improve rural communities and enhance rural development. As people and businesses have left rural areas for urban industrial centers, many towns have declined in importance as economic centers. Various loan programs are designed to improve rural communities by providing jobs and enhancing the quality of rural life.

Funds for the FmHA programs fall into two categories: guaranteed and insured. Guaranteed loans are made and serviced by a private lender, with the FmHA guaranteeing to limit any loss to a specific percentage. Insured loans are direct loans made and serviced by FmHA. Funds for the insured direct loan programs are available through the Federal Financing Bank which coordinates and consolidates the borrowing activities of about 20 different federal agencies.

The FmHA farm lending programs have been aimed at "high risk borrowers" who must be unable to get credit elsewhere at reasonable rates and terms. Borrowers typically include young farmers and those who do not meet usual credit standards such as capital position or repayment ability relative to loan size. FmHA is able to service this type of farm borrower without undue losses because it provides technical advice and loan supervision.

Life Insurance Companies

Life insurance companies play an important role in mortgage financing of rural real estate. Generally, reserves of a life insurance policy accumulate over time and provide insurance companies with large sums which can be invested for long periods. Long-term farm real estate loans are among suitable investments.

Although the life insurance industry is an important lender to the farm sector, the amounts loaned represent less than 3 percent of the industry's total investment portfolio. As recently as 1968, life insurance companies were the largest single institutional source of long-term real estate credit to the farm sector, but more recently their relative importance has declined. Relatively few insurance companies are active in the farm market. It is estimated that 21 life insurance companies account for some 96 percent of the farm mortgage lending. Further, the top eight companies hold 87 percent of the industry total.

Farm mortgage lending is relatively specialized, requiring agricultural specialists, and is costly. To minimize the cost of making loans, insurance companies generally concentrate their efforts in good commercial farming areas where many large loans can be obtained.

Life insurance companies are under no obligation to serve the farm sector. Should high return investment opportunities become available elsewhere, funds are diverted to these uses.

Other Sources of Farm Loans

Individuals

As noted earlier, individuals are the leading source of farm mortgage loans. Most of this lending occurs when a farmer retires and is willing and able to accept a down payment and carry the balance owed by the buyer on the purchase price of the farm. Land contracts are widely used, especially in time of tight credit, and in some areas may account for half or more of the financing of farm real estate transfers. The seller may find it easier to dispose of a farm if he or she is willing to accept a contract arrangement and may welcome the oportunity to leave more funds invested. When less than 30 percent of the value of the property is received as a down payment or in principal payments, the capital gain can be spread over a number of years, thereby reducing federal income taxes.

Vendors

Merchant and dealer credit is used by many farmers and has long been recognized as an essential part of a retailing operation. To supplement local credit sources, many firms in the major farm supply industries have their own credit affiliates or have working arrangements with financial institutions including banks to accept the notes of farm borrowers originating in the course of day-to-day business.

Financing of "hard goods" such as farm machinery and of "soft goods" such as fertilizer, feed or petroleum products presents a number of contrasts. Farm machinery has a typical productive life of several years, while fertilizer primarily contributes to a single year's output. Another difference is that a machine is a tangible piece of property that can be used as security for a loan and can be repossessed if necessary. Because of such differences, credit arrangements differ between suppliers of hard and soft goods.

In addition to individuals, merchants, and dealers, a number of institutions provide small amounts of credit to the farm sector. These include various kinds of investments and loan companies, credit unions, savings and loan associations, and the Small Business Administration. Data are not regularly collected on the farm lending activity of these institutions.

Choosing Credit Sources

Farmers make choices in their lenders depending on the degree of difficulty or ease in selling their debt. A number of things enter into this judgment.

First, the amount of the loan, and the length of the loan commitment, must meet the needs of the borrower. Second, familiarity with the lender often determines where money will be borrowed. It can be a traumatic experience for many farmers to borrow. If a farmer has had a long relationship with his banker or another credit source, he is likely to be much more comfortable in talking over a loan request.

Third, the cost of borrowing is always of concern. Nowadays, with disclosure requirements, it is fairly easy to compare the cost of money for a specific loan offered by one lender against another. All things being equal (which they rarely are), a borrower would choose the cheapest source of borrowing. Repayment terms, however, are important. This may be the number of years the loan can be stretched out, the possible deferment of principal payment for the first few years with interest only being charged; it could include a balloon payment at the end of the term which would permit smaller payments throughout the amortization period.

Fourth, documentation is considered to be a nuisance by most borrowers and the more docu-ments required by the lender, the more frustrated the borrower may become by the perceived "red tape." Documentation consists of filings with public authorities to record a lender's interest in a borrower's assets, trade credit reports, appraisal and field reports, and financial exhibits. Requirements to have a guarantor or cosigner often create the most unhappiness. Whether a loan is secured or unsecured by collateral also will affect the number of documents required to close the loan.

Fifth, the speed of response to a loan request can be important. Often quick decisions must be made. If a lender can only come to a credit decision in a three to four week period, even a favorable decision may no longer fill the needs of the borrower. Lastly, the accompanying benefits offered by one lender may make a difference. Some lenders offer all kinds of additional services that a borrower finds necessary and worthwhile. Other lenders have a single financial product to offer, such as equipment leasing.

Borrowers choose their credit sources by weighing all these factors for relative importance.

Crop dusters are often seen in farm areas of California, making aerial application of pesticides or fertilizers. Crops such as wheat or rice are also seeded by airplane. *Photo by Jack Clark.*

14

Technology and Services

The phenomenal increase in agricultural productivity in the period since World War II is attributable largely to the development and use of new technologies. While selective plant and animal breeding have been responsible for increased yields, equally significant have been the contributions of the mineral, chemical, and machinery industries to the old tasks of planting, cultivating, protecting, and harvesting crops. Concomitant with the development of often very specialized and expensive technologies has been the development of a custom-service industry to assist the farmer in their application. This chapter combines the perspectives of four specialists to review three basic technological inputs—fertilizers, pesticides, and machinery—and the service industry which has grown up around them.

A Revolution in Agriculture: Custom Services

By Edward A. Yeary

Less than a half century ago most California farmers owned their land and livestock as well as the machinery required to operate their farms. Production was very diversified. Feed crop production was balanced against livestock needs. Soil fertility was maintained by rotation crops as well as livestock waste. Rotation practices also helped maintain soil quality and minimized weed problems. Needed cash flow was generated from products sold. Tenant farmers and renters were generally accepted as being less stable and less reliable as credit risks than were the operators who owned their farms.

Today it is possible to operate a rented farm from a rented office using a leased vehicle to supervise the activities of contractors and custom operators who perform all of the tasks required for land preparation, planting, irrigating, cultivating and harvesting as well as packing and selling the crop. A telephone book and a check book are the only items needed by such a farm operator. This is not necessarily an ideal method to operate a California farm, but it is physically possible for many of the crops produced here.

Mechanical and Technological Change

During this brief span of years, major changes have occurred in our methods of farming. Two revolutions, mechanical and technological, have drastically changed the practices followed by most farmers.

The mechanical revolution put draft horses out to pasture, replacing them with efficient tractors of all sizes and a full complement of machines needed to perform farm operations. Some farm tractors now exceed 400 drawbar horsepower and are capable of utilizing equipment not dreamed of a few years ago. The machines they pull perform near miracles in such areas as soil modification, land preparation, weed control, planting and harvesting. The airplane and helicopter also have been harnessed to perform farm work, principally planting, fertilizing and applying chemicals. They are even brought into service occasionally for such tasks as removing rain water from ripe cherries and helping with frost control.

Machines take most of the hard physical labor out of harvest chores and in many cases electronically sort products into quality units with precision and speed that cannot be matched by hand labor. The newest mechanical tomato harvesters cost in excess of $140,000 (in 1980), but they pick and sort tomatoes at about 30 percent of the cost of using very capable hand labor. To do this, they must be used on as many acres as possible to operate at low unit cost.

This logic applies to most products of the mechanical revolution in California. Their functions are (1) to help produce more reliably at lower cost, (2) to relieve hand labor of some of the burdensome chores that were formerly required, and (3) to preserve and expand processing, transportation, sales and service employment that otherwise would be lost to other states and countries where wage rates are substantially lower.

The accompanying technological revolution in agriculture has provided chemicals for insect and weed control, nutritionally better feed mixes for livestock, manufactured fertilizers, and an array of other products now in daily use by California farmers. The technological revolution has furnished materials which nearly make unnecessary such older practices as maintaining soil fertility through crop rotation, driving to the mountains to gather ladybird beetles for insect control, and extensive hand hoeing for weed control.

A by-product of these two revolutions has been a rapid increase in the size of California farms. One manager is now able to operate and supervise much larger units than was previously possible. Pest and fertility problems formerly handled by crop rotation can now largely be solved by mechanical and chemical means. This has permitted more specialization—farmers can now make the best use of their physical resources and their individual skills by concentrating more heavily on the production of fewer types of crops and livestock.

Machinery Hire in California

Large and complex machines are very expensive and are often needed for only brief periods of time on any one farm. Agricultural chemicals may require very elaborate and expensive application equipment and may present some danger to the user unless they are properly handled. These expenses and complexities of modern technology have led to the development of a wide range of custom services which farmers may hire. Furthermore, good roads and excellent transportation equipment have made it possible to move machinery very quickly from farm to farm.

Typically, most farmers in California own their basic tillage and planting machines as well as water distribution systems—although even this is not true in every case. The nearly endless list of services and equipment which may be leased or hired on a custom basis includes land tillage, application of all types of chemicals, pollination of crops by bees, irrigation, and harvesting of all kinds of crops. Almost any process required today in farming can be obtained on a custom, lease or rental basis. Many crops are fully mechanized and every process is available in these programs. Hand work is only a supplement where unusual problems may occur, or operations are too small in size to justify bringing in equipment.

For example, an orchardist in the Central Valley some time between the world wars might have owned his farm and worked it with draft animals and a minimum amount of machinery. He and his family and the hired man would probably have done all of the work except at harvest time, when extra help would have been hired. Possibly he would have brought in some help for pruning, thinning, and harvesting, but the farmer would have worked with the crew and done as much of the work himself as possible.

Today with such an orchard the grower would probably own a tractor and some tillage equipment, but he could call in a custom farmer with specialized equipment to do the necessary tillage. He could, if he wished, rent an irrigation system for the season, which would be used with the farm well or surface water supply and operated by people in the employ of a labor contractor. Pruning could be accomplished by calling a labor contractor to bring out his crew to complete this job. If a dormant spray should be applied, the state abounds with highly skilled operators who have equipment and the expertise to safely handle and apply the needed chemicals. The grower could rely on a chemical company to see that fertilizer requirements were met. At blossom time, if there were need for a fungicide application, this could be accomplished by calling one of California's more than 200 aerial applicator businesses. (Approximately 750 active and fully licensed pilots are available in the state as well as about 300 more who move in and out of California.) If the fruit crop required thinning, the labor contractor might again be called upon. At harvest time it would be likely that the packinghouse which is to package and sell the fruit would send out the crew to do the harvesting, providing bins to contain the fruit and trucks to haul them to the packing shed. In literal fact, an orchard now can be operated by a manager who owns no movable equipment and performs no physical labor, but contracts for every operation.

Other Custom Services

In earlier years livestock operators sold cattle directly from the ranch or fed them in small farm feed lots. At present many cattle feeders are non-farm investors. They buy cattle through an agent who moves them directly to a large custom feed lot. A multi-million dollar computer decides what to feed them and how much it will cost. When they are ready for market they are sold and transported without any help from their owner at all.

Herds of dairy cattle can be leased or rented. They may be housed and managed on rented premises. In addition, services such as artificial breeding, hoof trimming and veterinary care may all be provided by contractors who function on a custom basis.

Cotton is a field crop that is almost 100 percent mechanized and in which every production process can be leased, rented or hired on a custom basis. Land preparation, planting, thinning, irrigating, pest and weed control programs and harvest operations are all available for hire. These services have become available because it is often far more economical to lease, rent or hire than to own needed equipment and to obtain the skills necessary to use it. Custom operated machinery has held the unit cost of harvesting cotton down to a very low level. At the end of World War II the cost of hand harvesting field run cotton was approximately 4 cents per pound; today it would be approximately 20 cents per pound. In 1980 the most sophisticated mechanical cotton harvesters cost approximately $145,000. By using such machines on the largest number of acres possible during a season, operators can harvest cotton for between 3 and 5 cents per pound. When this custom rate is available, very few cotton growers need to own their own harvesters. Custom harvesters who own fleets of machines can move to a cotton farm and very quickly remove and store all of the crop. It is common to see eight or ten

such machines at work on a reasonably large-sized cotton farm harvesting perhaps 200 acres per day depending upon the weather at picking time.

Operators of sophisticated equipment must be highly skilled. For example, an aerial applicator must have a commercial pilot's license plus many hundreds of hours of experience in flying. In addition, he must earn an agricultural pest control pilot's license issued by the California Department of Food and Agriculture.

Scale of Operations

These custom services are available to relatively small scale farmers at the same cost that is paid by operators of the largest units. Skilled services and the use of equipment which could not possibly be purchased by smaller farm operators often can be utilized through the custom programs available.

Nevertheless, very small farms may be handicapped in California. They cannot obtain custom services at the same *proportional* cost paid by larger farmers, and perhaps these services will not be available to them at all. Equipment must often be moved long distances and this must be charged for unless the farm unit to be serviced is very large. During the height of processing tomato harvest season, growers are often limited as to the amount of product they can deliver in any one day, generally proportional to their total amount of production. Very small growers are so highly restricted here that a custom harvester cannot provide his services at the same cost paid by larger operators because the harvester cannot run long enough each day on the small unit. Aerial applicators obviously would have difficulty applying chemicals to very small land areas, while custom operators who do heavy soil modification work such as deep ripping must charge owners of small units proportionally a higher fee than that normally received from an owner of a large block of land.

Large, complex, expensive machines tie up so much capital that many farm operators would be handicapped if they had to own such equipment. By availing themselves of custom rental and lease programs, capital can be used for such purposes as acquiring more land or adding permanent facilities such as building and wells. These custom programs greatly simplify the operation and management of large land units. At the same time, they may place a small unit at a competitive disadvantage. Small unit owners are further handicapped by the fact that they must use smaller equipment and improvise to accomplish needed results—normally at far higher costs per unit than paid by users of custom services. They must also acquire expertise not needed by managers of larger farms who hire work done.

The technical and mechanical revolutions have made American farmers the most efficient producers of food and fiber the world has ever known. Placing many of the products of these revolutions in the hands of custom operators has relieved farmers of many financial burdens, enabling them to expand to the limit of their abilities, free of the requirements of owning and managing complex, expensive machinery which may be applying potentially dangerous materials. The complexity and expense of many technologies have made it more attractive for farmers to hire custom work done than to do it themselves.

Fertilizers Past and Present

by Roland D. Meyer

Many early California settlers, such as John Sutter who established New Helvetia in 1839, dreamed of the potential for agricultural production they saw here. Although many came in search of gold, a number subsequently turned to farming endeavors, establishing orchards, raising livestock and producing vegetable and field crops. Coming from various parts of the United States and the world where soils had been under cultivation for decades and perhaps even centuries, they quickly realized the productivity of this virgin land with its beneficent climate. Early agriculturists soon became aware, however, that after repeated cropping, yield levels begin to decline and the question of maintaining or restoring soil productivity became paramount.

Early scientists attempting to explain the differences in the productivity of soils had difficulty identifying the elements which went into plant nutrition. They had, however, observed the stimulating effect on plants of various minerals; saltpeter was one of these, recorded by several people to give a very bounteous harvest. As the science of chemistry progressed, scientists became aware of the fact that vegetable substances consist largely of carbon, hydrogen, and oxygen.

Although they were aware that hydrogen and oxygen came from water, they were somewhat mystified as to the origin of carbon. Eventually the humus theory postulated that decomposing animal and vegetable material in the soil supply carbon to growing plants.

The founder of the modern fertilizer industry was a German scientist, Baron Justus Von Liebig. Beginning about 1840, he stressed the value for plant nutrition of certain mineral elements, and the necessity for replacing these elements to maintain soil fertility. He recognized the value of nitrogen and believed that plants could get most of their requirement from the air, while nutrients such as phosphorus, lime, magnesium and potash would need to be supplied by fertilizer sources. Liebig is probably best known for his "law of the minimum," which is still a very useful concept and forms the basis for the use of fertilizer in most agricultural production. This law states that if one of the nutritive elements of the soil or air is deficient or lacking in availability, plant growth will be poor even when all other elements are abundant. This concept has also been extended to include other factors affecting growth—moisture, temperature, the genetic capacity of plant varieties, insect control, weed control and many others. It has been the successive identifications of the most limiting factor (or integration of several factors) which have led to the continuing increase in crop yields characteristic of today's agricultural crop production.

Many people moving to California thought that the soils were of such unusual quality that they would continue to maintain a high level of production for an unlimited time. However, agriculturalists in this state as elsewhere soon learned that successive crops tended to have reduced yields after soils were in cultivation for some time. The soils supporting the bonanza wheat years of the late 19th century were maintained to some extent by the rotation of leguminous cover crops which biologically fixed nitrogen from the atmosphere. Fixed nitrogen supplied in rainwater (5 to 10 lbs/A/yr) and wastes from grazing animals also contributed a small part towards restoring the fertility of soils, but could never replace the large quantities of nutrients removed by harvested crops. The decline in yields of grain over a period of time made it clear that the productivity of the soil was being depleted faster than California farmers could restore it by these and other methods.

The Growth of Fertilizer Usage

Commercial Nitrogen Sources

Mined sources of nitrogen fertilizer such as Chilean nitrate or soda were applied to crops, and it soon became apparent that yields could be increased dramatically. These sources were usually in short supply and rather expensive. Ammonium sulfate was first produced by the evolution of ammonia in by-product coke ovens, and this soon became a growing source of nitrogen fertilizer. Direct synthesis of ammonia from nitrogen and hydrogen was first carried out successfully on a commercial scale in Germany in 1913. After the first world war several other countries built plants which derived their hydrogen-nitrogen synthesis mixtures from the reactions of coke with steam and air.

In the mid-1920s commercial nitrogen fertilizers such as sodium and calcium nitrates and ammonium sulfate were just beginning to be used in southern California. Most of these fertilizers were spread by hand in citrus groves at a rate of approximately five pounds or more per tree. In 1928 two brothers, Eugene and John Prizer, concerned about the labor required to spread this fertilizer, built an applicator to dissolve soluble fertilizers and introduce them into irrigation water.

About the same time, a Shell group of oil companies headquartered in Holland felt the best way to enter the fertilizer manufacturing business was through the production of synthetic ammonia from nitrogen and hydrogen. Their first plant was brought into operation in September 1929, and subsequently a second was designed for California. Because California produced large quantities of petroleum gases that could be "cracked" to get the needed hydrogen, the Shell company worked very closely with Southern California Gas company in Los Angeles to make refinements in the art of gas cracking. In August 1931 the Shell Point plant near Pittsburg, California, began to produce anhydrous ammonia, which in turn was reacted with sulfuric acid to produce ammonium sulfate. The young company lost a considerable amount of money during the depression but decided to continue operation. Like most early manufacturing plants in Europe, the plant had been designed with stand-by equipment for almost every part of the process; thus, with a few pieces of additional equipment, the production capacity could be doubled with very little added overhead. This greatly reduced the manufacturing cost.

Meanwhile, Ludwig Rosenstein, chief chemist of Shell, approached D.R. Hoagland of the University of California about the possibilities of using anhydrous ammonia directly as a fertilizer material. In February 1932 experimental work was initiated on the use of anhydrous ammonia as a source of nitrogen for application into irrigation water. Investigations dealt primarily with volatilization losses of ammonia from the irrigation water and from the soil following irrigation. It was learned that sufficient nitrogen to enhance plant growth could be applied while maintaining a concentration less than approximately 100 ppm. Other items investigated were the changes in pH of the soil, the depth to which the ammonia would penetrate into the soil, and the effect that cation exchange capacity had on the depth of movement.

In early 1934, the first commercial application of anhydrous ammonia was made to a hillside citrus grove in Tustin, California. At about the same time a number of companies were appointed as distributors for various parts of California. As the use of anhydrous ammonia injected into irrigation water became a widespread practice throughout California, it was apparent that other methods of application would be necessary on those soils where irrigation was not practiced, or where it was deemed necessary that nitrogen be added at a different time than irrigation. Thus the shank- or knife-type injection blade was developed to apply anhydrous ammonia below the soil surface.

Phosphates

The development of the phosphate fertilizer industry dates back to the early practices in Europe when ground bones were used as a source of phosphorus. Treatment of bones and later of rock phosphate with sulfuric acid was found to yield an effective phosphate fertilizer. In about 1840, the product containing 16 to 20 percent P_2O_5 resulting from the addition of sulfuric acid to rock phosphate came to be known as superphosphate. By the late 1800s concentrated or treble superphosphate, having a P_2O_5 content of 35 to 45 percent, was first produced in Germany. Treble superphosphate did not, however, become an important fertilizer until the 1950s, and the addition of ammonia to phosphate materials did not come into popular practice until the

1960s. Since then, these materials have become the leading forms of phosphate fertilizer in the world. Although rock phosphate was and still continues to be used to some extent in the U.S., many types of rock phosphate remain rather inactive and supply very little phosphorous to many crops unless they are applied on acid soils.

California has some very low grade rock phosphate deposits, but all current supplies come from mines in Florida, Utah-Idaho, North Carolina and several other parts of the world. Even though vast reserves of rock phosphates exist, the higher grades have been partially depleted and lower grades are now being utilized.

Potassium

The early sources of potassium were wood ashes, sugarbeet waste and saltpeter. Salt deposits in Germany were opened in 1860 and dominated the world market for about 75 years. Manure salts (20–25 percent K_2O) and kainite (19 percent K_2O) were the first low grade unrefined ores used as potassium sources. The development of refining methods gradually increased the grade of commercial products to the current 60–62 percent K_2O for potassium chloride, which is the main source of potassium fertilizer. Potassium sulfate and potassium nitrate are the other principal sources.

California had the first large scale production of muriate of potash, initiated in 1916 at Searles Lake. Some 10 to 15 years later the discovery and exploitation of the sylvinite deposits in New Mexico made North America self-sufficient in potassium, providing another major ore source for the world. The initiation of mining potash deposits in Saskatchewan, Canada, in the late 50s and early 60s provided the North American continent and the world with a rich and extensive new source of potassium. Although the brine sources of potash obtained from the Searles Lake area in San Bernadino County as well as the Salduro Marsh near Wendover, Utah, provide a source within 200 feet of the surface, the deposits in New Mexico are from 650 to 2,500 feet below the surface. The potash beds located in Saskatchewan are located at depths of approximately 3,000 to 4,000 feet, resulting in a number of problems with mine development and sinking of shafts.

Other Amendments

Another important soil amendment and fertilizer material mined in California is gypsum. Gypsum is a major source of calcium for reclamation of alkali soil, particularly in the San Joaquin Valley. It is also used in the northern intermountain areas as a source for the plant nutrient sulfur. (In most industrial countries, large quantities of sulfur from the combustion of fuels at one time passed into the atmosphere and were subsequently brought down by rainfall, supplying sulfur for the nutrition of various crops. As more and more sulfur is removed from stack gases in the effort to prevent atmospheric pollution, agricultural soils will no longer be fertilized in this manner. The build-up of sulfur wastes will also become more of a problem at the plant sites. Sulfur from the oil refinery process has, in fact, been collected and is currently marketed as a sulfur fertilizer.) Another major use of gypsum is to increase the salt content of water coming from the Sierra Nevada for irrigation, or to treat soils where this water is used.

In addition to gypsum, limestone and dolomite are currently increasing in use on soils to neutralize the acidification created by continued use of the nitrogen fertilizer ammonium sulfate combined with the extremely low salt cotent of Sierra-origin irrigation water.

A significant tonnage of micronutrients, which are normally required in rather small amounts by plants, is also applied to considerable acreages of California cropland. Zinc, iron, copper, boron and molybdenum are the most important.

Fertilizer Usage Since World War II

As the manufacture of fertilizing materials has progressed and larger supplies have become available, farmers have applied them to more and more acreage and at higher rates. Continued improvement in crop varieties with greater yield potential has provided growers an opportunity to achieve higher production if adequate fertility is ensured. The dramatic growth of the fertilizer industry is illustrated in Figure 1. Nearly a 20-fold increase took place in fertilizer use from 1940 to 1980.

Although cropland acreage in California has increased relatively little since 1940, yields of most crops have increased substantially. Much of this increase in production can be attributed to the more widespread use of commercial fertiliz-

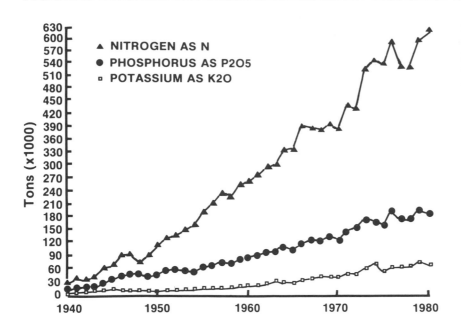

FIGURE **14.1** Fertilizer usage in California, 1940–1980. *Data from California Department of Food and Agriculture Fertilizer Tonnage Reports.*

ers. Not only is a larger percent of the total acreage in most crops being fertilized, but the rate of application, particularly of nitrogen, continues to increase as well.

The Need for Nitrogen

California soils almost always lack sufficient nitrogen to support intensive cropping, thus various sources of nitrogen fertilizer are applied to a large portion of all cropland. Anhydrous ammonia (82 percent N), which is stored as a liquid in high pressure tanks, converts to a gas under ambient conditions. The tanks which are often pulled behind field tractors prior to or soon after planting inject the liquid ammonia directly into the soil where it dissipates as a gas. Aqua ammonia (20–21 percent N), made from the addition of water to anhydrous ammonia, is another important nitrogen source. Other liquid nitrogen materials of importance include solutions of urea and ammonium nitrate either alone or in combination, dissolved in water. Dry granular forms of nitrogen fertilizer commonly used are urea, ammonium sulfate, ammonium nitrate and ammonium phosphate sulfate (16–20–0). Use of any particular form depends largely on the cultural practices utilized in the production of a crop, economics, and the personal preference of the grower. Ammonium sulfate has historically been the leading nitrogen source and continues to be used widely in California, but urea is increasingly used because of advantages in its manufacture and transportation. Because urea is 45–46

percent nitrogen compared with 21 percent for ammonium sulfate, it offers significant economy in application.

Two-thirds of all nitrogen fertilizers are manufactured using natural gas—thus the increasing cost of natural gas supplies has and will have a significant impact on fertilizer prices. In 1970 eight ammonia plants were operating in California; by 1979 six were closed because of the increasing cost of natural gas and the availability of cheaper off-shore ammonia from Trinidad, Canada, Russia, and Mexico. Special legislation permitted the remaining in-state ammonia plants to keep operating by guaranteeing a special rate for natural gas supplies for a limited period.

Field crops receive nearly half of the total nitrogen fertilizer applied in California. Rates of application range from 76 to nearly 200 pounds per acre, depending on the crop and situation. Overall, the large acreage of field crops in the state accounts for the large tonnage used. Vegetable crops receive a much higher applied rate per acre but represent a smaller acreage. Turf and ornamentals comprise an even smaller area but have the highest rate of nitrogen application.

Other Fertilizers

Although nearly half of the total phosphorus fertilizer is applied to alfalfa, irrigated pasture and rangeland, the highest rates of application are applied to vegetable crops. Vegetables are frequently planted during cool weather and phosphorus placement beneath the seed at planting stimulates seedling development and provides more uniform plant growth. Vegetables and turf account for more than half of the total potassium applied. Prunes and almonds, along with turf, have the highest rate of application.

Animal wastes continue to be used as fertilizer, but primarily near the places where they are generated—feedlots, dairies, and poultry houses—because of the difficulties inherent in collecting and distributing them. About 5 million tons of manure are used in agricultural production in California each year, which is reasonably close to the estimated amount of collectible manure produced in the state. A high proportion of this is used in hay-silage, pasture-range and vegetable production. This amount, however, represents only 3 or 4 percent of the total commercial nitrogen sold.

It is estimated that about 30 percent of total United States agricultural production is directly attributable to the use of fertilizers. Total United States use approaches 21 million tons annually, about one-half of this nitrogen, one-quarter phosphorus and another quarter potassium. In California the percentages are slightly different, since crops require more nitrogen and less phosphorus, and soils here have been less potassium deficient. About 68 percent of fertilizer usage in California is nitrogen, 23 percent phosphorus and 9 percent potassium. Average combined per acre rates of nutrient applications in California rose from 65 pounds in 1955 to nearly 167 pounds in 1975. Total fertilizer applications in 1979: nitrogen, 623,635 tons; phosphorus, 208,956 tons; and potassium, 79,311 tons. The value of commercial fertilizer in California approached nearly $500 million in 1979.

Outlook

Outlook for the future of commercial fertilizer is somewhat uncertain, since in an era of diminishing resources and increasing costs, expenses may be greater, supplies may be limited, or both. However, the economics of production usually dictate adequate fertility inputs; otherwise, crop yields are drastically reduced. Although the needs for potassium and phosphorus can probably be met domestically for an indefinite period, the manufacture of nitrogen will come into more competition with other uses for natural gas—residential and industrial heating, transportation and other manufacturing.

Research to improve biological fixation of nitro-

gen by legumes continues and, in addition, major efforts are underway to adapt the symbiotic fixation of nitrogen to other plant species. Growing of legumes is a cultural practice used by many agriculturalists to substitute for some of the nitrogen needed from commercial sources. Most of the current use of legumes in California is as a dual crop in vineyards, orchards, and rangelands, and not in a rotation as in the central U.S. Costs and energy requirements for growing and maintaining a second crop must be considered. If the secrets of the symbiotic relationship between bacteria and legumes can be transferred to other crops for more efficient production, perhaps food supplies can be maintained with smaller inputs of fertilizer nitrogen.

Pesticide Use and Regulation
by M. W. Stimmann

A Brief Review

In 1946, University of California plant pathologist Ralph Smith writing in *California Agriculture* said: "To grow plants entirely free from destructive enemies may seem a Utopian dream; yet such a condition actually existed in California within the memory of persons now living...." In 1858, a settler near Colusa declared: "I have apples, plums, nectarines, almonds, apricots, and peaches in bearing. We have no diseases in any of our fruit trees except almond, upon which there is a small reddish-colored insect which makes a web on the upper side of the leaf...." By 1890, however, the list of pests[1] which had been introduced into California from other parts of the world or had moved from their native hosts and habitats into California's crops was appalling—grape phylloxera, San Jose scale, pear slug, citrus mealybug, purple scale, corn earworm, Hessian fly, pear and apple scab, apricot shot hole, peach blight and many others were recognized as threats to the developing agricultural industry.

Until 1945, California farmers had utilized rela-

1. Pests are organisms which attack or otherwise adversely affect our crops, bodies, homes, pets, and livestock, or which compete with or attack our plants. Examples are some species of insects, weeds, nematodes, fungi, snails, slugs, rats, mice and so on.

tively few inorganic pesticides.[2] Those in use included Paris Green, lead arsenate and other compounds containing arsenic to control codling moths in apples, pears, and walnuts; Bordeaux mixture of copper and lime to control fungus diseases in grapes; and lime sulfur to control insects. Also available were a small number of organic chemicals: oils and solutions of soap, kerosene, tobacco, and gasoline. Additionally, a few botanical insecticides, such as ryania and pyrethrum were used. Early methods for applying these pesticides were crude; sometimes the pesticides were simply sprinkled on the crop or sponged on the foliage. Because these early pesticides and application techniques were relatively inefficient, it was often necesssary to apply them in tremendous quantities. By 1945, complicated devices had been developed to apply the few available pesticides to both field and tree crops. The application devices ranged from knapsack sprayers to gasoline-powered truck-mounted rigs. By then, the basic techniques of applying pesticides by aircraft also had been developed.

In completing his chapter on plant protection in *California Agriculture*, Smith discussed potential replacements for the older pesticides. His suggested substitutes were mostly inorganic materials such as cryolite and compounds of selenium and mercury. But Smith concludes that despite the fancy names and new combinations, "Bordeaux mixture, lime-sulfur, dusting sulfur, lead arsenate, petroleum emulsions, and nicotine sulfate still form the main basis of chemical plant protection." He went on to say "there has been surprisingly little development of new fungicides during the present century." Smith made no mention of any of the pesticides which were soon to prove so effective in controlling insects, mites, weeds, and plant diseases.

Today, surely, none can remember the time when California crops were grown without pests. Many, however, can recall when our crops were grown without the use of any of the modern chemical pesticides.

Following 1945, the most important pesticides in history were widely introduced into California agriculture. Included among these more complex compound chemicals were the insecticide DDT,

the herbicide 2, 4-D, and several new fungicides. These inexpensive and effective chemicals changed forever the methods which farmers used to control insect, weed, and disease pests. With the control of pests brought about by the new pesticides, agricultural productivity increased dramatically. Thousands of tons of pesticides are now used in California each year. These pesticides, their benefits and the problems associated with them, are one of the major issues facing contemporary California agriculture.

Many pesticides have been manufactured and sold over the past 35 years. Some have come onto the market to stay, others have been used for a while and then lost their effectiveness or been removed from sale for a variety of reasons. In all, approximately 1,200 active pesticide ingredients formulated into about 30,000 individual products are registered for use in the United States.

Current Usage

In 1980, agricultural pesticide applicators reported using some 112 million pounds[3] of active pesticide ingredients in California. The largest category of use was sulfur; twenty-seven million pounds were applied to control mites and plant diseases. The second most frequently used pesticide was petroleum distillates; 23 million pounds of these oils were applied for a variety of reasons including insect and weed control.[4]

The majority of the remaining 62 million pounds of pesticides were synthetic chemicals. Of these, ten insecticides/miticides, nine herbicides, three soil fumigants, two defoliants (including inorganic sodium chlorate) and four fungicides made up more than 55 percent of the state's reported synthetic organic pesticide use in 1980. (See Table 1.)

In 1980, two thirds of the pesticides used on California's agricultural crops were applied in just ten counties: those with the most intensive agriculture in the San Joaquin Valley; the Imperial Valley; and two coastal counties, Monterey

2. Pesticides are substances or mixtures of substances intended for controlling any form of life declared to be a pest. Included among the pesticides are insecticides, herbicides, defoliants, fungicides, nematicides, and rodenticides. The definition of a pesticide is not limited to any particular kind of chemical or pest.

3. Because of the reporting system, it is acknowledged that this is an *under* estimate of the total amount of pesticides used by California agriculture. The 121 million pounds reported in the 1980 Pesticide Use Report from CDFA is incorrect, having overestimated pesticide use in Santa Barbara County by more than 9 million pounds.

4. Both petroleum distillates and sulfur are considered to be relatively nontoxic to humans and to present little environmental hazard (although oils may make a significant contribution to air pollution).

TABLE 14.1
Reported Major Pesticide Use in California, 1980

Active Ingredient (Brand Name)	Lbs Applied**
INSECTICIDES/MITICIDES	
propargite (Omite)	1,798,000
methomyl (Lannate)	1,669,000
methamidophos (Monitor)	1,008,000
carbaryl (Sevin)	830,000
parathion	806,000
dimethoate (Cygon)	665,000
oxydemeton-methyl (Metasystox)	562,000
dicofol (Kelthane)	519,000
mevinphos (Phosdrin)	509,000
azinphos methyl (Guthion)	448,000
TOTAL	8,814,000
HERBICIDES	
nitrofen (TOK)	2,102,000
CDEC (Vegadex)	1,346,000
2,4-D (all forms)	890,000
DNBP (Dinoseb)	705,000
diuron	448,000
atrazine	430,000
paraquat	430,000
simazine	238,000
glyphosate (Roundup)	230,000
TOTAL	6,819,000
FUNGICIDES	
maneb	2,819,000
chlorothalonil (Bravo)	812,000
captan	790,000
benomyl (Benlate)	249,000
TOTAL	4,670,000
DEFOLIANTS	
sodium chlorate	3,749,000
DEF	745,000
TOTAL	4,494,000
FUMIGANTS	
D-D mixture (Telone)	13,395,000
methyl-bromide*	6,064,000
chloropicrin*	1,443,000
TOTAL	20,902,000
TOTAL of listed major use pesticides	45,699,000

*Soil and commodity fumigation
**Approximate values

TABLE 14.2
Major Pesticide Uses by County, 1980

County	Pounds Applied
Fresno	14,464,000
Kern	10,744,000
San Joaquin	9,092,000
Ventura	7,788,000
Monterey	7,484,000
Imperial	5,802,000
Tulare	5,773,000
Merced	5,521,000
Stanislaus	4,077,000
Madera	3,984,000
TOTAL	74,719,000

About 90 million pounds of pesticides applied to fifteen crops accounted for about 80 percent of all reported agricultural pesticides used in California in 1980. (See Table 3.)

and Ventura. The lowest rates of pesticide use were reported from some of the mountainous areas such as Alpine and Sierra counties where there is very little agriculture. (See Table 2.)

Such huge amounts of pesticides, some of them highly toxic, have not been used without causing human and environmental damage. Thus, the need for effective regulation has long been recognized at both the state and national levels.

Pesticide Regulation in California

Pesticide regulation in California is a three-tiered system: the federal and state government promulgates regulations and policies, the state implements a regulatory program, and the counties develop their own policies and regulations and enforce the program.

Before a chemical may be lawfully used as a pesticide for any purpose in the United States, it must be registered by the Environmental Protection Agency (EPA) under the Federal Insecticide, Fungicide and Rodenticide Act (FIFRA).

The conditions of California with its high pesticide usage, extensive agriculture, and highly concentrated population centers, create a unique set of pesticide-related problems. Thus, it is important that the state be able to regulate pesticide usage within its own borders. The federal law recognizes this and delgates to the state many pesticide regulatory activities. California pioneered development of legal controls promoting

TABLE 14.3
Pesticides Applied to California Crops, 1980*

Commodity	Lbs Applied**	Acreage Harvested	Lbs/Acre**
Grapes	14,300,000	605,300	24
Cotton	13,500,000	1,500,100	9
Sugarbeets	11,380,000	228,000	50
Citrus (all types)	10,047,000	265,000	38
Broccoli	6,100,000	73,600	82
Almonds	6,000,000	326,400	18
Lettuce	6,000,000	158,800	38
Alfalfa	4,700,000	1,030,000	5
Tomatoes	4,630,000	239,000	19
Strawberries	3,200,000	11,000	291***
Rice	3,000,000	548,000	5
Cauliflower	3,600,000	33,100	109
Carrots	2,400,000	37,200	65
Peaches	2,200,000	62,600	35
Celery	1,700,000	20,500	83

*Includes oils, sulfur and various fumigants
**Approximate values
***This represents the total amount of pesticides reported for the crop. Total values such as these tend to give a distorted picture of the amount of pesticides applied to the crop itself. For example, 291 lbs of pesticides/acre on strawberries represents the total reported pesticide usage/acre on California strawberry plantings for 1980. However, three *pre-plant* soil fumigants accounted for 90% of the pesticides reported. This leaves approximately 322,500 lbs of other pesticides which were applied to the 11,000 acres of strawberries grown that year (29 lbs/acre). Of this total, 151,000 lbs were sulfur and oils. It could be further argued that because of their relatively high degree of human and environmental safety, sulfur and oils can be removed from consideration. Thus significant pesticide usage on strawberries amounted to about 14 lbs/acre.

the safe use of pesticides in the United States. The first California pesticide laws were passed in 1901. Since that time, California has developed the most extensive system of pesticide control in the nation and the California pesticide regulatory scheme has, in many cases, served as a model for the federal law.

Under the California Food and Agricultural Code, the California Department of Food and Agriculture (CDFA) has primary reponsibility for implementing the regulation of pesticides; CDFA operates in cooperation with the County Agricultural Commissioners (CAC) who have the responsibility of enforcing the program. Within CDFA, pesticides are regulated by the Division of Pest Management. This Division is divided into a number of regulatory units, which are described in detail in Chapter 19.

Before any pesticide may be sold in California, each formulation must be registered with the United States EPA and the CDFA. The CDFA reports that more than 1,000 companies have registered more than 10,000 pesticide products. These are used on over 200 food and fiber crops grown in California and 300 non-crop targets to control some 5,000 species of pests. Data submitted by the manufacturer of each pesticide is reviewed and evaluated by CDFA to eliminate those which are dangerous to humans and the environment, misrepresented or not beneficial. The label instructions must specify which crops and pests may be treated with the pesticide (generally, crops not listed on the label may *not* be treated); what special safety and use precautions must be taken before, during and after use, and what special disposal and other criteria must be met to use the pesticide. Under FIFRA, the CDFA may not change an EPA-approved pesticide label; the state may only register or refuse to register the product in California. If registered in California, the state, using the permit process, can impose further restrictions on a pesticide by restricting its use to certain conditions.

Classification of Pesticides

Federal law requires that pesticides be classified for either "general" or "restricted" use. General use pesticides may be used by anyone. Restricted use pesticides may be used only by persons "certified" as competent to handle them safely.

California law is more strict in regulating restricted pesticides. California can require that *permits*[5] be obtained for any pesticide, regardless of whether or not the pesticide is restricted under FIFRA.

The pesticide classification scheme in California applies differing restrictions to the use of "restricted materials," "exempt materials" and unclassified "non-restricted" pesticides. The pesticides most hazardous to humans and the environment are classified as restricted materials. These may not be possessed or used until the applicator has obtained a permit.[6] Restriction of pesticides in California is based upon the following considerations: 1) danger of impairment of public health, 2) hazards to applicators and farm workers, 3) hazards to domestic animals (including honey bees) or to crops, 4) hazards to the environment from drift and other pesticide movement, 5) hazards related to persistence in soils, and 6) hazards to subsequent crops through soil contamination. About sixty active pesticide ingredients have been restricted in California. Some examples are parathion, aldicarb, methyl bromide, and 2, 4-D.

Pesticides may be designated as "exempt" materials if they have been found to need no restrictions other than registration and labeling. "Inert" ingredients and spray adjuvants are generally classified as "exempt." Examples of some active ingredients of pesticides which are exempt are sulfur, Bordeaux mixture, and *Bacillus thuringiensis*. All pesticides not classified as restricted or exempt are classified as "non-restricted." Generally, most non-restricted pesticides may be used without a permit, although under California law, counties may require a permit for any use of an agricultural pesticide.

County Agricultural Commissioners and the Pesticide Use Permit

The California system of County Agricultural Commissioners (CAC's) had its inception in 1881. Over the years the commissioners have acquired a wide range of responsibilities including plant quarantine inspection, nursery inspection, plant pest control,[7] vertebrate pest control and weed control. In fifty-four of the fifty-eight California counties, the CAC has the principal responsibility for pesticide regulation and enforcement. CAC's are employees of the county appointed by the County Board of Supervisors but their duties are prescribed by state laws. They are required to register persons in the business of selling pesticides (pesticide dealers), persons applying pesticides (pest control operators) and those giving pest control advice (pest control advisers). CAC's are also responsible for issuing permits for the use of restricted pesticides. While the broad outlines of pesticide regulatory programs are prescribed by the state, there is considerable variation in regulation and enforcement among the individual counties.

While pesticide users must report their use of restricted pesticides under a permit, and agricultural pest control operators must report *all* pesticides they apply, growers are not required to report non-restricted and exempt materials, and most urban uses of pesticides are unreported. Based on these reported uses, the CDFA compiles and publishes an annual "Pesticide Use Report." The California reporting system provides the most complete information about pesticide use in the nation.

Before issuing a permit, the CAC must consider conditions at the site of application including: 1) nearby dwellings, schools, or hospitals, 2) problems related to adjacent plantings of other crops, 3) the possibility of outbreaks of other pest species, 4) weather conditions which might adversely affect the application, 5) effects on honey bees, and 6) the provisions made for proper pesticide storage and disposal. The CAC may place other limitations on the permit depending on local conditions. By law, the CAC's are responsible for enforcing pesticide use and pesticide worker safety regulations and for investigating reported incidents of human illness or environmental damage.

California is the only state requiring medical doctors to report incidences of pesticide poisonings. Through this system, California physicians reported 1,477 cases of pesticide-related illness in 1979. Most pesticide illness involved individuals who were either mixing, loading, or

---

5. Permits are written permission to apply a pesticide to a specific crop and in a prescribed manner. Permits are issued by the County Agricultural Commissioner.

6. Some exceptions exist: certain restricted pesticides do not require a written permit when small amounts are used for non-agricultural purposes such as home use and structural pest control.

7. Pest control carried out by CAC's is associated with noxious weeds, and introduced pests which have not become widely established.

One of the techniques of Integrated Pest Management is to monitor insect populations closely. This technician knocks branches in an orchard so that he can calculate insect density. *Photo by Jack Clark.*

applying pesticides or working in crops such as citrus and grapes which may retain high amounts of dislodgable pesticides on the foliage.

In the event of a misuse of a pesticide, or the report of human illness associated with the application, the CAC will investigate and may issue a notice of violation, revoke a pesticide permit, or in more serious instances, revoke the applicator's or advisor's permission to operate within the county. In the most extreme cases, the law authorizes criminal prosecution.

Agricultural Pest Control Advisers

Most of the agricultural pesticides used in California are applied upon the recommendation of an Agricultural Pest Control Adviser (PCA). PCA's are responsible for recommending pest control methods to farmers. Although growers may make their own decisions regarding pest control activities, they may legally be given advice regarding these decisions by licensed PCA's or representatives of the University. In California, PCA's are required to meet certain minimum qualifications, including at least four years of relevant college education. They must pass an examination prepared by the state and register with the County Agricultural Commissioner of any county in which they wish to practice. Recommendations must be made in writing and must contain information on the crop to be treated, the target pest, and the type and amount of pesticide to be used. Currently, approximately 80 percent of the 2,500 active PCA's in California are employed by pesticide companies. Generally, these PCA's do not charge directly for their services, but are remunerated through pesticide sales. The remaining 10 percent are "independent" advisers who are not affiliated with pesticide companies. These independent advisers generally charge an acreage fee for the crops monitored.

Pesticide Applicators

LICENSED PEST CONTROL OPERATORS. Persons in the business of applying pesticides for hire must be licensed as Pest Control Operators (PCO's). Licensed PCO's must pass a state examination and be registered in each county in which they do business.

GROWERS. Farmers who apply pesticides to their own land are not required to pass a written examination, to pay any license fees, to obtain a written recommendation or to keep any records of their use of non-restricted pesticides. They must, however, pass an "oral examination" given by the CAC certifying them as a private applicator and obtain a permit for any application of restricted use pesticides they intend to make.

CERTIFIED APPLICATORS. California licensing and certification requirements meet federal law concerning certification of private applicators (growers) and commercial applicators (those who aply pesticides for hire).

Integrated Pest Management

Experience has shown that widespread and intense use of pesticides has led to many problems. Also, many pesticides gradually lose their effectiveness as certain pests, especially insects and mites, develop resistance and require greater amounts of pesticide for effective control. With a few pest species, control is very difficult, if not impossible, with any of the registered pesticides. This loss of effectiveness coupled with rising pesticide prices and public concern over health and environmental effects has stimulated increased research on alternative methods of controlling pests.

There is no single alternative to pesticides, and their need cannot be eliminated by current technology. Many people feel, however, that pesticide use can be reduced through the development of management systems which are better able to predict and evaluate pest outbreaks and economic damage to the crop, eliminate unnecessary chemical treatments, more precisely time pesticide applications and rely on other methods and factors in the environment. This approach to pest control is called Integrated Pest Management (IPM). IPM programs require an indepth understanding of crop and pest biology as well as careful field monitoring of pests, crop development and environmental factors. Within the past few years, the concepts of IPM have been embraced by many pest control scientists as well as by the state and federal governments. In August 1979, President Carter directed federal agencies having pest management research, control, and education programs to support and adopt IPM strategies whenever practicable. It is clear that governmental policy-makers are attempting to bring about a shift to IPM.

The developing shift from classical pest control to IPM is eagerly anticipated by those concerned about the deleterious effects of pesticides. However, the actual extent of change in our pest control methodology may prove to be limited by problems beyond our immediate control.

For example, an infestation of the world's most damaging insect pest, the Mediterranean Fruit Fly, was recently discovered in northern California. By the time the potential magnitude of this problem became widely appreciated, the Medfly had survived at least one winter in the San Francisco Bay area and moved into at least three Bay Area counties. If this infestation continues to spread in spite of intensive control efforts, the Medfly may become established in all of the major agricultural areas of California. Aerial applications of baited pesticides are the only proven method of controlling such widespread infestations of this pest. If an ongoing baited pesticide control program became necessary in major agricultural regions, it would have a severe impact on beneficial insect predators, parasites and pollinators, and, without a doubt, would have a devastating effect on dozens of current and planned IPM programs.

Summary

Our pest-free Utopia was lost long ago. The compounds of arsenic, selenium and mercury that were spread over California's fields and farms for decades have been replaced by more than 1,000 synthetic organic chemicals. With these chemicals has come the development of a regulatory system which seems to have become as complex as the pest problems themselves. Our federal, state, and county regulatory systems are massive, expensive, and burdensome. Yet, in spite of the size of the system (or perhaps because of it), practically no one is entirely satisfied with pesticide regulation in California. The fault lies not with the system; it is due to the very nature of the pest control problem. Pesticides are by necessity toxic materials which are designed to have a profound physiological effect on the pests they control. No pesticide is entirely selective, and the majority are toxic in minute amounts to pests and non-pests alike. It is this innate property of pesticides which leads to the fundamental dilemma. Pesticides are clearly of benefit—their use has resulted in greatly increased agricultural productivity and they have become a necessity for successful modern agricultural production in California; yet pesticides cannot be used without at least some risks to human health and the environment. The risks to health and the environment presented by legal use of many pesticides are very small; others, however, require extreme care in manufacture, transportation, and application. This benefit-risk dilemma will demand the attention of everyone involved in California agriculture for a very long time to come.

Farm Machinery in California

by Roy Bainer

An Historical View

Grain Harvesting

The single event that was destined to become one of the most important factors in the development of agriculture in California was the importation of the Hirum Moore combined harvester-thresher from Michigan in 1854. Moore had received a patent covering his invention on March 28, 1836, only five years after McCormick introduced his reaper. The machine was ground-powered and pulled by 16 horses, and cut a swath of 10 feet. It was shipped to San Francisco via Cape Horn and transported overland to the John Horner Farm near Mission San Jose. During the 1854 season it harvested 600 acres of wheat on the Horner and Breyfogle farms.

This successful harvesting experience stimulated combine development in California. Between 1858 and 1888, at least 21 individuals and companies built combines. These early machines were ground-driven and pulled with 20 to 40 horses or mules, cutting swathes of 20 to 30 feet. Steam engines for propelling and driving the harvester mechanisms started to replace animal power in the mid-1880s. George Berry, who grew 400 acres of wheat in Tulare County, is credited with being one of the first to build a self-propelled, straw-burning, steam-driven harvester for use in 1887. The machine cut a 22 foot swath, extended to 40 feet in 1888.

Stockton soon became the center for manufacture of combines as well as other agricultural machinery. Early manufacturers included Harris, Holt Brothers, Hauser-Haines and Shiplay, as well as Best of San Leandro.

The availability of agricultural machinery plus a favorable environment stimulated the development of a tremendous grain trade in the great Central Valley during the last half of the nineteenth century. Wheat production increased from 5.9 million bushels in 1859 to almost 41 million bushels in 1889. Between 1880 and 1900, however, other more valuable crops began to replace wheat. Eastern markets that could be reached over the new transcontinental railroads were gradually established. By 1899, 64 percent of the value of all crops in California came from noncereal commodities.

The new crops were vines, trees, vegetables, sugarbeets, cotton, and rice, all of which required a heavy input of hand labor. Domestic labor was augmented by Chinese who had been brought in to build the railroads, Filipinos, Japanese, Mexican nationals, and, later, the migrants from the dust bowl. This extra labor was still not enough to ensure adequate help. Other industries began to draw on the labor supply and force wages up. The labor problem became more critical with the involvement of the nation in two world wars, and action of the Congress which curtailed the importation of foreign labor. All of this led to a demand for labor-saving machinery.

Rice and Flax Harvesting

The binder-stationary thresher system for harvesting rice exited prior to 1929. In addition to a high labor requirement, the system left much to be desired; high head rice (whole kernel) yields were reduced during the milling operation due to "sun-checking," small fractures perpendicular to the axis of the kernel which develop during rapid drying in the field. Breaking along these fractures often occurs during milling. Laboratory studies demonstrated that sun-checking would be practically eliminated if rice taken from the field with a moisture content of 20 percent was dried slowly. Studies initiated in 1929 at the California Agricultural Experiment Station resulted in the substitution of direct combining and artificial drying for the binder-stationary threshing system for harvesting rice. This not only saved labor but resulted in greater recovery in the field and produced a superior final product. A study showed that the production of rice in Japan in 1948, where hand methods prevailed, required 900 man-hours of labor per acre compared to approximately 7 1/2 man-hours per acre under the highly mechanized system used in California. Airplane seeding of rice, first reported in 1929, plus direct combining and artificial drying prevails throughout rice-producing areas in the

United States today.

Rice combines, with headers in front to reduce field losses in opening irregular shaped rice checks, are now mounted on tracks for maneuvering in wet fields. The rice is handled in bulk. Bank-out wagons, also mounted on tracks, are used to transfer the rice from the harvester to transport trucks at roadside. Drying is done in three-stage, vertical-column driers.

The introduction of high-yielding Punjab flax into California from India in the early 1930s also presented a harvesting problem. When threshed with a standard spike-tooth cylinder, a portion of the seed was cracked and the straw was shredded, making separation of seed from straw difficult. The solution was to substitute a rasp-bar cylinder and mount a pair of rolls, one steel and the other rubber-covered, ahead of the cylinder. The rolls crushed the flax bolls thereby releasing the seed. The new cylinder left the straw intact which improved the separation of seed from straw, and cracked less seed.

Sugarbeet Mechanization

A cooperative project on sugarbeet mechanization was initiated in 1931 by the California Agricultural Experiment Station and the USDA. It was greatly expanded in 1938 as a result of substantial research grants from the U.S. Sugar Beet Association. At that time excessive amounts of multi-germ seed balls were planted to insure a continuous row of plants. Laborers using short handled hoes blocked the stand to 12-inch centers and used their fingers to reduce population on each block to one plant. During harvest, after the beet roots were loosened by a simple plow, laborers with hooked knives lifted each beet, cut off the top and threw it into a windrow. Later, the same crew picked up the beets from the windrow and tossed them onto a truck.

Labor used in beet production peaked higher in the spring during thinning than during harvest in the fall. This spring labor peak was reduced by the development of mechanical processes for reducing multi-germ seed balls to units approaching single germs, and by precision planting and thinning. Later, seed processing was replaced by mono-germ seed produced by plant breeders in the mid-1950s. Sugarbeet harvesters were developed in the late 1940s for reducing the back-breaking labor of hand topping and loading. Presently, the production of sugarbeets is done entirely by machines.

Three commercial farm shows each year display new machinery and technology to California farmers. This is a view of the Colusa Orchard and Farm Equipment Show which takes place in February.

Mechanical Cotton Picking

The development of a mechanical cotton picker challenged the minds of men for more than 100 years. Mechanical cotton harvesters are of two types, commonly known as pickers and strippers. Pickers are selective in that seed cotton is removed from open bolls by revolving spindles, whereas green unopened bolls are left on the plant to mature for later picking. They are used in Arizona, California, and the deep South, where cotton plants grow rank. The stripper, on the other hand, is a once-over machine. All bolls, open or closed, are removed from the plant in a single pass. These machines are used in the High Plains region where the plants are smaller. By 1962, 90 percent of the California cotton crop was harvested mechanically, compared to 10 percent in 1949. The rapid acceptance of mechanical cotton pickers was partially due to an economic study of 63 machines in operation in 1949. This showed that after charging the costs of labor, overhead, operating, field waste, and reduced sales value due to reduction in quality, against the machine, the saving of machine picking over hand picking was approximately $20 per bale.

Mechanical Tomato Harvesters

The rapid adoption of mechanical tomato harvesters was a direct result of labor shortage which resulted from the termination of Public Law 78 on December 31, 1964, drastically reducing the importation of foreign labor. At that time approximately 85 percent of the tomatoes for processing were picked by nationals from Mexico. The University, however, had launched a long-range research program in 1950 to mechanize the harvesting of tomatoes. During this period plant breeders developed a determinate-type tomato plant that set fruit which ripened over a very short period, permitting a once-over harvest compared to two or three pickings spread over a much longer period. Once the new plant was available, agricultural engineers were able to design a machine to lift the plants and remove the tomatoes in a single operation. The new system, somewhat imperfect at the start, became available at about the right time. The research behind it was truly interdisciplinary, involving plant breeders, food scientists, agricultural engineers, and finally manufacturers, farmers, and processors who were willing to risk their economic futures on mechanizing the crop. In 1963, about 1.5 percent of California's processed tomatoes were harvested by machine. Five years later, 1,300 mechanical tomato harvesters harvested approximately 95 percent of the 240,000 acres of tomatoes.

The tomato processing industry had threatened to move operations south of the border where labor was plentiful. This would have resulted in California losing an industry worth approximately $3 billion today (1980), including thousands of jobs. In 1963, the tomato harvest work force was made up of 37,300 foreign workers and 5,500 domestic workers; five years later, 95 percent of the crop was machine harvested using about 31,000 workers, most of whom were domestic. Back-breaking stoop labor had been eliminated as workers rode the machine. With the advent of electronic color sorting, the work force has recently been reduced to about 21,000 workers.

Other Developments in Mechanization

Labor shortages associated with World War II resulted in a demand for harvesting machinery for deciduous fruits and nuts, vine crops (wine and raisin grapes), and vegetables. In deciduous fruit, such as prunes, three men with the aid of two portable catching frames and an inertia-type rigid-boom shaker can now harvest 60 trees per hour.

In the case of mechanical harvesting of raisin and wine grapes, the canes are tied to a tight wire trellis during pruning. The grapes are removed by vibrating the wire with an impactor device on the machine as it moves down the row. For raisin grapes, a continuous paper tray is laid on the ground between rows to receive the grapes. After their day in the sun for two or three weeks, another machine picks up the paper and removes the raisins. Wine grapes, on the other hand, are collected as they are harvested, and placed in trailers or trucks for transporting to the winery.

In the early 1950s California farmers faced curtailment in acreage allotments for cotton, rice,

and sugarbeets. One of the substitute crops considered was corn (maize). Because of a history of limited production in this state, there was no restriction on acreage. California imported a considerable tonnage from the corn belt, and the freight differential gave an advantage to locally produced corn, but a problem was the lack of corn harvesting machines. Since many of the farmers owned grain combines, it appeared desirable to try adapting them for corn harvesting.

Cooperative field trials were initiated in 1954 between the International Harvester Company, Deere and Company, and the California Agricultural Experiment Station in the use of the combine for harvesting field corn. In these trials, snapping rolls replaced the header on their combines. Only unhusked ears were introduced into the machine. Tests run in Riverside, Kern and Yolo counties demonstrated that this system had merit. By 1970, approximately 70 percent of the corn produced in the five principal cornbelt states was harvested by combines equipped with corn heads.

Another important contribution from California was a track-type tractor that could cope with the loose peat soils of the Delta. Holt Brothers of Stockton introduced a steam-power track-type tractor in 1904 which was later, in 1907, powered by an internal combustion engine. Dan Best of San Leandro competed in this development. In 1925, the Best and Holt companies merged to form the Caterpillar Tractor Company, now known throughout the world. It is interesting to note that more track-type tractors are now used in industry than in agriculture.

The major suppliers of tractors and farm machinery today are large companies such as International Harvester, John Deere, Massey-Ferguson, Case, White, and Caterpillar. These companies as a rule are not interested in small volume output. Therefore specialized machinery such as tomato harvesters and sugarbeet harvesters are built by smaller companies like Blackwelder, FMC and Johnson Machine Company. The annual "farm shows" in Tulare, Stockton, and Colusa display great numbers of machines of all sizes and types manufactured by many companies in the U.S. and from abroad.

Timeliness is a very important factor in most farming operations. There is an optimum period for soil preparation, planting, and harvesting. The use of machinery permits staying within time limitations. For example, the harvesting of rice must be done within a certain time period or the crop may be lost as a result of inclement weather. The same is true in regard to other crops like sugarbeets, corn, cotton, and tomatoes. The application of machines to agricultural production has been one of the outstanding developments in American agriculture during the past century. The burden and drudgery of farm work has been reduced, and output per worker greatly increased.

References

Fertilizers

Hignett, T.P., et al. *Fertilizer Manual*. Muscle Shoals, Alabama: International Fertilizer Development Center, 1969.

McVickar, M.H., Bridger, G.L., and Nelson, L.B. (Ed.). *Fertilizer Technology and Use*. Madison, Wisconsin: Soil Science Society of America, 1963.

McVickar, M.H., Martin, W.P., Miles, I.E., and Tucker, H.H. *Agricultural Anhydrous Ammonia Technology and Use*. Madison, Wisconsin: American Society of Agronomy and Soil Science Society of America, 1966.

Rauschkolb, R.S., and Mikkelsen, D.S. *Survey of Fertilizer Use in California 1973*. University of California, Division of Agricultural Sciences, Bulletin 1887, February 1978.

Pesticides

Boraiko, A.A. "The Pesticide Dilemma." *National Geographic*, 157(February 1980):145–183. 1980.

California Department of Food and Agriculture. *Laws and Regulations Study Guide*. Pesticide Use Enforcement and Licensing, 1979.

California Department of Food and Agriculture. *Report on Environmental Assessment of Pesticide Regulatory Programs*. Vol. IV, State Component, 1978.

Dunning, H.C. "Pests, Poisons, and the Living Law: The Control of Pesticides in California's Imperial Valley." *Ecology Law Quarterly*, 2(Fall 1972):633–693.

Huffaker, J.D. "The Regulation of Pesticide Use in California." *University of California, Davis, Law Review*, 11(November 1978):273–299.

Hutchison, C.B. *California Agriculture*. Berkeley: University of California Press, 1946.

Tucker, W. "Of Mites and Men." *Harper's*, August 1978.

Farm Machinery

Bainer, R. *Harvesting and Drying Rough Rice in California*. California Agricultural Experiment Station, Bulletin 541, 1932.

Fite, G.C. *The Farmers Frontier 1850–1900*. New York: Holt, Rinehart and Winston, 1966.

Hedges, T.R., and Bailey, W.R. *Economics of Mechanical Cotton Harvesting*. California Agricultural Experiment Station, Bulletin 743, 1954.

Higgins, F.H. "97 Years of Combining in California." *California Farmer*. March 25, 1950, pp. 280–281.

Roske, R.S. *Every Man's Eden*. New York: Macmillan, 1968.

Sitton, G.R. "Mechanized Rice Production." *California Agriculture*, 7(February 1953):2.

Tavernetti, J.R., and Carter, L.M. *Mechanization of Cotton Production*. California Agricultural Experiment Station, Bulletin 804, 1963.

Walker, H.B. "A Resume of Sixteen Years of Research in Sugar Beet Mechanization." *Agricultural Engineering*, 29(October 1958): 425-430.

15

Marketing

Leon Garoyan

California's environment—favorable climate and temperatures, productive soils, and large water supplies—was recognized early on as conducive to the production of hundreds of different crops and animal products. Even during the state's development period, the ability to produce quality products in excess of California's requirements led to the development of a "marketing mentality" that still distinguishes agriculturists here from those of most states. The need to move surplus products from California's productive valleys to eastern population centers has always been a characteristic of the commercial aspects of California's agriculture.

A consumer in New York buys a pound of fresh peaches grown in Fresno County, little aware of the efforts of a host of people who have participated in making her purchase possible. The U.S. food production and distribution system appears to work so effectively that consumers expect to be able to satisfy their wants at all times—and usually they can.

Marketing is the process of identifying consumer requirements and arranging to fulfill them. It is more than selling a farm product to a shipper or a canned good to a consumer, although these activities are results of the marketing process. Marketing encompasses the assembly, processing, transportation, storage, finance, and other distributive activities that add value to farm products as they are converted to consumer goods. It really involves production also, because decisions as to what to produce and where to produce it are strongly influenced by market considerations.

Marketing has not always been as involved as it is now. As recently as the 1930s, it was possible for families to trade farm products, such as eggs and homemade butter, for groceries at neighborhood stores. But the system has changed. Population has increased substantially in California and throughout the nation. Consumer incomes have increased and people have moved to the cities. International markets have opened. The food system has responded. Production and distribution technology have improved, permitting profitable specialization and encouraging more concentration in production, processing, and marketing. This has increased the range of markets which can be served by California producers. As a result, farmers and processors have needed to meet requirements for large quantities of uniform products, often shipped thousands of miles from California's fertile valleys.

Functions of the Marketing System

Our modern agricultural marketing system has evolved to provide four interrelated functions:

1. *Encouragement of production decisions to meet food industry requirements for raw products.* The U.S. food system depends upon a large base of raw materials originating from the nation's millions of farms. The marketing system seeks out consumption information and transmits it so that farmers and ranchers can produce to meet consumer needs. These signals to producers are the start of a highly coodinated system of decisions and activities. The system produces, on the one

hand, livelihoods for farmers and others in the food system who carry out the marketing functions, and, on the other, products to satisfy consumer needs.

2. *Conversion of farm products into consumer goods.* When a consumer in Boston purchases a package of raisins, she receives a far different product than the fresh Thompson grapes from which they originate. Most agricultural products are processed in some way. Fresh products are cooled, graded, and packaged for consumers. Many of California's fruits and some vegetables are canned or frozen. Livestock are converted to cuts and consumer packaged ready for cooking. Wheat is ground and flour is baked into bread; and cotton is ginned, spun into thread and woven into fabrics. For many commodities, production is seasonal while consumption occurs year around, so that storage of commodities following harvest is required to maintain adequate supplies for consumers. For perishables, elaborate storage facilities maintain fresh quality. All of these activities are undertaken to meet market and consumer needs and therefore are part of the marketing process.

3. *Exchange of values—price making.* Pricing determines the market value of farm commodities in terms of money. Depending upon the demand for and the supply of a given farm product, its price may fluctuate hourly, daily or weekly. Buyers and sellers are concerned about the efficiency of price making. When the method of purchase and sale produces prices that erroneously reflect demand and supply situations, it can trigger unwarranted price differences between individual producers and regions.

Buyers and sellers must have knowledge of demand, supply, and prices—including the extent to which price uniformity exists for like grades. Such knowledge assures a more expedient buyer-seller transaction and more efficient marketing system. Clearly it is important for the marketing system to facilitate the flow of market information, because it improves price making and minimizes misallocations of resources.

The California Food and Agriculture Department assembles and disseminates market news to help producers and marketing firms acquire price and quantity information. In addition, the Department provides quality control services. Grading of agricultural products by uniform standards of quality assists producers to obtain fair and equitable prices. Grading also allows trading

to occur without direct inspection by distant buyers.

4. *Distribution of final product.* Distribution involves a vast network of railways, highways, water and air transportation. Wholesaling and retailing organizations perform the assembly, storage, and "break-bulk" services which are of gigantic proportions.

Farmers, processors, wholesalers, and retailers are continually involved in the effort to supply consumers with the "right merchandise or service at the right place, at the right time, in the right quantities, and at the right price." In a competitive marketing system like that of the U.S., distribution and merchandising include many related promotional activities, such as packaging, display and advertising.

Structure and Organization of the Market System

Californians have developed diverse ways to market their products. For the typical producer, the point of delivery of farm products is "the market." The so-called "cash market" is the original and typical method to pass title to farm products from producers to buyers. Typically in the past, local markets involved direct negotiations between sellers and buyers who were well acquainted. Prevailing prices were usually known, and only visual inspections were required to determine grades before title was transferred from producers to buyers. For a growing number of commodities, however, the number of buyers (markets) continues to become smaller, and producers have fewer alternative outlets for their output. Many of the remaining buyers are conglomerates, multinational firms, or large scale firms with corporate headquarters located in major cities outside California. Thus, local representatives have few key discretionary decisions left—in contrast to the situation even in the 1960s, when key decision makers were local residents and quickly available for talks with producers.

Producers' access to markets for some commodities has also become restricted because of the reduction in handling and processing firms. While there were 36 cattle (steer-heifer) slaughter plants in central and northern California in 1970, there were only 28 in 1977, and considerably fewer today. There is but one lamb slaughtering

Prunes are sorted for size and quality along an assembly line in a packing house. Standardization and packaging are part of the marketing process.

plant operating now in California. The number of fruit and vegetable canneries continues to decline as more processors found profits inadequate in the 1970s. From 37 canners of cling peaches in 1960, the number dropped to 15 in 1973, 11 in 1980. For canning pears, the number dropped from 26 to 17, and to 13 in 1980. As a result, agricultural products must be transported to greater distances, a costly process. More importantly, however, is the reduction in market alternatives available to California producers.

Market Coordination

The need to reduce uncertainty in marketing farm products has led to several methods to minimize producers' and processors' exposure to unusual changes in demand. Among the most common types of coordination are production contracts, vertical integration, joint ventures, and, in some cases, formation of producer cooperatives.

Contracts

Perhaps a handshake was all it took to close a sale in earlier years, but California farmers and marketing firms have long since used written contracts to express terms of their agreement. These agreements between producers and buyers may specify the type of agricultural commodity, how the commodity is to be grown or raised and harvested, and often the price to be paid to the producer.

Contract production is the most common method employed in agriculture to coordinate farm production with marketing activities. Since contract production is forward in nature (as distinct from a marketing or sales contract), and is entered into prior to planting, it serves three purposes. First, a production contract reduces producers' market risks. Second, it minimizes processors' risks by assuring a source and quantity of supply. Third, it helps to regulate product flow in conjunction with expected demands. Also, if price is stipulated in the contract, the producer's potential income can be more realistically projected, aiding in financial planning.

There is no guarantee that contract terms will be equitable since equity is predicated upon relative bargaining strength. Often, producers are in a better position to negotiate production contracts through a bargaining cooperative rather than as individual producers.

Output under contract production in the U.S. has increased moderately from 15 to 17 percent over the years 1960 to 1970, and may be higher in California. The percentage of the total production occurring under contracts is estimated as about 50 percent for fresh market vegetables and potatoes; 75 percent for citrus; 85 percent for turkeys; 90 percent for vegetables and potatoes for processing, and hybrid seed corn; and over 95 percent for broilers, fluid milk, sugarbeets, and vegetable seeds.

On the West Coast, contracting for fruits and vegetables for processing has been commonplace. Today most California fruit and vegetable processing cooperatives have renewable contracts providing for yearly withdrawal by members. At one time, one of the large wineries in California offered 15-year contracts to grape growers as an inducement for increased acreage plantings.

Overall, contract production is more concentrated in fruits and vegetables, cotton, dairy, and poultry, than in crops such as wheat, barley, and alfalfa. For many crops, planting without assurance of a home for the production is economically very dangerous.

Vertical Integration

Coordination of agricultural production with marketing activities may involve all stages in the furnishing of materials, the production of commodities on the farm, the processing of food products, and the distribution of those products at retail stores and eating establishments. In recent years, the initiative to integrate within the food system has come from food processors and marketing firms rather than from farmers.

Coordination can be achieved through "vertical integration," where one firm assumes two or more stages of production that have traditionally been handled by separate firms. The primary purpose is to improve profit potential by having more control over sources of supply or markets. Vertical integration may enable market firms to get a more steady flow of product or a more consistent quality than before. Another rationale for vertical integration is that in reducing risk and improving efficiency, it leads to reduced production costs.

Vertical integration through ownership can be achieved by farmers either as individuals or as a group. Group participation in vertical integration is seen in farmer-owned cooperatives. For example, when almond growers own a cooperative that processes, packages, stores, and markets almonds, they are vertically integrated forward, through ownership, toward the consumer. On the other hand, if a food processing firm decides to own farmlands to grow raw agricultural products for processing, it would be vertically integrating backward toward the farmer. Some large farming corporations have vertically integrated both forward and backward. For example, one large business started out as a broiler growing operation in California, then integrated forward by running its own processing plant wholesaling organization and also integrated backward into feed manufacturing. But one very large fruit processor that also owned California fruit farms has recently sold its farms, lessening the degree of vertical integration that previously existed.

Joint Ventures

Some coordination attempts consist of joint ventures of two or more companies that integrate all or part of their business for a single undertaking. They exercise control and share in the profits on some agreed basis. Small and medium-sized firms have been attracted to the joint venture scheme because they can compete more efficiently with large competitors in regional areas and stages of business often barred to them because of their size.

Some California farmer cooperatives have engaged in joint ventures to enter new activities. Joint venture organizations have been organized to expand sales to international markets, enter the food service area, and control the manufacturing of fertilizers, containers, or other needed materials. Some farmer cooperatives have formed joint ventures with regular corporations, to develop wider market participation and to strengthen their financing.

Institutions

Farm Cooperatives

Farmer cooperatives are businesses owned by and operated at cost for their members, who also strongly influence the management decisions through a board of directors elected by members. These unique characteristics—operations at cost for members, and influence and control of decisions by members—distinguish cooperatives from general corporations. In most other ways, cooperatives are similar to general corporations.

Traditionally, farmer cooperatives perform one or more of three functions: (1) marketing farm products, (2) purchasing farm production inputs, or (3) providing services. Cooperatives distribute their earnings in the form of patronage dividends that are returned to patron owners on a basis proportionate to their transactions with the cooperative.

The primary objectives of marketing cooperatives are to improve net returns to members, and to provide a reliable and consistent home for crops. In their pursuit of these objectives, cooperatives may reduce the prevailing marketing charges, change marketing practices to provide more orderly marketing, influence consumer or trade demand, and in the case of bargaining cooperatives, negotiate for prices and contract terms.

Agricultural producers also obtain services and credit through cooperative associations. For example, California's dairymen upgrade the performance of their herds through Dairy Herd Improvement Associations, organized specifically to provide dairymen with production records for individual cows. Also, many producers obtain capital through PCA's and the Federal Land Bank, both of which as cooperatives operate on a cost basis for member-producers (see Chapter 13).

While service and purchasing cooperatives are important forms of cooperation among farmers, it is marketing cooperatives that have been most important in California. One of California's oldest and most successful cooperatives evolved out of the need for safeguarding farmers' livelihood. In 1893, desperate to save their farms from economic ruin, a group of citrus growers formed a "pool" to collectively sell their fruit. They bypassed the commission agents and shared the expenses of packing and marketing, devising new ways to do business, with each member receiving his share of profits according to the amount and grade of fruit he shipped through the pool. The payment to each individual according to quality of product preserved individual initiative. The system worked and grew to become the first marketing cooperative, known today as Sunkist Growers, Inc.

The founding of Sunkist set the stage for the organization of many single-commodity California marketing cooperatives. Other prominent cooperatives can trace their start to the success gained by citrus growers: Blue Anchor, Inc. (1901), California Almond Growers Exchange (1910), Diamond Walnut Growers (1912), Sun-

TABLE 15.1
Commodity Groups Using Authorized Marketing Order Provisions

Marketing Order Provisions	In Effect Some Time During the Period[a]		In Effect In		
	1945–1975	1961–1975	1950	1960	1975
Quality regulation	30	28	17	22	21
Inspection	29	26	17	20	20
Container or pack	15	13	5	9	9
Volume control	19	11	5	7	7
Rate-of-flow	5	4	2	3	4
Research	44	43	5	17	33
Promotion or advertising	34	31	6	19	24
Trade practice regulations	4	3	1	1	2

[a] Includes only commodities for which authorized programs were actually implemented.
SOURCE: French, Ben C., N. Tamimi, and C. F. Nuckton, *Marketing Order Program Alternatives: Use and Importance in California, 1949–1975.* University of California, Division of Agricultural Sciences, Bulletin 1890, May 1978.

Maid Raisin Growers of California (1923), Sunsweet Growers, Inc. (1917), Rice Growers Association (1920), Dairymen's Cooperative Creamery Association (1909), and California Bean Growers Association (1916).

Agricultural bargaining cooperatives are a variation of marketing cooperatives, used more in California than elsewhere. Producers join to gain strength in negotiating with proprietary processors such items as price, quality, quantity, and delivery terms. Producers delegate authority to their bargaining association to establish common quality and common price, and to negotiate contract terms. Most bargaining cooperatives do not take possession of products or assemble, process, or distribute them. Since bargaining cooperatives wholesale their members' production to processors, their facilities are limited generally to an office and perhaps a quality testing laboratory, whereas members of processing cooperatives have substantial capital investments in property, plant, and equipment.

As returns on fruit and vegetable canning businesses declined in recent years, several proprietary firms announced plant closures. To maintain a home for their farm products, farmers who sold to these canners were compelled to purchase the facilities and organize them as cooperatives. Producers of livestock may be compelled to invest in slaughter plants if further closures of existing plants occur—for the same reasons that have compelled fruit growers to buy canneries.

Marketing Orders and Agreements

The kit of marketing tools used by California agriculture includes marketing agreements and orders available to producers and handlers since the mid-1930s. Federal and state marketing orders and agreement programs are designed to provide for an orderly approach to the marketing of food products, thereby improving prices for producers and reducing marketing costs. These are self-help programs operating within the legal frameworks authorized and established by legislation. These programs, self-financed, provide growers and handlers the opportunity to find tailor-made solutions to marketing problems with a minimum of external interference or support. Both federal and state enabling legislations have similar objectives.

Marketing agreements and orders have been authorized for some crops for almost five decades. They have been used by California farmers to obtain, through group action, orderly marketing, quality control, inspection, container regulations, product promotion and market development, prohibition of unfair trade practices, and financing of research. For example, through a marketing order, supplies can be allocated to different markets according to product demand. Examples are in-shell versus shelled walnuts, as well as domestic versus export sales of raisins. Or, an industry may wish to standardize grades so as to present a consistently uniform product to buyers. Many commodity groups sponsor and maintain such marketing regulations (see Table 1).

With stability in prices one objective, another

benefit of marketing order programs has been the encouragement of greater initiative and responsibility for marketing among producers. Whether a main goal is a price level (as for milk), or price stabilization, or improving demand for a commodity through promotion, marketing programs have made a significant improvement in the orderly distribution of products, and in the long run, have served to provide consumers with more stable supplies and prices.

CALIFORNIA MARKETING ORDERS. Because of the difficulties met in the enforcement of the early voluntary marketing programs and the need for a broader range of marketing program legislation in the state, the California Marketing Act of 1937 was enacted. Since a large number and variety of specialty crops are produced in California, most current marketing orders affecting California products have been established under this law. Other legislation specific to California has established marketing programs with somewhat different organization. Table 2 lists current state marketing programs.

The general purpose of the marketing orders developed under California legislation is similar to that under federal orders. California producers and industry leaders supported state programs because they were administered locally in the state rather than from Washington, D.C., and because they permitted market demand expansion programs such as promotion, advertising, and education.

Despite the benefits to growers and marketing firms, organized consumer groups have in recent years opposed many features of marketing orders. California's milk marketing order, for example, has been modified considerably since 1974 as a result of consumer advocate groups and political opposition to the program. Also, public representation on state marketing order committees has been initiated to widen consumers' influence on board decisions. One criticism of marketing orders by consumer groups is that volume control of some commodities can serve to keep consumer prices higher than they might be in periods of surplus production. For purposes of orderly marketing, the volume of fruits or vegetables shipped through particular channels during a given period may be sometimes controlled to regulate the rate of flow to markets, or certain amounts of a crop viewed as surplus may be diverted to alternative outlets such as processing or export sales.

TABLE 15.2
Active Marketing Orders, Marketing Programs and Marketing Agreements, December 1980

UNDER AUTHORITY OF THE CALIFORNIA MARKETING ACT OF 1937

Alfalfa Seed Production Research Board
Apricot Advisory Board
Artichoke Advisory Board
Avocado Advisory Board
California Brandy Advisory Board
California Celery Research Advisory Board
Citrus Research Board
Egg Advisory Board
Dried Fig Advisory Board
Honey Advisory Board
Iceberg Lettuce Advisory Board
Melon Research Board
Manufacturing Milk Producers Advisory Board
California Milk Producers Advisory Board
Processors Clingstone Peach Advisory Board
Producers Canning Cling Advisory Board
California Potato Research Advisory Board
California Prune Board
California Raisin Advisory Board
Rice Research Board
Strawberry Advisory Board
Processing Strawberry Advisory Board of California
Fresh Market Tomato Advisory Board
California Turkey Industry Board

UNDER AUTHORITY OF THE AGRICULTURAL PRODUCERS MARKETING LAW

Pear Program Committee

UNDER SPECIFIC ACTS OF THE LEGISLATURE

California Beef Council
Dairy Council of California
California Avocado Commission
California Iceberg Lettuce Commission
California Kiwifruit Commission
California Table Grape Commission

Controlling the volume of shipment may or may not affect the total quantity of shipments. Its purpose is to avoid the emergence of market gluts and subsequent possible severe price cutting. As in the case of fresh lemons, such volume control more or less assures a more stable level of prices during the entire season—an advantage for the producer, but also, over time, for the consumer.

Trade Associations

California's farm industries are well organized. Commodity trade organizations perform a wide array of economic, political, educational, public relations, and market development activities. Trade associations represent many of the state's agricultural producers in testimony before public bodies as well as to the general public. Not all trade associations are commodity oriented. For example, the Agricultural Council of California represents farm cooperatives. While technically not trade organizations, general farm organizations in California perform some of the same functions as commodity trade organizations, and are active participants in matters affecting the marketing of farm products. (See Chapter 12 for more information on farm organizations.)

Government Programs for Consumer and Trade Protection

No modern marketing system could function for long without market regulations to assure equity and protection for both consumers and producers. State and federal programs provide assurance that: (1) producers are paid for products delivered to buyers; (2) grades and standards are met; (3) transportation firms perform their services adequately; (4) provisions exist for inspection both at shipping and at receiving points where disputes might arise; and (5) provisions are made for honest weights, measures, and packaging. Inspection also provides customers with assurance of product wholesomeness, sanitation, and safety for animal and plant products. (See Chapter 19 for detailed description of state programs.)

In recent years, grades and standards for fruits and vegetables have been criticized by some consumer groups which argue that some nutritionally acceptable products are removed from market channels simply because they lack physical characteristics required by law. Such grading is claimed to result in higher prices to consumers. Grade standards, however, also enable greater efficiencies in the marketing system. Buyers in distant markets such as New York and Chicago can purchase commodities in California on the basis of common definitions, bypassing the need for personal inspection. Grades provide a basis for communication that otherwise cannot be achieved when buyers and sellers are geographically dispersed. Cotton, live cattle and hogs, butter and dairy products, eggs, hay and virtually all farm products now can be traded with some confidence because grades exist to describe their condition or characteristics. Com-

bined with telephones, television, and other telecommunication processes, descriptive grades provide the means by which food and agricultural commodities are traded throughout the U.S. and internationally.

Trade protection also is provided through government processing plant inspection services that certify conditions essential for wholesome and safe foods. Another set of regulations establishes acceptable terms of trade and identifies restraints or unfair trade practices by which firms may gouge customers or drive out competitors. These antitrust laws, based on state and federal statutes, cover both state and interstate trade.

Increasingly, agricultural firms have come under the closer scrutiny of those who enforce these antitrust laws. Besides laws that regulate general terms of trade, such as the Sherman and Clayton Acts, there are special acts tha apply only to agriculture. For example, USDA may bring action in the cattle and poultry industries through the Packers and Stockyards Act. The U.S. Warehouse Act, regulating warehouse licensing and inspection, provides assurances of safe storage of products. The Produce Agency Act prohibits fraudulent practices in consigned transactions for perishable fruits and vegetables, dairy and poultry products. The Commodity Exchange Act regulates trading and pricing on the nation's commodity exchanges.

Food Distribution and Retailing

Retail Groceries

Food retailing is a part of the miracle of the U.S. food production-processing-distribution system, unequaled in any other nation. Supermarkets and convenience stores, both chain and individually owned, and even some "Pop and Mom" stores, comprise a system that in 1979 accounted for $222 billion in sales.

As supermarkets have increased their volume of sales during the past 30 years, they have exercised marked influence over the entire system of food marketing. Supermarket chains have led in the improvement of food packaging, shelf display, showcasing of frozen foods in open counter freezing units or closed vertical freezing cases, and other methods of product merchandising. They have achieved high efficiency of operation—a necessity since some chains' net profit after taxes is less than one percent of sales

The marketing and distribution system which moves food from farm to urban consumer is massive, complex, and efficient. *Photo by Charles Papp.*

(although about 16 percent on investment). The distribution cost of food has often been reduced as a result of supermarket ability to purchase products in large volume, store them in central warehouses, and transport them to retail outlets at scheduled intervals.

Other changes will occur as retail food companies seek to adjust to high labor, energy, and transportation costs. These include computer-assisted checkout systems (CACS) designed to scan Universal Product Codes (UPC), thereby shortening checkout time, providing complete shopping records for consumers, and providing better inventory management. Labeling foods and other items generically (i.e., as a class of product rather than by brand) is a move by retailers to reduce prices to consumers. Consumer demand for generic-labeled products in supermarkets has now reached 13 percent of food store sales, and continues to increase. Generic sales, along with chain store labeled merchandise, have sharply cut into the sales of packer brand items.

Another cost-cutting development is the emergence of bulk or "box" stores, where merchandise is displayed in original shipping cartons on pallets, or in bins or containers, with no shelf stocking. Most "box" stores eliminate the need for costly shelf stocking, as well as individual pricing of each item. Because of economies resulting from labor, energy, containers, and investment in costly shelving, grocery prices can be 20 to 25 percent lower than in a typical supermarket.

Food-Buying Cooperative and Clubs

A food buying cooperative is a business enterprise formed by a group of people cooperating to serve themselves. It is set up on the basis of limited interest on capital or investment that is returned to the investors, one member-one vote, and net profits refunded to co-op members in proportion to their co-op patronage.

A food buying club, appropriately known as a "preorder" co-op, should not be confused with a food buying cooperative, which is structured as a business organization. A club operates by banding together neighbors and friends, consolidating orders and funds to purchase food directly from wholesalers, and dividing up food purchases among those involved. The work done in pooling orders and money and buying, transporting, and dividing the food is shared by the members.

Food buying co-ops in California account for over $125 million in annual store sales. Although this seems like a large volume, it represents less than one percent of statewide retail grocery trade.

Away from Home Food Sales

Until recently retail stores accounted for most food sold to consumers. As changes in life styles, incomes, and numerous other factors occurred, meals consumed outside the home became a growing part of total food sales. By the mid-seventies, about 25 to 30 percent of the consumer's food dollar was spent for food away from home. That figure is projected to increase to 35 to 40 percent by the mid-1980s, but really depends upon a vital economy to expand. Much of the away-from-home food dollar is spent in fast-food establishments. In 1980 there were some 50,000 fast food outlets in the country doing a $15 billion business. Approximately one out of every four meals is now eaten out, compared to a decade ago when one meal in six was purchased or consumed outside the home. The rapid increase in such prepared food has many implications.

Corporations and entrepreneurs representing various sectors of the food industry (food manufacturers, soft-drink bottlers, general merchandise retailers, airlines, bus companies, and others) have been acquiring restaurants and other food outlets. Restaurant franchisers and fast-food chain operators have taken advantage of new sales and profit opportunities in non-metropolitan areas. Fast food operations have been viable in smaller communities with as few as 3,000 residents, especially when such units belong to a nationally advertised fast-food chain. In revitalized and redeveloped urban centers, new opportunities have encouraged fast food operators to open up new units.

Concentration in chain food service firms and eating establishments is increasing rapidly. Perhaps this concentration will be analogous to the supermarket chain control of much of the grocery retailing industry. While customers will be assured of uniform products, consistent quality, and rapid service, they also will face the dilemma that a buying committee located in corporate headquartes will decide what products will be made available to them, in what quantity and how prepared, and how frequently a product will be served. For farmers, this means corporate decision-makers can influence per capita consumption of some foods very significantly. For example, meals eaten away from home seem to have affected the total demand for canned fruits. Canned fruit is seldom served on an airline meal, or in many other prepared meals served elsewhere. Producers and food processors will be challenged to develop formulations of such foods to facilitate their use in commissary-type meals.

Direct Marketing

As defined in the California Farmer-to-Consumer Direct Marketing Act of 1976, direct marketing is "the marketing of agricultural commodities at any marketplace established for the purpose of enabling farmers to sell their agricultural commodities directly to individual consumers in a manner calculated to lower the cost and increase the quality while providing increased financial returns to farmers." In simpler terms, direct marketing is a means to move food from the grower/producer directly to the consumer, bypassing the handler in the middle. The role of the "middlemen" in their traditional performance of the transportation, distribution, and food handling functions is virtually eliminated. Direct marketing may come in the form of a farmer's market, U-pick programs, or independent roadside stands, or perhaps bulk sales which shorten the marketing link between producers and consumers.

The main purpose of direct marketing is to open up new markets for small scale producers. Though direct marketing has certain advantages, there are also associated problems. Consumers may have to add an extra step in time and distance to complete their shopping, or producers must find a way to deliver produce to a centrally located direct market outlet. Produce supply is often geographically restricted and highly seasonal. Producers who do not have resources available to transport and market produce may find that it is not economical to coordinate both production and selling activities. Those farming substantial acreages may still have to market through traditional channels since the volume of produce which can be sold through a direct market outlet is limited.

Marketing Problems and Pressures

The marketing of California's food and fiber products involves complex interrelationships, which create complex marketing problems. These can have a detrimental impact on all who depend on agriculture, whether producer or consumer.

Equity Between Buyers and Sellers

Problems can be internal to the marketing system, such as the determination of equitable terms of trade between buyers and sellers. The role of farmers in establishing commodity prices through bargaining cooperatives will continue to remain problematical because of increasing concentration in the food industries. Many firms like to dictate prices to growers, and resent grower involvement in this process. However, since farmers have so much investment in their individual enterprises, it is unlikely they will give up their rights to participate in price determination through group bargaining. The process has become complicated as structural changes continue to reduce the number of processors in California, and those which remain seek to coordinate supplies through contracts and other devices. Concentration results in fewer buyers, each of whom represents greater market power. In some cases, producers now have but one outlet for their production. A growing number of producers have been compelled to purchase processing plants considered unprofitable by previous owners. While fruit growers have experienced this for at least a decade, the same pattern appears to be developing for the meat industry too.

Some firms have launched "participation" plans as a substitute for specific prices to be paid to producers of fruits and vegetables. Although these plans vary, growers usually do not receive full payment upon delivery, but must wait until the commodity is processed and marketed before receiving full returns. In essence, the producer in this case is partially financing the processor's operations. Unlike true cooperative operations, producers often do not share in the profits of the private firm, so that returns are often lower than for members of cooperatives. Nor do growers

influence decisions of the boards of directors of private firms.

The question of equity in pricing will remain a major issue during the 1980s. An alternative, which has not yet received wide consideration, is for organized farmers to negotiate prices for processed products with retail and institutional distributors, and then to have their products processed on a custom basis by existing processors. This proposal—in concept resembling a public utility—could reduce the necessity for producers to invest in processing activity.

Pricing

Another problem arises when the number of cash buyers decreases. As farmers organize into cooperatives to purchase unprofitable proprietary plants, how will the true cash value of farm products be determined? Members of cooperative processing firms have relied on a "cash" price against which they measure their cooperative's performance. Cooperatives advance part of the farm value at delivery time, process and market the products, and—after deducting processing and marketing costs—pay their members the difference between earnings and costs. In a private firm this difference is considered to be profit. To function in this traditional manner, cooperatives need not determine the farm value. However, because there have heretofore been sufficient private or cash buyers, there was a basis of comparison for returns to cooperative members— which have been in line with (often above) prices paid to other producers.

Handling Perishables

Agricultural products are bulkier and more perishable than most other products. The bulkier the product, the more space it occupies relative to its value, and thus unit transportation and storage costs for agriculture are high. Perishability also affects handling of agricultural products. Produce, fresh meat, and dairy products must move into consumption centers as quickly as possible to minimize loss of nutritional and economic value.

Agricultural products are ordinarily handled several times while en route from producers to consumers. The handling and rehandling process at each point in the distribution process is expensive because it requires a tremendous amount of human labor, plus many mechanical handling devices. Product delay also may affect quality.

Use of cargo containers and unit loads (such as bins) may help attain greater efficiency in the handling of perishables, thereby reducing costs. Palletization may result in saving both time and money, and in minimizing product loss.

Transportation: Rates, Availability, and Quality of Service

Because California produces large proportions of the nation's supply of many specialty commodities, consumers throughout the nation are affected by the availability and quality of transportation services. The very survival of many California producers and industries is also at stake. Transportation costs and transit time affect the location of agricultural production and the market area, the quantity and quality of products shipped, the size and form of products marketed, and the mode of transportation utilized.

Transportation needs of agriculture are unique in that agricultural production is diverse and seasonal in nature, and hence requires a variety of transport systems. Transportation users have been concerned about freight rates for different modes of agricultural transportation, in addition to rail car shortages, branchline conditions and abandonments, inadequate weight-bearing capabilities of rural roads and bridges, and numerous other problems related to waterways and air transportation.

Freight rates differ depending upon the commodity, shipment size, services performed, distance of haul, value of the transportation service to the shipper, competitive routes and rates offered by other railways or other modes of transportation, damage losses, and market conditions. Railroad officials have emphasized ratemaking freedom as critical to their financial recovery and stability. On the other hand, users have stressed the importance of regulation in controlling the dramatic rise of rates and the availability of rail cars. The established rates for each mode of transportation, whether they are "fair" or not, have influenced the geographic pattern of agricultural production and distribution.

Trucking freight rates vary depending upon the type of truck hauling agricultural commodities. "Common carriers" are for-hire trucks operating on regulated schedules and routes. The trucking industry is regulated by ICC, as were the railroads until 1980; there is expectation that deregulation of the trucking industry will also follow. Trucks privately owned by manufacturing or marketing firms for their own use do not fall under the ICC regulatory power. In addition, the ICC provides that motor carriers which transport fish, livestock, and all agricultural commodities are not subject to federal rate regulations. This exempt privilege is designed to provide farmers a low-cost and flexible transportation system in which rates are usually established by direct negotiation between truck owner and farmer.

Rising transportation cost is a problem that may inevitably make California less competitive in the marketing of fruits and vegetables and other specialty crops, despite its climatic advantages. Since transportation rates are often determined by distance, the state is at a disadvantage compared to states closer to market.

In the U.S. transportation system, motor trucks hauled about 90 percent of the nation's fresh fruits and vegetables in 1980. Railroads currently haul only a fraction of the total produce tonnage. Only five-tenths of one percent of fresh produce is moved by air transport. Ship transportation also accounts for only a very small percentage of domestic movement, although in some regions and some commodities, barge movement is significant. Railroads may increase their share of shipments of perishables in the future, with a reduction of truck shipping. A major concern in the trucking industry will be the expected increase in highway user tax as the source of funds used to repair and maintain the interstate highway system. Another concern will be the availability of supply and the price of diesel fuel. In 1979 the cost-per-mile to operate an over-the-road rig reached 87 cents, compared to 59 cents in 1976. Legislation is also being considered relative to backhaul regulation. Unless non-regulated agricultural trucks have greater freedom to backhaul regulated commodities, their ability to compete is jeopardized due to higher costs of equipment, fuel, and repairs.

One of the changes occurring in the American transportation system is the concept of intermodal systems. The piggyback trailer (TOFC—trailer on flat cars) has become more popular as a result of rising transportation fuel costs and the rail carrier's inability to purchase expensive rail equipment. Because of the unpredictable availability of diesel fuel and gasoline and our dependence on foreign oil, shippers will look toward utilizing the more energy-efficient rail system—with motor carriers furnishing service to and from the ramps.

International Trade

California's agricultural industries could not survive without exporting large quantities of commodities and food to consumers in foreign lands. Export sales, which once served as a buffer outlet when production exceeded domestic demand, are now a specific part of the sales program for many industries and firms. In fact, for some crops such as rice, almonds and cotton, substantial acreages planted in recent years have been in direct response to international sales.

With the excess capacity of U.S. agriculture and the inability of third world countries to produce sufficient food for burgeoning populations, the export market will continue to be an important part of American agriculture. Agriculture is now the largest export industry in the U.S. economy. In 1979 agricultural exports accounted for 20 percent of the total merchandise exports of the U.S., worth $32 billion. Feed grains, wheat, and soybeans continue to be the largest share of exports. In 1979 the U.S. agricultural trade balance was a record $16 billion (agricultural exports of $32 billion less agricultural imports of $16 billion). Although the U.S. exports about 5 to 6 percent of its gross national product, U.S. agriculture exports one-half of its wheat production and one-fourth of its feed grain production—the production from one out of every 3.5 acres harvested on U.S. farms. Agricultural exports in recent years have contributed between 14 and 17 percent of total national farm income.

The export market is significant for a number of California agricultural commodities. Income from rice, wheat, cotton, and almond exports, for example, accounted for over half of farm income for those commodities. The total estimated value of California agricultural exports in 1980 was about $4 billion. Table 3 gives information on exports as percentages of production values.

Sacks of rice are loaded into ocean vessels at the inland Port of Sacramento. Many California commodities are exported to foreign countries. *Photo courtesy of the Port of Sacramento.*

TABLE 15.3
California's Agricultural Export Profile, 1980

Commodity	Farm-Level Value of Exports as a Percentage of Total Value of Production	California Export Value as a Percentage of Total U.S. Export Value
	Percent	
FIELD CROPS		
Alfalfa hay	4.1	37.7
Cotton lint	87.5	43.3
Dry beans	41.9	51.2
Rice	79.8	62.7
Wheat	78.1	4.2
FRUIT AND NUT CROPS		
Almonds	60.7	100.0
Apricots	18.7	100.0
Dates	35.9	100.0
Grapes, fresh	18.1	99.0
raisins	23.0	100.0
Grapefruit	29.0	13.6
Lemons	31.4	80.5
Olives	3.2	100.0
Oranges	25.8	58.2
Peaches, clings	10.9	100.0
Pears	6.9	50.6
Plums	13.7	79.9
Prunes	31.3	100.0
Walnuts	30.8	100.0
VEGETABLE CROPS		
Asparagus	16.0	25.5
Broccoli	10.8	98.6
Garlic	12.9	100.0
Lettuce	5.1	73.9
Onions	29.1	48.8
Strawberries	6.9	76.0
Tomatoes, fresh	10.9	29.5
processing	2.1	87.9
LIVESTOCK AND POULTRY		
Beef cattle and products	1.9	3.5
Dairy products	3.5	27.0
Sheep	2.5	9.3
Chickens and eggs	6.9	9.2
Turkeys	1.1	7.2

SOURCE: California Crop and Livestock Reporting Sevice, *Exports of Agricultural Commodities Produced in California, Calendar Year 1980*, August 1981.

As advantageous as international trade is for U.S. and California agriculture, producers are faced with trade problems which may have impact on farm incomes, employment, and local economics. These may be very briefly summarized as follows:

TRADE RECIPROCITY. Some countries actively promote their own food exports but restrict food imports from the U.S. An example is the European Common Market.

POLITICAL STABILITY. Large disruptions in foreign trade may be imposed by political considerations such as the Russian grain embargo, leaving producers and firms vulnerable to quick market changes.

EXPORT PRICES. Highly variable international economic forces influence export prices. Producers with high fixed costs must sometimes cope with widely fluctuating returns.

BALANCE OF EXPORTS AND IMPORTS. During years of less than average production, continued large exports could result in domestic shortages.

References

Some general references on agricultural marketing include *The Agricultural Marketing System* by V. James Rhodes, Grid Publishing Inc., Columbus, Ohio (1978). This collegiate text emphasizes the management options of agribusiness and farmers relating to marketing, integrating organizational structure and applied economic theory. Emphasis is on procurement systems as well as marketing systems. A second text is *Marketing of Agricultural Products*, by Richard L. Kohls and Joseph N. Uhl, (5th ed.) published by Macmillan. The aproach in this text is mixed—some description of functions, institutions, market levels, and some description of commodity marketing, mostly crops produced in the Midwest.

Collections of papers given at conferences sponsored by the North Central Regional Research Committee (in which the University of California, Davis, is a participant) have been published which provide current research reports on major marketing issues. Some specific publications are (1) "Agricultural Cooperatives and the Public Interest," N.C. Monograph 4, September 1978; (2) "Coordination and Exchange in Agricultural Subsectors," N.C. Monograph 2, January 1976; and (3) "Pricing Problems in the Food Industry," N.C. Monograph 7, February 1979.

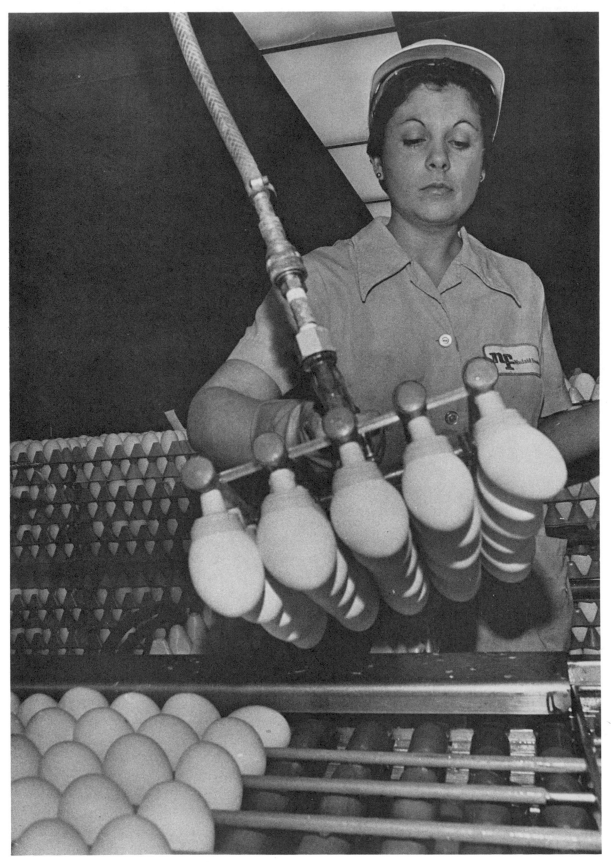

Mechanization has helped in many of the laborious tasks of preparing commodities for distribution. Here eggs are packed in flats after washing and grading.

16

Food Processing

Emil M. Mrak, Bernard S. Schweigert, and Ann F. Scheuring[1]

Since the discovery of agriculture, man has been able to produce food and in one way or another to preserve and process it for future use. Thousands of years of human history have seen the development of many methods of food processing. Today millions of urban consumers depend on a long chain of such steps in bringing food from distant farms to their tables.

Food processing should be called the twin sister of production agriculture since without it many products would not reach the consumer in safe, edible or palatable form. California's specialized agriculture, in particular, has historically been closely entwined with developments in the food processing industry. Only a small part of California's total annual agricultural production reaches the consumer in fresh form. By far the greatest part is processed and reaches the consumer in canned, frozen, dehydrated and other forms. Consequently, many highly perishable items are available for consumption throughout the year.

Most grain crops, and several other products such as olives and sugarbeets, must be processed to make them suitable for consumption. All of our meat products are processed to one degree or another. Even eggs are washed, graded, and packaged for protection. Much of our fresh produce receives some sort of treatment that in the broad sense may be called pro-

cessing. For example, chemicals retard or accelerate ripening of certain fruits after harvest, coatings preserve freshness, and precooling prevents enzymatic deterioration.

These food processing activities are the basis for a large business in California. According to the 1977 Census of Manufacturers, there were 2,583 food and kindred product establishments in the state, which provided employment for 162,000 people. These establishments purchased almost $14 billion worth of raw materials, primarily from California farmers, and added to them processing activities valued at over $6 billion. The total value of their processed product shipments was $20.2 billion.

California Processing Industries Since World War II

In the more than 30 years since the end of the Second World War, many changes have taken place in the food processing sector of California agriculture. These changes are partly contingent on the dramatic evolution of production agriculture in the state, and partly on improving technologies in the processing industry. The huge growth in population in California has produced vast new urban markets at home as well as elsewhere in the nation. Economic incentives have led some national companies to shift their manufacturing locales to California as production itself shifted toward the West (e.g., in the tomato processing industry). Then too, specialty crops produced largely or only in California logically

1. The authors gratefully acknowledge the assistance of A. Wade Brant, John C. Bruhn, Michael J. Lewis, Bor S. Luh, Robert F. Pearl, Robert J. Price, Lloyd M. Smith, and Eric Thor in developing this chapter. Comments by Kirby Moulton were especially helpful in preparing the final version.

TABLE 16.1
Trends in California Food Processing Industries

Item	Establishments			Employees			Value Added in Processing *million $*		
	1947	1963	1977	1947	1963	1977	1947	1963	1977
All food & kindred products	2,803	3,067	2,583	120,510	155,731	162,000	851.8	2,412.5	6,507.3
Meat packing plants	133	126	127	10,353	7,135	5,900	51.2	96.4	187.4
Poultry processing plants	39	89	32	927	3,822	3,600	3.3	34.9	109.9
Fish products (canned, cured, fresh, frozen)	60	58	57	8,286	6,181	9,000	53.4	72.0	261.5
Dairy products plants	176	431	263	4,554	19,124	10,000	32.1	262.3	414.1
Canned fruits, vegetables	218	192	145	26,379	25,060	26,300	161.1	321.3	847.3
Frozen fruits, vegetables	37	101	43	2,004	8,144	10,200	7.1	76.4	250.8
Dehydrated foods	87	120	84	1,701	6,360	6,400	9.9	67.0	246.1
Grain mill products	252	254	215	6,536	6,935	7,200	58.1	173.5	407.5
Bakery products	403	379	331	16,547	22,594	19,700	104.1	303.7	727.6
Sugar refining (beet)	10	11[1]	11	2,788	n.a.	3,100	21.4	n.a.	96.8
Cottonseed oil mills	n.a.	12	9	n.a.	1,033	900	n.a.	18.7	52.0
Malt liquors	16	15	11	4,144	3,740	2,600	48.4	90.1	184.5
Wines and brandy	218	131	165[2]	4,650	4,049	6,900	68.9	92.0	406.6

[1] Information from California Beet Growers Association.

[2] The Department of Alcoholic Beverage Control reports 333 winegrowers licensed to make and sell wine in 1977, plus 26 brandy makers. For 1980 the Department reports 435 licensed winegrowers plus 20 brandy makers. The discrepancy between Census and A.B.C. figures might be accounted for by size of operation reported; census figures are based on operations with 5 or more employees. Rapidity of increase may also be a factor.

SOURCE: U.S. Department of Commerce, Bureau of the Census: *Census of Manufactures*, Statistics by States, California (1947, 1963, 1977).

required local processing. New technology developed through public and private research has resulted in increased ability to handle volume, and developed many new and improved products appealing to consumers.

At the same time that volume and sales were generally increasing, changing consumer tastes had great impact on which products sold or failed in the marketplace. Economic pressures have forced a gradual consolidation of smaller operations into bigger ones. Table 1 gives a condensed record of some trends in various categories of food processing over the thirty year period since World War II.

Overall, value of production in the food processing industries of California has enormously increased. The total value of shipments as well as value added in manufacture of the raw product has increased consistently, and substantially more than the rate of inflation during the same period. Although the total number of establishments is down since the early 1960s, total employment is higher, even in a·period of greatly increased automation.

Within individual categories, trends differ. Canneries and bakeries have significantly declined in number; there are also fewer meat packing plants, grain mills, and breweries than 30 years ago. The frozen food industry, after phenomenal postwar expansion, experienced a significant amount of consolidation. Poultry processing and dairy products plants also reflected this surge and later decline in numbers. In contrast, in the last decade wineries have multiplied as acreage of grape varietals has expanded. Wineries may be one of the only processing categories where smallness of output is still prized and product differentiation is well rewarded. With a strong expanding market for California wines, the wine industry seems to be an exception to the trend toward more and more consolidation.

Fruit and Vegetable Processing: An Example of Change

California's specialty fruits and vegetables are often thought of as destined primarily for the fresh market, to be processed, if at all, by the consumer at home. Yet, as can be deduced from Table 2, a substantial portion of our fruits and vegetables is allocated to processing markets. Approximately 84 percent of the tonnage represented in Table 2 was processed in 1978.

Allocation between fresh and processing markets is explained not only by crop size and relative perishability, but by numerous economic and noneconomic factors including relative prices in the markets, institutional arrangements within the industry, and the number of alternative product forms (juice, preserves, fruit cocktail, prepared dishes, etc.).

Table 2 also indicates how the allocation to processing may vary between different crops. A high percentage of the state's total apple, apricot, fig, grape, peach and pear crops was sent to processing plants while the bulk of the California sweet cherry, nectarine, plum, and strawberry crops was sold fresh.

It should be noted that varieties of some fruits

TABLE 16.2

Utilization of Selected California Fruits and Vegetable Crops (1978)

Item	For Fresh Market	For Processing (canned, crushed, frozen, dried)
	Short Tons	*Short Tons*
FRUITS		
Apples	70,000	180,000
Apricots	6,500	116,500
Sweet cherries	11,000	3,000
Figs	3,700	26,700
Grapes, all	412,000	3,357,000
Nectarines	144,700	3,300
Peaches, all	114,300	637,700
Pears, all	46,500	241,800
Plums	151,000	3,000
Prunes	—	132,000
Strawberries	193,049	64,950
VEGETABLES		
Asparagus	26,450	12,750
Broccoli	126,550	131,349
Carrots	313,498	113,299
Cauliflower	68,600	84,950
Spinach	22,250	71,049
Tomatoes	376,748	5,289,650
Snap Beans	15,000	16,050
Cucumber	36,050	61,600 (pickles)
Green peas	1,800	10,050
TOTAL TONS (all fruits and vegetables)	2,139,695	10,424,798

SOURCE: *California Fruit & Nut Statistics, 1978–79* and *California Vegetable Crops, 1978–79*: California Crop and Livestock Reporting Service, Sacramento.

A variety of canned tomato products is displayed at the annual California Canners and Growers meeting.

and vegetables are bred specifically for processing. For example, the fresh market tomato is different from the processing tomato which is smaller, firmer, and has higher acids and solid content. Wine grapes are raised exclusively for processing while Thompson Seedless grapes may be shipped fresh, dried for raisins, or made into wine.

The Canning Industry

The national fruit and vegetable canning industry, widely dispersed geographically, is made up of approximately 1,000 canning establishments in the United States.[2] Industry sales of canned fruits and vegetables in 1978 amounted to $5.5 billion. A relatively small number of fruits and vegetables account for a very large proportion of the total volume canned. Five products account for about 70 percent of the fruit canned; four products account for over 45 percent of canned vegetables.

2. Information in this section is largely taken from an unpublished paper on the canning industry prepared by Dr. Eric Thor for the Assembly Select Committee on California Automotive Industry and Plant Closures: May 13, 1980.

In 1941, California canners packed 80 million cases of canned goods; by 1977 volume had increased to 153 million. California currently packs about 44 percent of all U. S. output of canned fruit, and 35 percent of canned vegetables.

The 1980 Directory of the Canning, Freezing, and Preserving Industries lists more than 250 individual processors in California, distributed widely throughout the state. These range from relatively small local family operations specializing perhaps in a single product (frozen fish cakes, olives, apples, etc.) up through the giants of the industry who pack millions of cases in a diversified line through half a dozen or more packing plants. For example, of Del Monte's huge organization, 30 U.S. factories besides packing facilities abroad, 10 are situated in California. Other national food processors, such as Hunt-Wesson and Libby, McNeill & Libby, also operate several California plants. Two of the world's largest cooperative canneries are California Canners & Growers and Tri-Valley, both organizations operating seven plants in 1980.

Because canning of fruits and vegetables is a relatively low profit industry, many of California's earlier small canning companies have either become cooperatives, become part of a multi-national firm, or gone out of business. Like many other segments of the food processing industry, for example, most olive canners began years ago as small family-owned operations. Their number has greatly decreased in recent years for a variety of reasons, including economic conditions in the industry and numerous consolidations involving mergers or the sale of one company to another. Thirty-five years ago there were 39 olive packing companies processing a harvest of approximately 25,000 tons. Today, the number of olive canners has dwindled to only 7. Their combined processing capability, however, has grown substantially during the same time period, and in 1980 the industry processed an olive crop of approximately 120,000 tons.

There are several reasons for the low profits canners receive: 1) lack of control over annual volume packed and year-end inventory; 2) increased marketing of canned fruits and vegetables as a commodity without the processor's label; and 3) competition between fresh and processed products. Canners do not have great control over the volume of various fruits and vegetables canned because of the annual variations in

Olives, delivered for processing to a Corning plant, are immediately dumped into water to clean away dirt and debris.

yield per acre at the farm level. Although some canners try to exercise control over the volume by purchasing only specific quantities from farmers, others have agreements to pack all that is produced on a specific number of acres. The major evidence that canned fruits and vegetables have become a commodity business is a decline in the importance of the processor's advertised label in comparison to the increase in volume marketed under the retail store's or distributor's private label. In addition, the percent of total volume packed for and sold to institutional buyers and further processors has increased substantially since the 1960s. During the late 1950s and early 1960s many retailers purchased their private label products from the smaller canners. During the 1960s many of these small processing plants either went out of business or consolidated with other firms into larger and more efficient processing plants. Meanwhile, rising costs of canned goods and greater consumer preference for fresh or frozen produce have combined to cut per capita consumption of canned fruits and vegetables.

Another economic factor leading to severe financial losses for processors has been overproduction in some crops—for example, peaches and pears. Also, costs of labor, transportation, warehousing, containers, financing, replacement of equipment and facilities, energy and other services have increased. Higher prices for canned commodities have led to consumer resistance and product substitution. Squeezed on both sides, fruit and vegetable canners have increasingly opted for some form of consolidation.

The Freezing Industry

The golden anniversary of the freezing industry was celebrated in 1980, exactly 50 years after Clarence Birdseye began freezing green peas in the New York-Boston area, as an offshoot of freezing fish. It was Birdseye who discovered how to protect the quality of frozen vegetables by deactivating the enzymes in blanching before freezing.

Until 1940 the frozen foods industry remained small. World War II, however, hit the canning industry hard because much of the supply of

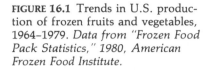

FIGURE 16.1 Trends in U.S. production of frozen fruits and vegetables, 1964–1979. *Data from "Frozen Food Pack Statistics," 1980, American Frozen Food Institute.*

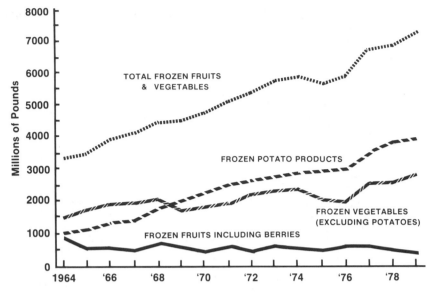

metal and tin was requisitioned, whereas the freezing industry could use cardboard containers. Thus a major boost was given freezers in competition with canners.

Nationwide, large numbers of firms entered the freezing business during the 40s and 50s as refrigeration facilities, trade associations, and marketing systems became established. There were quality problems in the adolescent industry, however, and by 1970 nearly 80 percent of the early processors had gone out of business. In spite of consolidation and adjustments, production of frozen fruits and vegetables doubled between 1964 and 1980. Although frozen fruits, including berries, experienced some decline, total frozen vegetable production soared from 3.8 billion to 6.8 billion pounds in the 15-year period. Production of frozen potato products jumped phenomenally during the same period, and in fact surpassed all other vegetables combined in total volume (see Figure 1).

Frozen fruit and vegetable processing is now a mature industry. In California in 1980 there were 11 major firms in the commercial freezing business. These included five stock companies with several plants and six independent firms processing full lines. Another half dozen firms processed single products or relatively narrow lines. In a comparatively low-margin business, the very large firms have experienced considerable attrition, while the independents and cooperatives have been better able to adapt to economic pressures because they do not have to pay dividends to stockholders and do not have the high over-

head of very large organizations. Over-production for current markets is a continuing problem, and further attrition among freezing firms may be seen in future.

California now produces about one-third of the nation's total supply of frozen vegetables, the Nothwest another third, while the remainder comes from across the country. The most important frozen vegetables packed in California are asparagus, green beans, lima beans, broccoli, Brussels sprouts, carrots, corn, green peas and spinach.

Energy expenditures in the frozen food industry are great, particularly for storage and transportation. A recent Cornell study comparing systems in the canning and freezing industries, however, concluded that production of frozen foods is actually somewhat less energy consumptive than that of canned foods, primarily because the energy cost of containerization (manufacture of wax-wrapped paper boxes or plastic pouches compared with tin cans) is significantly less.

The Dried Fruit Industry

For many years the Santa Clara Valley was a center of the dried fruit industry in California, until encroaching urbanization wiped out many acres of orchards and forced a shift in production to the Sacramento and San Joaquin Valleys. The primary fruits which are processed by dehydration are raisins and prunes, with about 200,000 and 150,000 tons respectively being processed annually. Small quantities of peaches, pears, figs, and apricots are also dehydrated. Though some small drying operations still exist, the industry has been characterized by the kind of consolidation seen in the canning and freezing industries. Sun Maid Raisins and Sunsweet Prunes, both cooperatives, are the giants among the dried fruit processors.

Some Trends in Meat, Poultry, Dairy and Seafood Processing

Meat and Poultry Products

Cattle and calves are the second most important agricultural commodity in California, while chickens, turkeys, and eggs combined are fifth in importance. These commodities, with the exception of eggs, require the conversion of a living animal into human food. This is food processing in its most irreplaceable form. It is simply not possible to pluck these products from the field or tree and eat them raw as can be done with many of the fruits and vegetables.

With few exceptions, federal laws require that all slaughtering and processing of livestock and poultry be conducted under inspection by the U.S. Department of Agriculture. There are over 700 establishments in California now receiving such inspection.

A major trend in shipping fresh beef as "boxed beef" rather than in carcass form has occurred throughout the United States in recent years and this trend is expected to continue. In this system, carcasses are cut into primal and sub-primal cuts at the slaughtering plants, packed in cartons, and shipped to retail outlets. Final cutting and wrapping are performed at this level. The excess fat and bones are utilized at the slaughtering plant while costly transportation charges are avoided.

While some pork is produced in California, most is shipped in from the Midwest for sale as fresh product or, more importantly, used in the form of cured and processed products (ham, bacon, sausages, luncheon meats).

Animal and poultry products are highly perishable. At ambient temperatures meats remain edible and safe for only a day or two. Numerous methods of preservation, such as refrigeration, heating, canning, curing, salting, and drying, are some of the most intricate and demanding in the entire food field. California is particularly noted for its production of Italian Dry Salami. The Uni-

versity of California has contributed in a major way to the understanding of this ancient art. Microbiology of meat and meat products has also received much attention by University researchers.

Over 50 percent of all turkeys now reach the consumer in a form requiring the meat to be removed from the bone. This remains a labor-intensive process, but productivity has been measurably increased through equipment and work method innovations. Many of the traditional red meat products such as frankfurters, bologna, pastrami, salami and ham have been successfully produced from turkey meat and to a lesser extent from chicken meat. These products have met with wide consumer acceptance.

The washing of eggs has become a standard practice, improving the microbiological condition and salability of California eggs. Through diligent attention to efficiency and the rapid incorporation of new technologies, poultry and eggs remain one of the least expensive sources of animal protein for consumers.

Dairy Products

California is the second leading dairy state in milk production (12 billion pounds annually), following Wisconsin with about 20 billion pounds. California cows produce more milk than those in other states (about 14,500 lbs/cow), and California continues to produce increasingly more milk annually while maintaining a fairly stable cow population (Figure 2). California also has the highest herd sizes (est. 350 cows/herd) in the nation; these herds are predominantly located in the San Joaquin Valley (50 percent of milk produced) and Southern California (35 percent).

The number of processing plants, manufacturing both fluid milk products and by-products such as cottage cheese, butter, skim milk powder, continues to decline. Much of the loss has been among small (less than 5,000 gallons a day) processors, many of which were located in Southern California, although some large processors have also closed obsolete facilities. In some instances other major California processors have purchased the closed facilities and/or markets. The volume of milk (sales) processed by the remaining plants has consequently increased.

Fluid milk processing plants in California are principally located in the San Francisco Bay area and in southern California. Nearly all of the

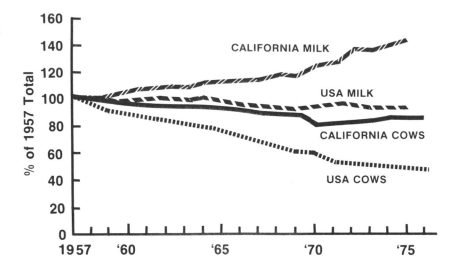

FIGURE 16.2 Trends in cow population and milk production, U.S. and California, 1957–1977. *Source: J. Bruhn.*

manufacturing product plants (e.g. butter, powder, cottage cheese) are located in the San Joaquin Valley; for example, five cottage cheese plants in this area produced about 90 percent of the state's total production and 12 percent of all U.S. production during the 1970s.

In California more milk is sold through supermarkets than in any other state and the proportion sold through these outlets continues to increase. Home delivery and cash-and-carry operations (drive-in dairies) handle less milk than a decade or two ago. Accompanying the increased supermarket sales has been a marked change in fluid milk sales by carton size. In 1953, half-gallon sales represented 16 percent of the total, while sales in quarts were 70 percent. Today only about 5 percent of sales is in quarts, while 70 percent is in half-gallons. Sales in gallon units are also steadily increasing.

With the increased milk supply has come a marked increase in the production of manufactured products. Where California once had to import butter from other states, four principal plants now manufacture the entire state's supply (130 million pounds annually).

Although until quite recently California was not an important cheese manufacturing state, "jack" cheese originated in Monterey County about 1892, as a way of disposing of surplus milk. Monterey jack cheese is still a state specialty. Since the 1960s there has been rapid increase in all types of cheese manufactured in California. In 1969 nearly 14 million pounds of cheese were manufactured in the state; by 1979 production had surpassed 150 million pounds. Although two cheddar cheese plants were installed in the 1970s, Italian cheese, principally Mozarella, has consistently accounted for most production. California has a very limited production of other specialty cheeses, including a very fine Camembert.

The California Seafood Industry

Over 170 species of fish and shellfish are processed in California. The processed value of these products in 1977 was over $647 million, ranking California number one among the states for value of seafood products. The major products produced are canned tuna, industrial products (including fish meal, oil and solubles), canned pet foods, shrimp products, fish sticks and portions, and fresh and frozen fish fillets and steaks.

During 1977, over 1,045 million pounds of seafoods were processed into pet foods, industrial products, and food for human consumption, an increase of 31 percent since 1970, but far below the record 1.76 billion pounds processed in 1936. The total amount of fish and shellfish processed declined steadily between 1936 and 1968, but has been increasing since 1969. The fluctuations in the amount of seafood processed during the last 40 years have been due primarily to variations in the amount of fish processed into industrial products, since production for human consumption has remained relatively constant since 1950.

Although California is a net exporter of canned tuna, pet food, squid and sea urchin roe, its fishermen supply less than one-half the state's total consumption. The state is a net importer of haddock, cod, halibut, lobster tails, shrimp and scallops. Of the major species consumed, only in canned tuna, sole, rockfish, fresh salmon and Dungeness crab are California suppliers a dominant factor.

According to *Current Fishery Statistics* (No. 8000: U.S. Dept. of Commerce), in 1978 there were 184 plants processing and wholesaling seafoods in California, with about 12,272 employees. Between 1970 and 1978 the number of processing plants decreased by 9 percent while the number of employees increased by 33 percent.

Major Methods of Processing Agricultural Products: A Review

Having considered some of the evolution of processing industries in California, we now turn our attention to descriptions of basic processing methods. Although food preservation and conversion is millenia old, in the last 50 years it has reached a level of sophistication in technology never before known. Simple processes such as dehydration, fermentation, and milling have become increasingly complex. The following sections briefly describe some of the major developments in food processing, particularly with reference to California.

Dehydration

Drying of fruits in California has been important since the early days of the state's agriculture. The basic forms of drying used for fruits are sun drying and hot air dehydration. Cut fruits, such as apricots, peaches and pears are generally still dried in the sun after they have been sulfured. When the sun drying process is used, the trays are stacked after two or three days so that the drying process is completed in the shade.

Sun drying of light colored fruits tends to darken the fruits because of enzymatic action and chemical deterioration. This can result in off-tastes, the loss of nutritional value (principally vitamin C) and a poor appearance. Sulfuring inhibits this process. However, the fruits do retain varying amounts of sulfur dioxide after drying.

Most grapes are still dried in the sun and produce the familiar dark raisin. Raisins are dried on paper trays placed in vineyard rows. The drying time may be two weeks or more which subjects the crop to considerable risks from bad weather.

Some fruits are dehydrated using forced draft air. Prunes are the primary example in Califor-

nia. The dehydrating process uses hot air which is forced over trays of fruit at the rate of several hundred feet per minute. The drying process takes place at a higher temperature and in a shorter period of time than sun drying. Figs are also dried in dehydrators, without pretreatment. Some grapes are also dried in dehydrators, after being dipped in a mild lye solution and then sulfured. The required drying time is about 18 hours. The end product is "golden bleach" raisins which are very light in color.

Apples are commercially dried in evaporators which are buildings with slanted floors that have heaters beneath. Hot air moves up through the room where the apples are several inches deep on the floors. The apples are shoveled over during the drying process, which may take 24 to 36 hours.

A completely different situation exists in the case of vegetables. Important treatment prior to drying of most vegetables (not onions or garlic) is blanching in steam or hot water. This is usually done by exposing the peeled, washed, cubed, sliced or shredded vegetable to steam over 200° F for several minutes. This inactivates peroxidase and other enzymes and helps prevent subsequent deterioration and development of off flavors. The vegetables are then generally sent through an air blast dehydrator similar to those used for fruit. The moisture content of the final product, however, is much lower. A few vegetables such as cabbage, potatoes and carrots are briefly immersed in a sulfite solution after blanching and prior to dehydration in order to improve their storage characteristics.

Onions and garlic are generally dried immediately after slicing without any other pretreatment. After that, they may be ground into onion or garlic powder or at times mixed with salt.

Most vegetables are dried to about 5 or 6 percent moisture content. Drying down to a 2 percent level greatly enhances their storage life but is very costly and entails the risk of heat damage.

Many other foods are dried in one way or another. Milk is spray-dried to make powder; cereals are dried before storage to prevent natural fermentation; even yeast is dried.

Concentration is also a type of moisture removal process. It may be used in combination with freezing. Orange juice concentrate is probably the best known example, but other products

Owners of an apricot orchard inspect trays of fruit laid out to dry in the sun. Dried fruits were one of the earliest of California's agricultural exports.

such as tomatoes and evaporated milk are also processed by concentration.

Fermentation

Like dehydration, fermentation is an ancient natural process used to convert some foods to a more useful form. Controlled fermentation is used in the production of such foods as cheese, pickles, sauerkraut, olives, vinegar, certain sausages, and yeast.

Fermentation is also the underpinning of the great California wine industry. Some of the advances in wine-making which have been pioneered by the Department of Viticulture and Enology at the University of California at Davis are described in Chapter 7.

Several California commodities have also been used for brewing beer, including barley, hops, and rice. A miniature brewery in the Department of Food Science and Technology at Davis has conducted studies on the brewing properties of these California crops.

The Use of Heat

CANNING. The first food cannery in the United States was started near Boston over 130 years ago. Spoilage of canned goods, however, was commonplace in the early days. As the industry grew, researchers helped solve some of the problems encountered in the canning process. Research showed that bacteria caused spoilage of canned foods and that heating to a temperature above the boiling point of water was needed in order to kill the bacteria. Research also showed the necessity for thorough heat penetration in the canned product, and the importance of cooling canned foods after heating. The spoilage known as "flat sours" was found, to a large extent, to be caused by improper cooling after processing.

Eventually engineering and technology made great advances. The development of the thermocouple for measuring temperature became an important item in studying heat penetration, and the development of more accurate and reliable thermal processing times took place. The spiral type continuous cooker was developed for both pressure and non-pressure cooking of fruits. Later an aseptic canning procedure was developed for soups, milk products and pumpable materials. A more recent development is the so-called "flame" sterilizer, which uses direct flame heat on the cans as they are being rotated,

thereby achieving more rapid processing. The new "steri-vac" process permits the canning of products without added liquid, under vacuum using flame sterilization. These new processes still under development may very well result in improved nutrient retention and better texture and taste characteristics.

Problems still being studied today include the better retention of nutritive values and avoidance of lead contamination. Research has been done on two piece cans, which eliminate the soldered seam.

PASTEURIZATION. Pasteurization is a heat treatment of food intended to destroy organisms dangerous to health, or to destroy those which cause food spoilage or interfere with desirable fermentation. It does not mean that the products are sterile, but that undesirable organisms are to a large extent destroyed and the product can be handled and held for a considerable period of time when kept under refrigeration.

THE ASEPTIC PROCESS. This is a variation of pasteurization although it really involves sterilization. It is the so-called "high temperature-short time" (HTST) treatment used for liquids, particularly milk products. The liquid is exposed to a very high temperature for a very short period of time and then rapidly cooled and filled into containers under aseptic conditions. The temperature may run well over 230°, 240° or even up to 250° F. The advantage of this process is that though it does sterilize the product, the period of heat treatment is so short and rapid that it does not materially alter flavor and other characteristics of the material being sterilized.

Refrigeration and Freezing

Natural refrigeration, such as cold water, ice, snow, or naturally cold atmosphere, has been used by human beings for eons. The actual invention and exploitation of practical means of producing mechanical refrigeration, however, took place in the last century. Improvements during the past 50 years have been dramatic. The development of the frozen food industry was at first slow because it entailed not only the production of frozen items, but installment of complete distribution systems including refrigerated trucks, freight cars, cold lockers and cabinets in grocery stores, and refrigeration units for the home. Mechanical refrigeration has, however, come into common household use, and refrigerated transportation has revolutionized the national food system. This has been especially important as a means of preservation for dairy products, meat, eggs, fish, and other animal products, but it has also been important for fruits and vegetables. The development of reliable mechanical refrigeration systems for ocean transport was a great advance for global food distribution. Refrigerated cargo ships have, for example, enabled the shipment of chilled and frozen beef from Argentina and Australia to Europe and elsewhere.

STORAGE OF ANIMAL PRODUCTS. Low, but not freezing, temperatures are used extensively for holding of animal products. Interestingly, if a carcass is stored under refrigeration for a period of time, a natural tenderizing process seems to take place.

Storage life of meat products varies depending on the nature of the product, degree of sanitation, and the temperatures used for storage. Beef, lamb or chicken can be kept for a year or more when properly frozen and protected from dehydration. Frozen fresh pork has a shelf life of approximately six months. Meats kept under refrigeration only have a much shorter life. Cooling techniques also permit storage of shell eggs for many months if the humidity of the storage is not too high. This enhances the ability of egg users to maintain inventory against production needs. Refrigeration is also an essential part of the production and distribution systems for milk and dairy products. They are often shipped and marketed at a temperature of about 32° F and have a shelf life of one to three weeks or more under normal conditions.

The development of efficient production and handling facilities for frozen animal products has greatly enlarged the range of markets which California producers can serve. For example, the ability to preserve turkeys through freezing has permitted California producers to smooth out their production cycle and to serve distant markets in competition with producers from out of state and more central locations.

STORAGE OF FRESH PRODUCE. Many fresh fruits also benefit from the use of refrigeration, though some (for example, bananas) may be damaged by chilling. Large quantities of apples, pears, grapes, citrus and other fruits, however, are stored under refrigeration after harvest, for shipment and distribution at later times. Some fruits are benefited by controlled atmosphere (oxygen and carbon dioxide levels) in the storage room.

Ventilation in storage is important, since fruits naturally emit ethylene gas, which is a ripening agent. If this natural gas accumulates, ripening may take place even at lower temperatures, with subsequent deterioration as a result. Plant physiologists have recently found means of reducing ethylene production in cases where it may be deleterious. In other cases, such as melons shipped to the east coast, controlled ethylene gas may be used in refrigerator cars to accelerate ripening.

Fresh vegetables with high respiration rates such as asparagus, peas, corn and lettuce are benefited by chilling for holding following harvest. This minimizes loss of moisture and certain chemical changes. Large quantities of onions and potatoes are also held in cold storage for long periods of time in the U.S. each year.

Packaging of fresh produce sometimes poses a dilemma. Packaged produce stored in bins over mild refrigeration holds up well, while unpackaged produce tends to wilt and lose moisture and quality. Consumers, however, tend to prefer bulk purchasing over packaged items to enable them to pick and choose individual items. Some retail markets now use packaging primarily for items like cabbage where single heads may be enclosed in plastic material, or use waxing to retain moisture in rutabagas or cucumbers.

FREEZING VEGETABLES AND FRUITS. Vegetables are blanched before freezing in order to inactivate certain enzymes. The product is then cooled, placed in packages and frozen by the air blast method. Packaging is extremely important, since it must be relatively moisture proof and protect the product from oxidative deterioration.

Some fruits are frozen after peeling, pitting, and packaging, or covering with a syrup. Some are frozen in cubes without any additions and used by manufacturers of jellies and jams. Some may be packed with sugar and subsequently used by pie bakers.

Milling

The conversion of grain into a relatively fine meal for cooking is one of the most ancient of food industries. Today's highly automated American milling industry has evolved from these age-old processes. The American industry is centered close to the great grain-producing regions of the central and plains states. Though the milling industry in California is relatively small by midwestern standards, grain mills here process wheat brought in from out-of-state as well as the grains grown on over a million acres of California farmland.

WHEAT MILLING. Seventy-five percent of wheat grown in California is hard red wheat while about 100,000 acres of land in the Imperial Valley are used for the production of durum wheat. Wheats differ in milling and baking properties. The protein content of wheat determines its suitability for use. Most California wheat is of a lower protein content than the varieties grown in northern and midwestern states. It is preferred for making cookies, cakes, and other bakery products, while wheat with higher protein content is preferred for making bread.

In the manufacture of white flour in a modern mill, the aim is to separate the endosperm of the grain from the bran and germ, followed by pulverizing. The steps involved in wheat-to-flour production are wheat selection and blending, cleaning, conditioning or tempering, breaking, bolting or sieving, purification, reduction and bleaching. Wheat is conditioned by adding water to the grain several hours before it goes to the rolls. This toughens the bran and mellows the endosperm. The milling process consists of a gradual reduction in particle size through a series of rolls. Some of the endosperm is lost and some bran and germ are present in certain flour streams. The variation in purity give rise to different grades of flour. Vitamins and minerals may be added for enrichment purposes. The problem for the flour mill is how to most efficiently blend the various flour products to achieve a profitable product mix. To improve their capabilities, modern mills in California have adopted computer analysis to optimize their production streams and have continually improved the equipment in their mills.

RICE MILLING. Rice is a very important product in California and provides substantial income to both growers and rice millers. In contrast to wheat milling, where the object is to grind the kernel to produce flour, the objective in rice milling is to reduce kernel breakage to a minimum.

Paddy rice, as received from the field, is put through precleaning machines to remove impurities such as straw, soil, stones, weeds and insects. Aspirators are used to remove the hulls separated from the kernels by the shellers. The kernels are then separated into large and small sizes and sent to the pearler, which removes most of the bran by abrasive action. The rice kernels can be further polished to remove all traces of bran to produce a shiny white surface. Since milling removes the germ and bran layers, polished rice is reduced in nutritional value. This loss may be offset by means of vitamin, mineral and protein fortification.

A new rice processing technology, known as solvent extraction rice milling, is now in commercial use. Bran removal is conducted under more gentle conditions than the usual pressure machinery method. This creates much less rice kernel damage than does dry milling and yields are considerably higher. The products obtained by the solvent process are white rice, defatted bran, and rice oil. The white rice has improved color but no differences in eating quality. Defatted rice bran is free-flowing and light colored with a bland flavor. It has potential for food uses. Rice oil has been used for quality cooking and salad oil.

CORN MILLING. Relatively little corn is grown in California, and most of what is produced is used for feed. However, there is a growing interest in corn products for various purposes, and this demand has led to considerable advancement in the techniques of corn processing.

Almost all corn prepared for human consumption is dry milled. This involves cleaning, tempering, degerminating, drying, cooling, grading, aspiration, grinding, sifting, purification, packing and oil extraction. The corn kernel, because of its size and shape, is more difficult to mill than wheat. In the early days, corn was ground on stone mills and the meal produced often was not sifted and therefore contained the germ. Recent changes require the removal of germ and the production of low fat (less that 1 percent) grits or flour of specific granulations depending on use.

A wet milling process is used to obtain starch and oil. The corn is steeped in water to about 45 percent moisture in the presence of sufficient amounts of sulfur dioxide or sodium acid sulfite to inhibit microbial growth. The softened corn is passed through a degerminating mill which separates the germ by a tearing action rather than by grinding.

In addition to traditional corn products, many new corn products such as snack foods, breakfast foods and industrial items are being produced in increasing amounts. Although this trend has not yet had much impact in California, the increase in corn syrup production is more significant. Already, corn syrups have captured a large share of the market for sugars used by the food processing industries. To the extent that such processing of corn for sugar becomes profitable in California, it could provide an incentive for considerable expansion of the state's corn acreage.

Sugar Refining

Sugar refining in California is important to the growers of California sugarbeets. Farmers plant approximately 250,000 acres of sugarbeets annually in California and vary that acreage considerably depending upon the contracts offered by sugar refiners. Although the sugarbeet industry has been protected through various government programs in the past, it is subject to intense competition on the world market. The resulting refined sugar prices place a great emphasis on efficient refining techniques.

On arrival at a sugar refinery, sugarbeets are washed and sliced into long slender shreds or "cossetts." The cossetts are whirled through a steam bath and dropped onto a conveyor belt which passes to a diffuser. The diffuser uses a continuous stream of water at 70 to 80 degrees centigrade to remove the sugar from the cossetts.

The remaining beet pulp is converted into cattle feed while the sugar solution is sent on to the carbonation and settling processes which remove most impurities and leave a thin golden juice containing about 10 to 11 percent sugar.

This solution is further concentrated in a series of evaporators and the resulting liquor is filtered and pumped into boilers where it is boiled at reduced pressures and temperatures to prevent caramelization. Sugar crystals are formed after this solution has cooled and they are separated from the liquid in a centrifuge. Finally the damp sugar crystals are dried, sorted, and packaged. The total process requires about six hours.

Over half of the beet sugar produced is granulated, suitable for table use, canning, jelly making, preserving, cake baking and candy making. Other forms of sugar are liquid sugar for canners, brown sugars, cubes, confectioners powdered sugar, superfine and bakers' special sugars.

Oil Processing

The edible oil industry of California has a long history and a wide variety of sources have been used. These include olives, avocados, cottonseed, safflower seed, sunflower seed, grape seed, almonds, walnuts, rice, and sesame. Edible oil is also extracted and/or refined from imported sources, such as coconut, palm, palm kernel, and soybean. For some crops, such as safflower and sunflower, oil is the principal product for which the crop is grown. For other crops, oil production is incidental to the primary utilization.

The most important California source is cottonseed which is processed by several plants in the San Joaquin Valley. Interest in the processing of flax (linseed oil for paint), and of safflower and sunflower seeds for edible purposes during the past 50 years has varied with market prices of competing crops such as wheat, corn, and sugar beets. Processing of oilseeds to produce cooking, salad or margarine oils requires a complex series of steps including seed preparation, oil extraction, and refining of the crude oil. Special techniques are available for processing each type of oil.

The California industry has pioneered advances in the processing of cottonseed oil as a byproduct of cotton, a major crop in the state. Cottonseeds are removed from the fibers by cotton gins and are transported to the oil processing plant. The seeds are cleaned and hulled and then subjected to continuous screw pressing and/or a solvent extraction process. The crude oil is degummed and refined by washing with alkaline water solutions to convert the free fatty acids into oil-insoluble soaps. After recovery by centrifugation, the refined oil is heated and mixed with adsorbent clay or carbon to produce a light colored oil. Steam is blown through the oil to remove volatile odors and flavors. Cottonseed oil to be used as salad oil is chilled or "winterized" to remove higher melting glycerides. It is partially hydrogenated to give better keeping quality and physical properties when used as a shortening or for margarine.

Other Forms of Food Preservation

USE OF GASES. The injection of various gases into closed containers is another method of food preservation, which has been used to ripen such fruits as pears, melons, and persimmons. The procedure is very old, the Chinese using it perhaps a thousand or more years ago, especially for persimmons.

Studies are presently underway on extending the life of fresh products by holding them in various mixtures of gases under reduced atmospheric pressures. This is not entirely new, for apples have been held in an atmosphere of CO_2 for a long time, as have fruit juices.

Fumigation is used to protect foods against insect damage. Dried fruits and grains, for example, are fumigated in bulk with gases such as methyl bromide. This compound has unusual penetration properties and is very effective in the control of insects and rodents.

RADIATION. The potential use of radiation as a means of food preservation appeared shortly after World War II. The procedure involves the killing or inactivation of micro-organisms by ionizing radiations, particularly beta and gamma rays. Radiation also includes the use of ultra-violet light. Unfortunately, while the process seems to look somewhat promising, it is clear that commercial application will not be simple and extended use may never take place.

Conclusion

The food processing industries in California represent a major segment of the food system today. As food processors adjust their operations to stay viable in the marketplace, they must respond to varied pressures. Historically, safety and stability in processed products have been the processors' chief concern. More stringent requirements for sanitation and in tolerances for additives and pesticides continue to keep this concern high on the list of priorities. Other issues, however, increasingly bear on the viability of the food processing industries. Foremost amongst the economic pressures is the low profitability of first stage processing as compared to more complex processes producing differentiated and higher valued products. Consumers, concerned with nutritional values, at the same time often demand ease of product use, and the sometimes elusive factor of "acceptability" is the subject of much market research. Inflation erodes both sales and profits as the costs of processing labor, energy inputs, containerization and transportation continue upward, while consumers resist higher product prices by substituting lower-priced items. Like other industries, food processors find they are confronted with an increasing number of burdensome regulations imposed by a proliferation of government agencies. The question of cost-benefit ratios arises with regard to the ability of regulations to protect the consumer, versus the ultimate costs of enforcement which must be reflected in the price of food.

Nevertheless the food processing industry remains an indispensable part of California agriculture. As part of the complex system which moves food from the field and feedlot to the consumer's table, the processing sector contributes significantly to the income of the state and to the diets of the nation.

References

Altschul, A.A., and Wilcke, H.L. *New Protein Foods—Animal Protein Supplies*. New York: Academic Press, 1978.

Chou, M., and Harmond, D.P., Jr. (Ed.). *Critical Food Issues of the '80s*. New York: Pergamon Press, 1979.

Coulston, F. (Ed.). *Regulatory Aspects of Carcinogenesis and Food Additives: The Delaney Clause*. New York: Academic Press, 1979.

Herschdoerfer, S.M. *Quality Control in the Food Industry*. New York: Academic Press, 1972.

Kosikowski, F.V. *Cheese and Fermented Milk Foods*. Ann Arbor: Edward Bros, 1977.

Luh, B.S., and Woodroof, J.G. *Commercial Vegetable Processing*. Westport: Avi Publishing Co., 1975.

Luh, B.S. *Rice: Production and Utilization*. Westport: Avi Publishing Co., 1980.

Mountney, G.J. *Poultry Products Technology*. Westport: Avi Publishing Co., 1976.

Price, J.F., and Schweigert, B.S. *The Science of Meat and Meat Products*. (2nd edition). Westport: Food and Nutrition Press, 1978.

Stewart, G.F., and Amerine, M.A. *Introduction to Food Science and Technology*. New York: Academic Press, 1973.

Tannahill, R. *Food in History*. New York: Stein and Day, 1973.

Woodroof, J.G., and Luh, B.S. *Commercial Fruit Processing*. Westport: Avi Publishing Co., 1975.

17

The Changing Consumer

Sylvia Lane and Desmond Jolly [1]

In today's market economy, producers, if they are to remain in business, ultimately depend upon consumers. Modern agriculture has moved far beyond subsistence farming to a highly specialized industry tailoring its production toward the needs and demands of the market. The producer-consumer relationship is symbiotic: the producer provides the product which the consumer needs or wants, but the consumer is also influenced by the efforts of the producer, or his agents, to sell his product. The consumer, however, is complex and occasionally unpredictable—tastes change as times change.

California's highly diversified agriculture is part of a complex production and distribution system. The system is influenced by numerous economic and social forces and must respond to often elusive shifts in consumer preferences. This chapter concentrates on consumption patterns and trends and how they might affect the evolution of California agriculture. It begins with a general description of current consumption patterns, continues with a discussion of important trends, and concludes with a summary analysis of major consumer concerns which may help shape the future of agricultural production in California.

Trends in the Consumption of California Agricultural Products

California produces a wide variety of agricultural products, among which are many important fruits and vegetables. The consumption of these commodities has changed significantly since

1. With grateful acknowledgment of the assistance of Lynn Horel, Kirby Moulton, and Ann Scheuring.

1960, as indicated in Table 1. (Trends implied in the table are amply supported by data from intervening years.)

Overall per capita consumption of fresh and processed fruits, many vegetables and potatoes has declined. On the other hand, per capita consumption of frozen vegetables, canned vegetables, lettuce, canned tomatoes and tomato products, fresh onions, avocados, orange juice, and strawberries have all increased. Frozen fruit and juice consumption has increased, while frozen vegetables have increasingly been substituted for fresh vegetables. Fresh potato consumption per capita (retail weight) declined from 80.4 pounds to 51.8 pounds. The consumption of processed potatoes since 1950 in the form of potato chips, dehydrated and frozen potatoes, and potatoes used in prepared package food, however, has grown; per capita frozen potato consumption increased from 2.6 pounds per capita in 1960 to 16 pounds in 1977. (See Chapter 16 for discussion of food processing industries.)

Increased or decreased per capita consumption of other food products produced in California are also of interest. Per capita consumption of beef increased from 64 pounds per person in 1960 to 96 pounds in 1976. By 1979, however, it was down to 78 pounds per person. Lamb and mutton declined from 4.3 pounds per person in 1960 to 1.5 pounds in 1977. In contrast, pork, fish and poultry have enjoyed increases in per capita consumption; pork from 60.3 pounds in 1960 to 65 in 1979, fish from 10.3 to 13.7, and poultry products from 76 to 108. Within poultry products, however, eggs suffered a sharp decline from 42 to 36 pounds per person in part as a result of consumer health concerns. In the dairy product category, consumption of fluid whole milk has declined, while consumption per capita of cheese has increased.

Factors Influencing Consumption Patterns

To understand why per capita consumption of food has changed and will continue to change, it is necessary to understand the effects of various factors in the marketplace on the buying decisions of consumers.

Product Availability

Food consumption patterns are strongly influenced by the availability of commodities. For example, food consumption patterns vary *seasonally*; summer diets contain more fruits and vegetables because not only are these lighter, fresher foods preferred in hot weather, they are also in more abundant supply. Weather variations also influence product availability, and lead to day-to-day or week-to-week changes in per capita consumption which are not related to underlying demand shifts.

California agricultural products are often available to consumers in the United States during the winter and spring months when other regions are not producing. As more products have become available for longer periods of time during the year (lettuce from Imperial Valley) and in some cases throughout the year (frozen vegetables), their consumption has increased.

Moreover, as new items become available and consumers try them, they may become part of regular consumption patterns. New products or varieties have been an important stimulant to consumer demand for California agricultural products. Introduction of these innovations has been encouraged by the adaptive nature of California's consumers and the influence they

TABLE 17.1

Trends in Per Capita Consumption of Selected Commodities, United States

Commodity	1960	1977	Increase or Decrease 1960–1977
	Retail Weight in Pounds		
All fruit (fresh)	90.0	80.1	−9.9
All fruit (fresh & processed)	119.5	105.9	−13.6
Melons	23.2	19.3	−3.9
Oranges (fresh)	18.6	12.6	−6.0
Grapefruit (fresh)	9.7	7.6	−2.1
Apples	18.3	16.8	−1.5
Avocados (fresh)	0.8	1.2	+0.4
Peaches (fresh)	8.7	5.1	−3.6
Strawberries (fresh)	1.2	1.8	+0.6
Fruit Juices (canned)	13.0	15.8	+2.8
Frozen juice* (other than orange)	1.9	2.9	+1.0
Orange juice*	15.6	26.9	+11.3
Tomatoes (fresh)	10.5	10.6	+0.1
Tomatoes (canned) and tomato products	17.6	22.6	+5.0
Vegetables other than tomatoes (fresh)	85.5	82.5	−3.0
Vegetables other than tomatoes (canned)	28.1	34.3	+6.2
Carrots (fresh)	7.0	5.0	−2.0
Corn (fresh)	7.7	7.0	−0.7
Lettuce (fresh)	17.5	22.7	+5.2
Onions (fresh)	11.5	12.2	+0.7
All vegetables (frozen)**	9.7	26.3	+16.6
Potatoes (total)	87.9	75.3	−12.6

*Single strength equivalent
**Including potato products

SOURCE: USDA, ESCS Food Consumption and Expenditures, Supplement for 1977 to Agricultural Economic Report No. 138, Washington, D.C., September 1979.

have on consumers of other regions. Avocados, for example, were first introduced and accepted in California, and are now sold nationwide. Many health/natural foods (alfalfa sprouts and frozen yogurt as two examples) were first accepted by Californians, westerners and then the rest of the nation. California's wines, increasingly popular in many markets, still sell best at home.

The factors influencing California's role as a food innovator appear durable and therefore likely to influence future consumption patterns. Two of these factors are the California climate and the California "image." Favorable climatic conditions in most of the state encourage outdoor activities which influence what Californians eat and how they entertain, while the California image has been promoted by the motion picture and television industries which often feature California scenes and activities.

The volume and character of the state's population growth have also contributed to California's role as an innovator. From 1970 to 1977, California's population increased by nearly two million persons. Almost half of the increase was due to immigration from other states or countries at a rate almost twice the national average. Immigrants not only increased the demand for goods, they also introduced new tastes, foods, cuisines and patterns of consumption. (A *Sunset Magazine* survey found that nine out of ten western households eat and enjoy foreign and ethnic foods.)

Californians are also relatively affluent and sophisticated. Income per capita in California was $700 above that for all of the United States in 1978. The median number of completed years of schooling for Californians was over 12.7 as opposed to 12.5 for the nation in 1978, while 61 percent of the population was employed, compared with a national average of 59 percent. Over 90 percent of California's population is urban. The majority of women over 18 are employed outside of the home. Between a quarter and a third of all alternative lifestyle groups in the nation live in California. It is these factors among others that contribute to the California "lifestyle." Typical Californians seem less constrained by tradition, more willing to take risks, more interested in new experiences, and less inhibited. They are good prospects for new products, and particularly the specialty products of California's agriculture, and, in turn, they influence other consumers throughout the nation.

Consumer Income

Although food may be viewed as the most essential consumption good, it is not the largest item in the consumer's budget in the United States. According to the Bureau of Labor Statistics, food accounted for 18.2 percent of the expenditures by urban consumers in 1978 while housing absorbed 44 percent.

TABLE 17.2
Selected Foods Consumed by a Higher Percentage of Western Than United States Households

	% Household Usage	
	WESTERN	U.S.
Natural breakfast cereals	15.2	10.7
Honey	66.2	52.9
Jam	37.9	28.7
Wheat bread	57.8	39.1
English muffins	60.5	49.4
Milk, 2% lowfat	44.8	34.7
Milk, dry	37.0	27.8
Cheddar cheese	55.2	39.4
Monterey Jack cheese	41.1	11.9
Cottage cheese, small curd	66.0	52.6
Yogurt	44.5	33.9
Guacamole/avocado dip	13.6	4.2
Apricots, fresh	23.6	9.2
Avocado, fresh	47.7	17.4
Cantaloupes, fresh	60.8	47.8
Cherries, fresh	35.1	24.9
Lemons, fresh	40.5	36.8
Pineapples, fresh	32.1	20.1
Strawberries, fresh	59.6	50.0
Tomatoes, fresh	76.1	69.8
Watermelon	51.8	46.7
Link sausages	50.4	44.1
Dry soup mix	63.0	53.8
Salad dressing, dry mixes	34.7	20.1
Dry gravy & sauce mixes	42.1	33.3
Hot sauce	21.7	16.0
Soy sauce	35.7	25.4
Orange juice, frozen	76.1	70.0
Other juices, frozen	19.0	12.3
Tomato & vegetable juices	61.3	55.1
Pop/party wines	14.4	11.4
Domestic dinner/table wines	41.0	30.6
Imported dinner/table wines	17.5	15.6
Champagne, cold duck, sparkling wines	32.2	25.0
Cordials & liqueurs	22.3	18.2

SOURCE: *Sunset* Magazine Research Department, data compiled under title "1978 Target Group Index."

BUDGET SHARES. The budget share of food consumed at home was 12.6 percent in 1978. This percentage has been decreasing over the years. By contrast, expenditures for food away from home have been expanding, thus increasing demand for restaurant-type foods and ingredients in prepared foods (meats, buns, lettuce, tomatoes) served by other retail outlets, especially fast food outlets. A breakdown of consumer expenditures for various food categories appears in Table 3. Fruits and vegetables, California's major specialty crops, constitute only 1.8 percent of the current budget for all urban consumers in the United States; fresh fruits, one of our major specialty crops, only .4 percent.

In most countries, the proportion of expenditures allocated to food is inversely related to consumer incomes. As per capita income has increased in the United States, France and Egypt, for example, the proportion of income allocated to the purchase of food has declined. This almost universal relationship (formalized as Engels' Law) suggests that future budget shares for food will be no more than current levels in developed countries and probably significantly less in developing countries. The total value of food consumption, however, may rise as the result of price increases and consumer shifts towards "better" and presumably more expensive foods.

The aggregate per capita consumption of food, particularly in developed countries, is not likely to increase; hence, increased food consumption will be caused primarily by population growth, although the pattern of consumption will change as income increases. It is these changes in consumption patterns which are the most promising determinants of demand growth for California specialty products such as fruits and vegetables. The export-oriented cereal crops may also benefit from income-induced demand changes in less developed countries.

Population growth in the United States and in most developed countries, however, has slowed. Birth rates have been declining, although the United States is still many years away from zero population growth due to a large proportion of women being in their child bearing years into the 1990s.

INCOME ELASTICITIES. Various agricultural commodities have different "income elasticities of demand." Economists define the income elasticity of demand as the responsiveness of quantity purchased to changes in income. Food has a relatively low but positive income elasticity of demand—estimates range below .2 in the United States for food at home. This means that as income increases by one dollar, less than 20 cents (on the average) of that additional dollar may be spent on food. While overall budget shares allocated to food remain steady or decline as incomes increase, the particular mix of products purchased will reflect to a limited degree the rising affluence of consumers. Although they will probably not actually consume more food, they will tend to adjust their diets to accommodate more expensive tastes and preferences.

Commodities with positive income elasticities of demand are termed "normal" goods. Most foods are in this category (meat, milk, fruits and vegetables). But some have negative income elasticities of demand. Economists term them "inferior" goods—which has no connotation concerning their nature or quality. Margarine, evaporated milk, and beans are examples of such foods. As incomes increase, people in the United States eat less of these and more of other foods.

Food products may be classified according to their degree of dietary necessity. Basic (or staple)

TABLE 17.3
Selected Food Items in the Consumer Price Index*, December 1978

Commodity	Percentage
All items	100.00
Food and beverages	18.2
Food at home	12.6
Meats, poultry, fish and eggs	4.4
Fruits and vegetables	1.8
Dairy products	1.7
Cereals & bakery products	1.5
Fresh fruits and vegetables	.9
Processed fruits & vegetables	.9
Fresh fruits	.4
Processed vegetables	.4
Oranges	.1
Potatoes	.1
Lettuce	.1
Tomatoes	.1
Frozen fruit and fruit juice	.1
Frozen vegetables	.1

*In composing the CPI, weights are assigned to various food items thought to be reflective of their representation in a "typical market basket" of consumer expenditures. These weights may be considered somewhat like "budget shares."
SOURCE: U.S. Department of Labor, Bureau of Labor Statistics, "Consumer Price Index for all Urban Consumers," 1979.

foods are essential components of the standard American diet. They include bread, milk, meat, eggs, and potatoes. As diets are expanded to provide more varied sources of nutrients, products such as oranges, beans and carrots might be included. Finally, specialty crops such as artichokes, pomegranates, and pistachio nuts will be added as incomes permit greater diversity. Even though many specialty fruits and vegetables are nutritious, they may not be perceived as essential to the basic diet. Consumers buy them for variety and because they enjoy them.

This classification is useful because it tends to separate products with ready consumer acceptance and fairly consistent consumption from those whose demand is sensitive to price and especially income changes, often requiring stimulation through advertising and promotion. During hard times or in anticipation of hard times, consumers usually return to basics and purchase fewer exotic or special items. Many of California's out-of-state shipments are the non-staple items whose sales are heavily dependent on consumer incomes. Thus, a prolonged recession in the U.S. would doubtless have an impact on the demand for many of California's products.

Prices

Consumers make their purchase decisions based in large part upon product prices. How much they will increase quantities purchased when prices fall, or decrease quantities purchased when prices increase, is indicated by the "price elasticity of demand" for different products. This, like the income elasticity of demand, is a measure of the responsiveness of quantity purchased to economic changes.

PRICE ELASTICITY OF DEMAND. The price elasticity of food in the United States has been estimated to be about .46. (If prices increase by 10 percent, the quantity purchased will decrease by 4.6 percent.) For specific products price elasticities vary widely. If a product has few or no substitutes in the diet, consumer demand will tend to be inelastic and the quantity purchased will not vary much if prices increase or decrease. The demand for staples is relatively price inelastic (on the order of -0.2 in the U.S.), hence variations in supply create large fluctuations in the price of commodities like rice, wheat and corn. These price movements often lead to volatile political situations in areas where such grains are important

dietary items. On the other hand, if a product has many substitutes to which consumers can turn if the price of the product increases, quantities purchased may fluctuate widely depending on price. Many fruits and vegetables, with price elasticities ranging between $-.07$ and -1.0, fall into this category. Consumers buy available fruits and vegetables and meats, and canned and frozen foods too, for that matter, whose prices are relatively low compared to other products in the same category, and vary menus in order to maximize the utility from their food dollars.

For low income consumers who spend a higher proportion of their income on food, this may be especially important, particularly when prices are increasing and their incomes are not. Researchers have found they tend to cut down on more expensive foods and substitute lower-priced foods, but this often means that their diets contain a higher proportion of starchy (cheaper) staples.

COMPLEMENTS AND SUBSTITUTES. Some foods are customarily often eaten together—bacon and eggs, for example—and if the price of one, bacon perhaps, goes up, the quantity of bacon sold decreases and so does the quantity of eggs. Bacon and eggs are "complementary" products; so are strawberries and whipped cream. In the case of products which can be substituted for each other—for example, rice for potatoes—if the price of one increases and consequently the quantity purchased decreases, the purchased quantity of the substitute increases. Factors leading to product substitution also include changed population characteristics, altered lifestyles, shifting occupation and habits, and increased consumer concern about the wholesomeness and safety of food.

RESERVATION PRICES. Consumers also have "reservation prices." They will only buy some commodities when prices are below a certain level. This again is particularly true of products which have available substitutes and determines part of the seasonal consumer purchasing pattern for fruits and vegetables.

BUYING HABITS. Since food products are purchased frequently, in many cases consumers tend to form food buying habits. Once they are satisfied with a product they will continue buying it until circumstances or the product itself or its price changes enough so that this is no longer appropriate behavior. For example, a household shopper may buy the same type, brand and size of bread from the same store every week or two or three times a week. Habits are efficient. They allow consumers to make fewer decisions and so save time and energy. For some foods (fruits and vegetables) consumers may habitually buy in the same category, but they may substitute one kind for another depending upon the price.

TASTES AND PREFERENCES. Tastes and preferences largely underlie consumer purchase decisions, although budget constraints often operate to modify them. Consumer preferences are not entirely predictable. They shift, especially over time. Many of the factors affecting changing consumer purchasing patterns have been identified. They include changing demographics and changing popular values, as well as advertising and promotion.

Changing Character of the Population

As the demographic composition of the population changes, food purchases change. Different age and sex groups have different food preferences. Teenagers often prefer hamburgers to stew and french fries to salads. More men than women list steaks as their favorite food.

The population of the United States is growing older and, as time goes on, consists of more women than men. The percentage of the resident United States population over 65 was 9.2 percent in 1960; 9.9 percent in 1970; and 11 percent in 1978. The number and percentage of the elderly will grow in the 1980s and the 1990s; they eat less than all other population groups except small children. Children, who eat the least until they are teenagers, are a decreasing proportion of the population, although their absolute number is not changing markedly. Women, who outlive men in the United States, are increasingly becoming a larger proportion of the population. Women eat less than men and prefer different foods. They eat more fruits and vegetables and less meat, tending to favor salads, for example, over sandwiches for lunch.

The urban population has been increasing, from 69.9 percent of the population in 1960 to 73.5 in 1979, though there is now some evidence that this trend is reversing itself. Rural households tend to eat more than urban households, but urban households tend to purchase more specialty items and rely less on staple foods.

Meanwhile, the area of the country that is growing the fastest has changed. The Sunbelt states have had the fastest growth rate in recent years and this is expected to continue. Because of the hot climate of the Southwest, diets tend to be lighter—less meat and potatoes and more fruit.

The ethnic mix of the population has changed and is changing. Black and Spanish-origin groups make up a larger share of the population. Blacks have increased from 9.9 percent of the nation in 1950 to 11.7 percent in 1978, while the population of Spanish origin increased from 3 percent in 1970 to 4 percent. The latter group particularly differs in dietary habits from other groups, and has larger sized households, 3.58 persons in 1978, in comparison with black households with an average of 3.1 persons. (The average size household in the United States as a whole that year was 2.8 persons.) Mexican-Americans have had a marked influence on the cuisine generally available in the United States. Mexican food has joined Italian, French, Chinese, and customary American foods (hamburger, french fries, and fried chicken) on the list of foods available in every sizeable urban center and even in many nonurban areas in the United States.

The changing occupational structure of the population has also affected our eating patterns. The proportion of white collar workers in the population, for example, has been steadily increasing (white collar workers were 48.3 percent of employed persons in 1970, 51.3 in 1979). Blue collar workers have remained a nearly static proportion of the population (31 percent in 1970, 32 percent in 1979), but with improved technology many jobs are not as arduous. As workers move increasingly into more sedentary occupations, they require fewer calories.

CHANGING TYPES OF HOUSEHOLDS. The number of households in the nation has increased from 53 million in 1960 to 63.5 million in 1970 and 76 million in 1978. In large part this is because the population per household has declined. Two-person households increased from 8.3 percent of total households in 1970 to 9.8 in 1978; single (never married) males increased from 18.9 percent of the population in 1970 to 22.5 in 1978; single females from 13.7 percent to 16.4 percent. Single persons living alone and those in two-person households tend to eat less than persons in larger households, although they spend more on food per capita. They do not prepare as many meals at home and they spend less time on food preparation. In part this helps explain the trend toward frozen and pre-prepared foods.

CHANGING LIFESTYLES. Indicators of changing lifestyles in the United States include many of the demographic trends mentioned above: more elderly persons and fewer children, smaller households, more households headed by women (21 percent in 1970, 25 percent in 1978), and more women working. Of women over 16 in the United States, 49 percent were in the civilian labor force in 1978; more notable, 48 percent of married women living with husbands were in the labor force, and 55 percent of these women had children under 18. In 1970, only 41 percent of married women were in the labor force.

Working women and single household heads spend less time in food preparation than the traditional stay-at-home wife, and various members of the household may eat at different times. Meanwhile, fewer people are working regular 9 to 5 hours as jobs tend to be characterized by more flexible days and staggered shifts. More leisure-time activities, sports and recreation, and organizational activities away from home have meant fewer traditional family meals. The result is greater use of pre-prepared foods or foods requiring little or no preparation, and of microwave ovens which speed the process of food preparation. In part, this is because people's time is worth more, but the desire to save ever more costly energy (gas, electricity) has also made for less cooking at home.

Fast food eating places increased dramatically in numbers during the last decade in part because people drove more, but also because they provide fast, relatively inexpensive and convenient meals that do not have to be cooked at home. For some households a large number of breakfasts, lunches, snacks, and dinners are now eaten away from home. The share of food eaten away from home can only be expected to increase as our lifestyles become less home-oriented, with fewer people at home as many hours, and families and other households becoming less cohesive as greater demands are made on the time of household members.

On the other hand, entertaining and gourmet cooking are becoming increasingly popular. So is eating in fine restaurants. More time and money are being spent on gourmet meals and beverages, and this growing consumer interest is reflected in demand for fresh, high-quality ingredients and specialty food items.

The values that underlie our changing lifestyles are continuing to evolve. According to Arnold Mitchell, Director of the Values and Lifestyles Program at Stanford Research Institute, in the next decade we can expect a near-doubling of the groups which are "inner-directed" and espouse self-expression. By 1990 these groups will comprise about one quarter of the population. Of the groups which embrace the ecology movement and advocate social responsibility, many are also vegetarian in their food preferences. According to the May 1980 *Consumer Reports*, approximately 7 million Americans now practice vegetarianism. Typical items in their diet are alfalfa sprouts, soybean curd, and trail mix (peanuts, raisins, sunflower seeds and other ingredients). Mitchell also anticipates a decline in the number of consumers espousing traditional values. He predicts they will decrease from 70 to about 60 ercent of the population. The remainder of the population, whom Mitchell terms "money restricted," are the poor and near-poor. Though their proportion will decrease, their number, because of a still slowly rising population, will stay about the same.

Advertising and Promotion

California agricultural products are advertised by proprietary firms, cooperatives, voluntary organizations, commissions, councils, boards, and by individual producers, all of whom attempt to increase or maintain consumer demand. In economic terms, promotion represents an attempt to shift the demand curve upward so that at every price, consumers will buy greater quantities of the product. An effort to make the demand curve more price inelastic may also be involved, so that if prices increase the change in demand associated with the percentage change in price will be smaller. In some cases advertising may also try to increase income elasticities, so that sales of the product will increase more than proportionately, as incomes increase. Advertising may also attempt to capture a larger share of the market by differentiating the product (i.e., a Sunkist is not just any orange but a better orange), and by increasing product substitution.

In group advertising, producers generally assess themselves voluntarily in accordance with the quantity produced. Farmer marketing cooperatives have been especially large advertisers. Sunkist, Diamond Walnut Growers, Calavo, and The California Prune and Apricot Growers Association are examples of cooperatives with sizeable budgets for advertising and promotion.

Generic advertising (by the California Raisin Advisory Board, for example) promotes the product irrespective of who the grower is, or the brand, if the product is branded. Brand advertising promotes the particular brand (Sun-Maid raisins). Proprietary advertising (advertising by a particular firm) is generally brand advertising (Hunt's, Del Monte) but it may be generic, simply building the image of the firm and not any particular product (Hunt-Wesson Foods).

There is no definitive study on the results of the advertising and promotion of California agricultural products, but sales increases following advertising campaigns are usually attributed to the advertising effort (although other factors are nearly always involved), and lack of sales increase is taken to mean that the campaign was unsuccessful. An example of a seemingly unsuccessful campaign was the Prune Association's effort in 1979. Though the radio spots were amusing and widely enjoyed, no subsequent increase in prune sales could be detected that could be attributed solely to the radio spots. The avocado, orange, wine, and almond campaigns known to many are, among others, examples of successful campaigns.

In some cases, advertising campaigns are, in part, defensive. The Dairy Council of California spent over $5 million on market development, education and promotion in 1973 in the effort to slow the decline in fluid milk consumption per capita, a decline that has been apparent since 1947. Whether the expenditures acted to slow the decline or not is unknown in view of other factors that may have affected the outcome (advertising of substitute products, for one).

Consumer Concerns

As consumers' incomes and levels of education increase, they tend to become more discriminating and begin to demand attributes in food that relate to health, convenience, and aesthetics. Consumer demand for food products has been markedly affected in recent years by concerns over nutrition, food safety, and environmental contaminants. Although consumer organizations have not always been effective, they are increasingly a force to be reckoned with, and some of their concerns have been institutionalized in public agencies and regulations.

Nutrition

Notions of a nutritionally beneficial diet change over time, especially as we learn more about nutrition and health. According to a 1980 survey conducted by *Woman's Day* and the Food Marketing Institute, all those interviewed agreed, "Americans are more interested in nutrition than they were two years ago." Growing concern over diet's role in health and longevity has been demonstrated by governmental involvement in nutrition education as well as by booming sales of diet and fitness books and by the proliferation of health food outlets.

Nutrition education has become a high priority on both the federal and state levels. Public concern is evident in the policy-oriented publication of the U.S. Senate, "Dietary Goals for the United States," (1977). Specific response to the issues raised in this report came in the authorizing of the Secretary of Agriculture to carry out a program of nutrition information and education by establishing a system of grants to state agencies. The Child Nutrition Facilities Act of 1975 in California gave a significant boost to nutrition education in the state's public elementary schools by providing special educational projects relating to nutrition. Federal funding has stimulated the California Department of Education to develop a comprehensive nutrition curriculum for kindergarten through 12th grade, focusing on integrating classroom instruction with school food service programs. Nutrition education has also been carried on under the auspices of Cooperative Extension in its Expanded Food and Nutrition Education Program (EFNEP), operating since 1969 in approximately 17 counties to provide instruction and assistance in planning nutritional diets for low-income consumers.

Public awareness of the importance of proper nutrition has also been heightened by a spate of special-diet books during the decade of the 70s, many of which have emphasized a greater proportion of fruits, vegetables and whole grains in the diet while cutting intake of cholesterol-laden meats, fatty foods and refined sugars.

Health and Safety

Bacterial contamination of foods has been the safety concern given the highest priority by government officials and academic researchers. While food-borne illnesses continue to occur, necessitating continued monitoring of hygiene and manufacturing practices, the concerns of consumers have gravitated to other issues related to health and safety. Specifically, consumers have become more aware of the use of potentially hazardous chemical compounds in the production, processing and fabrication of foods. Public and private research efforts often with the aid of highly sophisticated measuring techniques have identified potentially damaging effects on health from use of some of these chemicals. In some cases, the research results are highly tenuous; in others, they are relatively solid (ordinary sugar and cavities; salt and high blood pressure). The cumulative effect of the publicizing of the research has been to heighten the anxiety of food consumers, and their representatives, about the short-term and long-term effect of these chemicals, and to increase the consumption of "natural" foods. The allegations of mutagenic and carcinogenic properties and links to allergies, heart disease, sterility, miscarriages and other problems, have had significant psychological impacts on consumers.

Environmental Contaminants

Environmental contaminants, such as pesticides (including herbicides, insecticides, fungicides, nematocides, *et al.*) which have increased crop yields by reducing losses to disease, insects, damaging organisms and weeds, pose risks if they are toxic to benign insects, organisms or weeds, or other plants or animals, or workers. Residues beyond acceptable levels can also collect in soil or water or on the product that reaches the consumer. State and federal regulations exist to deal with these risks, as described in Chapter 14. Enforcement of regulations cannot always be complete, however, and new problems arise, making this a highly controversial issue. Without pesticides, prices for fruits and vegetables would increase. But the effects of these increases have to be weighed against actual and potential risks to the environment and human health. The increase in the consumption of "organic" foods is in part a response to concern over pesticides.

Food Costs

A principal concern of consumers is, of course, inflation in general and food costs in particular. As the costs of transportation, housing and services increase, they tend to reduce the available funds which can be allocated to the food budget, even while food costs themselves may be rising.

There is pressure to reduce the overall consumption of foods as well as the "quality" of the diet. For some households, increases in food costs have marginal nutritional and psychological impact. For others, inflation means major changes in their consumption patterns.

Food prices in the United States have been relatively low in recent decades. From 1970 to 1977, however, the food price index rose 77.3 points in comparison with the overall consumer price index increase of 56.2 points. Basic concern over food prices and additional concerns over quality have given rise to renewed interest in institutional alternatives such as food buying cooperatives and direct producer-to-consumer marketing in local farmers' markets and roadside stands. Although none of these are new either in concept or in practice, their return appearance in altered forms on the California scene has satisfied a desire of some citizens for alternatives to the conventional food marketing retail system.

Consumer Organizations

The oldest, largest, and most influential of the food buying cooperatives in California is the Consumers Cooperative of Berkeley, begun in 1937 by church and university people. From a storefront operation staffed by volunteers the Berkeley Co-op has grown to a 1980 membership of 97,000, controlling 12 stores which retail not only food but pharmaceuticals, hardware, garden supplies, and recreational equipment. Other major activities of the Co-op include nutrition and consumer education through a staff home economist, and testimony before legislative hearings on such consumer issues as food labeling and grade standards. The Co-op claims to achieve considerable savings over conventional retail prices for its members, some of this coming in the form of annual rebates on total purchases.

Other nonprofit groups have organized to achieve food savings. Some cultivate community gardens, and some community groups "glean" recently harvested fields for agricultural surplus. Many of these "gleanings" have been carried out by, or in behalf of, senior citizens, a sector of the population with a higher than average proportion in poverty.

Other California consumer organizations voicing concerns about agricultural products are the Consumers Federation of California, a coordinating group for a number of smaller organizations which serves as their spokesperson at state and federal hearings, and the San Francisco office of the national Consumers Union, which also lobbies for legislation.

Within state government, the Department of Consumer Affairs has actively assisted in the formation of food cooperatives, with a special program for inner-city areas, and in the establishment of a consumer "hotline" for direct marketing information within the Department of Food and Agriculture. Representatives of the Department have testified before legislative committees concerning the impact on consumers of milk pricing in California, the certification of farmers' markets, and item pricing in supermarkets. The CDFA (partly in response to DCA pressures) now requires the appointment of public members to state marketing order boards.

On the federal level the 1976 creation of a special position in the U.S. Department of Agriculture, Assistant Secretary for Food and Consumer Affairs, marked the growing interest of government in consumer issues. The assistant secretary held hearings in California on the labeling of ingredients in food products as well as on USDA grading standards for meat and poultry.

Summary

California has developed innumerable innovations in foods which have been adopted elsewhere, and has contributed a wide variety of agricultural commodities to the market baskets of the nation and the world. Consumption of staple commodities will probably increase as population increases, but many California agricultural commodities are not staples, and their sales will be contingent on rising affluence. Should income levels decrease, consumption of most nonessential commodities would also decrease—and more rapidly than consumption of staples.

The 1970s have seen development of the consumer movement. Concerns over nutrition, health and safety are causing some shifts in consumption patterns. Additionally, as consumers experience the pressures of inflation, they seek alternative means of meeting food needs. Consumer concerns have been institutionalized in the form of new regulations on production and marketing, as well as in formal representation in public agencies.

California agriculture can operate profitably only when consumers will buy its products in sufficient quantity at high enough prices. Thus, the knowledge of factors affecting consumption is essential to the various sectors of the industry. For agriculture to prosper, it must adapt, as it has in the past, to changing conditions in the marketplace, and to the changing consumer.

References

The key reference work on the economics of consumption is still Marguerite Burk, *Consumption Economics*, John Wiley & Sons, Inc., New York, 1968. For an overview of the entire food economy including consumption and consumer concerns see *Your Food: A Food Policy Basebook*, Cooperative Extension Service, The Ohio State University, 1975. Current conditions in the food economy, and research findings concerning food expenditures, food marketing, food consumption and food safety and quality are contained in the *National Food Review*, a publication of the Economics and Statistics Service of the United States Department of Agriculture, issued quarterly. On the changing demand for food, an interesting reference is Letitia Brewer and Michael F. Jacobson's *The Changing American Diet*, published by the Center for Science in the Public Interest, Washington, D.C., 1978. Food cooperatives and their growing importance are discussed in Art Danforth's "Food Co-ops: Their Significance Today." These were introductory remarks for the Food Co-op Workshop Conference on Emerging Cooperatives held in St. Louis, Missouri, May 16–20, 1976. For a succinct overview of the history of the consumer movement, see Robert O. Herrmann, "The Consumer Movement in Historical Perspective," reprinted in David A. Aaker and George S. Day, *Consumerism*, 3rd ed., The Free Press, New York, 1978, pp. 27–36.

Students gain hands-on experience in activities at Cal Poly's Beef Cattle Evaluation Center. California Polytechnic State University works closely with industry in such programs as the Annual Bull Test and Sale. *Photo courtesy of CPSU Public Affairs Officer, San Luis Obispo.*

18

Education and Research in Agriculture

Loy L. Sammet[1]

A complex agriculture with highly developed technology owes its existence to the research and experimentation which are its underpinnings, and to the availability of competent people. California is almost unequalled in the research and educational system which has encouraged and supported its agricultural growth. From modest beginnings in the 19th century, this system has been expanded to a multi-level network which serves a twofold function: contribution of new knowledge and new methods in conjunction with the training and education of those who work in agriculture.

In a historically short span of years a very large and complex system of education and research in agriculture and related fields has evolved in the U.S., and probably nowhere in degree equal to that in California. This chapter provides a historical perspective on this development, with attention to the philosophical divisions that had to be bridged when the University of California and its program in agriculture were created. The perennial reappearance of the basic founding issue—the reconciliation of both classical and applied education within a single institution—and its role in the development of a four-tiered system of agricultural education in California are traced. The chapter concludes with a brief description of contemporary programs of agricultural research and education in California and an appraisal of the issues—some deeply rooted in the past—that are shaping these programs.

1. Acknowledgement and appreciation are due many individuals who provided data and critiqued the initial draft of this work, in particular Elizabeth M. Wauchope, who made a substantial contribution to the section on contemporary programs.

Prologue

Among early proponents of education in the United States, some were strong advocates of a classical education as essential to social and political development and to the preservation of religious and moral values. Others stressed practical training in support of economic progress for the nation and economic and social mobility for the individual.

This mix of educational philosophy was brought by eastern migrants to California in the mid-1800s and is reflected in a provision of the State Constitution of 1849, declaring that the Legislature "... shall encourage by all suitable means the promotion of intellectual, scientific, moral, and agricultural improvement." The first public school law in California was passed in 1851, a year in which the State Superintendent of Instruction reported a total enrollment of 1,846 students. There were less than 20 schools of any kind in the state at that time, and no public high school until 1856.

Early leaders included the Reverend Samuel H. Willey, a graduate of Dartmouth, who came to California in 1849 as a representative of the American Home Missionary Society, and Henry Durant, an 1827 graduate of Yale and a Massachusetts teacher and minister, who arrived in 1853. Strong advocates of higher education, both worked with a committee of clergymen for the establishment in 1853 in Oakland of a Contra Costa Academy. Despite financial stress, the Trustees of the Academy in 1855 petitioned for and were granted a charter for a College of California. Pressures of a rapidly expanding City of

Oakland soon led the College Trustees to seek a new site. This was formally accomplished at a dedication at Founder's Rock in 1861 on the present site of the UC Berkeley campus.

In a document called the "Organic Basis" the Trustees of the College of California declared in 1859 a common interest "... in seeking the highest educational privilege for youth ... in the promotion of the highest welfare of the state ... as fostered by the diffusion of sound and liberal learning." "Laws of the College," adopted by the Trustees, provided for a strongly classical curriculum, closely parallel to many New England schools.

As the College of California struggled to establish a school in the classical tradition, a national movement was developing in support of higher education in agriculture and the mechanic arts. The proponents strongly believed such training to be essential to economic growth and development, and the movement drew considerable strength from populist concepts advocated in defense of the common man.

Populist grievances shared with an emerging farmer organization—a secret society known as the State Grange of the Patrons of Husbandry—included suspicion of big business generally and opposition to trusts, combines, pools, and monopolies. Both the Grange and the more broadly based populist movement were suspicious of gouging through excessive rail freight rates, and both were concerned about access for farmers and small businesses to adequate and reasonably priced credit. Positions of particular interest to the Grange involved support for cooperative systems of commerce, for the introduction of European laborers and for the elevation and increase of mechanical industry. A position the Grange fervently endorsed was that education in agriculture should be practical and should include actual work experience on farms. By formal resolution in 1874, the Grange complained that the University's instruction was science oriented and that "... no instruction has ever been given (either in agriculture or the mechanic arts); nor has the manual labor system required by law ... been encouraged or practiced."

Recurrent expressions of these themes have had an important effect on the development of higher education and, in particular, research and teaching in agriculture in California—even to the present.

Legislative Initiatives and The University of California

The Morrill Act

A national campaign for federal support of higher education resulted in the passage in 1862 of the Morrill Act, which has come to have enormous influence on higher education in the U.S. The Act provided land grants to each state (150,000 acres to California) for "... the endowment, support, and maintenace of at least one College where the leading object shall be, without excluding other scientific and classical studies, and including military tactics, to teach such branches of learning as are related to agriculture and the Mechanic Arts, in such manner as the legislatures of the states may respectively prescribe, in order to promote the liberal and practical education of the industrial classes in the several pursuits and professions of life."

In some states, implementation of the Morrill Act was accomplished simply by assigning the benefits to an existing institution. Following this opportunistic line, the College of California Trustees in 1864 created a Mining and Agriculture College, an aspiration never given substance, and advised the Legislature of their readiness to accept the land grant. Instead, the Legislature enacted in March 1866 a proposal to create an Agricultural, Mining, and Mechanical Arts College. A major practical problem was securing a site, and one of several sites considered was that north of Oakland, recently acquired by the College of California. A practical difficulty of a different sort was the financial straits of the College of California. These difficulties, and rather strong disagreement over educational philosophy, were in part resolved and in part embodied in an "Organic Act" of the Legislature, signed into law by Governor Haight on March 23, 1868.

The University of California was thus created. Its character reflected a precondition imposed by the College of California that a "complete University" be established to teach the humanities as well as practical subjects, and it was constituted—in conformance with popular sentiment in the state and with the Morrill Act—to offer instruction in agriculture, mining, and the mechanic arts. The blend of commitment to both the liberal arts and applied subjects in a publicly supported institution proved later on to be of enormous significance in the development of higher education in the U.S. How much so, and the extent to which this uneasy compromise would both strengthen and strain the University in the future, may not have had much notice when the University of California opened in Oakland in 1869. However, controversy was already brewing by the time classes began at Berkeley in 1873.

A beginning in the formation of a College of Agriculture in the University was made in the appointment in 1869 of Ezra Carr as Professor of Agriculture, Agricultural Chemistry, and Horticulture. Dissension soon arose from Carr's inability to satisfy desires from within the University that the work in agriculture be conducted at a scientific level, and from the public demands for training at a practical level. Carr shared the latter view, and in fact worked very closely with the Grange movement in California to assure its accomplishment. These efforts became politicized and divisive to such an extent that Carr was dismissed in August 1874. His successor, Eugene W. Hilgard, was appointed in the same year.

Meanwhile, in 1870, it was announced that the University proposed "... to furnish the facilities for all needful experiments; to be the 'Station' where tests can be made of whatever claims attention; to become the exponent and repository of our progressive knowledge." In 1873 President Gilman reiterated that the University domain was being developed with a view to illustrate the capabilities of the state for special cultures, whether forests, fruits, or field crops, and the most economical means of production. He described the University as a place where new plants and processes will be tested and the results made available to the state.

Hilgard—Philosophy and Influence

Hilgard was gifted with intelligence, persistence, training in both the sciences and humanities, and vision. He was able to establish with his University colleagues the credibility of the academic program in agriculture, while also persuading the farming interests of the practical potential of the scientific work he was beginning.

Within a year of his appointment, Hilgard undertook the first field experiment in the University, an experiment to compare the effects of shallow and deep plowing. This work began under control of the Secretary of the Regents. But four years later—and after much dispute—

control was vested in the Professor of Agriculture, and $250 annually was allocated by the Regents to support the work. Thereafter, continued support was provided by the Legislature through specific annual appropriations. This development was of far reaching significance, for it established the precedent for public funding of research in the University and for its control by the faculty.

Meanwhile, there was growing pressure nationwide for federal support of research in agriculture. Hilgard was active in this effort, and published a timely and very influential paper on the subject in the *Atlantic Monthly* (May 1882). The movement was supported by the Teachers of Agriculture, an association formed at the University of Illinois in 1880 and by the Association of American Colleges and Experiment Stations established in 1887. (The latter organization eventually became the National Association of State Universities and Land-Grant Colleges.) Support also was provided by George L. Bailey, Commissioner of the U.S. Department of Agriculture, by numerous professors of agriculture at eastern and midwestern colleges, and by congressional leaders.

The Hatch Act and the Agricultural Experiment Station

A consequence of the national movement was the passage of the Hatch Act in 1887, which provided the basis for cooperation in agricultural research between the federal government and land-grant institutions and for the allocation of federal funds in support of Experiment Station research. The objectives of the Act of 1887 were to "... aid in acquiring and diffusing among the people of the United States useful and practical information on subjects connected with agriculture, and to promote scientific investigation and experiment respecting the principles and applications of agricultural science ...." The Act also provided for the establishment of a department "... to be known and designed as an 'agricultural experiment station' in each of the land-grant institutions established under the Morrill Act of 1862." Section 2 of the Hatch Act of 1887 specified broad aspects of agricultural science to be investigated and provided for variations reflecting "... due regard to the varying conditions and needs of the respective States or Territories." The University's announced intention in 1870 to be a "station" for research is the basis for credit to California as the first state in the

U.S. to establish an Agricultural Experiment Station.

Numerous subsequent Congressional acts extended the coverage of the Hatch Act of 1887 and increased the amount of funds appropriated. Of particular interest is the Hatch Act of 1955 which consolidated the provisions of the previous legislation. This Act stated the policy of Congress to be "... to promote the efficient production, marketing, distribution and utilization of products of the farm as essential to the health and welfare of our peoples and to promote a sound and prosperous agriculture and rural life as indispensable to the maintenance of maximum employment and national prosperity and security." The 1955 Act also calls on the State Agricultural Experiment Stations "... to conduct original and other researches, investigations and experiments of a permanent and effective agricultural industry of the United States, including researches basic to the problems of agriculture in its broadest aspects, and such investigations as have for their purpose the development and improvement of the rural home and life and the maximum contribution to the welfare of the consumer, as may be deemed advisable, having due regard for the varying conditions and needs of the respective states." The Act places the foregoing charge in the context of "expenditure of funds hereinafter authorized ...."

Other significant enactments were the McIntire-Stennis Act of 1962 in support of research on matters concerning forests and related rangelands, and a Congressional Act of 1966 that established a National Sea Grant Program.

Early Policy Conflicts and Their Resolution

The establishment of a system of cooperative federal-state research in agriculture was a victory for those who wished to emphasize the role of science in education and research in agriculture. This success did not, however, still the popular clamor for practical training.

Pressure from the farm interests for greater emphasis on the practical, coupled with low enrollment in formative years of the College of Agriculture at Berkeley, let President Gilman to direct the Professor of Agriculture to spend time lecturing in all the agricultural counties and centers of population. Hilgard complied, and with notable success. A high point in these efforts was the organization of a series of Farmer's Institutes, the first in March 1871. These highly suc-

cessful institutes and Hilgard's persistent efforts in extending to farmers the results of scientific investigations in the University led to specific state funding for such activities in annual appropriations to the University; this put California in a state of readiness to cooperate in a federal program of agricultural extension education authorized in the Smith-Lever Act (U.S. Congress, 1914).

The Smith-Lever Act created a cooperative partnership between the U.S. Department of Agriculture, the land-grant colleges and universities, and county governing boards. The stated purpose was "... to aid in diffusing among the people of the United States useful and practical information on subjects relating to agriculture and home economics and to encourage the application of the same." This objective was expanded in subsequent legislation to provide funding for new programs in areas related to agriculture, such as 4-H Club work, rural health and sanitation, and—much later—for such activities as special efforts with disadvantaged farm families and the dissemination of information on the uses of solar energy in agriculture.

Another manifestation of the concern for practical education in agriculture was a drive to establish a university farm, where practical instruction could be given. Public interest was whetted by Peter J. Shields, a Sacramento attorney, who lobbied for the proposal with great determination and skill. Legislative approval and an appropriation of $150,000 were granted in 1905. Soon thereafter, 779 acres adjoining the townsite of Davis, California (now the site of the Davis campus) were purchased, and formal instruction at a non-degree (high school) level was begun there in 1909. In a parallel development outside the University of California, a state school designed to emphasize vocational training in agriculture and engineering was established at San Luis Obispo in 1901 (later to become California Polytechnic State University).

A further consequence of the difficulty of combining within the University of California all aspects of instruction in agriculture, ranging from the scientific to vocational, was the introduction of education in agriculture at the high school and secondary school levels. The first such program appeared in the Kern County Union High School in 1905. The California Legislature in 1907 specified that all elementary schools give "... instruction in nature study with special reference to agriculture," and in 1917 the U.S. Congress passed the Smith-Hughes Act which offered federal funding—on a matching basis—in support of vocational education. A developing state effort in vocational education in agriculture in the public schools was thus augmented through federal support.

As vocational instruction in agriculture at the secondary and high school levels was being established, instruction in this field was also being introduced in the junior colleges (later established as the California Community Colleges, under a separate Board of Governors). The first such program appeared in the Bakersfield Junior College in 1915, and a steady growth—accelerated following World War II—has ensued.

Agriculture in Transition in the University

With the enactment of the Smith-Lever Act in 1914 and the Smith-Hughes Act of 1917, the basic elements of California's diverse system of education and research in agriculture were in place. However, as the state's population and economy grew, dramatic changes were induced in this system.

Riverside and UCLA

Early evidence of the evolutionary process was the pressure developed by growers of horticultural crops in southern California for the establishment of a local research station.

The Legislature responded in 1905 by authorizing a pathology laboratory and Citrus Experiment Station in southern California, the appropriation of funds to the University, and the appointment of a three-member Commission to direct the enterprise. The Commission recommended the siting of a pathology laboratory at Whittier, and a Citrus Experiment Station at Riverside. The Regents responded in February 1907 with approval of a lease of 30 acres for use by the Citrus Experiment Station at the base of Mt. Rubidoux in Riverside. From this small beginning the Citrus Experiment Station grew steadily in terms of size and variety of programs. In 1912 a graduate degree program in tropical agriculture was established in cooperation with the UC Berkeley campus. In 1954 a regional headquarters for Cooperative Extension and a College of Letters and Sciences were established at Riverside,

and in 1960 the Riverside units became a part of a general campus of the University. At the same time the Regents authorized the establishment of a College of Agriculture on the Riverside campus. In 1961 the Citrus Experiment Station was renamed the Citrus Research Center and Agricultural Experiment Station and it became more firmly established as a major unit of the California Agricultural Experiment Station. Through several reorganizations, the College at Riverside has achieved an integration of basic and agricultural science departments unique in the University and is now known as the College of Natural and Agricultural Sciences.

As the University's research and educational development at Riverside progressed, much public discussion grew over plans for the expansion of the Los Angeles State Normal School (established as a branch of the Normal School at San Jose in 1881) and its possible affiliation with the University. This culminated in a Legislative Act in 1919 creating a branch of the University of California at the site of the Los Angeles Normal School. The branch campus later (1927) became the University of California at Los Angeles. The present site in Westwood was occupied in 1929. The new campus included a College of Agriculture with related activities in the Agricultural Experiment Station. These flourished for a time, but with increasing urbanization in the Los Angeles area, the Regents—in voting to create a general campus and College of Agriculture at Riverside—also voted to terminate the College and Agricultural Experiment Station activities at UCLA, contingent on the availability of facilities on other campuses to continue the UCLA programs. This was accomplished over time by transferring some faculty and staff to agricultural programs at Berkeley, Davis and Riverside.

Development at Davis

Practical instruction in agriculture was one of the principal motivations for the 1905 acquisition of the university farm at Davis. This objective was attained in 1909 with the institution of a three-year non-degree program of instruction at the high school level. In 1922, the Regents authorized the introduction of university courses for freshman and sophomore students at Davis—a move to forestall pressure from some farming interests to separate the College of Agriculture (Berkeley) and the Davis campus from the University and create a separate institution. The new program at Davis grew rapidly and, by sharing faculty with the Berkeley campus, Davis soon was able to offer courses in nearly all fields usually found in a four-year program. As time passed, administrative and staffing ties to the Berkeley campus diminished, a College of Agriculture was established, and the Davis campus began granting the B.S. degree. A School of Veterinary Medicine was established in 1946. As enrollment pressures increased, a College of Letters and Science was created, and in 1959 the Regents established a general campus at Davis.

The period of rapid population and economic growth following World War II and the Korean War was also one of rapid expansion in student enrollment at Davis. It was also a period in which urban impingement on agriculture's traditional open-space preserve was being keenly felt, and in which natural resource and environmental issues were being pressed with increasing vigor. These new influences led in 1964-67 to a substantial restructuring of the programs of instruction and research in agriculture at Davis and in the renaming in 1967 of the College of Agriculture as the College of Agricultural and Environmental Sciences.

Change at Berkeley

Under the broad perspective expressed by President Gilman in the early 1870s, and with the vision and leadership of Dean Hilgard and later Dean Edward J. Wickson, the College of Agriculture at Berkeley experienced continuing development, including the introduction of numerous new fields of specialization. An especially significant event was the approval in 1913 of the first Professorship of Forestry and the appointment to that position in 1914 of Walter Mulford. This followed several decades of discussion and proposals concerning the introduction of a curriculum in forestry in the state. Under Mulford's persistent leadership, and with the support of Dean Claude B. Hutchison, the work in forestry was first organized as a Division, later as a Department within the College, and finally as a School of Forestry.

Separately administered programs of education and research in agriculture and forestry continued at Berkeley for several decades. By the early 1970s, however, continuing developments within the University and increasing societal concern about the status of the natural resource base and environmental quality—along with the evi-

dent student interest in such issues—stimulated a substantial change in curriculum and organization. The programs in agricultural sciences and forestry were brought together in a new College of Natural Resources, with planned emphasis on the biological, social, and management sciences as they apply in the areas of agricultural and forestry production, natural resource utilization and conservation, and environmental and ecological protection. Multidisciplinarity, especially in undergraduate instruction, was a major planning objective. In the Regents' approval of the new college it became the first in the University of California formally to be chartered as both a scientific and professional college.

Cooperative Extension

As campus education and research in agriculture grew and became more diverse, a similar development was occurring in Agricultural Extension. Specialists in various agricultural sciences were appointed on each of the agricultural campuses to work in close association with related campus departments, and professional staffs were established in Extension offices in virtually every county in the state. In March 1974, to reflect close working relationships with counties and with the Federal Agricultural Extension activities, Agricultural Extension in California became known as Cooperative Extension.

Differentiation of Campus Programs

Growth in the size and diversity of California agriculture and recognition of the limitations for field experiments at Berkeley were the basis for expansion of resources and programs at Davis and Riverside and, for a time, at UCLA. Decentralization of the University's program in agriculture was accelerated in the post-World War II years and numerous fields of specialization at Berkeley were transferred to the other campuses, primarily Davis. A statewide coordinating function was provided through formation of the Division of Agricultural Sciences, within which statewide administration of the Agricultural Experiment Station and Cooperative Extension are still centered.

To a significant degree after the fact, the policy of decentralization was formally recognized in the academic plan for the University published in 1961 which declared that the programs on each campus would be developed in accordance with the designated role of the campus in agricultural

Rice breeding and management trials attract large numbers of visitors on a Cooperative Extension field day. University research on agricultural problems has contributed significantly to productivity.

teaching and research, "… to be determined in part by the particular needs of California agriculture related to the geographic needs of each campus." The 1961 academic plan specified that, "The program at Berkeley should continue to emphasize teaching and Experiment Station research in the basic physical, biological, and social sciences, taking advantage of the vast array of scientific resources on that campus to add to the pool of fundamental knowledge upon which advances in agricultural technology depend. To insure the excellence of the program at Davis, which is the principal center for agricultural instruction and research, encouragement must be given to the developing of outstanding teaching and research programs in Letters and Science."

Emergence of Other Segments

The establishment of a vocational school in agriculture and engineering at San Luis Obispo in 1901 was preceded by the founding of a State Normal School at San Jose in 1857 and a state-supported school at Chico in 1887. These institu-

tions were the forerunners of the present 19-campus system of the California State University and Colleges. As of 1971, four of these—the campuses at San Luis Obispo, Chico, Fresno, and Pomona—had degree programs in agriculture. By decision of the Coordinating Council for Higher Education, offerings in agriculture in the state universities and colleges system are limited to those campuses. In addition, at Humboldt State University, degree programs in Wildlife and Fisheries were begun in 1946, and in Forestry in 1957.

State Universities and Colleges

The phenomenon of evolutionary growth in higher education in California is exemplified in the five institutions mentioned above. From its founding as a non-degree vocational school in agriculture and engineering, the California Polytechnic State University, San Luis Obispo, progressed to the offering of a cooperative B.S. degree program in the 1930s (with the University of California and Utah State University) and thereafter established its own B.S. degree prgram in 1942. It became the base campus for a branch at Pomona when, in 1938, it took over responsibility for Voorhies School for Boys. The Pomona branch became a separate institution within the state college system in 1966, and since 1972 has had standing as the California State Polytechnic University at Pomona. The beginning at Fresno in 1910 was as a junior college, which a year later became a state normal school. Its program and organization approached university status in 1965–68 and it became the California State University at Fresno in 1972. At Chico, instruction in agriculture was not offered until 1953, when a two-year program was begun. That program achieved four-year baccalaureate status in 1957. The campus became the California State University at Chico in 1972. At Humboldt, the beginning in 1913 was as a teacher's college. By 1935 the institution had developed as a liberal arts college, and it achieved its present status as Humboldt State University in 1972.

A common experience of each of the state university programs in agriculture and forestry is their establishment as places of learning in the vocations of agriculture and forestry and their commitment to the provision of hands-on instruction in the related arts and crafts. Each now continues applied instruction but with growing emphasis on work at the professional level. Each also now offers the Master's degree.

Community Colleges

Vocational instruction in agriculture appeared in a different setting with its introduction in the Bakersfield Junior College in 1915. A second program was begun in Modesto Junior College in 1921, and in Reedley Junior College in 1939. A specialty in ornamental horticulture was established at San Francisco City College in 1938. Growth prior to World War II thus was very slow, but the pace quickened in the post-war years. By 1952 the number of junior college programs in agriculture had grown to 12. In 1968, the number reached 36, and the present total is 56. Throughout most of this period, the junior college system was under the jurisdiction of the State Department of Education, but in 1969 a separate governing body, the Board of Governors for the California Community Colleges, was established.

Vocational Education in High School

Further evidence of strong interest in the state in vocational instruction in agriculture was the introduction of the first such program at the high school level in 1905, in the Kern County Union High School. By 1910, six of the 215 high schools in the state offered work in this area. While growth continued to be slow in California, significant impetus was provided nationally through passage in 1917 of the Smith-Hughes Act (Federal Vocational and Education Act of 1917). The Act appropriated funds in support of vocational instruction and teacher training in the states and required state matching funds on a one-for-one basis. It also required the creation of a State Board for Vocational Education. The Legislature promptly passed an enabling act and the State Department of Education established required plans and regulations.

The Smith-Hughes Act was at first interpreted as limiting its programs to regularly enrolled high school students. However, its language was later interpreted as permitting extension of the program to out-of-school youth and to adult farmers in evening classes. Over the following 50 years, federal funding for vocational education grew substantially. There was commensurate growth in California in vocational instruction in agriculture during this period—from 22 schools and 337 students in 1918–19 to 283 schools and 23,556 students in 1968–69.

In deference to legislative intent in the Smith-Hughes Act, the first programs under it provided for part-time day vocational courses in agriculture involving not less than three class hours per day on farm project work ("supervised farm practice") and related instruction, including farm mechanics, and not less than three hours per day in class or school. Provision was also made for related subjects in science and English. Evening and part-time courses in vocational agriculture were introduced in 1926.

Collateral program developments in the formative years included the institution of judging contests, held at the University Farm, Davis, for students in vocational agriculture (1922) and establishment of a camp and judging contests at the California State Fair (1923). The Future Farmers of America program was launched in 1928. It became an important adjunct of the program of the California Polytechnic School with the appointment there of a one-half time Executive Secretary of the FFA in 1932.

Vocational instruction in agriculture was officially extended to the junior colleges in 1933 through revision of the State Plan for Vocational Instruction. This was followed by further revision of the state plan to accommodate the introduction in 1938 of a curriculum in Agricultural Education in the University's College of Agriculture at Davis.

Despite momentous changes in the society—the Depression of the 1930s, World War II, and in the structure of California agriculture—the nature of vocational agriculture in California changed relatively little through the three decades preceding the 1960s. Early in the 1960s, however, a statement of expanded objectives was prepared in the Bureau of Vocational Education which introduced a concept of training for occupations requiring a knowledge of agriculture in addition to the traditional emphasis on training for farming.

The new direction involved recognition of the changes in agriculture itself—reflected in the coined term "agribusiness"—which included rapid advances in the level of agricultural technology (including increased mechanization), increases in size of individual farm and agricultural marketing firms, and a rapidly expanding role for industries supplying goods and services required in this new era in the production, processing, and marketing of agricultural commodities. It was supported by the Vocational Education Act of 1963, which quadrupled the federal funds available and emphasized vocational education as one of the basic contributions to social adjustments required by technological change.

The period of program development and growth was also one of change in the provision of teacher training and in the structure of state administration in vocational instruction in agriculture. Teacher training in the early part of this period was largely the responsibility of the University, and the resulting instruction in the vocational programs was widely thought to be too scientific and academic, and seriously deficient in practical training. Disagreement between the Bureau of Agricultural Education and the University led to discontinuation of the position in teacher training in the University and the reestablishment of this function in the Bureau.

The Master Plan

A near century of growth and sometimes reluctant adjustment in the system in higher education in California was drawing to an end in the period immediately following World War II. This also was a time of great growth in the state's population and economic development, and of unprecedented growth in enrollments and costs in its system of higher education. The climate thus was near-optimal for the introduction of proposals to rationalize the structure of higher education in the state. Such a proposal emerged in the late 1950s, initially as a constitutional amendment, but finally enacted by the Legislature and signed into law in April 1960. This legislation, known as the Donohoe Act, established the basis for a master plan for higher education in California and established a coordinating council.

The Master Plan formalized and restructured the existing three-tiered system of higher education in California. The junior colleges were reconstituted as "community colleges," removed from the jurisdiction of the State Board of Education, and placed under the authority of a new Board of Governors. Provision was made for the offering of vocational instruction, and also for college-level instruction through the 14th grade and for transfer privilege to the other segments for qualifying students. An associate degree in the arts or sciences was authorized.

The Plan removed state colleges from the administration of the State Board of Education, reconstituted them as the California State Colleges, and placed this system under the authority of a new and separate Board of Trustees. The degree-granting authority of the former system was continued, and authority was granted to conduct research consistent with the primary function of the state colleges. The Plan permitted award of the Ph.D. degree jointly with the University of California.

The governmental structure of the University of California was unchanged in the Plan, but some functions were redefined. The University was authorized to continue instruction in the liberal arts, sciences, and professions (including teacher education) and to have exclusive jurisdiction over graduate instruction in law, medicine, dentistry, and veterinary medicine. Sole authority for award of the doctoral degree was reserved to the University—excepting the joint program with the state University and Colleges. And the University was designated as the primary state-supported academic agency for research.

In subsequent legislation the state college system was designated the California State University and Colleges, and authority for the Ph.D. was extended to include joint programs with accredited private universities, provided the doctoral program was approved by the California Postsecondary Education Commission.

Contemporary Programs

The 20 years since the introduction of the Master Plan for higher education in California have brought major changes in the content of instruction and research in agriculture. These reflect dramatic advances in science and technology, and major shifts in social, economic, demographic, environmental, and other factors affecting agriculture and society in general. As the following brief sketches of current programs indicate, however, institutional adjustment has been faithful to the policy guidelines established in the Master Plan.

Campus Instruction

University of California

As an outgrowth of historical precedent and its charge under the Master Plan, the University's programs of instruction in agriculture are strongly focused on the related biological, physical and social sciences, and include extensive instruction toward the M.S. and Ph.D. degrees. Instruction is offered in the College of Agricultural and Environmental Sciences (Davis), the College of Natural Resources (Berkeley), and the College of Natural and Agricultural Sciences (Riverside).

By far the most comprehensive of these programs is at the Davis campus. The College continues to emphasize the traditionally important program in food and fiber production. It also is engaged with broad issues arising from social and technological change, a concern reflected in programs in environmental protection, improving nutrition in major segments of the population, developing and utilizing human and renewable natural resources, and in many related fields. Campus instruction is organized within a total of 39 undergraduate majors and 28 fields of graduate study. These majors include curricula in various fields of agricultural production, plant

and animal protection (from pests and disease), food technology, consumer science, applied behavioral sciences, agricultural and managerial economics, environmental toxicology, land, water and air resources, and so on. In a separate school, the Davis campus offers the state's only program in Veterinary Medicine.

The degree programs at Berkeley have as a central theme the use of natural resources in ways that are at once productive, conservative of those natural resources, and protective of environmental quality, while meeting the accelerating rise in demand for essential food, fiber, timber, wood products, and recreational use of open space. The College offers 17 undergraduate majors, and 16 fields of graduate study. These include several majors in applied behavioral sciences, and majors in forestry and wood science, conservation genetics, entomology, plant pathology, plant and soil biology, nutrition and dietetics, conservation of natural resources, genetics, agricultural and resource economics, and other related fields.

At Riverside instruction in agricultural sciences is closely integrated in a single College that includes the biological and physical sciences. The College offers 20 undergraduate majors and graduate instruction in 15 fields. Agricultural applications are centered in such areas as biochemistry, plant science, environmental science, entomology, plant pathology, nematology, and soils and plant nutrition.

State Universities and Colleges

The four campuses in California's State University and College system with degree instruction in agriculture and a fifth campus with a program in forestry and natural resources share a common commitment to undergraduate instruction, with emphasis on the application of science in vocational instruction—including hands-on production experience. In support of this philosophy, the programs on the agricultural campuses maintain extensive facilities—including well-equipped university farms of substantial acreage—for student experience in agricultural production and for applied research. An additional important curriculum of the four agricultural campuses is the teaching credential program in vocational agriculture. Award of the credential requires achievement of the B.S. degree, except in unusual instances when the candidate has extensive occupational experience in a particular skill. A restricted credential may, under this circumstance, be awarded which is valid only in that particular area of expertise.

While the five State University programs share a basic educational philosophy, they differ in enrollment, choice of curriculum, and areas of emphasis.

Humboldt State University at Arcata is unique in the State University system with its extensive degree program in natural resources. The B.S. degree is offered in forestry, fisheries, oceanography, range management, wildlife management, watershed mangement, and resource planning and interpretation. The M.S. degree is offered in natural resources.

California Polytechnic State University at San Luis Obispo has the largest agricultural enrollment in the State University system (54 percent of the four-campus total). It offers the B.S. degree in 13 areas: agricultural engineering, agricultural management, agricultural sciences, animal science, crop science, dairy science, fruit science, food science, mechanized agriculture, natural resources management, ornamental horticulture, poultry industry, and soil science. In 1979, nearly 70 percent of the enrollment was in four majors—agricultural management, ornamental horticulture, animal science, and natural resource management. Approximately 3,700 acres of the 5,176-acre campus are utilized for instruction in agriculture. Twelve hundred students each year participate in a hands-on program of instruction financed by the University Foundation.

The California State Polytechnic University at Pomona offers the B.S. degree in seven departments: agricultural business management, agricultural engineering, agricultural science, animal science, food and nutrition, ornamental horticulture, and plant and soil science. The major in animal science—with specialities in animal production, animal agribusiness, and preveterinary studies—accounted, in 1979–80, for one-third of the agricultural enrollment at Pomona. The extensive facilities include an 800-acre university farm and rangeland unit and a 53-acre citrus and avocado center, which are used to provide production experience to every major.

The State University at Fresno offers three B.S. degree curricula in agriculture: agricultural business, agricultural education, and a general agricultural science program, with 16 fields of specialization. The campus has the only recognized enology program in the State University. Its M.S. degree programs in agriculture include an option in Agricultural Chemistry. A 1,200-acre university farm is maintained for instruction and applied studies.

The State University at Chico offers the B.S. degree in agriculture with seven options: agricultural business, agronomy, horticulture, range management, agricultural mechanics, animal science, and general agriculture. The M.S. degree in agriculture also is available. The University maintains a farm of 803 acres of irrigated farmland, an additional 95 acres of rice land, and 240 acres of rangeland that are available for student crop production experience.

California Community Colleges

At present (1980–81), 56 of the California Community Colleges—over half of the total of 106—offer a two-year program of instruction in agriculture and natural resources. The number of separate programs and subjects varies among campuses (ranging from one to over a dozen choices on a single campus). The fields of specialization systemwide include agribusiness, agricultural production, agricultural supply and services; agricultural mechanics, mechanized agriculture; ornamental horticulture, landscape installation and maintenance, nursery management; animal science, animal health technology; food processing; natural resources, wildlife technology, forestry, forestry/timber technology; and plant science, soil science, and crop production. In 1978 more than 60 percent of statewide enrollments were in two curricular areas—agricultural production/agribusiness, and ornamental horticulture. Approximately 65 percent of current enrollments are part-time (less than 12 credit hours per term).

All of the community college programs emphasize practical instruction and most of them provide hands-on vocational training in preparation for employment in agriculture and natural resources. Twenty campuses operate laboratory/farm training stations to facilitate practical instruction.

High Schools

Vocational instruction in agriculture continues to be an important curricular component in California's high schools, with 400 schools engaged in this field statewide in 1980–81. The programs serve high school students who plan to enter farming or agribusiness occupations as well as students planning advanced study in the agricultural colleges. Practical instruction is offered in such fields as soils, animal and plant production, insect control, selection, operation and maintenance of farm machinery, and management. Practical applications are designed for realism and frequently are conducted as a part of commercial farming operations. An important adjunct is the Future Farmers of America, which provides student leadership activities as an integral part of high school instruction in vocational agriculture.

Extension Education

All four segments of agricultural education in the state engage in various forms of extension education and public service. The most comprehensive of these is conducted by Cooperative Extension, a part of the University of California's Division of Agricultural Sciences. This program, a long-established activity in land-grant universities, receives substantial annual federal funding under the Smith-Lever Act and also is supported by county governments. It conducts educational and demonstration programs over virtually the whole range of subjects dealt with in the University's teaching and research programs, and itself supports applied research of its academic staff. A major activity of Cooperative Extension is its 4-H Club Program for youth. Initially established as a farm-related educational service for rural youth, this program now has a substantial urban clientele.

Cooperative Extension operates extension education and public service offices in 54 California counties. Each office is staffed with advisors in particular fields of expertise as appropriate within each local geographical area. In addition, specialists in various scientific disciplines are stationed on each of the Berkeley, Davis, and Riverside campuses and at the San Joaquin Valley Agricultural Research and Extension Center at Parlier, where they work closely with counterparts in the Agricultural Experiment Station. The specialists work primarily in extension education in cooperation with the county-based advisors, as well as with state agencies and various consumer and commodity groups. A vital function beyond the extension education role is feeding back to the teaching-research faculty information concerning problems in need of research.

Each state university campus with programs in agriculture and natural resources also is engaged in various kinds of outreach activity. The specific nature of these programs varies with institution and location. The means include conducting field days and judging events, seminars, short courses, workshops, and other educational and public service activities. An important part of these activities involves extension education for adults and continuing education for high school and community college teachers of vocational agriculture.

Research

While several public agencies conduct research in agriculture and related natural resource fields in California, the Agricultural Experiment Station of the University continues as the state's major center for such research. This work is fully integrated with the campus teaching programs at Davis, Berkeley, and Riverside. This is accomplished by means of a staff appointment pattern in which nearly every investigator's appointment pattern is split, with part time as a member of a college teaching faculty. On the average, statewide, about one-third of such appointments are budgeted in teaching and the remainder in research.

With the close tie between research and teaching, most of the work in the Agricultural Experiment Station is conducted on the three campuses, although extensive field research is performed at nine field stations maintained by the Experiment Station. Field research also is conducted in cooperation with other public agencies, and with individual farmers throughout the state. A substantial effort in applied research also is carried on in Cooperative Extension, frequently in cooperation with staff of the Agricultural Experiment Station.

The University's research in agriculture continues to emphasize varietal, cultural, and other forms of technological change that will increase agricultural productivity. However, major changes in the economic and social structure of the U.S., and new categories of problems arising in part from demographic and technological developments affecting agriculture, have led to a remarkably broadened and more complex research agenda. Much effort now is expended on issues concerning structure and economic performance in agriculture production and marketing, natural resource management and conservation, environmental protection, human nutrition, family and consumer welfare, farm labor, youth development, and other related areas. The variety and range of this program is implied in the fact that the Agricultural Experiment Station supports work in more than 1,200 separate research projects.

As provided for in the Master Plan, applied research consistent with their primary functions also is conducted in the five State University campuses with specialized programs in agriculture and natural resources. Unlike the University of California, no state funds are appropriated for this purpose. At the Humboldt campus, applied research is conducted on problems in resource management. A major focus of this work is in the California Cooperative Fishery Research Unit (established by agreement with the University, U.S. Department of Interior, and California Department of Fish and Game) where emphasis is given to environmental problems affecting fishery and aquatic resources. The Humboldt campus also cooperates in a Sea Grant program administered by the University of California. At San Luis Obispo, applied research is conducted on problems related to crop and livestock production, soil science, natural resource management, and agricultural engineering, while applied research in animal science and other fields in conducted at Pomona. At the Fresno campus applied research is facilitated by two Centers, one in viticulture and the other in irrigation technology. Emphasis in this work is given to the resolution of problems in such areas as viticulture, enology, raisin production, pest management, and animal science. Various applied research studies, particularly varietal testing of grapes, walnuts, almonds, cotton, and wheat, are in progress at the State University campus at Chico.

Two other state agencies, the Resources Agency and the Department of Food and Agriculture, expend a small fraction of their total effort on specialized applied research in agriculture and natural resources, and also sponsor research within their statutory responsibilities through contract with other agencies, frequently the University and the State University and Colleges.

In addition to the research performed within California state agencies, a major program in agriculture and natural resource studies is conducted in the state by the U.S. Department of Agriculture. This activity is organized within an

administrative grouping known as the Science and Education Administration (SEA) established in conjunction with the Food and Agriculture Act of 1977. Research specific to forestry is organized within the Forest Service of the USDA.

The USDA agencies directly concerned with agricultural research in California are the following: (1) SEA-Cooperative Research which administers the Hatch and McIntire-Stennis funds made available in support of research in the Agricultural Experiment Station; (2) SEA-Agricultural Research which conducts all USDA research in California concerning food and fiber production and processing; (3) SEA-Economics and Statistics Service, which conducts economic studies concerning agriculture and natural resource issues; and (4) the Forest Service's Pacific Southwest Forest and Range Experiment Station.

The locations of USDA research are widely scattered throughout the state. Research funded under the Hatch and McIntire-Stennis Acts is conducted in the Experiment Station (primarily at the Berkeley, Davis, and Riverside campuses). The agricultural research activity of the U.S. Department of Agriculture has its largest operation at facilities in Fresno and Albany, but also conducts research at locations in Brawley, Shafter, Salinas, Indio, and Pasadena. This unit also carries on research on or adjacent to the University of California campuses at Berkeley, Davis, and Riverside.

Research of the Economics and Statistics Service of the USDA almost without exception is carried on in close collaboration with Agricultural Experiment Station researchers at Berkeley, Davis, and Riverside, often under Cooperative Agreements whereby University and USDA resources are pooled in jointly conducted projects.

The Pacific Southwest Forest and Range Experiment Station is headquartered in Berkeley and conducts a significant part of its programs there. It also operates an Institute of Forest Genetics at Placerville, the Central Sierra Snow Laboratory at Soda Springs, and Redwood Science Laboratory at Arcata, and the Forest Fire Laboratory at Riverside. Several field research units also have been established, specifically for the field evaluation of chemical pesticides at Davis, for field research on the silviculture of Sierra Nevada conifers at Redding, for research on range wildlife and endangered species at Fresno, and for work with the management of chaparral and related ecosystems at Glendora.

While a substantial part of the USDA research in California is specific to California climatic, soil and cultural conditions, the programs also are regional—and in some respects national—in terms of subjects of inquiry and applicability of findings. A major element involves crop production and protection from agricultural pests, post-harvest physiology and storage of crops, resource problems associated with irrigated agriculture, and with the production of perishable fruits and vegetables. Forestry and range management problems under highly variable coastal, mountain, and semi-arid conditions are additional areas of focus in California-based USDA research on problems of regional significance. A major facility, the Western Regional Research Laboratory in Albany, is concerned with utilization and food safety research, nutritional quality of foods, toxicology and mutagenicity, energy conservation in processing, and related problems. Economic studies conducted by the USDA in California emphasize such areas as world trade in fruits and nuts, the pricing of agricultural products, and economic aspects of pest management technology, and of water and related land resources conservation.

Quantitative Measures

A few statistics concerning agricultural education and research in California are of interest in suggesting the magnitude of current programs. In the fall of 1980, for example, there were over 84,500 regularly enrolled students seeking certificates or college degrees in the field of agriculture, forestry, and related natural resources. Of these, 7,600 were enrolled in the University of California and 9,100 in the State University and College system. Community College fall 1980 enrollment in agriculture and natural resources was approximately 22,000, while at the high school level there were nearly 46,000 students receiving vocational instruction in this field. Regarding degrees awarded in agriculture and natural resources in 1980, there were 2,400 granted by the University (1,790 B.S., 370 M.S., and 240 Ph.D.), and 1,790 by the State University and College system—roughly 110 of which were the M.S. degree.

Professional staff engaged in public-supported teaching and research in agriculture and natural resources totaled nearly 3,300 in 1979–80. Of these, 746 held research and teaching titles in the University and 484 were in academic appointments in Cooperative Extension. Faculty listings in agriculture and natural resources in the State University approximated 380 in 1979–80, and in the community colleges there were about 660 instructors, about 60 percent of whom were employed part time. In 1980–81 there were 630 high school teachers in vocational agriculture at work in 400 schools. In addition to state-employment personnel, approximately 370 scientists were assigned by the U.S. Department of Agriculture to research activities in California.

Outlook

While the organizational structure of agricultural research and education in California has not changed substantially in recent years, there have been major adjustments in the content of instruction and in research agenda. These flow from numerous highly interdependent factors, some arising from past agricultural research and development and some a reflection of change in society.

Demographic change, for example, is a key societal influence. Important developments include changes in population age distribution and, particularly in this country, increasing concentration of population in urban areas. Worldwide, the dominant demographic factor is accelerating population growth. In the U.S., increased concentration of population in cities is resulting in the transfer of some of the best producing land from agricultural to urban use. When coupled with increased affluence and greatly increased population mobility, a consequence is heavy new demands on the traditional open-space habitat of agriculture and forestry. The impact includes land use transfers at the margins of expanding cities and, perhaps more important, a pervasive intrusion of recreational and residential land use throughout the open space. This impairment of the basic land resource runs counter to the need for greatly increased agricultural production implicit in accelerating growth in world population.

While urban population has been expanding rapidly, the number of U.S. farmers and farms has decreased dramatically. A similar increase in concentration of food processing and distribution in few firms has occurred, along with increased integration of producer-processor-distributor operations. The consequent reduction in the number of open markets and auctions for farm products is the basis of diminishing confidence among consumers that prices are competitively determined and that decisions as to food safety and nutritive qualilty are being made in the public interest.

Recognition of inevitable increases in future world food needs must be linked with the fact that constraints on increased agricultural productivity are multiplying. These include continuing soil fertility depletion and erosion, progressive depletion of easily accessible supplies of strategic natural resources, and serious side effects from some of the new output-increasing technology (health hazards to applicators and consumers from chemical pest controls, pollution of groundwater aquifers and surface waters, etc.). Water—both as to quantity and quality—is of particular concern, and is regarded by some experts as a greater constraint on increased agricultural output in California than is land.

Positive factors in regard to maintaining, or increasing, agricultural productivity are found in recent scientific developments. These include advances in research techniques in plant breeding and reproduction, in knowledge of the basic process of photosynthesis, in the nitrogen fixation processes of certain plants, in animal and avian physiology and reproductive processes, in gene manipulation through recombinant DNA techniques, and in advances in computer capability and in the use of computers in data storage and operations control.

A suggestion as to the changing direction of agricultural research in California is given in the statement of research priorities made in the University of California Academic Plan for Agriculture, 1980. These priorities involve research contributing to:

• Achieving a long-term balance between world food and fiber needs and agricultural and forestry production capacity; evaluating strategies as to food production levels and the relief of nutritional deficiencies both domestically and internationally.

• Maintaining—or increasing—per unit productivity (e.g., per acre of crop, range, and forest lands, or per animal unit) through the improvement of plant and animal strains and improvement of cultural and disease control practices.

• Compensating for the effects of resource depletion—especially energy, water, and land; development of policy in regard to the technical, institutional, and management aspects of natural resources, both among competing uses and over time.

• Developing new sources of energy, with special concern for solar and biomass source.

• Development of more efficient and water-conserving water management systems, and dealing with water quality degradation resulting from over-pumping of groundwater and the leaching of mineral fertilizers and natural salts into underground aquifers or surface waters.

• Protection of environmental quality, in particular from toxic materials used in the control of pests and diseases.

• Protection of farm workers and consumers from farm chemical application hazards and residues; and providing ecological and consumer protection from the accumulation of toxicants in the food chain through recycling processes.

• Improving knowledge and education in regard to human nutrition-health interrelationships, and of farm production and food processing methods as they affect the nutritional quality of foods; development of nutritional standards and a food and nutrition policy for humans; improving knowledge and practices in regard to animal and plant nutrition—and in particular in regard to the role of trace elements in plant nutrition.

• Assessment of the social, as well as economic, impact of technological development.

• Development of information and policy analysis with regard to rural community development and improvement in the status of the rural work force.

The above priority listing indicates greatly increased attention to issues concerning resource development and conservation, environmental protection, and the use of less ecologically disruptive and less toxic methods in the control of pests. Concurrent goals include adaptation of plant and animal species to make them more disease and pest resistant, more tolerant of biological stress, and, as well, to improve productivity and nutritional quality. Other issues involve human nutrition, consumer information and protection, farm labor, small farm viability, and rural community development. The list implies the need for increased attention to problem solution through multidisciplinary effort.

Recapitulation

From the early and sometimes turbulent developments in education and research in agriculture in California, a clear pattern of philosophical division and institutional evolution is evident. At the outset, in the mid-1800s, there was the strong public commitment to education, initially a commitment most urgently pressed at the college rather than public school level. Support at the college level was sharply divided, with some insisting on education in a classical, liberal arts tradition, and others demanding more practical education in agriculture and the mechanic arts. It was in the midst of this controversy that Professor Hilgard exercised his talent in establishing credibility for education and research in agriculture at a scientific level in the University, while also persuading farmers of the validity of the scientific approach and providing them with practical instruction in lectures throughout the state and in the Farmer's Institutes. But this reconciliation was not easy. On more than one occasion the pressure for more practical emphasis threatened to split or restructure the University.

Accommodation to opposing views and adjustments designed to serve perceived needs was achieved through an evolutionary process and through the creation of new institutions with more narrowly focused objectives. This was accomplished within the University of California through the development of academic curricula with strong scientific orientation designed to utilize strength in the basic sciences in the University as a whole. These scientific curricula were closely coupled through joint appointment of faculty in both teaching and research in the Agricultural Experiment Station. An applied element was retained in the research of the Agricultural Experiment Station and in the public service activities of the teaching-research faculty. Public service involving extension education in agriculture became a special function in the University with the creation of the Agricultural Extension Service. The establishment of vocational instruction at community college and high school levels, represents another response. Through these varied means, the state has come to offer a broad spectrum of education and research in agriculture such that within one or another of its several segments, a desired level of program may be found.

The Wheel Turns

As the institutional and technical setting of agricultural production has become more complex, so also have the relationships of academic institutions to their several constituencies. At the outset Professor Hilgard had only to deal with academic colleagues determined that the University engage in work of classical and scientific merit, and with the farm interests who for the most part wanted emphasis on practical training. It was then universally accepted that to improve agricultural productivity would make farming more profitable and at the same time benefit consumers.

Now the dominance of few and large firms, particularly in processing and distribution, has inspired a constituency concerned about purity and nutritional quality of product, and suspicious that equity is not achieved in the pricing of products. Environmentalists seek objectives that inhibit free choice by producers in the utilization of production technology. Conservationists press for reduction in the level of energy use and recreationists seek to reduce the commitment of natural resources to the production of commodities as opposed to recreational use. Positions are strongly argued and there is frequent resort to legal process.

Academic institutions attempting in their teaching and research to span the wide range of issues involved find that they have not one but many publics—publics often in conflict and easily disturbed if a strongly held point of view is challenged, even though in a scrupulously objective analysis. A consequence has been prolonged controversy over policy with respect to agricultural research, focused primarily on the University.

During the past decade or so, observers close to the producer sector have sensed some anxiety there about a perceived lessening of commitment to the resolution of pressing problems in traditional areas of production and marketing. A more visible cadre of critics has appraised California's research and educational programs from a different viewpoint. These critics, described in one commentary as a "...mixed lot, not numerous as yet (including) some scientists, farmers and farm workers, environmentalists, nutritionists, social scientists, a few politicians..." claim that the work of University researchers has conveyed a special economic benefit to narrow, private agribusiness interests at the expense of farm workers, small family farms, taxpayers, and the quality of rural life.

Mechanization, with attendant displacement of farm workers, has been a primary target. More generally, it is argued that in concentrating research and education on output-increasing and labor-saving technology, specific obligations under the Hatch Act have been ignored. It is said, for example, that University research has been inattentive to the Hatch Act (1955 Amendment) commitment to the promotion of a "...sound and prosperous agriculture and rural life as indispensable to the maintenance of maximum employment and national prosperity and security—and the maximum contribution of agriculture to the welfare of the consumer..."

University researchers in agriculture are disposed to refute such claims, arguing that only a small fraction of their work (roughly 3.5 percent) is likely to produce results useful primarily to large operators, with the remainder of their research size-indifferent. Many are puzzled by ambiguity as to what values in rural life are to be protected, and as to the role agricultural research should play in relation to change stimulated in large part by massive shifts in the nonagricultural sector. Achievement of maximum employment in agriculture is seen as a questionable alternative to productivity objectives; and it is suggested that the Hatch Act exhortation on maximizing employment was stated in the context of national policy with respect to employment in the nation as a whole—not specifically agriculture.

Counter arguments notwithstanding, the critique has been strongly pressed in hearings before the University of California Board of Regents. A group known as the California Agrarian Action Project Inc. filed suit in January 1979 against the Regents seeking redress of grievances and revisions of University policy in regard to University research in agriculture. Counsel in this filing was the California Rural Legal Assistance, itself a federally funded program.

Meanwhile, a group of students on the Davis campus of the University were proposing the introduction of a new course of study, its aim being to develop a program of education and research in support of an "appropriate competitive and sustainable agriculture." Its features include decreasing dependency on scarce resources and on the use of environmentally harmful

chemicals—objectives to be sought in part through the establishment of "experiential/experimental" farms at Davis and elsewhere in the state. The perceived clientele: students wishing to prepare for careers as small farm operators.

The continuing critique of education and research in agriculture has been accompanied by some marginal adjustments in University research and extension, for example, in the areas of small farm viability, farm labor, and the social and economic implications of technological change. Major new research has been launched in the area of integrated pest management.

The 1970s controversy over agricultural education and research revives issues intensely argued in the founding years of the University and periodically since. There is visible in the current critique some of the humanist fervor in defense of the common man against the trusts that characterized the populist movement of the 1860s, regard for farming as a way of life and for toil on the land as espoused by the Grange, and a recurring clash concerning practical versus scientific approaches to education in agriculture.

Today's debate, however, is taking place in a vastly different setting. Policy with respect to how California's carefully segmented system of education and research in agriculture should adapt to current and future circumstances will remain an object of lively debate.

References

Boyce, A.M. "History of the Citrus Research Center and Agricultural Experiment Station." Proceedings, First International Citrus Symposium, Vol. 1, 1969.

Carr, E.S., *The Patrons of Husbandry on the Pacific Coast*. San Francisco: A. L. Bancroft and Sons, 1875.

Casamajor, P. (Ed.). *Forestry Education at the University of California, the First Fifty Years*. Berkeley: California Alumni Foresters, 1965.

The Centennial Record of the University of California. University of California Press, 1968.

Hicks, J.D. *The Populist Revolt*. University of Minnesota Press, 1931.

Hutchison, C.B. "The College of Agriculture, 1922–52." An interview conducted by Willa Klug Baum, Regional Oral History Project, Berkeley, 1961.

Knoblanch, H.C., Law, E.M., Meyer, W.P., et al. *State Agricultural Experiment Stations: A History of Research Policy and Procedure*. Cooperative State Experiment Station Service, U.S. Department of Agriculture, Misc. Pub. No. 904, May 1962.

"Memorial to Congress on an Agricultural College for California, 1853." *Agricultural History*. XL, 1, Jan. 1966.

Postsecondary Education in California: Information Digest-80. California Postsecondary Education Commission, Sacramento, 1980.

Rochester, A. "The Populist Movement in the United States." New York: International Publishers, 1943.

Stadtman, V.A. *The University of California 1868-1968*. New York: McGraw-Hill Book Co., 1970.

Stimson, R.W. and Lathrop, F.W. *History of Agricultural Education of Less than College Grade in the United States*. U.S. Office of Education, Vocational Division Bulletin No. 217, Agricultural Series No. 55, 1942.

Sutherland, S.S. "A History of Agricultural Education in the Secondary Schools of California, 1901 to 1940." State Bureau of Agricultural Education, State Department of Education. Typescript. July 1, 1940.

Sutherland, S.S., and Burlingham, H.H. "A History of Agricultural Education in California Secondary Schools, Vol. III." Bureau of Agricultural Education, California Department of Education, 1975.

Wickson, E.J. "Beginnings of Agricultural Education and Research in California." Report of the College of Agriculture and the Agricultural Experiment Station, University of California, July 1, 1917, to June 30, 1918.

19

State Government's Role

Jerry Scribner

Earlier chapters have described the rapid growth of California agriculture, one of the largest and most complex agricultural economies in the world. This chapter focuses on the important role of state and local government in the shaping of the farm scene as we know it today.

Over the years, most state regulation of agriculture in California has been sponsored by farmer-legislators with the active support of agricultural constituencies. More recently, agriculture's impact on the environment and interaction with other parts of the state's economy have resulted in an increasing amount of regulation sometimes at odds with farmers' interests. Not only is California a large agricultural state, it is also a large industrial state with powerful labor unions, strong environmental and consumer movements, and a progressive political tradition. In addition, because California citizens have felt geographically and politically isolated from eastern power centers and other western states, problems which elsewhere have been viewed as federal or regional have in this state been viewed as appropriate for local action.

Historically, California farmers turned to government for help in three critical areas: (1) resource development (land, water, transportation, and labor), (2) marketplace regulation, and (3) assistance in preventing the spread of plant and animal pests and diseases. This early pattern of government regulation continues to the present. Most of the governmental activity in disease prevention and marketplace regulation affecting agriculture has gradually been consolidated into one state agency, the Department of Food and Agriculture, and in the offices of county agricultural commissioners and sealers of weights and measures.

Government's role in resource development meanwhile has grown and shifted dramatically in focus. The promise of never-ending progress based on unlimited resources has given way to a realization that resources are limited and unchecked growth can have profound environmental consequences. With this realization has come a demand for government resource management with a strong emphasis on conservation. As the largest user of limited water supplies and as a major user of land, energy, and labor, agriculture has found itself increasingly regulated by agencies other than the Department of Food and Agriculture. Some of these are: (1) the agencies overseeing labor (the Agricultural Labor Relations Board, the Employment Development Department, and the Department of Industrial Relations); (2) the agencies overseeing natural resources (the Department of Water Resources, the Water Resources Control Board, the Air Resources Board, the Energy Commission, and the Department of Conservation); and (3) the agencies supervising business and transportation.

State regulation of agriculture is also strongly influenced by the political philosophy and party affiliation of those in the governor's office and legislative leadership positions.

The following pages describe the evolution of state regulation of agriculture as implemented through the executive and legislative branches in the last two decades; the history and current structure of the Department of Food and Agriculture; and the role of other state agencies in issues which affect agriculture. In addition, brief reviews of the scope of state government and of the legislative process are included as chapter appendices.

The Executive Branch

Policy and Decision-Making

The people of California exercise their political power by selecting 120 legislators and the top leadership of the executive branch. The governor, lieutenant governor, treasurer, controller, secretary of state, attorney general and superintendent of public instruction are elected every four years. Each of these elected officials has by statute the right to appoint, on a noncivil service basis, various deputies and officials to assist in the administration of the laws charged to their particular office.

By far the most important office is that of governor, since he appoints virtually the entire executive leadership. Not surprisingly, the governor's appointments typically reflect his political philosophy and party affiliation. The governor also appoints judges to vacancies which occur during his term; it has often been remarked that because most judges serve for life this is frequently the most enduring impact of a governor's tenure.

Day-to-day decision-making in the executive branch revolves around the governor and his chief of staff. Weekly cabinet meetings are usually held to discuss major administration goals and policies in some depth. Conflicts between agencies are typically handled by direct face-to-face meetings with agency heads and a member of the governor's staff or the governor himself. Relations with the legislature are coordinated through the governor's secretary for legislation.

The governor's key agricultural policy and political advisor is the director of the Department of Food and Agriculture, who is a member of the cabinet and equivalent in rank to an agency secretary. The governor also appoints the 15-member State Board of Food and Agriculture which advises the director and governor on matters of importance to the state's agricultural interests. The governor also meets and confers by telephone with the leaders of agricultural organizations on particular legislative or administrative issues.

Many issues involve more than one department—for example, water. The federal government, local water agencies, and the State Department of Water Resources have primary responsibility for water management. However, the State Water Resources Control Board has responsibility for water rights, water quality, and waste treatment. The Fish and Game Department is concerned about any adverse impact of water decisions on fish or wildlife or their habitat. The Health Department must be included in decisions involving drinking water, swimming or boating, mosquito control or possible disease implications of whatever might be happening with the water. Most bodies of water are also used for recreation, so the Department of Parks and Recreation may need to be consulted or the Department of Boating and Waterways. The Department of Food and Agriculture will be concerned about decisions of other agencies which make it more difficult for farmers to obtain or use water.

Issues which involve major disputes within the executive branch or are likely to anger important political constituencies in the state are often bucked up to higher levels for a decision. Most executive branch decisions, however, are made well below the director's or governor's level. They involve the day-to-day implementation of laws passed by the legislature and are reached after a formal and informal interchange of information and views between the various state agencies. Many advisory committees within the executive branch help facilitate decision-making by keeping neighboring departments informed of actions which may cross administrative lines. The Department of Food and Agriculture, for example, maintains a Pesticide Advisory Committee composed of representatives from Health, Industrial Relations, Fish and Game, the Agricultural Commissioners Association and other agencies to assure that all views are considered in reaching pesticide regulatory decisions.

Major executive branch decisions are issued as administrative regulations. The California Administrative Code currently contains over 30,000 pages of regulation. These regulations can only be adopted or amended after a public hearing and other procedural steps. At the hearings interested members of the public, legislators and other administrative units offer comments and suggested changes in proposed regulations. The regulations which are adopted must be based on statutory authority and can be challenged in court if they go beyond the authority of the agency issuing them.

The policies and decisions of the executive branch, though governed by statute, are directly influenced by the political persuasion of the governor. Naturally, there may be opposition to the general change in direction and philosophy which follows from a change in administration. In general, however, the agricultural sector has enjoyed bi-partisan support in the executive branch regardless of who was elected governor.

Changes in the Executive Stance Toward Agriculture Since 1960

Fifteen years of political change, from 1960 to 1975, including reapportionment and a change of governors, left California agriculture relatively unaffected. The two-term administration of Governor Edmund G. Brown, Sr. was marked by rapid state development and a strong pro-agriculture, pro-water development posture. Governor Reagan's two terms from 1966 to 1974, on the other hand, brought a more conservative fiscal policy to the state and saw the creation of a number of new resource-oriented agencies. Although reapportionment changed the composition of the legislature in the 1966 and subsequent elections, the presence of a conservative Republican governor served to mask the liberal drift and increasing urbanization of the legislature, and agriculture retained a strong voice in the Capitol. Reagan's appointments as director of the Department of Food and Agriculture were generally agreeable to the agricultural community.

Governor Edmund G. Brown, Jr.'s election in 1974 brought these previous conditions to an abrupt turning point. Most significantly, he was a liberal Democrat, elected with the active support of the United Farm Workers and their urban liberal allies. One of his earliest appointments was that of Rose Elizabeth Bird to be Secretary of the Agricultural and Services Agency. She was the first woman ever to be accorded cabinet status, and the first ever to be directly responsible for communications between the governor and agriculture. Moreover, Bird was the first such appointment to represent agriculture from outside the farming community.

Political evolution also involves the matter of style. Edmund G. Brown, Sr. was an ebullient, outgoing governor who had, before his election

to governor, been the state's attorney general and prior to that the district attorney of San Francisco. He had a wide circle of friends and political acquaintances, and was approachable. Governor Reagan, elected in 1966, was less accessible than Brown and relied almost totally on his staff for policy decisions. Governor Jerry Brown employed neither the multi-channeled, broad-based political strategies of his father, nor the modern managerial techniques of Governor Reagan. He and his staff took a very intense, personal interest in a great many details that had not been of interest to previous governors. They also were uncomfortable with the old-boy, club-member approach to political networking that is a hallmark of American political tradition. As a consequence, traditional farm groups found themselves virtually cut off from communication with the new administration.

The first major policy question the new Brown administration grappled with was the enactment of a farm labor law. The Agricultural Labor Relations Act of 1975 was drafted by the Agricultural and Services Agency Secretary and her staff, negotiated with key legislators and agricultural representatives, and signed into law on June 5, 1975. The governor's initial appointments to the five-member Agricultural Labor Relations Board were viewed by agriculture as pro-United Farm Workers and anti-agriculture. Furthermore, the board's unprecedented workload caused it to go bankrupt approximately eight months into its first year of operation. Agricultural interests blocked refunding for four months, reflecting their hostility toward the administration. For nearly two years, relations between the executive branch and farm groups were strained and tense.

Communication between the governor and agriculture improved, however, in April 1977 with the appointment of Richard Rominger, a fourth generation Yolo County farmer, as director of the Department of Food and Agriculture, and the simultaneous restructuring of the executive branch to make the director a member of the cabinet. Although the governor actively courted agricultural interests in his second-term reelection drive and in his two brief presidential campaigns, his support of farm labor and of environmental protection measures undercut gains with agriculture on other fronts.

The Legislative Branch

Reapportionment: Implications for Agriculture
Prior to 1966, the California Assembly, like the
U.S. House of Representatives, was divided into
districts of roughly equal size based on population, and the state Senate, like the U.S. Senate,
was divided into districts based on geography.
Senate districts could not be smaller than one
county nor larger than three counties, and counties could not be split. Thus most senators represented one county with a few representing two
or three to make up the 40-member Senate. Under this system, nearly half of the state's population residing in Los Angeles County had one
senator, while sparsely settled northern and rural
counties held a majority of seats in the state Senate.

In 1965, the California Supreme Court in re-
ponse to federal decisions requiring state legislatures to be elected on the basis of the "one
man—one vote" principle ordered the Senate
and Assembly reapportioned. The effect on the
Assembly was negligible, with one seat from San
Francisco being dropped and an additional seat
being added in the Orange County/San Bernardino area, but the effect on the Senate was
dramatic.

Many of the senators representing smaller
northern California counties were forced to run
against each other in much larger geographical
districts. At the same time, in the populous
southern half of the state, new senate districts
were created and urban-oriented legislators elected to fill them. One example of the impact of
the court's decision was that of the First Senatorial District which went from seven senators
representing 15 counties to one. Simultaneously,
Los Angeles County went from one senator to
13. The 15 counties in the first senatorial district
contain a significant portion of California's watershed, timber, cattle grazing, orchards, and rice
growing. The result of the 1966 election was that
the once prestigious Senate Agriculture Committee had to be reduced in size (to seven members)
and even then it was not possible to find enough
interested members to fill all the seats. Although
the 1966 elections resulted in a loss of agricultural background and experience in the Senate, it
did not result in the strong switch to a liberal-
urban orientation that many had predicted. In
fact, the Senate may have ended up slightly
more conservative as a consequence of the replacement of a number of moderate to liberal
Democratic senators from the north by suburban
Republicans or conservative Democrats from
southern California.

In the meantime, the Assembly Committee on
Agriculture, which had declined in influence in
the early sixties under the urban-oriented speak-
ership of Jesse Unruh, experienced a resurgence
as northern rural Democrats, and later valley Republican interests, strengthened and revitalized
the committee.

By the late sixties and through the seventies,
agriculture's hammerlock on state policies
viewed as inimical to their interests was gone.
Although the urban-liberal shift had been muted,
the loss of agricultural power and expertise in
the Senate, combined with the increasing concern with the environmental consequences of
growth, opened the door to enactment of a number of sweeping environmental protection measures with significant impact on agriculture. For
example, in 1963 a bill to ban DDT was quickly
killed in committee. By 1969, DDT had been ban-
ned and California had undertaken the most
comprehensive regulation of on-farm pesticide
use in history. This regulatory leadership continues today with California far out in front of all
other states, the federal government, and foreign
countries. During the seventies, the California
Energy Commmission was created, the powers
of the State Water Resources Control Board
greatly expanded, and the Air Resources Board
was given wide powers to require industry to install expensive emission control equipment. The
Wild Rivers Act was also passed and water development generally frustrated. It is noteworthy
that nearly all these legislative actions took place
during the two terms of Governor Ronald Reagan, who enjoyed the support of agriculture
and opposed much of this legislation.

*Agricultural Issues: Land Development,
Farm Labor*
Land use legislation aimed at balancing the need
for urban expansion with the need to preserve
land for agriculture and open space occupied the
attention of the legislature nearly continuously
from the late fifties and early sixties through the
seventies. In 1961, the legislature placed a proposition on the ballot to constitutionally authorize
local assessors to assess farm land on the basis of
existing use rather than highest and best use.

However, opposition by urban interests and some assessors concerned with loss of local revenue led to its defeat.

In 1963, the Assembly Agriculture Committee undertook a study in cooperation with the State Department of Agriculture and others which led to the enactment in 1965 of the California Land Conservation Act, popularly known as the Williamson Act. The Williamson Act provides for voluntary contracts between local landowners of "agricultural land" (as defined in the law) and local assessors—under which the land is taxed at a reduced rate in return for the landowner's agreement to restrict the land to agricultural uses for a period of 10 years. The contracts are automatically renewed each year, unless cancelled, in which event there is a nine-year wait before it can be developed.

There are currently about 16.2 million acres of land in California under Williamson Act protection, roughly one-third of the privately owned land in the state. However, much of the land under protection is in the Central Valley, far from the spreading urban fringe. Moreover, landowners in the path of urban development have by and large opted not to put their lands into agricultural preserves.

In response to the weaknesses in the Williamson Act, efforts to enact mandatory agricultural land use controls were mounted sporadically during the seventies. In 1976, major legislation to comprehensively regulate prime farm land passed the Assembly and one Senate committee. The bill narrowly failed passage in the Senate Finance Committee after intense lobbying against it by farm groups and real estate and development interests.

Believing that land use legislation might be successful in the 1977 session, and desiring to take a more constructive approach to the problem, farm lobbying organizations joined together in a series of discussions which led to the introduction in early 1977 of a bill to protect farm land, which was supported by agriculture. Rival bills were supported by environmental interests. Two of these bills were extensively debated in the Senate in 1977 and ultimately passed and sent to the Assembly in early 1978. There, along with a rival Assembly bill, all died in a crossfire of opposition. Farm groups, local government, and the real estate lobby all insisted that there be "local control" of the process while environmentalists feared that without strong state over-

sight local control would inevitably lead to the continued piecemeal urbanization of California.

Although all parties continue to support in principle the idea of agricultural land preservation, the reasons underlying the demise of the '77–'78 legislation persist. With increased fiscal conservatism and a backlash against government regulation at all levels, the drive to protect California's agricultural land from urban development has been at least temporarily dampened.

Farm labor was a major issue in the Legislature prior to reapportionment and many observers expected the change in representation to virtually guarantee that legislation on farm worker protections agreeable to labor would soon be enacted despite agriculture's opposition. This did not happen, in part because the nonagricultural senators swept in by reapportionment were more suburban and conservative than urban liberal, and partly because Governor Reagan consistently vetoed those bills on this subject which did reach his desk.

However, the farm labor struggle set the stage for agriculture's belated recognition of the need to recruit nonagricultural legislators to their point of view. If agriculture's declining influence in the legislature reached its low point in the controversy surrounding the Agricultural Labor Relations Act in 1975 and 1976, then its defeat in November 1976 of Proposition 14 (a farm worker-backed initiative to further strengthen the ALRA) marked the end of the decline and the beginning of a new era—in which agricultural influence would be exercised by increased lobbying efforts and major increases in campaign contributions through farm bloc political action committees.

From 1966 forward, agricultural leaders have frequently spoken out about the need for agriculture, which now represents only five percent of the population, to "tell its story" to the urban public. Agriculture's future may well rest on its ability to do that.

The Department of Food and Agriculture

Background

Settlers who flocked to California during and after the 1850s brought with them not only their extensive knowledge of horticultural practices and problems, but years of experience in the uniquely American interrelationship between government and the economy. Concessions and encouragements to agriculture in the form of bounties and land grants, the maintenance and enforcement of standards in marketing, the use of import and export duties, and favorable taxation policies, were already well established in eastern states before the settling of California. In agricultural matters as well as others, California's founding fathers borrowed heavily from the constitutions and laws already in use in other states.

Between 1850 and 1870 the state attempted to encourage new immigrants with farming skills by giving bonuses or premiums for the production of new crops, and tax exemptions for growing crops and vines. This same period saw the decline of the cattle industry and the emergence of grain and specialty crop agriculture. Central to this evolution was a change in the Trespass Law which, prior to 1850, favored the cattle owner. By the 1870s the law had been reversed, making the cattleman responsible for keeping his stock off the lands of others. Cattlemen, however, did secure favorable legislation regulating the use of brands and provided for state registration, a program which continues up to the present day.

As California crops became more diversified, farmers found they needed more help with disease problems, marketing problems, and other types of public assistance. In 1880 the Board of State Viticultural Commissioners was authorized and appointed. Its enabling act provided for pest control, instructing the Board to "devote special attention to the study of phylloxera and other diseases of the vine." In 1881 County Boards of Horticultural Commissioners were established; two years later the State Board of Horticulture was authorized and took over pest inspection functions, employing an inspector of fruit pests. 1899 saw the enactment of the first basic Plant Quarantine Act providing for interstate quarantines. California's quarantine laws were compiled and reenacted in 1911 and rodent control and weed control responsibilities were added in 1917.

In 1919 the State Department of Agriculture was created, superseding the State Commissioner of Horticulture (established in 1903), the State Dairy Bureau (1895), and the State Veterinarian created in 1899. Other miscellaneous functions incorporated into the new Department were the enforcement of the Johnson Grass Law, the Insecticide and Fungicide Board, the regulation of commercial fertilizers (a law enacted in

1903), the Cattle Protection Board created in 1917, the control of predatory animals, control of the walnut codling moth, and the supervision of fairs and exhibits.

In 1921 the Department assumed responsibility for the enforcement of the Weights and Measures and the Net Container laws, as well as the Weighmaster Act, Hay Baling Act, Control of Dairy Containers, and the State Bread Act. This regulatory pattern continued to be expanded with the addition of new responsibilities and the broadening of previously enacted provisions. By 1929 the Market News Service had been established, as well as the federal/state Crop Reporting Service. The State Board of Agriculture had been created and the County Horticultural Commissioners had become County Agricultural Commissioners.

In the early 1930s the California Agricultural Code was drafted, consolidating all the previously passed enactments into one code. The California Marketing Act of 1937, which paralleled the depression-spawned Federal Marketing Act, significantly broadened the state's marketing program. Meanwhile, the legislature passed four milk stabilization laws sponsored by the dairy industry, and charged the Department with their administration and enforcement. The first of these was the Young Act, effective in 1935, authorizing the establishment of minimum prices which distributors must pay producers for fluid milk. In 1937 the Desmond Act authorized the director to establish minimum wholesale and retail prices for fluid milk. In 1947 the Dairy Industry Advisory Board was created (changed in 1963 to the Dairy Council of California), and the legislature also passed the Unfair Practices Act to regulate the sale of dairy products in connection with the sale of fluid milk. These four enactments combined with the Gonsalves Milk Pooling Act in 1968 to provide a comprehensive regulation of the dairy industry until the mid-1970s.

In 1961 the present organizational structure of the Department of Agriculture was established by consolidating the various bureaus into a system of seven divisions. A second reorganization took place in 1968 when Governor Reagan reorganized all state cabinet-level departments into agencies. The Department of Agriculture then became one of ten departments in the Agricultural and Services Agency. Concern that agriculture was losing its traditional seat at the cabinet table was handled by elevating Director of Agriculture Earl Coke to the position of Secretary of the Agency, with the tacit understanding that this position would continue to be filled by someone with strong ties to agriculture.

In 1972, the Department was renamed the Department of Food and Agriculture to reflect a growing consumer interest in agricultural matters. In 1977 the 1968 reorganization was effectively rescinded when Governor Brown pulled the Department out of the Agency structure and once again gave the Director a cabinet-level post. Another major change in 1977 was the creation of a new division—the Division of Pest Management, Environmental Protection and Worker Safety. The rapid expansion of this division reflected growing public concern with toxic substances in general and pesticides in particular.

Current Roles

The primary statutory charge of the Department of Food and Agriculture is to promote and protect California agriculture in the interests of the general welfare. Its present activities are described in the following catalog of its divisions. Figure 1 shows an organizational chart of the Department as of 1980.

The Executive Office

The executive office of the Department consists of the director and chief deputy director, appointed by the governor, and two deputy directors appointed by the director. One deputy serves as legislative liaison and as executive secretary of the State Board of Food and Agriculture. Other executive office personnel include a legal advisor; an information officer; a long range planning unit; three assistants on export trade, environmental concerns, land and water use, department hearings and special projects; and two regional coordinators. The executive office oversees the activities of seven operating divisions plus all administrative services.

Division of Animal Industry

This division provides inspections to assure that meat and dairy products are safe, wholesome, and properly labeled, and surveillance programs to prevent animal diseases from causing serious financial losses to producers. Detection and diagnosis of animal diseases is provided by state veterinary diagnostic laboratories. The division also helps protect California cattle producers from loss by theft or straying.

BUREAU OF ANIMAL HEALTH. In the past few years, the bureau has kept under control such animal diseases as bovine tuberculosis and brucellosis. Through enforcement of import regulations and controlled destruction of ship and aircraft garbage, exotic diseases such as foot and mouth disease, and African swine fever have been kept out of California.

VETERINARY LABORATORY SERVICES. Accurate and prompt diagnoses are required for livestock and poultry disease control. Laboratories, located in Petaluma, San Gabriel, Turlock, Fresno, and Sacramento, conduct tests, perform autopsies, and provide diagnoses for state and federal animal health veterinarians, private veterinarians, poultrymen, meat inspection and dairy service personnel, and UC research and extension veterinarians.

MEAT INSPECTION. A comprehensive state and federal inspection program assures that meat products sold in California for human consumption are wholesome, clean, truthfully labeled, and rigorously monitored for pesticide residues, antibiotics, and disease-causing organisms. The U.S. Department of Agriculture performs most of the meat inspection in California. The Department's role is to cover the small operations which are exempt from USDA inspection.

LIVESTOCK IDENTIFICATION. The high mobility of cattle limits the protection that can be provided by cattle owners and local law enforcement agencies. Statewide enforcement is, therefore, required. Under the program, livestock brands are registered and cattle are inspected for ownership prior to transportation, sale, or slaughter. Owners pay the total cost of the program through inspection and brand registration fees.

MILK AND DAIRY FOODS CONTROL. This bureau administers state laws relating to inspection and sanitation of the more than 10 billion pounds of milk produced by California's dairy cows each year, and sold fresh or converted to such items as butter, cheese, and ice cream. Milk products are inspected at various points to assure they are safe, wholesome, unadulterated, and correctly labeled.

Division of Plant Industry

The Division of Plant Industry's entomologists, plant pathologists, biologists, plant nematologists, and other scientists along with their counterparts in county agricultural commis-

DEPARTMENT OF FOOD AND AGRICULTURE ORGANIZATION CHART

DIRECTOR
CHIEF DEPUTY DIRECTOR
DEPUTY DIRECTOR
DEPUTY DIRECTOR

COUNTY AGRICULTURAL COMMISSIONERS
COUNTY SEALERS OF WEIGHTS AND MEASURES

REGIONAL COORDINATORS
SANTA ANA
SACRAMENTO

STATE BOARD OF FOOD AND AGRICULTURE
EXECUTIVE SECRETARY

ASSISTANT EXECUTIVE SECRETARY

ADVISORY BOARDS AND COMMISSIONS

STAFF TO DIRECTORATE
ADMINISTRATIVE ADVISER
INTERNATIONAL TRADE AND AGRICULTURAL ECONOMICS
ENVIRONMENTAL RESOURCES
EXECUTIVE ASSISTANT
AFFIRMATIVE ACTION
WOMEN'S PROGRAM COORDINATOR
PUBLIC INFORMATION OFFICE
LONG-RANGE PLANNING

DEPUTY DIRECTOR

ADMINISTRATIVE SERVICES
Departmental Services
Financial Services
Personnel Services
Training and Development

ASSISTANT DIRECTOR MARKETING SERVICES

DIVISION OF MARKETING SERVICES
Agricultural Statistics
Market Enforcement
Market News
Marketing
Milk Marketing Enforcement
Milk Pooling
Milk Stabilization
Nutritionist

CHIEF MEASUREMENT STANDARDS

DIVISION OF MEASUREMENT STANDARDS
Metric Conversion Council
Metrology
Petroleum Products
Quantity Control
Weighing and Measuring Devices
Weighmaster Enforcement

ASSISTANT DIRECTOR PEST MANAGEMENT

DIVISION OF PEST MANAGMENT, ENVIRONMENTAL PROTECTION AND WORKER SAFETY
Worker Health and Safety
Pesticide Enforcement
Pest Management and Environmental Monitoring
Pesticide Registration and Agricultural Productivity

CHIEF DEPUTY DIRECTOR

ASSISTANT DIRECTOR INSPECTION SERVICES

DIVISION OF INSPECTION SERVICES
Chemistry
Egg and Poultry Quality Control
Fruit and Vegetable Quality Control
Grain and Commodity Inspection
Feed, Fertilizer and Livestock Drugs

ASSISTANT DIRECTOR PLANT INDUSTRY

DIVISION OF PLANT INDUSTRY
Exclusion and Detection
Control and Eradication
Nursery and Seed Services
Laboratory Services

DEPUTY DIRECTOR

LEGISLATIVE ASSISTANTS

ASSISTANT DIRECTOR ANIMAL INDUSTRY

DIVISION OF ANIMAL INDUSTRY
Animal Health
Milk and Dairy Foods Control
Livestock Identification
Meat Inspection
Veterinary Laboratory Services

Approved Milk Inspection Services
County Veterinarian

CHIEF FAIRS AND EXPOSITIONS

DIVISION OF FAIRS AND EXPOSITIONS
Fair Budgets
Exhibits and Premiums
Engineering Services
Support Services

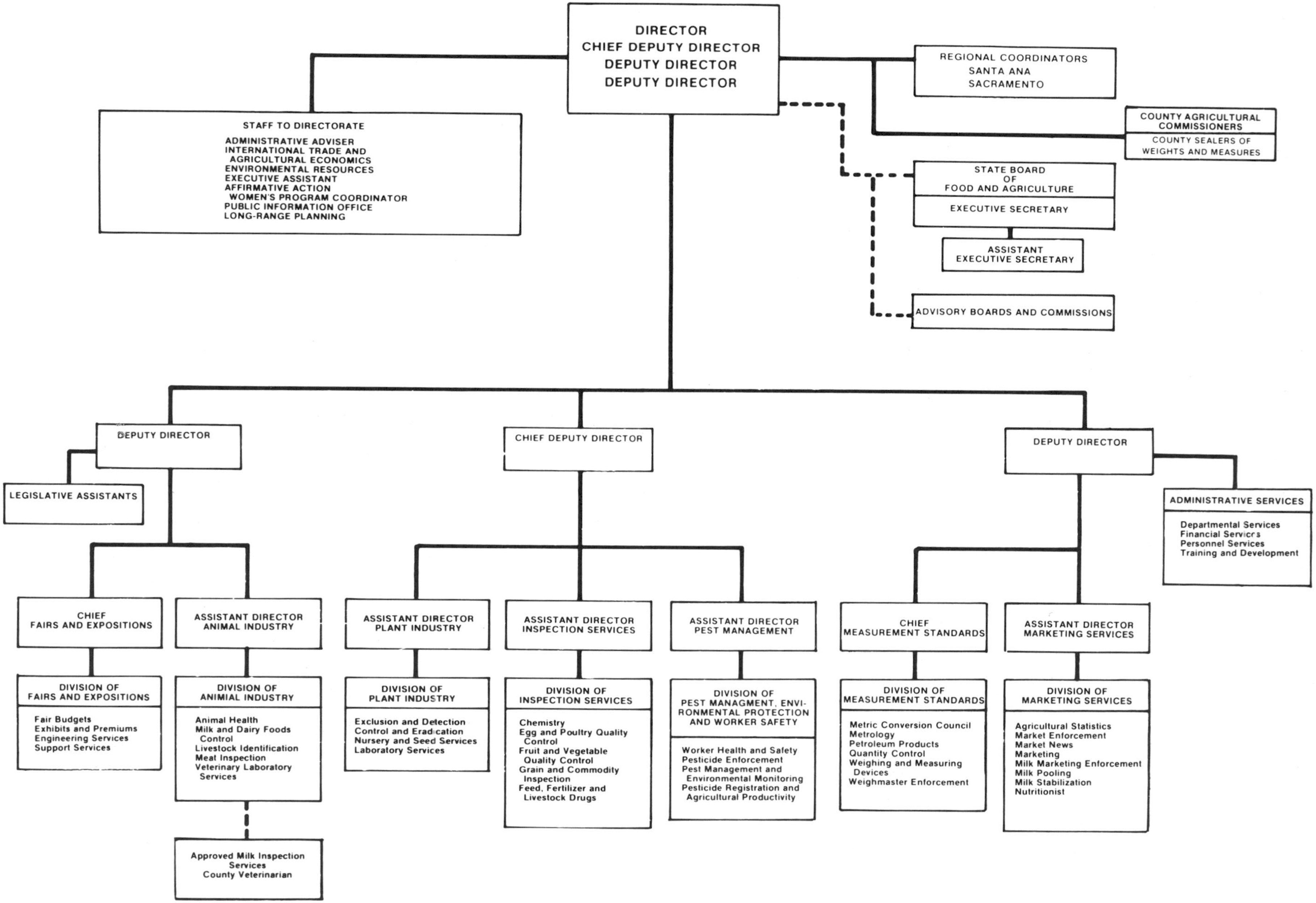

FIGURE 19.1 Department of Food and Agriculture organization chart, 1980.

The State of California uses border inspection stations to control potential pest and disease problems brought in on agricultural commodities transported from other states. *Photo by Charles Papp.*

sioners' offices combine their knowledge to protect California's home gardens, farms, forests, parks, and other outdoor areas from the introduction and spread of harmful plant, weed, and vertebrate pests.

LABORATORY SERVICES. This unit identifies pest organisms for governmental agencies, the agricultural industry, and private citizens. It has four labs: entomology, plant pathology, nematology, and botany-seed, plus a warehouse and greenhouse.

CONTROL AND ERADICATION. This unit, in its mission to control or eradicate major diseases and pests in the state, maintains a cooperative relationship with the county agricultural commissioners, USDA, and the University of California; gives technical advice; and helps in planning and coordinating the pest management work of the county departments of agriculture.

EXCLUSION AND DETECTION. Personnel at border agricultural inspection stations inspect inbound vehicles to protect against hitchhiking pests. They also inspect commercial truck shipments to prevent the entry of infested or substandard fruits, nuts, and vegetables. The unit also works with county commissioners for early detection of plant diseases, insects, and weeds.

NURSERY AND SEED SERVICE. This unit sees that the agricultural industry and the public receive high quality seed and nursery stock that is truthfully labeled and free from harmful insect pests and plant diseases. All persons selling nursery stock are licensed. Nurseries meeting cleanliness standards are given a California Nursery Stock Certificate, which allows nursery stock of suitable quality to be shipped interstate or intrastate without inspection at destination.

Division of Inspection Services

This division provides consumer protection and industry grading services on a wide range of agricultural commodities. These services involve the operation of chemistry laboratories; regulating the manufacture, labeling, and sale of fertilizing materials, livestock feed, and drugs; inspecting and certifying fresh fruits, vegetables, nuts, honey, shell eggs, grains, rice, and other agricultural commodities for grade, quality, condition factors, and weight.

CHEMISTRY LABORATORY SERVICES. The division's Sacramento laboratory has sophisticated electronic equipment to analyse and quantify a wide range of chemical substances. The division also operates four branch laboratories and two mobile laboratories. Each year more than 40,000 samples are processed from which over 100,000 specific analyses are made.

FERTILIZING MATERIALS. More than 4.5 million tons of fertilizer materials are used annually in California. Inspectors visit manufacturing plants, distribution outlets, and retail stores to check accuracy of labeling. Fertilizers are randomly sampled and laboratory tested to determine if label guarantees are correct.

COMMERCIAL FEED AND LIVESTOCK DRUGS. Producers need accurate and complete labeling in purchasing feed and drugs for the safe and economical production of meat, milk, and eggs. Livestock drugs are registered after label review. Feed manufacturers are licensed and inspected and feeds are randomly sampled and laboratory tested to determine label compliance and to ensure against harmful levels of toxic substances.

EGG AND POULTRY QUALITY CONTROL. California has, for many years, been the leading state in shell egg production. To assist in orderly marketing, the Department serves as a third-party agency to assure that eggs comply with the state's minimum standards for quality, size, and labeling. County agricultural commissioners are responsible for the day-to-day enforcement activities.

GRAIN AND COMMODITY INSPECTION. Certifications of grade and quality are the basis of an orderly grain and commodity marketing system. This unit has been designated by the U.S. Department of Agriculture as the official agency to inspect, weigh, and certify grains and other agricultural commodities at export terminals.

FRUIT AND VEGETABLE QUALITY CONTROL. This unit is comprised of three major program elements: (1) Fruit and Vegetable Standardization, (2) Shipping Point Inspection, and (3) Fresh Products for Processing.

Fruit and Vegetable Standardization is a cooperative activity with county agricultural commissioners to enforce state minimum standards for maturity, quality, packing, and marking. Under this program are inspections for fresh fruits, vegetables, nuts and honey, made at production,

wholesale, and retail locations and at inland and border highway stations.

An optional inspection and certification service is performed by Shipping Point Inspection upon request of anyone with a financial interest in fruit, vegetables, and nuts grown in California and destined for commercial resale or usage.

Fresh Products for Processing provides a neutral, third-party inspection service which certifies the condition and quality factors of fresh products transported from grower to processor, such as canning tomatoes, wine grapes, canning cling peaches, and garlic and onions for dehydration.

Division of Marketing Services

This division provides crop and livestock reports, forecasts of production, market news information, and other marketing services for agricultural producers, handlers, and consumers. The division also oversees the operation of over 30 marketing orders and administers the state's milk marketing program.

AGRICULTURAL STATISTICS. Also known as the California Crop and Livestock Reporting Service, "Ag Stats" is a federal-state program operating under a cooperative agreement between the Department and the USDA Statistical Reporting Service. Funding is approximately two-thirds federal and one-third state.

MARKETING. This bureau assists grower and handler groups in developing and operating self-help marketing programs (marketing orders, commissions and councils), and helps those in a particular agricultural industry work together to handle their problems. Marketing programs cover a wide range of activities including advertising and sales promotion; production; processing; pest control; marketing research; quality improvement; surplus control; and control of unfair trade practices. These are financed by the growers of a particular commodity voting to assess themselves to provide the necessary funds.

Additional activities include operating the farmer-to-consumer direct marketing program; analyzing various marketing problems; assisting in the organization of cooperative bargaining and marketing associations; and developing economic material for use by the director, the Department, and the agricultural industry.

MARKETING ENFORCEMENT. The Bureau of Market Enforcement is designed to protect producers of farm products from financial loss due to dishonest or unfair business practices by middlemen

Many inspection and regulation activities are carried on by trained technicians within the Department of Food and Agriculture. Here an employee sets up seed germination trials. *Photo by Charles Papp.*

and processors. Under California law, all persons, firms or corporations handling or purchasing farm products on a wholesale basis must be licensed. All sales must be recorded and the licensees must make proper payment to the sellers in accordance with contract terms or within 30 days as required by law. Investigations are made when complaints are received from producers of licensees. Where possible, bureau personnel settle the complaint, but formal hearings are held if necessary. Licenses may be suspended, revoked, or the licenses placed on probation.

MILK MARKETING ENFORCEMENT. This unit audits the records of milk processors, distributors, and retail stores, and enforces the laws on minimum producer prices and practices in the processing, handling, and marketing of milk and dairy products.

MILK POOLING. A statewide pooling plan for fluid milk has been in operation since 1969 to develop and maintain market stability and equalize distribution by producers. Production base and pool quota are assigned so each producer is guaranteed a share of the market. Each handler reports his total milk receipts and disposition monthly. Based on this and other information, pool prices are calculated each month. Producers are paid according to their pool quota, adjusted for differences in location.

MILK STABILIZATION. This bureau administers the California Milk Stabilization Law, which establishes minimum prices paid to producers for market grade milk. Class 1 prices are established by a formula which considers the costs of production, consumer ability to pay, and the value of milk used for manufacturing. The formula is subject to amendments by public hearing.

Division of Pest Management

The Division of Pest Management is responsible for regulating the registration, sale, and use of pesticides. It works with growers, the University of California, county agricultural commissioners, state, federal, and local departments of health, the federal Environmental Protection Agency, and the pesticide industry.

PESTICIDE REGISTRATION AND AGRICULTURAL PRODUCTIVITY. All insecticides, fungicides, disinfectants, rodenticides, herbicides, and other materials used around homes, industry, or farms for the control of pests must be registered with the Department of Food and Agriculture before being offered for sale in California. The Pesticide Registration Unit determines which pesticide products can be registered and how they are to be used in California. This is done through an evaluation of data gathered from chemical companies; from other units within the division and from other agencies of state, local, and federal governments; from universities; and through the public input process of notices and hearings.

PESTICIDE ENFORCEMENT. Working along with county agricultural commissioners, this unit develops and enforces regulations governing pesticide use, storage and disposal, as well as investigative techniques and worker health and safety procedures. Violation notices, investigative interviews, commissioner and district attorney hearings, accusations against licensees, and criminal prosecution are avenues of enforcement.

Through its licensing and certification program, the Department assures the competence of pesticide applicators, agricultural pest control advisers, and pesticide dealers. When a private applicator applies for a restricted materials permit, an oral examination by the county agricultural commissioner is required, which includes knowledge of label directions and restrictions, pesticide poisoning symptoms, and awareness of surrounding environmentally sensitive areas. Commercial applicators must pass a written examination to demonstrate their ability to conduct pest control operations and their knowledge of the environmental and human health effects of materials used. Agricultural pest control advisers must pass a written examination for each category in which they wish to make recommendations, and meet minimum education and experience qualifications. In addition, in order to renew their licenses, the adviser must complete a minimum of 40 hours of instruction relating to pest management within each two-year period. Pesticide dealers must also pass a written exam-

ination on the regulations governing the sale and use of pesticides.

This unit is also responsible for residue monitoring to detect and prevent the sale of any pesticide-contaminated farm produce from other states or foreign countries. Laboratories are located in Downey, Fresno, Berkeley, and Sacramento, and mobile labs are available for pesticide residue tests in the field.

WORKER HEALTH AND SAFETY. This unit, composed of four interacting programs, works to protect farm workers who mix, load, apply, store, or otherwise handle pesticides or who work in treated areas.

The medical support program provides medical support to agricultural commissioners and the public on pesticide illness investigations, evaluates scientific literature, and conducts health monitoring studies on specific pesticides. Program personnel help to educate health clinic and hospital emergency room staff to recognize pesticide illness symptoms and give proper treatment.

The pesticide residue program studies pesticides to determine when it is safe for workers to reenter treated fields.

The volatile pesticide monitoring program conducts on-site monitoring of volatile pesticides to measure possible exposure to the applicator and to people in the surrounding area.

The hazard evaluation program reviews and evaluates data regarding guidelines for safe use, pesticide formulations, and changes in pesticide labeling.

ENVIRONMENTAL MONITORING AND PEST MANAGEMENT. The primary responsibility of this unit is to assemble information and data on integrated pest management, biological control, fate and behavior of pesticides in the environment, air pollution effects, and effects of air pollution-pesticide interactions on agricultural production and the environment. Information is gathered through field investigations and studies, literature searches, contracting for research, and interacting with the University of California, the Environmental Protection Agency, the United States Department of Agriculture, county agricultural commissioners, representatives of the agricultural industry, and pest control advisers and operators. The unit issues "Pest Management Directories" which tell growers what pesticides are registered for use on particular pest problems.

Division of Measurement Standards

Working with county sealers of weights and measures, the division oversees the accuracy of weighing and measuring devices, from those used in supermarket checkout stands to the giant scales used at highway check stations. All new weighing devices, liquid measuring devices, and liquefied petroleum gas devices are inspected, and certificates are issued if requirements are met.

Maintaining the state's primary weights and measures standards and certifying county field standards is the job of the Metrology Laboratory. Standards are regularly checked against those of the National Bureau of Standards.

The Metric Conversion Council is responsible for organizing and coordinating the voluntary conversion to the metric system of measurement in California.

The Petroleum Investigation Unit is charged with enforcing quality control standards on such petroleum products as gasoline, diesel fuel, motor oil, automatic transmission fluid, antifreeze, and brake fluid. The program is funded by a special tax on motor oil.

The Weighmaster Enforcement program minimizes inaccuracies in weight and measure transactions through a licensing and bonding program of all weighmasters.

The Quality Control Unit works to assure that all statements of quantity on packaged goods be accurate, understandable, and easily read. It enforces provisions of the California Business and Professions Code and trains and assists county personnel in package control work.

The Division of Measurement Standards works closely with county sealers of weights and measures and local consumer affairs offices. It is also involved with a number of state and federal agencies, such as USDA, Food and Drug Administration (FDA), Federal Trade Commission (FTC), Environmental Protection Agency (EPA), Commerce, Transportation, and the National Bureau of Standards.

Division of Fairs and Expositions

Local fairs throughout the state exhibit and promote agricultural products while providing education and recreation for the community. This division assists the state's 80 district, county, and citrus fairs in upgrading their services and exhibits in response to changing conditions. The staff works directly with local fair managers and boards of directors in areas of planning, budgeting, exhibits, vocational education, events, construction, and maintenance. It distributes an administrative manual and a master premium list to all of the local fairs, reviews their financial reports, and approves their budgets. Local fairs are partially funded through horseracing tax revenues; this division distributes approximately $11 million received from parimutuel racing.

Agencies Protecting and Planning Resources

In the early days of statehood, government officials were primarily builders who viewed their role as one of facilitating the growth and development of California's economy. Nevertheless, even then there were strongly expressed environmental concerns which led to the outlawing of hydraulic mining and other extremely destructive economic activities. Not until relatively recently, however, have such concerns become institutionalized in the form of state agencies specifically mandated to oversee and protect natural resources.

The rapid urbanization in California following World War II and continuing through the 1950s gave way in the early 1960s to a growing disillusionment with the environmental consequences of growth. At the same time, a radical change in political power at the state level began taking place as population continued to shift to urban areas. Growing environmental concern combined with increased legislative power from urban areas resulted in the creation of a number of new agencies charged with protecting the environment for the common weal. These agencies included the California Air Resources Board created in 1967, the California Energy Conservation and Development Commission created in 1974, and the California Coastal Commission created in 1976. In addition, preexisting state agencies, like the Department of Water Resources, the Water Resources Control Board, and the Solid Waste Management Board, significantly expanded their environmental activities.

The Department of Water Resources

Most of California agriculture depends on irrigation, and the state has the most comprehensive irrigation system in the world. The key state agency responsible for building and maintaining much of this complex irrigation system is the

Department of Water Resources. Next to the Department of Food and Agriculture it is the state agency of greatest importance to California farmers. Water is regulated within the state also by federal agencies which operate the Central Valley Project, by more than a thousand special water districts, and by municipal water departments. (See Chapter 4.)

The Department of Water Resources is responsible for protecting and managing California's water resources and implementing the water resources development system—including the State Water Project and additions to the project such as the long-discussed Peripheral Canal—as well as safety inspections and supervision of dams and drinking water projects. The Department also furnishes technical assistance to other agencies. The California Water Commission, consisting of nine members appointed by the governor and confirmed by the Senate, serves as an advisory body. Within the department is also the Reclamation Board, consisting of seven members appointed by the governor, whose primary responsibilities are the construction and maintenance of levees on the Sacramento and San Joaquin river systems. Ninety percent of the Department's $350 million annual budget comes from various bond acts passed over the years for water development and recreation.

State Water Resources Control Board

The State Water Resources Control Board consists of five full-time members appointed by the governor to four-year terms. The board has two major responsibilities, the control of water quality and the administration of water rights law. The former responsibility is carried out through nine Regional Water Quality Control Boards which oversee water pollution control programs in accordance with federal, EPA, and state board guidelines. The water quality control function has been greatly augmented by federal initiatives to clean up rivers and promote better waste water treatment facilities, as well as by state clean water bond funds. Approximately 90 percent of the budget of this agency comes from bond funds or from federal grants. Concern over water quality has significantly affected agriculture in the 1970s in the form of increased costs for dealing with the effluent from canneries and food processing facilities as well as run-off from dairies, feedlots, or pesticide applications.

Department of Fish and Game

The Department of Fish and Game administers laws and programs protecting fish and wildlife resources and assuring a harvestable surplus of game species for hunters and fishermen. About 60 percent of the budget of the Department is derived from hunting and fishing licenses, and the rest from general fund appropriations, federal funds and miscellaneous sources. In carrying out its responsibilities to protect both game and nongame species of fish and animal life, the Department of Fish and Game works closely with the Department of Food and Agriculture to ensure that necessary pesticide use by farmers and others does not result in the killing of fish and animals. The Department of Fish and Game, in cooperation with the federal Fish and Wildlife Service and with the Department of Food and Agriculture, also carries out the predatory animal control program. The Department of Fish and Game also administers the Wild and Scenic Rivers Act of 1972 which declared legislative intent to preserve five north state rivers in essentially their natural condition.

Air Resources Board

Air pollution has been a serious problem for California agriculture, particularly in the Los Angeles basin. The Air Resources Board, the Department of Food and Agriculture and the University of California at Riverside have cooperated for the past five years in assessing crop damage caused by air pollution in the Los Angeles basin. Estimates are that over $55 million a year in damage is done to the crops still grown in the Los Angeles basin. Many varieties of vegetables are no longer grown commercially there because of their susceptibility to air pollution damage. In other areas, some segments of agriculture still depend on burning as a way of removing large quantities of agricultural waste, and of controlling pests and diseases. This creates another type of air pollution problem.

The California Air Resources Board was originally created in 1967 and reorganized in 1972. It is now comprised of five part-time paid members appointed by the governor. The Air Resources Board works closely with the federal Environmental Protection Agency which, under the Clean Air Act of 1970, has national responsibility for promoting clean air. Main areas of activity are in motor vehicle inspection and pollution control requirements and in stationary-source pollution control. The latter program affected agriculture in the 1970s much as did water quality control programs, in that canneries and other processing industries were required to install expensive air pollution control equipment to reduce their emissions.

County agricultural commissioners are also often the county air pollution control officers and thus are the local enforcement officers for both Food and Agriculture regulations and Air Resources Board rules.

Solid Waste Management Board

The Solid Waste Management Board is responsible for implementing a comprehensive litter clean-up program throughout the state, promoting recycling, and developing projects for the recovery of energy and resources from solid wastes. The federal Resource Conservation and Recovery Act requires each state to develop its own plan for solid waste management and the Solid Waste Management Board is currently in the process of developing such a plan for California. The Board has devoted considerable attention over the last several years to the problem of rice straw disposal, an ongoing problem.

Energy Resources Conservation and Development Commission

Next to water, and in the short term perhaps even more critical to California agriculture, is the assurance of a reliable supply of energy at reasonable costs. Prior to 1975, energy was a privately owned, publicly regulated commodity and the principal state agency involved in this regulation was the Public Utilities Commission. In 1975 the State Energy Resources Conservation and Development Commission came into being as a full-time, five-member commission with responsibility for developing energy conservation measures, forecasting supplies, and certifying power plant locations. The commission is funded, in part, from a special surcharge on consumers' electric bills.

Agriculture has been particularly interested in three programs administered by the Energy Commission. A $10 million demonstration grant program was authorized in 1979 to develop projects to demonstrate the feasibility of converting agricultural or forestry wastes into energy. Also authorized in 1979 was a grant program to investigate the practicability of alternative motor fuels. These funds were appro-

priated to the Business and Transportation Agency which in turn made available $2 million to the Department of Food and Agriculture for loans to farmers and farm organizations for small ethanol production plants. The third program has been the Energy Commission's Fuels Allocations Office, which has helped correct imbalances in the availability of diesel and gasoline to the state's farmers and agribusiness industries.

Department of Conservation

The Department of Conservation primarily administers the mining, geology, and oil and gas laws in the state of California. The Department is also responsible for administering the open space subvention program, and has a minor soil resource and planning program. Under the soils program, the Department prepared and released a draft report (April 1979) entitled *California Soils: An Assessment*. In addition, the Department has, in cooperation with the U.S. Soil Conservation Service, participated in a computer-mapping project to identify the location of open space and agricultural lands subject to the Williamson Act.

Other State Agencies Affecting Agriculture

The Agricultural Labor Relations Board (ALRB), created by the legislature in June 1975, opened its doors on August 28, 1975. In the first five months of operation, the Board exhausted its first year budget—a budget that had been prepared based on experience of the National Labor Relations Board, which conducted 31 elections in its first 10 months with a 30 percent rate of election objections. The ALRB in just half of that time received 604 election petitions, and conducted 423 elections. Objections were filed in 80 percent of the elections. In addition, the Board received 988 unfair labor practice charges during its first five months. The work overload drained funds and in April 1976 the Board ceased operation. Legislative debate finally resulted in refunding and the ALRB was reactivated on July 2, 1976. It has operated continuously since then.

Generally the ALRB's responsibilities are to guarantee agricultural workers the right to vote to join employee organizations in order to bargain collectively. The budget is currently slightly over $8 million, the staff about 200.

It is likely that farm labor organizing and election processes will continue to provoke controversy for the next several years. In 1980, the legislature refused to confirm two of the governor's appointments to the ALRB, evidencing continued dissatisfaction by the agricultural community with its rulings.

Department of Industrial Relations

The Department of Industrial Relations is charged with the responsibility for promoting and developing the welfare of working men and women in California, providing safe working conditions and administering the state worker's compensation laws. It is responsible for administering the California Occupational Safety and Health Act and federal Occupational Safety and Health requirements as they apply in California.

Of specific interest to the agricultural community are the laws and regulations with respect to farm worker health and safety, and the requirements for safe working conditions under CAL/OSHA. Worker health and safety related to pesticide illness or injury is primarily the responsibility of the Department of Food and Agriculture; however, it is required under state law to consult with the Department of Industrial Relations and the Department of Health Services on these issues. In addition, farm worker training programs are funded through the Department of Industrial Relations.

Employment Development Department

The Employment Development Department, formerly called the Department of Employment, is responsible for providing job placement services for both employers and employees, and for maintaining the unemployment insurance and disability insurance benefit payment system. The Department administers a total of $2.5 billion, of which $1.8 billion are federal funds and $634 million come from unemployment compensation funds.

Prior to 1975, farm workers were excluded from coverage under the unemployment insurance program, but since then they have been covered. The Department keeps statistics on the number of hired farm workers and the number of farmers and farm family members who work on farms in California. It also provides job referral services to match farm workers with agricultural labor requests.

Office of International Trade

Although small, the Office of International Trade is very important to California agriculture. As part of the Department of Economic and Business Development, the office, in cooperation with CDFA, has assisted in increasing agricultural exports, especially to Pacific rim nations and to the European Community.

Department of Housing and Community Development

The Department of Housing and Community Development assists in the provision of affordable housing for California residents. Particularly important to agriculture have been the efforts of this agency to improve farm worker housing.

Conclusion

The needs that led California farmers to seek governmental involvement in 1850 have changed over time but remain with us.

Without resource development and management, farmers would not have the water they need to grow crops, nor the network of highways and railroads to move products to market. Without marketplace regulation, markets could be destroyed overnight by the unscrupulous. Without continuing protection from exotic foreign pests and diseases, crops which have flourished for decades could no longer be profitably grown in California.

As our society has become more complex, so has government. In the future as in the past, the appropriate degree of government and the best mix of private activity and public good will be determined by the people through our democratic processes.

Appendix A: The Scope of State Government

Figure 2 gives a breakdown of total state expenditures and revenue sources. Nearly three-fourths of the state's expenditures are for education or health and welfare programs. Less than .5 percent of the budget is allocated to programs dealing directly with agriculture.

State government can be divided into five major categories. They are:

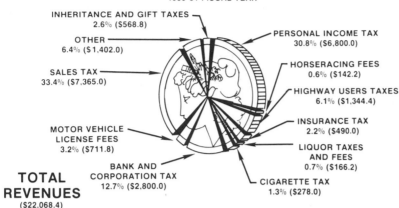

REVENUE DOLLARS
(amounts in millions)
1980-81 FISCAL YEAR

INHERITANCE AND GIFT TAXES
2.6% ($568.8)

OTHER
6.4% ($1,402.0)

SALES TAX
33.4% ($7,365.0)

MOTOR VEHICLE
LICENSE FEES
3.2% ($711.8)

BANK AND
CORPORATION TAX
12.7% ($2,800.0)

PERSONAL INCOME TAX
30.8% ($6,800.0)

HORSERACING FEES
0.6% ($142.2)

HIGHWAY USERS TAXES
6.1% ($1,344.4)

INSURANCE TAX
2.2% ($490.0)

LIQUOR TAXES
AND FEES
0.7% ($166.2)

CIGARETTE TAX
1.3% ($278.0)

**TOTAL
REVENUES**
($22,068.4)

FIGURE 19.2 California state government revenues and expenditures, 1980. *Source: State Department of Finance.*

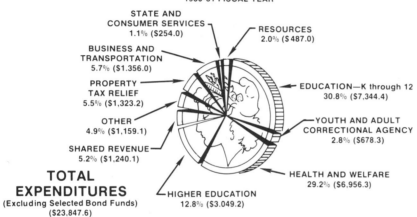

BUDGET EXPENDITURE DOLLARS
(amounts in millions)
1980-81 FISCAL YEAR

STATE AND
CONSUMER SERVICES
1.1% ($254.0)

BUSINESS AND
TRANSPORTATION
5.7% ($1,356.0)

PROPERTY
TAX RELIEF
5.5% ($1,323.2)

OTHER
4.9% ($1,159.1)

SHARED REVENUE
5.2% ($1,240.1)

RESOURCES
2.0% ($487.0)

EDUCATION—K through 12
30.8% ($7,344.4)

YOUTH AND ADULT
CORRECTIONAL AGENCY
2.8% ($678.3)

HEALTH AND WELFARE
29.2% ($6,956.3)

HIGHER EDUCATION
12.8% ($3,049.2)

**TOTAL
EXPENDITURES**
(Excluding Selected Bond Funds)
($23,847.6)

1. Health and Welfare, which includes the Departments of Health Services, Social Services, Rehabilitation, Mental Health, Employment Development, and others;
2. Education, which includes the Department of Education, apportionments to local school districts, and the university, state colleges and community colleges systems;
3. Resources, which includes the Department of Water Resources, the Energy Commission, the Air Resources Board, the Departments of Parks and Recreation, Forestry, Fish and Game, and others;
4. Business and Transportation, which covers the Departments of Transportation (formerly the Department of Highways), Economic and Business Development, and Real Estate, Motor Vehicles, and the Highway Patrol, as well as the Departments of Insurance, State Banking, Savings and Loans, and Corporations; and

5. General Government, a category which includes the cabinet level departments, such as Finance, Food and Agriculture, and Industrial Relations, as well as the independent commissions such as the Public Utilities, State Lands, Fair Political Practices, California Arts Council, and the Agricultural Labor Relations Board. Also included in this category are the general housekeeping departments, including the Department of General Services, the Personnel Board, Veterans Affairs, Public Employees Retirement System, and the Department of Consumer Affairs, plus the court system and constitutional offices.

To fund these activities, the state budget for 1980–81 was just under $24 billion. Of 221,000 state employees in 1980, 181,000 were civil service employees and 88,000 were employed in higher education. The remainder were in miscellaneous categories, such as legislators, judges, and constitutional officers.

Figure 3 is an organizational chart of the executive branch of California state government.

Appendix B: The Legislative Process

This brief description gives the path a bill must follow as it moves through the legislative process. The first step is the recognition that a new law (or clarification or repeal of an existing law) is needed. This can come from a constituent, a farm organization, or another organization such as the Sierra Club or consumer group, or from the legislator's reading of newspapers, magazines or government reports. Executive branch agencies often submit suggestions for making laws work better or for repealing statutes which are no longer needed. The next step is getting the bill drafted by the legislative counsel's office and introduced.

Once a bill is introduced, it is assigned to committee by the Speaker in the Assembly and the Senate Rules Committee in the Senate, depending on its primary subject matter. Bills affecting agriculture are most often assigned to the Assembly Agriculture Committee or the Senate Agriculture and Water Committee. However, environmental legislation affecting agriculture is generally assigned to the Senate Natural Resources and Wildlife Committee or the Assembly Resources Land Use and Energy Committee because the primary subject matter is environmental protection or resource management. Similarly, farm labor legislation is usually assigned to the Senate Industrial Relations Committee or the Assembly Labor, Employment and Consumer Affairs Committee.

The key step in the process is the hearing of the bill in committee. Both proponents and opponents will generally try to talk to members of the committee prior to the hearing to make their points and answer questions. Typically, at the hearing, the author has prepared a statement of why the bill is needed, what it does, and who supports it. After hearing from the author and supporters and asking a few questions, the committee chairperson will ask for opposition testimony. After hearing both sides, the committee usually votes on a "do pass" motion to send the bill either to the floor or, if there are budget implications, to the fiscal committee (Senate Finance Committee or Assembly Ways and Means

Committee). The bill must receive a majority vote of the entire committee (even if not all are present) in order to continue forward. If it fails, the committee may vote to allow the author to have the bill "reconsidered" which means the hearing process will be repeated at a subsequent hearing and the author will get a second try.

The process in the fiscal committee and on the floor of each house is similar. Once through one house, the bill goes to the other house and through the full process there. If a bill, after it passes one house, is changed as it is going through the other, it must come back to the floor of the first house for "concurrence" in the changes. If there are no changes, it goes directly from the second house to the governor.

The committee system is the key to the legislature. In the case of agriculture, the Department of Food and Agriculture (the key executive branch agency) and the two committees on agriculture work very closely. When the executive branch is considering new regulations or other actions opposed by one or more farm organizations, it is not uncommon to have members of the agricultural committee call the Director of the Department of Food and Agriculture and express their concern or even opposition. They may also consider holding legislative oversight hearings on the matter or maybe testifying themselves at the administrative hearing. Likewise, the Director of Food and Agriculture will consult from time to time with agriculture committee leaders before proposing new legislation or before advising the governor on major agricultural issues.

Figure 19.3 (opposite) Executive branch organization chart, California state government, 1980.

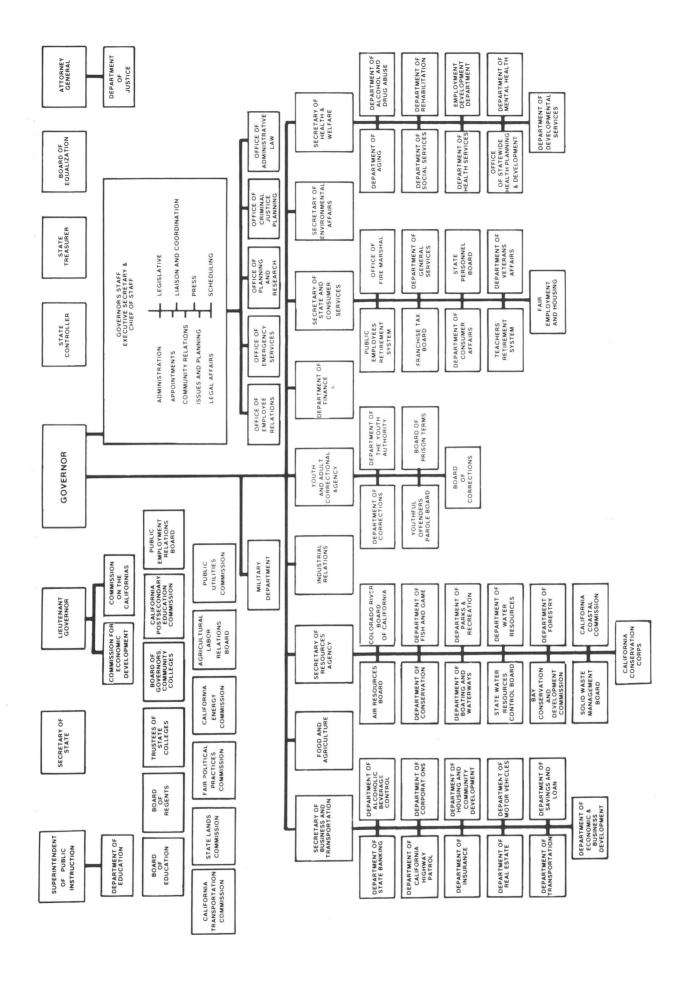

The Soil Conservation Service shares expenses with farmers on land improvement to help conserve resources. S.C.S. provides consulting service and must approve both plans and execution on such projects as land leveling.

20

Federal Programs and California Agriculture

B. Delworth Gardner and Carole F. Nuckton[1]

California lies a continent away from Washington, D.C., and many of the institutions described in previous chapters have evolved from the necessity to deal with problems unique to the state's location and resources. Nevertheless, the history of agriculture in California, as elsewhere, is entwined with the U.S. government—from early policies on land distribution to the most recent decisions on commodity supports or international trade embargoes. Tax laws, resource development projects, environmental legislation, or employment programs: all have potential impact on agriculture, helping to shape what it will become in the future.

The purpose of this chapter is to summarize some of the broad areas in which the federal government influences agricultural development in California. Because of the diversity of the state's agriculture, federal programs and policies have probably been more varied in their impact here than in many other states. Policies designed in Washington, D.C., to benefit agriculture in other regions may not always be suited to California. Some federal programs benefit certain of the state's farmers, others may leave them totally unaffected. Still others may even be detrimental to some segment of the industry—for instance, support prices for grain keep feed costs high for the state's livestock producers.

Some federal programs are directed specifically at farm problems; some, like federal tax laws, affect agriculture indirectly; still others, such as federal water projects, benefit not only agriculture but also municipalities and nonagricultural

industries. The heterogeneity of California means that the state's farmers have little in common that can be effectively represented to federal policy makers. A major worry for a vintner may be labor; for a drylot dairy, waste disposal; for a rice farmer, permission to burn stubble; for a citrus grower, frost; for a poultry producer, disease; and so on. For nearly all agriculturalists, however, at least in the southern two-thirds of the state, a major concern is water—in sufficient quantities at an affordable price. Another pervasive issue, perhaps soon to eclipse other matters, is energy.

Probably one of the strongest impacts of government policy on agriculture has been seen in the results of public support of agricultural research in the USDA and in the land grant institutions in each state. The governmental encouragement of research and education has been a significant factor in the development of agricultural technology. Successful research efforts have resulted in technological changes in the form either of new inputs or qualitative improvements in traditional inputs, bringing greater efficiency in agricultural production. Among the many significant research achievements through public investment are the development of hybrid corn, of medicated feed allowing the concentration of birds in the poultry industry, and of the mechanical tomato harvester and a tomato capable of being thus harvested.

While the land grant system operates at the state level, it was designed by federal legislation and considerable federal funding is provided on an ongoing basis. Although in California the state contribution to agricultural research and

1. The authors wish to thank Chester O. McCorkle and Alex F. McCalla for most helpful comments on an earlier draft. In addition, the assistance of Ray Huffaker is gratefully acknowledged.

extension represents a higher fraction of the total investment than in most other states, the federal share currently runs at approximately 22 percent (including grants and contracts). (Chapter 18 describes the state's research and education system in agriculture.)

Another highly significant federal contribution to agricultural development has been the Farm Credit System, described in detail in Chapter 13. Financial institutions directed by the federal government are the largest lenders of funds to rural America. The Farm Credit System provides credit amounting to nearly $60 billion annually to individual farmers and cooperatives.

The discussion in this chapter is developed under four main headings to describe significant areas of federal influence which have not been mentioned elsewhere in this volume—laws or programs relating to (1) natural resources, (2) hired farm labor, (3) federal taxes, and (4) agricultural prices.

Natural Resources

Land Use

The early land-use policies of the federal government were based on Jefferson's application of the Lockean principle that property rights were among the natural rights of man. Legislation such as the Homestead Act of 1862 and the Reclamation Act of 1902 reflected the pervasive national ethos favoring widespread private land ownership. Once the nation was settled, however, federal attention shifted away from such developmental policies. Under the Constitution, the primary authority over private land use was delegated to the states. Today, the major direct involvement of the federal government in land use is in the management of millions of acres of publicly held land. The Bureau of Land Management (BLM) and the National Forest Service (NFS) administer systems of grazing permits in the West that allow livestock grazing on the public lands.

California has approximately 36 million nearly snow-free acres of range lands. Much of this land is under the control of either NFS or BLM. In 1979, there were 493,100 animal-unit-months[2] of livestock grazing on California's NFS land

2. An AUM (animal-unit-month) represents the equivalent of one mature cow grazing for the period of one full month. Sheep and goats convert to cattle at the rate of 5 to 1.

(USDA, 1979). The BLM estimates that approximately 750 ranchers graze livestock on 6–10 million acres of public domain lands in California. This grazing is administered under approximately 700 allotments providing approximately 400,000 AUM's.

A continuing controversy rages over how much grazing should be allowed on public lands. Both the NFS and BLM have drastically cut livestock permits since the 1940s, and many range specialists believe further cuts are necessary to improve range conditions. Ranchers have few good substitutes for federal grazing in most instances, however, and thus the cuts have been very costly. Ranchers also argue that livestock losses to predators greatly increased when certain poisons were outlawed by the federal government in the early 1970s.

Recently, both agencies have also had to heed the demands of those promoting the recreational uses of public lands. It is conceivable that the two uses—agricultural and recreational—could conflict in given instances. Ranchers, for example, may not like recreational users to be around their livestock, and people seeking recreation may not enjoy camping next to cattle and sheep. It is not all clear how this and many other public land issues will be resolved in the years ahead.

Water Development

California's water problem may be described as a geographic and seasonal maldistribution of moisture in relation to human needs. The problem is not so much a paucity of supply as it is a natural deposition of too much water in the "wrong" places at the "wrong" times to meet the state's needs. An equitable and regulated redistribution has had to be man-made. Both the state and federal governments have been deeply involved in planning, constructing, and maintaining one of the most elaborate, complex, and costly water systems ever devised (see Chapter 4, "Water in California Agriculture"), but the focus in this section is on the federal involvement.

The Central Valley Project (CVP) under the Bureau of Reclamation, Department of Interior, is the state's largest integrated water and power project. CVP was conceived during the great drought of the 1920s, when Central Valley farmers had to resort to large-scale groundwater pumping to save themselves from ruin. In 1933, a project was pushed through the state legislature and a $170,000,000 bond proposition was endorsed by California voters. With the Great Depression in progress, however, the state could not find buyers and the bonds were not, in fact, ever put up for sale. The federal government was called on for financial aid and through the public works program of Roosevelt's New Deal the CVP became a federal undertaking.

Over the years the project has supplied irrigation water to farmers at an affordable price. The federal government has charged no interest on the money invested in the project and, additionally, has used excess revenue from the sale of electricity generated by the project to subsidize irrigation costs and reduce the price of water. Who receives the benefits of these water "subsidies"—consumers of feed and fiber, land owners, or land renters—is far from clear and is the subject of continuing research.

There were some strings attached, however, to receiving federal water. Farmers who contracted for CVP water thereby put themselves under the provisions of the Reclamation Act of 1902. The Act provided that no surface water from a federal reclamation project shall be delivered to a land parcel exceeding 160 acres to any one individual landowner (or 320 acres for a farmer and spouse) and no such water shall be delivered unless the farmer is a bona fide resident on the land or a neighboring resident thereof. The original intent was to give encouragement to settling and developing the West.

Probably there are few examples in American jurisprudence in which a law has been so variously enforced and administered. Since 1933, the Imperial Valley has been altogether exempt from the provisions of the law. The Westlands Water District is in the process of compliance, but most farms there are still very large by any standard, and full compliance will be costly to some farm operators. To further complicate matters, by court order a moratorium has been placed on land sales that are required to comply with the 160-acre limitation.

Recently, attention has been called to the discrepancy between the law and its actual implementation. Many bills have been introduced in Congress representing alternative approaches ranging from strict enforcement, through increased acreage limits, to complete repeal. The controversy continues. Given the heterogeneity of California growers, it would be a mistake to assume them to be unanimously hostile to the limitation. The Grange, an organization of small farmers, has traditionally defended it.

Soil and Water Conservation Programs

There are currently more than 34 programs within the USDA, administering 143 laws dealing with soil and water conservation, according to the U.S. Chamber of Commerce. The Soil Conservation Service (SCS) and the Agricultural Stabilization and Conservation Service (ASCS) are two agencies within USDA that are directly concerned with the protection of the nation's land and water resources. SCS offers technical assistance in such matters as erosion prevention, soil improvement, and water management. ASCS committees are county-based and look after fiduciary arrangements between the federal government and local farmers, such as price-support payments. ASCS cost-sharing programs are available to prevent resource depletion and to restore deteriorating resources. An example is cost sharing in the installation of drip irrigation systems, the most efficient in terms of water use.

In 1977, Congress passed yet another law—the Soil and Water Resources Conservation Act. The 50-year plan, as interpreted by SCS, was estimated to cost about $3.9 billion a year in implementation of programs affecting energy conservation, wildlife conservation, water pollution, and land use as well as the more traditional areas of soil and water conservation. It may well be that this Act represents a shift in emphasis away from managing resources in order to increase agricultural production to preserving resources even at the expense of agricultural production in the short run.

Environmental Protection Agency Regulations

While federal EPA regulations and, of course, state environmental regulation are often necessary for public health and comfort, they have added substantially to the costs of production in many industries. Agriculture is no exception. The balance of costs against potential profits plays an important role in agricultural decision-making, such as in cropping choices. The Clean Air Act Amendment of 1977, for example, calling for the attainment of National Ambient Air Quality Standards by 1982 (unless a waiver is obtained until 1987) has put pressure on farmers to curtail the practice of open field burning of agricultural residue (particularly rice straw). Proposed state legislation has exerted additional pressure. Regulations which threaten to reduce profit margins to an untenable point have obvious impact on what, where, and when farmers plant.

Hired Farm Labor

California, specializing in labor-intensive crops, paid 20 percent of the nation's hired farm wage bill in 1978, according to the Census of Agriculture. Eighty percent of the workers on California farms were employed less than 150 days on any one farm, indicating the seasonal nature of the demand for labor. The state has been the national leader in promoting better working conditions for its farm work force, although these conditions still lag behind those of most nonfarm employees. In this chapter, however, the focus is on programs generated at the federal level to address the problems of the farm worker population. A few of these are directed specifically at farm workers; others include them as a part of a generally disadvantaged population. Here space permits only a brief discussion of four of the more important: (1) the Comprehensive Employment and Training Act (CETA), Title III, Section 303, (2) the U.S. Employment Services (ES), (3) the Occupational Safety and Health Administration (OSHA), and (4) the National Labor Relations Act (NLRA). (The social history of farm labor in California is given in Chapter 1; the history of farm labor unionization, in Chapter 13; and a description of state farm labor programs, including the Agricultural Labor Relations Board, in Chapter 19.)

The Comprehensive Employment and Training Act

Title III, Section 303 of CETA, administered by the Department of Labor (DOL), was designed to provide training, job placement, and other employment services specifically to migrant and seasonal farm workers. California was the top state recipient of funding for this program, having received $14,900,000 for the 1980 fiscal year.

In general, however, Section 303 has reached only a small percentage of the target population. There was a rapid turnover of program offerings. Services lacked permanence and dependability partly because of the annual competitive bidding system among grantees. Follow-up is needed once a worker has been placed, but expensive longitudinal studies have not been done. The emphasis in some programs has been to prepare workers for nonfarm work. It is possible that opportunities in farming were overlooked. A notable exception is the Section of 303 that gave technical assistance to the production cooperative movement in California whereby former farm workers entered farming as self-employed entrepreneurs, cultivating labor-intensive crops (mainly cherry tomatoes and strawberries). Section 303 also provided support for a "peer training" project in which successful cooperative members trained others so that they could then establish other cooperatives.

The U.S. Employment Service

One part of the ES program, also under DOL, was to be the maintenance of placement offices with migrant and seasonal farm workers receiving special consideration. Various complaints about ES—for example, that farm placement offices were employer-oriented—culminated in a public suit filed against DOL in 1971. In August 1974, Judge Richey of the U.S. District Court signed a consent order negotiated by the parties to the suit, requiring DOL to take specific steps in providing more equitable services, benefits, and protection to migrant and seasonal farm workers. Various follow-up actions culminated in a settlement agreement, January 4, 1980. Regulations to be followed by state ES offices were spelled out in detail in the January 11, 1980, *Federal Register*. It remains to be seen whether deficiencies in ES services to migrant and seasonal farm workers have been rectified.

U.S. Occupational Safety and Health Administration[3]

US/OSHA, also of DOL, was given authority in 1971 to regulate employee working and living conditions on farms. Congressional funds, however, were not sufficient to hire enough inspectors, so US/OSHA concentrated mostly on nonfarm employment.

In 1973 US/OSHA promulgated regulations on pesticide usage, but the Environmental Protection Agency claimed jurisdiction over pesticides and US/OSHA retreated. The proposed regulations were less stringent than those already enforced by the California Department of Food and Agriculture. US/OSHA was successful, however, in making requirements for tractor rollover bars,

3. This section and the next, on the National Labor Relations Act, are from Sosnick.

protective shielding on certain equipment, and sanitation facilities in temporary labor camps. In 1976, when US/OSHA turned to field sanitation requisites—toilets, hand-washing facilities, and potable water—employers' protests won out and such matters were left up to the individual states to determine.

California was already ahead in the field sanitation area, for, after a decade of study in the 1950s, the first order was issued in 1961 by the California Industrial Welfare Commission. The regulations applied just to employers of women and any persons under 18. In 1965, the state legislature passed a Health and Safety Code requiring toilet and hand-washing facilities (not drinking water) in the field wherever five or more people work in fruit, nut, and vegetable crops.

By 1973 jurisdiction over field sanitation was transferred from the Industrial Welfare Commission to CAL/OSHA, the California state agency, funded by the federal government. Requirements identical to those in factories were to be maintained in the fields. Compliance with increasingly stringent regulations, however, can only be accomplished over time. By 1977 CAL/OSHA had a budget large enough to employ 180 investigators for nonfarm and farm working conditions which allowed spot checking of field sanitation.

The National Labor Relations Act

The NRLA governs the relationship between employers and labor in all industries engaged in interstate commerce except agricultural field workers. The original 1935 bill, known as the Wagner Act, was intended to cover farm workers as well but after being redrafted in committee, specifically excluded agricultural workers. Growers argued that they are more vulnerable to strikes because of the perishability of their products, that they are unable to control product prices and so cannot easily raise wage rates when demanded, and finally that they perform society's most necessary function, raising food and fiber, and must therefore have an adequate supply of labor. The exclusion actually was based on the argument that the administrative burden would be too great were agriculture to be included.

In 1966 Cesar Chavez spoke against the exclusion of agriculture from NRLA coverage, but three years later led the United Farm Workers

(UFW) in opposition to coverage. Growers also vacillated on the subject. UFW saw the procedure for union elections under the National Labor Relations Board (NLRB) as far too slow for seasonal farm workers, many of whom would move on between the filing date and the actual election. "Cease and desist" orders for employers' unfair practices take even longer under NLRB. More important, UFW wished to avoid NLRB restrictions such as those against member-only hiring halls or against demanding that an employer discharge workers for anti-union activity. Probably most important of all to UFW was that under NRLA, boycotting activities could not be conducted. Although used only as a last resort because it is costly, only partial in its effect, and often injurious to third parties, the boycott has been sometimes utilized by UFW, an organization officially committed to nonviolence. Boycotts may work when strikes fail, for strikebreakers are relatively accessible to growers, and fields are difficult to barricade.

California growers in the 1970s began pushing for coverage under NRLA so that UFW would be subject to the above restrictions, especially the restriction against boycotting. The resulting state law, the Agricultural Labor Relations Act of 1975, represented something of a compromise, affording growers some protections against picketing and boycotts and giving the UFW the right to obtain recognition through election.

Other Federal Programs

Other federal programs directed specifically at the hired farm worker population have been (1) Title I of the Elementary and Secondary Education Act of 1965, administered through the Office of Education, to meet the special educational needs of children of migrant agricultural workers, (2) Section 319 of the Public Health Service Act of 1975, administered by the Department of Health, Education and Welfare (HEW), to provide migrant farm workers with health care, (3) Title III, Section 304 of the Rehabilitation Act of 1973, also administered by HEW, to partially support the vocational rehabilitation of handicapped agricultural workers, (4) Section III on farm housing, of the Housing Act of 1949, administered by the Department of Housing and Urban Development, to give financial assistance for low rent housing to groups of farm workers, (5) Community Development Corporations, funded under Title VII of the Community Services

Act of 1974, to provide assistance to small groups of farm workers to establish producer cooperatives.

Changes in federal administrations naturally have significant impact on legislation and budgets. Not all of these programs have been consistently funded, and some have been discontinued.

Federal Tax Impacts

The federal tax structure has had tremendous influence on American decision making. A small investor may have purchased common stock *because* the first 100 dollars of dividend income was tax free. A high income individual may have bought municipal bonds *because* interest income is not subject to federal tax. Another may have decided to invest in gas-oil exploration *because* of the large "write off." Divorce settlement amounts may hinge on the fact that alimony is deductible for the donor and taxable to the recipient. Even gift-giving to one's children is often done for tax reasons. Tax-credits have recently encouraged people to insulate their homes. Thus, the pervasive influence of the federal tax system is an inescapable part of life.

Agriculture is by no means an exception. In fact, agricultural decision making may be even more subject to such influences because of certain special tax provisions allowed farmers. A few of these tax advantages, such as those which encourage orchard and vineyard development, have been especially relevant in California. The state tax structure is, of course, also important in agricultural decision making; in particular, high property taxes have levied a heavy burden on farmland. (See Chapter 19 for a description of the California Land Conservation Act, one effort to alleviate the property tax burden on farmland.) In this section, however, attention is confined to various aspects of the federal tax structure in relationship to California agriculture.

Income Taxes
While usable by all taxpayers, two general income tax provisions—income averaging and loss carry-over—are particularly well suited to agriculture where incomes can vary dramatically from year to year. It is said that the progressive nature of the income tax induces risk-taking behavior. A person in a high tax bracket is more willing to invest his marginal dollar in a risky venture since a good part of that dollar belongs to the government anyway. In farming this may mean the planting of high risk crops—more acreage in fruits and vegetables, fewer in field crops which have price support programs. In addition to the above-mentioned general features of the income tax structure with effects on farming, some tax provisions have been made specificially for agriculture.

CASH VS. ACCRUAL ACCOUNTING. Since an administrative decision in 1915, farmers have been given a choice between cash and accrual accounting methods for income tax purposes. By ignoring inventories, cash accounting allows timing of product sales and input purchases to minimize the total tax obligation. Fertilizer purchase, for example, could be made one year, expensed against that year's income, for use the next year. By careful planning, therefore, one presumably could alternate "good" and "bad" years, putting as many input purchases as possible into the "good" year.

While cash accounting has been helpful to farmers in general, it has also been attractive to investors who work outside agriculture but who may be attracted to investment opportunities in agriculture. Cattle feeding operations were one of the most popular tax shelter activities in the early 1970s. Youde and Carman estimated that 60 percent of all cattle being fed in California in 1972 were owned by nonfarm investors. End-of-year feeder cattle purchases can be written off against investors' other income. One consequence may be that such activity distorts the market, raising the otherwise normally low year-end prices for feeder cattle.

DEVELOPMENT COSTS FOR ORCHARDS AND VINEYARDS. Probably the provision with the most California-specific impact stems from a 1919 Treasury regulation allowing farmers to "expense" the development costs of an orchard. Costs which add value to an asset (land)—which normally in other businesses are capitalized and depreciated over several years—can be deducted against the current ordinary income of a farmer or a nonfarm investor. Then, upon sale of the developed orchard several years later, the difference between the land-purchase (without trees) and land-sale price (with trees) is subject to the lower capital gains tax rate.

This tax provision may have led to overplanting in the 1950s and 1960s. Citrus and almonds were very popular during this period. In 1969, however, the tax law was changed by the Tax Reform Act to require citrus and almond development costs to be capitalized within four years after planting. Carman estimated the immediate and longer-term effects of this capitalization requirement on acreage, production, and price. The estimated impact was to reduce acreage in citrus and almonds substantially. For example, Valencia acreage was estimated to decrease 10.1 percent by 1973, 17.4 percent by 1978, and 19.0 percent by 1985. A price increase as a result of decreased production was anticipated. The apparent shift in investor interest to other crops was shown in significant estimated increases in walnut and grape acreage—an estimated 99,000 acres greater in 1978, than would have been the case without the change in the tax law. Thus, when tax laws have differential impacts, the effects on crop mix can be dramatic.

The Economic Recovery Act of 1981 will undoubtedly also have structural impacts on agriculture. One provision allows the buyer of any developed orchard to fully depreciate it within five years. This change may carry differential impacts since various types of orchards had different lengths of "useful lives" under the old law. Almonds, for example, had a 40 year depreciation period; nectarines, only 15.

CORPORATE FARMING. There are several reasons why an increasing number of farms have incorporated their operations in recent years, but one of the strongest of these is the tax structure. In 1958, Subchapter S of the Internal Revenue Code eliminated the double taxation disadvantage for all small qualifying corporations. (Larger companies must first pay corporate income tax and then shareholders pay income tax on dividends received.)

Corporate tax rates are considerably lower than individual tax rates for incomes above $25,000. Subchapter A of the Internal Revenue Code of 1954 allows a farming corporation to pay the lower corporate rate on any earnings retained within the firm. This provision has worked out especially well for farmers wishing to expand operations.

COOPERATIVES. Under the Capper-Volstead Act administered by the U.S. Department of Justice, preferential tax treatment has been extended to agricultural cooperatives in both the input and the product sectors. Only income retained with the cooperative is taxed (at the corporate rate).

All other income is distributed to members (who then, of course, pay individual income taxes on it). The avoidance of double taxation borne by noncooperative corporations has given a boost to cooperatives. Producers may have benefited from lower input and higher product prices than would have been the case without cooperatives. Cooperatives have enabled small producers to compete more favorably with large agricultural private firms in the marketplace.

Estate Taxes

Since estate taxes are progressive, large farms are more burdened than small ones, unless through careful planning intergenerational transfer can be eased. One way is by incorporating the operation and then making annual gifts of corporate shares to potential heirs. In 1981, the annual gift tax exclusion was increased from $3,000 per recipient per year to $10,000.

The Tax Reform Act of 1976 differentiated between farm and nonfarm estates in two important ways. First, qualifying farmland can be valued for tax purposes at its use, rather than its fair market value, reducing estate tax liabilities substantially in most cases. (The maximum reduction in the value of the farm estate was increased from $500,000 to $750,000 in 1981 over a three-year phase-in period.) Once use value is established, land sales are restricted to only family members for 15 years. While the intent of this provision is to prevent the loss of the farm to the family because of taxes on high land values, the effect may have been to erect yet another barrier to beginning farmers. The 15-year restriction serves to reduce the availability of land on the market, in effect "locking up" land ownership.

The second difference between farm and nonfarm estates added in 1976 was that farm estate taxes can be paid in installments, amortized at four percent over a 15-year period. The intent of this provision was to discourage the sale of farms at probate, making it easier to keep the farm in the family—but again making it more difficult for others to buy.

Probably the most significant of the tax law changes embodied in the Economic Recovery Act of 1981 are the phased-in increases in the size of the exemption from federal estate taxes (up to $600,000 in 1985) and the unlimited marital deduction. It may no longer be necessary after 1981 for any widowed spouse to liquidate the family farm to meet the estate tax burden.

Agricultural Prices

Federal Commodity Programs

Because of the volatility of prices and generally low level of agricultural income during and after the Great Depression, government programs with the objective of increasing farmer incomes were initiated. The mechanism used was price supports based on parity. (Parity means that the ratio of prices received by farmers to prices paid would be the same as in some base period. The period chosen was 1910–1914, the so-called "Golden Age," when prices received stood in a favorable position relative to prices paid.)

By the mid-1950s, however, research-induced productivity gains in combination with government price support incentives resulted in enormous over-supplies of most grains. The sheer quantity of stocks became a burden to taxpayers and presented serious storage problems in spite of various disposal schemes including "dumping" abroad in the form of Public Law 480 food aid programs. The problem lay partly in the fact that the support price became the market price and the government had to maintain it by various supply removal programs.

The system that was devised in response to these costly surpluses was ingenious—when it works well, an economist's delight. The market price was freed from the support price, which then could be established at market-clearing levels. Specific policies are different for nearly every commodity. Space, therefore, limits the discussion to highlights.

The Agricultural and Consumer Protection Act of 1973 established the concepts still in use today as administered under the Food and Agricultural Act of 1977.[4] For wheat, the feed grains, and cotton, market prices are relatively free to seek equilibrium levels, except that a lower limit is effectively established by a price called the *loan rate*. This is so designated because a farmer with commodity stocks on hand can use these stocks as collateral to obtain a government loan, the amount of which would be determined by the per unit loan rate. Thus, if a farmer has 10,000 bushels of wheat stored and the loan rate is

4. The Reagan administration, however, has sought to modify some programs and move them toward a free-market orientation. In general, there is some hostility in this administration toward target prices and the disaster programs of USDA.

$2.50 per bushel, he can get a government loan for $25,000. This loan has a *nonrecourse* feature which establishes the loan rate as the effective minimum for the market price. If the market price exceeds the loan rate, the farmer can sell his wheat in the market, repay the loan, and keep the excess to meet production costs and living expenses. If the loan rate exceeds the market price, the farmer has the option of exercising the nonrecourse feature of the loan, and can turn his wheat over to the government in full payment for his loan. Thus, the government will accumulate wheat until the market price rises above the loan rate. The loan rate tends, therefore, to be a floor under the market price. Of course the market price may be significantly higher than the loan rate, depending on demand and supply.

As a floor on the market price, the loan rate is very important in determining stocks in government hands and levels of exports. If the loan rate were raised above the world price, importing countries would turn elsewhere for needs and U.S. exports would fall. Also, the higher the loan rate the larger the stocks that must be acquired by the government—if market prices fall below loan rate levels, inducing farmers to exercise their loan options. The Secretary of Agriculture has discretion to set the loan rates and he must do so with great care in order to keep exports and government stocks at acceptable levels.

Under this same program, farm incomes are supported by another price called the *target price* that replaces the old support price. The government guarantees to pay the farmer the difference between the target price and the prevailing market price based on the quantity that is the average yield for the area. This payment is called a *deficiency payment* and is given only to those farmers who enroll in the program. In exchange for this price protection, the farmer has to commit himself to reducing his acreage by a certain amount (called a *set-aside*). In the years 1978 and 1979 this set-aside requirement generally averaged from 10 to 20 percent, depending on the crop and the year. At this writing (1981), set-aside requirements have been dropped entirely. In the case of rice, no set-aside has been required but acreage is determined by historical allotments, and participation rates in California are higher than for wheat, cotton, and the feed grains.

The Food and Agricultural Act of 1977 provided for a target price for feed grains for the years 1975–1981 at the 1978 level of support adjusted for charges in costs of production, including variable costs, machinery costs, and general overhead costs. The Emergency Assistance Act of 1978, however, provided the Secretary of Agriculture with discretionary authority to raise target prices if economic conditions seemed to warrant this action. Thus, if market prices rise, suggesting more supply is desired, the Secretary may raise the target price, inducing more supply, thereby putting downward pressure on the market price. At the same time, the costs to the taxpayers will rise as deficiency payments will likely increase. Target prices, therefore, influence market prices for program commodities, farmer incomes, and also taxpayer costs.

In an effort to limit deficiency payments to large farmers, a payment ceiling of $45,000 per farm was imposed for the 1979 and later crops for wheat, feed grains, and upland cotton. For rice farmers, the limit is $50,000.

Farmers who enroll in commodity price support programs also are eligible for protection by the government's disaster program. If natural disasters occur (flood, drought, hail, etc.) and plantings are prevented and/or yields are low, the program permits government payments. For prevented plantings of wheat, feed grain, cotton, and rice the payments are one-third of the target price times 75 percent of normal yield. Low-yield payments are made when yields fall below 60 percent of normal for wheat and feed grains, and 75 percent of normal for cotton. Low-yield payments are one-half of the target price for that part of production below 60 percent of normal for wheat and feed grains, and one-third of target price for that part of production below 75 percent of normal for cotton. Disaster payments are not limited by ceilings.

The price support program for manufacturing milk works quite differently. In the old price support fashion, the government announces that milk prices will be supported as a certain percent of parity at the discretion of the Secretary of Agriculture. Then, to maintain manufacturing milk prices at the support level the government simply purchases butter and powdered milk (the main products of manufacturing milk) until the milk price rises to the support level. Of course, high support prices are associated with large government stocks of butter and milk powder and vice versa. The large (and costly) stocks in government hands led the Reagan administration

to cancel increases in milk support prices scheduled for April 1981.

Although it is probably safe to say that no federal influence is so direct or important as are price support policies on U.S. agriculture as a whole, the impact on California farming is probably more moderate. Most specialty crop growers, for example, remain totally unaffected by price support programs. Price supports have been used only for the following commodities: wheat, feed grains, sugar, rice, cotton, tobacco, peanuts, soybeans, manufacturing milk, and wool. Of these, only milk, cotton, rice, wheat, and sugar beets are major commodities in California, ranking first, third, eighth, twelfth, and seventeenth, respectively, in terms of 1980 sales value.

Further, as has already been mentioned, participation rates in various commodity programs are much lower in California than the nation. For example, in 1979 only 5.2 percent of the USDA Agricultural Stabilization and Conservation Service sorghum acreage in California was enrolled in the program compared with a national average of 54.4 percent. Set-aside costs of past years are higher in a state like California where many farmers simply cannot afford to lay fallow high-priced, irrigated land in exchange for some protection against fluctuating market prices. As economists say, the opportunity cost of set-aside acreage is very high in California.

Marketing Orders

Federal and state marketing orders have been used in California for the past 40 years. They have played a role in the fruit and vegetable industries somewhat analogous to that of price supports in the grains. Both programs have the same overall goal of price stabilization, but use different means to reach it. Marketing orders provide a means for growers to organize legally to regulate various aspects of marketing a particular commodity and to collect funds for research and promotion. Federal and state orders authorize similar kinds of programs, but differ in the extent of implementation of the various provisions. In general, state orders tend to emphasize quality control and inspection, research and promotion. Federal orders stress volume control and rate of flow to market, helping to coordinate marketing of commodities produced in more than one state. (For more on marketing orders see Chapter 15.)

Tariffs and Nontariff Restrictions

Domestic policy involving any degree of protectionism, such as the market price support program of the dairy industry, unavoidably involves trade barriers against other nations also producing those products. Tariffs and nontariff barriers may be formulated by the USDA, the State Department, the White House, or Congress, but the recommended tariffs or restrictions are generally implemented by the U.S. Treasury. The Sugar Act of 1948, amended at various times, but in effect until the end of 1974, provides an important example of nontariff restrictions. The purpose was to maintain a domestic sugar industry by limiting imports, with an intricate system of quotas placed on producing countries.

Perhaps more prevalent or at least more obvious than nontariff restrictions are tariffs placed on thousands of imported items, including many agricultural commodities. Since the imposition of these duties is intended mainly to protect domestic producers, commodities not produced here, such as coffee and fresh bananas, come in duty free, whereas champagne and other sparkling wines have a tariff of $6 a gallon, and still wines, $1.25 a gallon.

By charging differential tariffs, depending on whether a country is developed, less developed, or least developed, U.S. policy presumably is constructed to "give a break" to countries in the latter two categories. For example, fresh or dried dates (Chiani type) with the pits removed are charged 3¢ per pound if coming from a list of "least developed countries," 6¢ per pound if from those designated as "beneficiary developing" or as a U.S. insular possession, and 7.5¢ per pound if coming from anywhere else.

Of particular importance to California specialty crop producers is the fact that some tariffs are imposed only seasonally when the domestic crop is ready to harvest and then lowered during the remainder of the year. Thus, consumers can have, for example, fresh vegetables from Mexico at lower prices during the U.S. off-season.

Conclusion

Much more could be said about the influence of the federal government on California agriculture. Since legislative programs are forged in the crucible of public debate, they reflect values and priorities which may vary widely from time to time. While some enactments have benefited the state's agricultural sector directly (water subsidies), others have been costly (environmental regulations). Some policies aimed at helping a target group (family farmers) may have unintentionally helped another group (nonagricultural investors in farm land). Legislation aimed at aiding depressed farming areas may be totally inappropriate for more affluent ones.

Generally it may be true that California agriculture operates with less dependence on the federal government than some other regions. One reason may be that the state has historically been underrepresented on federal agricultural committees.[5]

Grain, cotton, sugar, and milk are, of course, also highly important in the California farm economy, but the state's farm operators tend to participate less in federal programs (where applicable) than their counterparts in other states. In 1979–80, University of California, Davis, researchers surveyed a large group of set-aside crop growers in nine California counties. Less than half of those surveyed participated in federal price and income programs when required to set aside a portion of their acreage. A majority

5. There are four congressional committees dealing directly with agriculture: (1) the Senate Committee on Agriculture, (2) the Senate Subcommittee on Agricultural Appropriations (a subcommittee of the Senate Committee on appropriations), (3) the House Committee on Agriculture, and (4) the House Subcommittee on Agricultural Appropriations (a subcommittee of the House Appropriations Committee). Over the years California's representation in these important committees has been relatively meager. The exception was Representative B.F. Sisk's chairmanship of the powerful House Subcommittee on Cotton during the 93rd Congress (1973). While California has averaged four to five places on the 36-member House Agricultural Committee over the past twenty years, it has been represented on neither the Senate nor the House subcommittees on agricultural appropriations for well over two decades (using the 97th session (1980) as the base year). California has had only four representatives on the Senate Committee on Agriculture since achieving statehood in 1850. The state had no representation on the Senate Committee on Agriculture for over two decades prior to the appointment of Senator S.I. Hayakawa (Republican). Representative Gene Chappie was appointed to the House Agricultural Committee in 1981.

viewed the set-aside cost as excessive when compared to program benefits. Many California farmers have made extremely large investments in machinery and irrigation equipment. While this has reduced their risk of crop loss, it means higher costs of set-aside in terms of foregone income. California yields are significantly higher in many crops than in other states (see Chapter 2) and also more stable, partly because of a more dependable climate. Thus, the incentive for California farmers to participate in price support or disaster programs is less than in many other places—rice being a notable exception.

The benefits of federal attention, however, have been felt in many aspects of California agriculture—from construction of water systems to tax write-offs for land development and technological innovation. Agriculturalists nevertheless increasingly chafe at the federal rules and restrictions which are also part of federal attention to agriculture. The following chapters discuss some of the major policy issues regarding California agriculture in the 1980s and beyond.

References

Carman, H.F. "Coming: More Corporate Farms in California." *California Agriculture.* 34(January 1980):9–10.

Carman, H.F. "Taxation as a Factor in Economics of Scale in Agriculture," in *Farm-size Relationships, with an Emphasis on California.* Ed. Carole Frank Nuckton, University of California, Davis, Giannini Foundation of Agricultural Economics, 1980.

Carman, H.F. "The Taxing Effects of Tax Reform." *Western Fruit Grower.* 101(September 1981):12–14.

Durst, R., Rome, W., and Hrubovcak, J. *The Economic Recovery Act of 1981: Provisions of Significance to Agriculture.* USDA, Economic Research Service, Staff Report AGES 810908, September 1981.

Kramer, R.A., Pope, R.D., and Gardner, B.D. "Participation in Federal Farm Commodity Programs." *California Agriculture.* 35(May–June 1981): 4–6.

Luttrell, C.B. "Farm Price Supports at Cost of Production." *Review.* 59(December 1977): 2–7. Federal Reserve Bank of St. Louis.

Moore, C.V. *Effects of Federal Programs and Policies on the Structure of Agriculture.* USDA, ERS, unpublished paper, January 1977.

Peterson, W.L., and Hayami, Y. "Technical Change in Agriculture," in *A Survey of Agricultural Economics Literature*, Vol. 1. Ed. Lee Martin, Minneapolis: U. of Minnesota Press, 1977.

Rochin, R.I., and Nuckton, C.F. "Farm Worker Service and Employment Programs," in *Seasonal Agricultural Labor Markets in the United States.* Ed. Robert D. Emerson, University of Florida, Institute of Food and Agricultural Sciences, Florida Agricultural Experiment Station, Food and Resource Economics Department, 1980.

Sosnick, S.H. *Hired Hands: Seasonal Farm Workers in the United Gtates.* Santa Barbara: McNally and Loftin, 1978.

Youde, J.G., and Carman, H.F. "Tax-Induced Cattle Feeding." *California Agriculture* 26(June 1972):13–14.

U.S. Chamber of Commerce. "USDA to Protect Resources, But Who'll Protect Farmers?" *Washington Watch.* Vol. II, No. 4, April 1980.

U.S. Department of Agriculture. *Certification Report*. Issued on 1979 Feed Grain, Wheat Acreage Compliance, No. 37–80, January 4, 1980.

U.S. International Trade Commission. *Tariff Schedules of the United States, Annotated, 1980*. USITC Publication No. 1101. Washington, D.C.: U.S. Government Printing Office, 1979.

PART V   Challenges: Today and Tomorrow

Environmental and land use issues have become increasingly controversial in recent years. The Los Angeles basin represents both kinds of problems; here a layer of smog floats above urbanized areas. *Photo by Jack Clark.*

21

Policy Issues

Warren E. Johnston and Harold O. Carter

It is the task of this chapter to examine some of the public policy issues which will shape the future of California agriculture. A general characteristic of the state's agriculture has been its propensity for technical innovation. Another common thread, increasingly visible in today's fabric of production and marketing, is the industrialization of traditional processes which used to be smaller in scale and performed by many more entrepreneurs. Agriculture is now also highly integrated with other sectors of the economy—an evolution which has accelerated since World War II and which will continue to have profound impact.

This chapter will discuss four major areas of concern relative to agriculture today: (1) growing concentration, (2) the availability of adequate water, (3) land use, and (4) environmental problems. All of these have drawn increasing attention in the 1970s, and in the years to come will continue to be debated as policy issues in a complex and interdependent society.

An Overview of California's Industrialized Agriculture

California farm output in terms of sales is concentrated towards larger units, the top 10 percent of the farms accounting for almost 80 percent of the state's value of production. Individual family farms still predominate in numbers (80 percent of the total), with less than 5 percent of all the state's farms identified as corporations in 1978. It is the largest farms, however, which dominate production. Furthermore, commercial farms are

becoming fewer and larger, continuing past trends, though at a decreasing rate.

Wide-spread mechanization is characteristic of industrialized agriculture. Agriculture has become progressively more energy-intensive; besides large machinery fitted for use on larger farms, increased usage of fertilizers, pesticides, and irrigation water has also demanded more energy inputs. The historical substitution of fossil fuels for human and animal energy was in response to relatively cheap fuel supplies, abundant land, and relatively high-cost labor. (Nevertheless, hired labor is still important in California. Farmers in this state hired more workers in 1978 than any other state and paid 20 percent of the nation's hired farm wage bill. They continue to employ an average of over 200,000 workers annually.)

Integration with Other Sectors

The system by which agricultural products are assembled, processed, and distributed to consumers also reflects the industrialization process. New technology and changes in organization have gradually revolutionized the food system. The impacts on farming have been both direct, via constraints imposed on the farmer's production choices, and indirect, through effects on size and location of marketing firms. The technology of assembly has been altered substantially in recent decades by the development of large-scale bulk handling methods for grains and milk, as well as many kinds of fruits and vegetables. These have changed the manner in which farm products are transported to processing facilities and to markets. The ability to handle greater volume has, in turn, led to demands for greater volume—and to the attrition of smaller producers, processors, and distributors.

Decisions of processors influence planting schedules, cultural practices, applications of inputs, and harvest schedules to meet their needs for certain raw product characteristics. Changes in raw product specifications and the use of substitutes have also affected the geographic location of production and possibly the nature of contractual arrangements between processors and growers. In these regards, much of California's commercial agriculture is influenced by industrial arrangements, in contrast to systems in other countries where processors and consumers have more limited choice of commodity characteristics, including quality. In effect, many California products are grown "to order"—thus producers must collaborate with processing and marketing firms.

For many commodities, free market conditions also have been affected by integration through common ownership of production and marketing facilities. These vertical arrangements generally have been developed in an effort to reduce both processor and grower uncertainty and to provide closer coordination between the raw product requirements of marketing and processing firms and the production decisions of farmers. A consequence is that the number of buyers for many agricultural commodities has declined over time. Hence, producers now find that there are fewer market alternatives for their products. Many of the existing firms that purchase the state's agricultural production are either larger farmer cooperatives or conglomerates and multinational firms, often with corporate headquarters located in major cities outside California. For the latter, local representatives may have relatively few decisions left to them, in contrast to those common in previous marketing structures.

California's agriculture has also taken on important international dimensions. The state's outlets for food and fiber have expanded over time—first from local to regional, then to national and subsequently to international markets. The volume of farm exports has increased significantly and now contributes to offset this nation's deficit nonagricultural trade balances. In the aggregate, agricultural exports utilize one-third of the harvested cropland acreage in California.

Three major crops (almonds, cotton, and wheat) had exports of more than 80 percent of their production in 1979. Almost one-half of the state's rice production and one-third of its lemon output were shipped to foreign markets. A wide variety of speciality commodities—nuts, fruits, and vegetables—are also important components in California agricultural exports.

Markets for California speciality commodities are largely found in high income countries. Larger populations and greater affluence in other nations have encouraged the expansion of trade, providing more outlets for the production of American and California farms. Such trade is strongly affected, however, by government policies both at home and abroad, which are always subject to change. Tariffs, nontariff barriers, variable levies, export incentives (subsidies), concessional sales, and domestic agricultural programs all affect agricultural trade. These nonmarket forces promulgated by governments can cause price instability in international markets for food commodities.

In summary, all parts of the food system have undergone structural changes, generally in the direction of concentration. Prices are increasingly being determined less by traditional, competitive, open market arrangements, and more by contracts and arrangements between highly integrated subsectors of the food system. The extent of the transition is, of course, highly variable by commodity group and region.

Having presented this characterization of a highly industrialized agriculture in an increasingly urban society, we cannot overlook the concerns expressed by some that concentration has progressed beyond the level required for economic efficiency. For example, the concentration of farms into fewer but larger units may impair the economic base of many rural towns and cities, and indeed eventually result in a decline in the quality of community life. Fewer processing and distribution firms may have increased market power, and the ability to control or unduly raise consumer prices.

Technological advances have contributed to the concentration of production and processing in the state. Although private and public research have developed these new and innovative technologies, it is research and development interacting with the social, economic, and political environment that has resulted in the current levels of concentration in agriculture. In this regard, agriculture is responding to forces much broader than itself—and paralleling trends in other sectors of the society.

Despite the concentration in commercial agriculture, another consideration is the fact that nonfarmers, and part-time farmers, are of growing importance in rural areas. Consequently, there is a blurring of lines between agricultural and nonagricultural interests in rural locales. In 1930, less than one in six farms was part-time. Now two of three farm families receive more than half of their income from nonfarm sources. Increased nonfarm incomes have reduced rural migration to urban areas, allowing a "middle ground" between staying in agriculture and exiting completely. Meanwhile nonmetropolitan areas have been experiencing faster growth rates than metropolitan areas as numerous people opt to live in less congested areas. If this trend continues, it will profoundly influence the size and composition of rural populations. It already represents a reversal of one of the nation's traditional long-term population trends.

Who Will Control Agriculture?

Is it necessary or inevitable that trends toward concentration in the food and agriculture complex must continue as they have over the past half century? Who controls the destiny of agriculture in California? Traditionally, farmers made their own decisions about input use and markets. Time and events have contributed to the steady shifting of control away from the farmer, as we have seen. This gives us cause to ponder what new goals are being pursued—that is, how are we to determine the kind of agriculture best for California over the succeeding decades?

When the land grant college system was established in 1862, the farm population in the U.S. was almost 60 percent of the total; now it is less than 3 percent, and declining. For California, less than one-half of one percent of the state's population are now commercial farmers. Even when counting people in related industries, the relative percentage of those engaged in agriculture is small. Surprisingly, the number of people on food stamps is three times the number of people on farms. Even less than 50 years ago a majority of the House of Representatives was from farm districts. By 1976, only one-fifth came from so-called farm districts. Thus we have seen steady attrition of both farm population and their political representation.

Meanwhile, environmentalists concerned about ecology, and consumers concerned about food prices and food safety, have increased influence in policy discussions on how agriculture is to operate. Control of practically all aspects of the food and agriculture system has been diffused among a spate of governmental and private institutions. On one hand, we appear to have inexorable forces in the nonfood section affecting agriculture—and simultaneously we have a diverse set of governmental and regulatory policies also exerting control. Clearly, the forces bringing change, both up to now and in the future, are no longer principally internal to agriculture.

The Availability of Water

Water is essential for agricultural production. Of the total acres cropped in California, about three-quarters are irrigated. In 1978 a University of California-sponsored task force (made up of over 200 study group participants representing every segment of California agriculture, including farmers, government personnel, processors, and consumers), considered and debated major policy issues facing California agriculture in the 1980s. The consensus, after a year of study, was that the single most limiting resource likely to constrain food and fiber production in California was water.

Natural Distribution and Human Intervention

As other chapers have documented, water in California, while generally plentiful, is not always present at the right time and place. Large water storage and diversion systems have been designed to overcome these temporal and spatial imbalances. Surface runoff provides about 60 percent of the water used in California, with the remaining coming from groundwater supplies. The spatial distribution of surface runoff actually presents a more difficult problem than does seasonal flow variation. Roughly two-thirds of the annual precipitation occurs in the northern one-third of the state with the runoff flowing toward the west and northwest—away from where it is needed in the southern portions of the state. Hence, the need for water transfer facilities.

Groundwater helps mitigate the imbalances between surface supplies and use. The bulk of groundwater supplies are found in the inland and central valleys of California. The total storage capacity of California's groundwater basins has been estimated at 1.3 billion acre-feet of water with the largest quantities in the Central Valley. However, usable groundwater storage is only estimated to be 200 million acre-feet, or about three times the existing total surface storage capacity. Of particular concern is the fact that average withdrawal rate exceeds overall recharge by more than 2 million acre-feet annually. Thus, withdrawals are currently depleting groundwater basin water supplies. Overdrafting of groundwater can, in addition, lead to a variety of problems, such as land subsidence, saltwater intrusion into fresh water aquifers, increased energy consumption, and even disruption of social and economic activities in extreme situations.

Critical overdrafting of groundwater supplies has been identified in 16 of 200 inventoried groundwater basins, and these overdrafted basins account for about 20 percent of the state's usable groundwater storage capacity. For the near future, water will probably be pumped from increasingly greater depths. With greater energy costs, however, the extraction of groundwater will ultimately become uneconomical for irrigation. Some critical overdrafted basins in the San Joaquin Valley have already reached the stage where pumping costs are approaching levels too costly for normal agricultural use.

Basic water costs for much of the state's agriculture currently range from $2 to more than $85 per acre-foot, depending on the source of water. Increasing water costs can cause farmers to change their cropping patterns, reducing or eliminating low value crops, or to change technologies in water systems. In extreme cases they may be forced into dryland farming or may even abandon farming operations on some land altogether. Naturally, net returns to farm operators and owners of farmland are gravely affected.

Where Do We Go From Here?

What is the outlook for increased water availability? In some areas groundwater basins have been overdrafted in anticipation of additional imported surface supplies. Conflicts between environmentalist and agriculturalist groups have, however, slowed new water supply development, and in some cases stopped it. The time lag between authorization and completion of new facilities requires at least 10 years; currently there are no authorizations for new major surface water augmentation in the state.

The water issue is more, however, than simply adequate supplies at reasonable costs. It is pervasively intertwined with energy, land, and environmental quality issues. Land quality, for example, can be severely affected if salts from irrigation water are accumulated in surface and subsurface profiles. Such salinization may lead to reduced crop yields, and at the extreme may even preclude economic production from severely affected soils. In many areas, groundwater supplies are poorer in quality than local or imported surface water supplies. Thus, high quality irrigation water is desired, and up-graded water management practices, including drainage, will be required to maintain productivity and land value.

Future debates, according to the Task Force report, will center on the available water supply, efficiency of use, allocations within agriculture, and allocation among agricultural, municipal-industrial, and environmental uses. The total amount of water available in the 1990s will depend to large extent on public action to generate new transfer and storage facilities and to promote incentives for conservation and redistribution. Ultimately these might include legal transfer mechanisms to permit reallocation of water supplies—and possibly water rights among classes of users. Competition for available water between agriculture and other sectors will undoubtedly become more intense. Since about 85 percent of the water used in California is currently received by agriculture, the stakes in the outcome of these policy debates are high.

Land Use

Future demands for agricultural commodities, both domestic and foreign, seem sure to increase because of rising populations and incomes in the U.S. and abroad. Nevertheless, urbanization continues to reduce California's agricultural resource base by converting agricultural lands and increasing nonagricultural demands for water. These two competing sets of forces make it imperative to monitor, manage and conserve our primary resources so that they will not be irretrievably lost—and, if possible, to expand the productivity of the resource base upon which California agriculture depends.

Urbanization and "Best Use"

As our economy has become more developed and our citizens more urbanized, debate has intensified about the highest and best use for agricultural land. Conflicts may revolve around preserving, protecting, or reserving land from or for particular uses. An implicit assumption is often made that the market system for allocating land to its highest and best use is short-sighted and inadequate without governmental intervention. Others would argue that the market system is not truly allowed to work because of customs, laws and regulations.

Of California's 100 million acres of land, 34 million acres produce food and fiber. Of that, about 7.5 acres are irrigated and another 3 million produce dryland crops, while the rest is grazing land. Urban land now occupies 5 to 7 percent of the state's most productive land. With population pressures and the fact that much agricultural land lies in the path of urban development, concerns are expressed by farmers and others that prime lands are being irrevocably lost for agricultural use. While some of these concerns relate to our future food supply, other factors are also cited—for example, the need for open space in or near urban areas, and the need to control urban sprawl.

Legislation and Conflicting Values

The primary authority regarding the use of private land is granted by the Constitution to the states. Pressures for the establishment of an explicit, comprehensive national land use policy have, however, persisted. During the 1970s land use legislation was introduced, though not passed, by Congress. Within the past two decades in California, the California Land Conservation Act (CLCA) of 1965 and the Coastal Zoning Act of 1976 both addressed land use issues, each dealing with separate concerns. The objectives of the CLCA were to preserve prime agricultural land and to deter urban sprawl. In contrast, the Coastal Zoning Act had aesthetic and open space objectives; it limits private development of land in the coastal zone. The former is based on changes in tax policy, whereas the latter is a zoning initiative which offers no economic incentive to property owners. Both have been controversial, and are likely candidates for revision— which is indicative of the complexity of land use problem resolution.

Conflict inevitably results both from redistribu-tive gains and losses from policy decisions, and from differences in the value systems of citizens. People tend to be ambivalent. Generally they can support efforts to achieve "good" land use to the extent that it does not impinge on their own personal freedom in managing private property.

Environmental Problems

Future food production is also likely to depend on society's success in dealing with two broad categories of possible environmental problems— those related to expanding and intensifying the use of resources, and those related to the increased use of inputs such as fertilizers and pest control chemicals.

Maintaining Inherent Resource Productivity

Among the first group are problems of deteriorating soil structure and fertility, problems of soil erosion and desertification, and problems related to irrigation. Unless they are ameliorated by technological improvements, upgraded management, and concerted conservation efforts, these problems may lead to a gradual deterioration in resource productivity and declining levels of food production. Deteriorating soil fertility may result from the interplay of many forces, including improper irrigation, compaction by cultural and harvest practices, intensified cropping, reduced incorporation of residual organic crop materials, or other factors. Thus far, improved plant varieties, increased chemical inputs and management practices have compensated for basic declines in natural soil conditions, but there is rising concern that the continued development of these enhancing sources of increased agricultural productivity may be slackening worldwide. Soil loss through wind and water erosion also has implication for future crop production. A recent U.S. study concluded that soil losses would need to be cut in half if crop production were to be maintained indefinitely at present U.S. levels. Current and anticipated levels of food and fiber exports to foreign countries may exact a heavy toll on our topsoil resources. Meanwhile, overgrazing and other forces could reduce the natural productivity of rangelands which support parts of California's very important livestock industry.

Loss and damage of irrigated lands are of significant concern because these lands are the most productive in terms of yields per acre. Unfortunately, irrigated lands are also vulnerable to factors which may sharply reduce agricultural productivity. About half of the world's irrigated land is already said to be damaged to some extent by salinity and waterlogging. Similar conditions are evolving in parts of California. While it may be possible to restore damaged lands, it is costly and slow. Thus it is rational to try to mitigate or avoid possible adverse conditions. Careful attention should be focused particularly on recently developed areas in the San Joaquin Valley and on lands in the Sacramento Valley and in the Delta.

Problems from Use of Modern Inputs

The second group of problems includes potential pollution from fertilizers, pesticides, and other materials; and increased susceptibility to diseases and pests of high-yield varieties grown in monocultures. Despite recognition of initial productivity increases from the application of chemical inputs, some critics argue that there may be longer-run, more adverse consequences. For example, contamination of groundwater can occur if fertilizer application rates exceed the absorption capacity of the soil and the nutrient need of plants. Unfortunately, the current information base on fertilizer and chemical pollution is fragmentary and limited. It is recognized, however, that misuse of such inputs may generate unanticipated environmental problems over time.

Destruction of insect and predator parasite populations and the increasing resistance of pests to heavily used pesticides are among the types of problems encountered. On California farms, for example, 17 of 25 major agricultural pests are now resistant to one or more types of pesticides, resulting in significant crop damages. Nationwide, in 1976, 364 species of insects were identified as having developed resistance to classes of insecticides. These considerations have led to increased emphasis on integrated pest management (IPM) strategies in agriculture.

Environmental degradation may well affect the productivity and economics of agriculture in future years. As with concentration, water, and land use, some environmental problems may affect more than agriculturalists; for example, water pollution from agricultural practices may have adverse effects on municipal or industrial users. On the other hand reciprocal effects are also to be noted—as occurs with damages to crop yield and/or product quality due to increases in air pollution from urban and industrial activities. The loss for California agriculture alone from 1972 levels of air pollutants was estimated to be at least $125 million. Trace elements from stack emissions, and pathogens and heavy metals from sludge application to agricultural lands, may also enter food chains. There is ample concern that these complex interactions need to be better understood. Our often fragmentary and meager information base must be expanded, to bring better understanding of changing environments and their effects on future agricultural production.

Conclusion

Universally acceptable resolutions to the economic and environmental issues that will shape the future of California agriculture are not to be easily won, or settled to the satisfaction of agriculturalists. Controversies in these areas clearly extend beyond the agricultural community. Debate will continue as enlightened and concerned California citizens consider the agricultural sector, and try to determine how the resources that underlie it can be better managed to meet a variety of private and public goals. Satisfactory resolution will require the informed and active participation of all groups in the debate.

References

Carter, H.O., and Johnston, W.E. *Farm-Size Relationships, with an Emphasis on California.* Giannini Foundation Special Report, December 1980.

Carter, H.O., and Johnston, W.E. "Some Forces Affecting the Changing Structure, Organization, and Control of American Agriculture." *American Journal of Agricultural Economics.* 60(December 1978):738–748.

Paarlberg, D. *Farm and Food Policy: Issues of the 1980s.* University of Nebraska Press, 1980.

U.S. Congress, House of Representatives. *Agricultural and Environmental Relationships: Issues and Priorities.* Joint Committee Print, 96th Congress, 2nd Session, June 1979.

U.S. Department of Agriculture. *Structural Issues of American Agriculture.* Agricultural Economic Report 438, November 1979.

University of California Agricultural Issues Task Force. *Agricultural Policy Challenges for California in the 1980s.* Division of Agricultural Sciences Special Publication 3250, 1978.

Wood, W.W., Jr. "Planning and Use of Agricultural Land—In Whose Interest?" Unpublished paper. University of California, Riverside, 1980.

A field of sugarbeets near Salinas demonstrates the conversion of water, soil nutrients, and sunlight into food and fiber. Plants are biological factories. *Photo by Ansel Adams.*

22

An Ecological Overview

R.S. Loomis

The preceding chapters provide remarkable detail on the evolution and present state of California's agriculture. The changes have been dramatic and continuous since the first mission settlements by Spanish colonists, not only in terms of the landscape itself but also in the technological and cultural context in which modern farming takes place. Let us now look more to the future, considering California's agriculture in a somewhat different way, as an ecological system. How does California's agriculture compare as a system with those found at other times or places? What environmental problems exist or may occur? What are the energy equations in food production? What have we learned, and what do we need to know?

The important thing about looking to the future is to gain a more rational basis for our present thinking. For that, we need a systems view of the forces, limits and controls which influence our agriculture. Agriculture involves thousands of entrepreneurs independently approaching complex decisions about crops, acreage, technologies and markets. The emergent patterns of successful strategies tell us a great deal about controlling factors. Being quantitative about the total system is more difficult, but some sense of actual numbers may be possible.

What follows represents a personal analysis and opinion. Whether or not the reader agrees with these conceptions of present and future, our purpose is accomplished if understanding and concern for agriculture are enhanced. This ecological overview will center on four basic views of California agriculture: its unique diversity and productivity; its resources of land, water and plant nutrients; its energy system; and its information systems.

The Nature of California's Agriculture

California's agriculture differs greatly from what one finds in other parts of North America and the world. California is divided into a series of provinces, each with somewhat different suitabilities for agriculture. Favorable climates with long growing seasons, fertile soils, irrigation, and high levels of production and processing technology are factors favoring highly diversified and flexible cropping. Production and quality of specialty crops are high. These, plus our access to large markets with an affluent population, lead to emphasis on high value fruit, nut and vegetable crops—of which California now produces half of the nation's supply.

Diversity

The unusual diversity of our cropping is summarized in Table 1. The total area cropped is only 10 percent of the state's land. This is somewhat larger than the irrigated land area since some acreage, particularly for wild hay, barley and wheat, is still cropped without irrigation.

In the 1860s and 1870s, cereal grains grown in dryland culture dominated the farming systems of California. The combine harvester made this compatible with the scarce labor supply within the state. Though there was only a small internal market, dry grain is a compact commodity and can be shipped long distance by sea at low cost. For a time, California was the world's leading exporter of wheat. Later, dramatic increases in crop diversity came with irrigation, processing and refrigerated rail transit. Until recent decades, however, yields of many crops were relatively low because agriculture depended on internal sources and recycling of the essential nutrient, nitrogen. The native supply of nitrogen in California's desert soils was very low. Inputs of nitrogen were gained by rotation of field and vegetable crops with a leguminous, nitrogen-fixing, forage crop—alfalfa. Before 1950, this commonly led to

TABLE 22.1
Use of Land for Crop Production in California

Crops	Harvested Area
	1000 Acres
FIELD CROPS	
Food grains: wheat	1,150
rice	550
Sugarbeet	230
Dry beans	220
Feed grains: barley	710
corn	270
sorghum	150
oat	70
Silage (corn + sorghum)	190
Hay: alfalfa	1,030
other	520
Irrigated pasture	750
Cotton	1,490
Seed crops	200
	7,530
VEGETABLES	
Tomato	240
Lettuce	160
Broccoli	70
Potato	50
Other	250
	770
TREE AND VINE CROPS	
All nuts	610
All grapes	690
Deciduous fruit	340
Citrus	290
Other	90
	2,020
TOTAL ALL CROPS	10,320

The numbers vary slightly from those of previous chapters. Field and vegetable data are drawn from 1980 reports while the tree and vine data represent averages for 1977 to 1979. An additional 36 million acres are classified as rangeland. The total area of California is 102 million acres. For comparison, the State of Iowa harvested crops from 25.6 million acres with 87% of that accounted for by two crops, corn and soybeans. (These numbers may be converted to 1,000s of hectares, the metric unit, by multiplying by 0.40.)

a lock-step rotation in California agriculture—legume, high-value processing crop, low-value grain and then back to legumes. Yields were increased, but the amount of nitrogen left in the soil by the legume still fell short of allowing subsequent crops to achieve their full potential. The advent of low-cost nitrogen fertilizers in the 1950s eliminated the need for these fixed rotations, overcoming the cropping limitations set by nitrogen. (This was not accomplished without some adjustments, however. It was 1970 before stiff-strawed wheat strains were available which would support the heavy yields of grain which are now possible.) Farmers today have much greater flexibility in cropping. And their output is more responsive to current markets since they now can decide each year which crops seem to hold the greatest promise for returns. Acreages of wheat, corn, barley, sugarbeet, tomato and other crops fluctuate accordingly.

An Island Economy
A surprising inference which can be drawn from Table 1 is that California probably does not produce sufficient food for its present population of more than 22 million people. California now has about 2.2 people per acre of cropland before correction for export to out-of-state markets. Roughly 80 percent of the vegetables, tree and vine crops and nearly 100 percent of the rice grown in California are exported. (In addition, the food returns from cotton are limited even though the seed oil appears in vegetable shortenings and the oil seed cake flows to animal feeds. Let us assume that the effectiveness of cotton acreage for human and animal feeds is 50 percent.) These adjustments reduce the area that produces food for Californians to about 6.7 million acres, yielding a ratio of 3.3 people per acre. While this is less than in densely populated countries like Egypt, The Netherlands and Japan, where ratios range from 6 to 11 people per acre of crop land, it is twice the world average and four times the U.S. average (currently near 0.7 people per acre after correction for exports). Even though California crop lands are supplemented by grazing on an additional 36 million acres, it is not surprising that California depends upon very large imports of grain, oil seeds and meat from other parts of the U.S. and the world. In 1971–72, for example, our yearly imports averaged 3 million tons of grain, 1.7 million live cattle (including 300,000 finished animals ready for slaughter) and 1.4 million finished hogs. In addi-

tion, the equivalent of 1.4 million dressed carcasses of beef and 4.6 million hogs were imported annually. These amounts grew steadily throughout the 1970s; in 1980, 2.6 million dressed carcasses of beef were imported. All of this, plus large amounts of vegetable oil, represent the produce of perhaps 4 million acres of land in other states.

This is not a unique situation—many other regions of the U.S. and the world are similarly dependent upon imports. Sometimes, as is the case for California, the economic and nutritional values of specialty agricultural exports tend to balance with imports. What is unique for this state, however, is the very great land distances over which those goods are exchanged with other parts of the country. Agriculturally, California takes the form of an "island economy" with the degree of insulation from midwestern grain and meat supplies depending on transportation costs and other factors. Many forces might operate to change the present relationships. Increasing pressures on land, water and food supply by California's expanding urban population might call for more imports from the east. Conversely, sharp increases in the costs of transportation or of imported grains—due, for example, to declining water supplies in the Great Plains' aquifers—might favor an expansion of feed grain production in California or a trend toward less meat consumption.

Complete substitution of imported meat by in-state production is unlikely; and in any case it would be an immense task. In 1980, the equivalent of 3 million pounds of dressed meat were imported. Since feeding efficiencies require about 10 pounds of grain and forage per pound of meat, we would need an additional 30 billion pounds (15 million tons) of feed. That equates to high levels of production (10 tons of hay or 5 tons of grain per acre) from 2.5 million acres of crop land.

But these issues are subject to modification by even more fundamental events. For example, supply-demand relations are closely linked with the level of productivity—which in turn is open to very large changes.

Production Levels—Here and Elsewhere
We tend to think of California's agriculture as oriented to near-maximum production per unit of land area. But even with nitrogen fertilizer, the yield levels of this state's crops fall well short of their biological potential for our climate. Compar-

ison of California's average yields with high-yield records for corn (7,000 vs. 18,000 pounds/acre), alfalfa (12,000 vs. 30,000 pounds/acre), and rice (6,000 vs. 12,000 pounds/acre) show clearly that our agriculture is far from "intensive." Japan and The Netherlands, for example, operate much closer to their yield ceilings. The difference reflects the levels of inputs and the intensity of field management which are only moderate in California. California's farming is quite efficient in its use of labor, capital and resources of water and nutrients, but simply is not driven to its biological and climatic limits.

California thus has considerable reserve for increases in production. That would require additional inputs of energy (mostly for nutrients) and more "labor" in the sense of more knowledge, more time in management, and more care in operations. Labor is currently the most costly input for American agriculture. Particularly with field crops, farming practices here tend to optimize production per unit of labor rather than per unit of land or capital as in other parts of the world. But mainly, increased production would require more water.

But even greater reserve potential exists in other regions. Africa, Asia and South America all show enormous potential for increased production, both in land area and production level. Indonesia, for example, averages less than 3,000 pounds of rice per acre compared with its potential for 10,000 to 12,000 pounds. The shortfall is not so much due to management skill and labor supply, which seem abundant, but to a shortage of plant nutrients. The low level of nutrient recycling in Indonesia can be traced to poor incentives and a low social priority for agriculture. Given sufficient capital, as well as social and economic organization, many of the developing countries have the potential to become exporters of food. The recent emergence of Brazil as a major exporter of soybeans is but a single example. The dominance in world food supply which the U.S. has held during the last 100 years, therefore, may diminish in the years ahead. Our country has remained competitive in the world market mainly through rapid increases in labor productivity. Little room remains for further increases in that area except with fruit and vegetable production. It is quite possible, despite the expanding world population, that the next 20 years will see a return to the land and production surpluses which plagued America in the 1950s and 1960s. The pace at which such changes might occur, and how they might affect California's agriculture requires some rather complex crystal gazing. Much depends on population growth and the priorities given to agriculture here and abroad.

A similar analysis can be made for production levels on California's timber and grasslands. In both cases, productivity is well below potential. Our grasslands are limited mainly by low rainfall and low temperatures during the wet season. But utilization is also restricted by nutrient deficiencies, the presence of plant species of low quality for grazing, predators such as coyotes, and inadequate fencing.

The situation for forests is more complex. Multiple-use concepts limit the harvest from public forests to less than their potential growth rates. Those forests tend to be dominated by slow-growing older trees. Outstanding performance is achieved on a few of the smaller private holdings and by Christmas tree growers. But in general the small private holdings are poorly managed for timber production. The forest plantations of the larger timber companies are much better managed for production than are the public forests but they also fall well short of what might be achieved. As with the grasslands, low returns limit the application of technology which would increase the volume of production. Terracing, fertilization, thinning, control of weed species, and the introduction of genetically superior timber species are among the practices employed much more extensively in Europe and Japan to come closer to potential growth rates.

Resources: Uses and Constraints

Land is a basic resource for agriculture. More than just land area is involved, however, since climate, chemical and physical characteristics of the soil profile, and the availability of water all determine the type of cropping and yield levels which may be possible. Rather marked changes occur to soils under cultivation and exploitation by cropping. Some of those changes can degrade the resource. The quality of the land for agriculture, however, may also improve under tillage. Drainage and salt removal are key types of improvements found in arid western states like California. Land leveling and additions of plant nutrients and organic matter are also important.

A basic point to recognize is that soil formation and change continue under agriculture.

As principal concerns for California's soil resources, we may list, in decreasing order of importance, water, nutrient depletion, salinity relations, and erosion.

Water Use in California's Agriculture
The "developed" water supply of California amounted to 40.5 MAF (million acre feet; 1 acre foot = 1,234 m³) in 1980. Around 14.7 MAF is pumped from groundwater while 25.8 MAF is surface water from streams and reservoirs. Agriculture currently uses about 85 percent of this; the actual numbers, however, are complicated by the amounts of return flows and deep percolation to groundwater and by the amounts used for salinity repulsion in the Delta and as flows to salt sinks. Of concern to agriculture is the fact that groundwater is being overdrafted by about 2.2 MAF each year and that Arizona will soon begin claiming a major share of the Colorado River supply. Much of the overdraft occurs in the agricultural areas of the San Joaquin Valley, an area already well short of the amount of water it could use in a fully developed agriculture. To compensate for the loss of Colorado water, Southern California will begin to draw its full entitlement from the California Aqueduct, further reducing the present San Joaquin supply.

Any change in water supply or cost will lead to a change in agriculture. The reason is that there is a fundamental physical relationship between the transpiration of water from plant leaves (accounting for 80 percent of water use by crops) and photosynthetic production. Transpiration and photosynthesis both are energy (sunlight) dependent, and both involve the exchange of gases—water and CO_2—through the stomatal pores of the leaves. Those pores must open for CO_2 uptake to occur—but water loss also occurs whenever the pores are open. It turns out that there is a more or less constant ratio between the production of biomass and the loss of water. That ration is Water Use Efficiency:

$$\text{WUE} = \frac{\text{dry matter production}}{\text{transpirational water loss}} \quad (pounds/pound)$$

WUEs with healthy crops vary between 3 and 5 pounds of production per 1,000 pounds (125 gallons) of water. Since only 30 to 50 percent of plant production flows to economically useful portions like grain, efficiencies per pound of food or feed go as low as 1/1,000.

Water use efficiency varies with environment and plant species. As a result, there are many adaptive strategies for maximizing production with limited supplies of water, and most of them are already employed to varying degrees in California's agriculture. WUE is greater during the cool season, with crops such as wheat and broccoli, than during the summer. Also, some summer crops such as corn and sorghum have intrinsically higher efficiency than others such as sugarbeet and tomato. Crops should be healthy, weedfree and adequately nourished; otherwise, WUE may drop well below the basic values. We also can change total water use by varying the acreage and growth duration of the crops. In most of inland California during the growing season, solar energy evaporates 0.2 to 0.3 inches (5 to 8 millimeters) of water per day from a crop or free water surface. A crop operating at a potential rate of 0.3 inches per day for 150 days would consume about 45 inches of water. Replacing it with a crop which matured in 100 days would reduce water use to about 30 inches, but at lower yield. Shortages of water are a major reason why we still have reserves of undeveloped farm land in California and why crop land lies fallow so much of the time. The climate would permit longer season crops and a greater amount of double cropping—the water supply does not.

There also is room for conservation. The main losses may be from reservoirs and canals. In addition, about 20 percent of the water applied in fields is lost to evaporation from the soil or by deep percolation. Those losses could be reduced to some extent by greater care and expense in irrigation. Methods include shorter irrigation runs (more labor), sprinklers (more energy and expense) and "trickle" techniques (more labor and expense, particularly for annual crops). On the average, however, agronomic assessments lead one to conclude that irrigation practices in California are already reasonably efficient in the use of water. This means that changes in water supply will lead to parallel changes in production.

There are certain major exceptions to our assessment that irrigation practices are relatively efficient. These occur principally with rice in the Sacramento Valley, and in some locations along the eastern fringe of the San Joaquin Valley, where water applications may considerably exceed evapo-transpiration requirements. The "excess" water, however, accounts for a major portion of the groundwater recharge in the state, and return flows to streams are used downstream for irrigation and for salinity repulsion. In these cases, the efficiency of individual fields may be low but that of the whole basin remains high.

Salinity

Soils in semiarid regions are subject to salinization through the accumulation of soluble salts such as sodium chloride. In low rainfall areas, many soils are saline in their pristine condition and require reclamation. Basically that involves adding water to leach the salts from rooting zones. But irrigation water is not free of salt, and under agriculture those salts can become concentrated in the soil as pure water is transpired to the atmosphere. So some extra water, a "leaching requirement," must be added from time to time to move these new salts away from the roots.

All of this is no great problem if the irrigation water is relatively pure and abundant, and if room exists in the deep profile for drainage. Problems arise if the soil profile fills or if the downward movement of salt is too concentrated and thus degrades the groundwater. The Imperial Valley represents an extreme case with very poor water quality—about 3,000 pounds of salt are added with each acre foot of water (1 gram per liter)—and a limited profile. Underground tiles or plastic drain lines have been laid in this area to intercept the leachate and carry it away as waste water. Eventually, large areas of the San Joaquin Valley will need similar drains if the land is to remain in production.

Drain lines need an outfall. In the Imperial Valley, the Salton Sea serves as the saline sink. In San Joaquin Valley, lower lying lands are beginning to serve in that role. Those lands might be sacrificed as salt marshes to absorb the saline wastes of the higher lands. That has attractive possibilities for wildlife but very serious consequences in the movement of salt to groundwater and in the loss of several hundred thousand acres of farmland. Thus the plans and debate regarding a drainage canal to remove salt from the San Joaquin.

Breeding for increased "salt tolerance" in crop plants may not be a very viable alternative for commercial agriculture. Increased tolerance could

A satellite view of the Salton Sea, center, shows the Coachella Valley to the north, the Imperial Valley to the south, and surrounding desert and mountains. The Mexican border can be clearly distinguished where the darker checkerboard of irrigated fields gives way to a different farming pattern. *Photo from the Remote Sensing Research Program, University of California Space Sciences Laboratory, Berkeley.*

reduce the amount of water allocated to drainage, and thus by a slight amount, the total of salt to be moved. But a drain would still be necessary and the costs in reduced production and diversity of crops, and soil tillage problems resulting from the higher salt level, are apt to be large.

Drains, dams, canals and the price of water promise to continue as major themes in California agriculture.

Nutrient Supply Depletion

Agriculture is an extractive activity. Certain chemical elements are derived from air and water so their supply is essentially limitless. Other nutrients are extracted from the soil. Table 2 provides some information on the nutrient composition of plant material. Several other elements including magnesium, boron, copper and chlorine are required by plants, but soil supplies

TABLE 22.2
Estimated Content of Selected Nutrients in Corn and Alfalfa

Element	Corn Grain		Total Corn Crop[1]		Alfalfa Hay	
	% Dry Wt.	*Lbs/Acre*	*% Dry wt.*	*Lbs/Acre*	*% Dry Wt.*	*Lbs/Acre*
Carbon[2]	43	3440	40	6400	40	8000
Hydrogen	7.0	560	6.5	1040	6.3	1260
Oxygen	47	3760	46	7360	41	8200
Nitrogen	1.63	130	1.33	210	3.07	610
Sulfur	0.14	10	0.11	18	0.30	60
Phosphorus	0.31	25	0.23	37	0.26	50
Calcium	0.03	2.4	0.33	53	1.43	290
Potassium	0.38	30	1.15	180	2.68	540
Iron	0.003	0.24	0.02	3.2	0.05	10
Zinc	0.00001	0.0008	0.00002	0.003	0.00002	0.004

[1] Calculations for corn assumed 8,000 lbs grain and 8,000 lbs residue per acre, and for alfalfa, 20,000 lbs hay per acre. Those yields are above the current averages in California but are still well below record yields.

[2] Carbon, hydrogen and oxygen calculated from the proximate analyses for protein, lipid and structural and nonstructural carbohydrates. The composition for other elements are representative values from the National Academy of Sciences tables on feed composition. (To convert pounds per acre to kilograms per hectare, multiply by 1.12.)

are either so vast (as with calcium) or plant use so small (as with copper) that they seldom if ever become limiting to plant growth. In California, it is nitrogen supply which strongly limits production on most crop and rangelands. In addition, phosphorus and sulfur deficiencies are fairly common, particularly on the older terrace soils surrounding the Central Valley. Phosphorus is of particular concern because the world's supply of high-grade ores is limited. Some areas also encounter zinc shortages.

The parent materials and secondary mineral clays of our soils serve as one source of nutrients. Taking five feet as a typical depth of rooting, crop plant roots on a acre of land will interact with as much as 20 million pounds of soil (dry weight). In semiarid areas such as California, the soils are generally quite high in base minerals (calcium, magnesium and potassium). So even very slow rates of weathering and dissolving will supply the needed amounts of the base elements for a very long time. This is one reason for the high fertility of California soils. In addition, the soil accumulates and stores nutrient elements which it receives as rainfall, dust, crop and animal residues, and fertilizer. Corn and alfalfa, for example, in going largely to feed livestock, may potentially be recycled to the land as manure. But the net export of nutrients as human food, and thus to sewage effluent, is also large. Over the longer term (hundreds to thousands of years), supplies of certain elements de-

rived from the clays may become limiting. Nutrient replacement then must be increased and our agriculture may assume some of the character of hydroponic culture now practiced in glasshouses.

Special Concerns for Nitrogen

A main concern at present is with replacing nitrogen which is derived from soil organic matter rather than mineral clays. Organic matter is the end product of fungal and bacterial decomposition of plant residues; in fact, humus consists mainly of degraded bacterial protein. Humus nitrogen is released slowly in the mineral forms (nitrate and ammonia) used by plants. Additional small supplies—10 to 15 pounds of nitrogen per acre (11 to 17 kg per hectare) each year—come by rainfall, fixation by "free-living" bacteria and other natural sources. A series of elegant studies by Hans Jenny of U.C. Berkeley has shown that, left undisturbed, soil-vegetation complexes will come to a point where such inputs equal outputs, and thus to an equilibrium content of soil organic matter and nitrogen. (With the high summer temperatures and low rainfall of California, the equilibrium content is small in absolute terms—in the range of 3,000 to 6,000 pounds of nitrogen per acre.)

It turns out that natural inputs of nitrogen are so small relative to nitrogen extraction by crops (130 pounds per acre per year for corn) that the soil nitrogen pool is quickly (50 to 70 years) re-

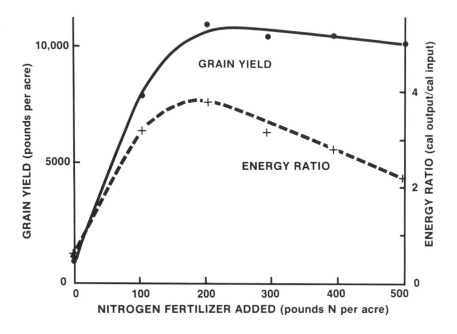

FIGURE 22.1 Responses of corn grain yield and energy output-input ratio to additions of fertilizer nitrogen. These experimental results were obtained on an irrigated soil near Fresno, California. The energy ratio is the heat of combustion of the corn grain divided by the energy content of the fossil fuel needed for fertilizer manufacture, farm operations, and machinery manufacture and repair. Note that the energy ratio is less than 1.0 without added N. (Pounds per acre multiplied by 1.12 equals kilograms per hectare.) *Data adapted from unpublished paper by P. R. Stout, U.C. Davis.*

duced by cropping to a new equilibrium. That is what has occurred in most primitive agriculture systems of the world. Grain yields have dropped to the level possible with natural inputs. An acre of corn receiving only 15 pounds of nitrogen per year yields around 900 pounds of grain per acre (rather than 8,000 to 16,000) containing 1.6 percent nitrogen (see Figure 1 and Table 2). Agriculture can be sustained indefinitely at that level, but at an enormous cost of land and labor for the necessary human food. The situation can be helped some by manure recycling and considerably by legume rotations, but yields still fall well short of biological potentials. Maximum crop production therefore requires large inputs of fertilizer nitrogen.

Fortunately, our atmosphere is 78 percent dinitrogen gas (N_2) by volume, so the supply for recycling is not limiting. This is the source material for nitrogen-fixing bacteria as well as fertilizer factories. Both use hydrogen to reduce N_2 to ammonia (NH_3) which can be used by crop plants. Skill is required for proper application of fertilizers. In soil, ammonia is converted rapidly to nitrate. Excessive applications, improper timing or water-logged soils can lead to loss of nitrate by gasification back to N_2, or by leaching which leads to contamination of drainage and groundwaters. In general, current practices are rather efficient—with 50 to 90 percent of applied nitrogen recovered directly by crops. Most of the rest is captured by bacteria and converted to humus, and is thus available to crops in subsequent

years. But the occurrence and threat of water contamination represent serious side effects of fertilizer application.

Erosion and Desertification

Erosion of surface soil by wind and water can be an insidious companion of agriculture. Loosening of the soil by tillage and removal of plant cover by tillage and grazing increases the susceptibility of soil particles to movement. On occasion, this contributes to disastrous losses in top soil.

Recent state and federal studies receiving wide attention by the public media have cited California as a region with a high rate of land degradation due to erosion and "desertification." The term desertification, referring to man-induced decline in the biological productivity of semiarid and arid lands, was born out of the apparent increases in the desert area of Africa in the 1960s and 1970s, thought to be due to overgrazing. Though it now appears that the desert did not expand and that grazing and other human activities were not involved, the term was popularized, and is retained to describe the consequences of over-use in such regions (including the salinization which may occur in the San Joaquin Valley without development of drainage).

The erosion question is worth considering in some detail. Steep landscapes and high geologic activity contributed to high natural rates of erosion and thus to the great depths of alluvium which have accumulated in our valleys since the last glacial period. Things have changed markedly under management by western man. A sharp reduction in the frequency and extent of wildfire is one important factor. Wildfire results in a severe reduction in cover and marked increases in soil loss, a phenomenon well known to residents near the shrublands of Southern California. At Muir Woods, north of San Francisco, one can examine a redwood log showing fire scars on the average of every 29 years for over 1,000 years prior to 1850. Since then, no fires have occurred in the area. Sierran forests, subject to even more frequent burning, were open and relatively free of shrubby undergrowth, in contrast to today's scene with heavy, fire-prone, but erosion-resistant, undergrowth. While fire decreased, grazing and timber harvest have increased and that could cause an increase in erosion.

Current estimates of erosion in California are based mainly on calculations with the "Universal Soil Loss Equation." Although there are serious questions about its applicability to California, that equation has been used to estimate that California "loses" an average of more than 12,000 pounds of soil per acre (0.04 inches or 0.5 millimeter) each year. At that rate, one inch is "lost" each 25 years. ("Lost" is the word used although it is not accurate, since the soil may only move to an adjacent area.) We know even less about the other side of the balance, i.e., the rate of soil formation and thus the tolerable rate of loss. Tolerable losses may exceed 12,000 pounds per year on some deep valley and mountain soils but many sites have much lower tolerances. Although soil organic matter can be restored in tens of years, clay minerals may require hundreds. (Soil can, however, become too "old." The terrace soils around the Central Valley are extremely old and heavily leached of sulfur, phosphorus and other nutrients. A bit of erosion to expose new parent materials might bring sharp improvement.)

Water erosion is essentially absent in the valleys. (But not wind erosion—which is a perennial problem for the organic soils of the Delta and occasionally elsewhere.) Most of California's erosion, therefore, is confined to hill lands; let us first consider the grasslands.

Pristine California had very few large herbivores and grasslands were under little pressure from grazing. The species associations found at that time seem to have evolved in adaptation more to fire than to grazing. The combination of fire suppression, grazing increases and the introduction of new species has led to dramatic changes in species composition. Moderate grazing leads to average lower fuel supply for wildfire; that, along with fire suppression, results in sharply lower erosion. (But heavy overgrazing leads to the same erosion susceptibility as does fire.) The opinion of most California range experts is that erosion is less under present management than in the pristine condition. By their criteria, only a small percentage of such lands—mostly around waterholes and salt licks—are subject to overgrazing. The uncertainty of annual range production due to variable rainfall, and the high cost of alternative feedstuffs lead to understocking. And the low levels of production, 1,000 to 2,000 pounds forage per acre each year, work against capital investments in fencing, waterholes and nutrient additions needed for more intensive ranching. Undergrazing is the general condition on California rangelands, and residual vegetative material present when fall rains begin is generally more than adequate for soil protection.

There is another view of rangeland conditions, however. It is based on a "climax" concept of plant species composition and has its roots in studies by Nebraskan botanists over 50 years ago. One conclusion applied to California is that perennial grasses—which were more abundant prior to western settlement—represent a form of higher "quality" than the introduced annuals which are now dominant even where grazing is excluded. This change occurred over 100 years ago. Based on experience in more humid parts of the U.S., this is now assumed by some to be a sign of overgrazing. The U.S. Soil Conservation Service follows this view and consistently rates California's grasslands as being in "poor condition," which also is taken as a sign of desertification and leads to use of a high erodibility factor in the Universal Loss Equation.

California range authorities reply that dominance by annual species does not indicate a poor condition in Mediterranean-type grasslands. They find many of the native species highly unsuited for grazing and land stabilization, in contrast to the excellent properties of introduced annuals such as soft chess, wild oat and subterranean clover. By the climax definition, the introduction of clovers to provide an internal nitrogen source for the rangeland system represents degradation!

The proof of rangeland viability lies in the productivity and stability of the system over time. By those criteria, our rangelands have improved considerably over time. Erosion rates have probably declined, since undergrazed annuals without fire give good protection while erosion rates around perennial bunch grasses can be quite high. Unfortunately, there are not enough long-term quantitative data to fully prove or discredit either view. While it is well to be cautious, alarmist concerns for "desertification" are not borne out of the real world.

Similar arguments and uncertainties exist for the forest lands. Fire has been reduced, but timber harvest and particularly road construction increase erosion. The roads will eventually stabilize and can be used in the next cutting cycle, but whether erosion is greater or less over 50- to 100-year cycles than under natural conditions, with fire, is unknown.

Attention has been drawn to California's north coastal area where the argument of erosion control was used recently to justify the expansion of the Redwood Park. Fire was a common event there. That, plus high rainfall and easily erodable soils and land forms, led to high natural rates of erosion. But the load of sediments in streams and rivers may now be greater than prehistorically. In harvested forest areas this appears due mainly to mass soil movements such as "creep" and "slump," and to subsequent channeling by streams, rather than to sheet and rill erosion. Mass movements are correlated with an increase in soil wetness when seasonal evapotranspiration is reduced after conversion of forest to grassland and, briefly, following timber harvest. Reforestation practices on commercial timberlands are now reasonably effective in rapid re-establishment of cover—as is required by California law—and nutrient and soil losses following harvest are minimal. The problem seems to arise more from earlier conversion of forest to grassland than from present logging practices.

The Energy Question

All organisms ultimately depend upon energy stored in the carbon compounds produced by green plants through photosynthesis. In photosynthesis, carbon dioxide is "reduced" to high-energy organic compounds. Sunlight is the source of energy for that process. In this way,

solar energy is converted to chemical bond energy. Plants use most of this reduced carbon to form their biomass; the remainder is respired back to carbon dioxide. The energy released in respiration meets the cost of plant metabolism for growth and for maintenance of the living structure. In addition to expenditures for growth and maintenance, animals also expend energy in doing work, i.e., in exercise.

Plants are rather poor converters of solar energy. Under the best conditions (nutrients and water nonlimiting), the maximum conservation of chemical bond energy equals only about 5 percent of the sunlight energy. That contrasts to some solar cells which approach 20 percent efficiency.

Eventually, all organic carbon recycles to carbon dioxide although large amounts may be sequestered for extended periods in biomass, soil organic matter, carbonates and various fossil forms. For the earth as a whole, most carbon is found as carbonates in the oceans and soils, with smaller, but still very large amounts in fossil fuels. Living organisms, soil organic matter, and the atmosphere each contain less than one-tenth of the amount found in the fossil pool.

Energy Analyses

Tracing carbon and energy pathways through food chains is an important aspect of ecosystem analysis. Ecologists commonly analyze plant and animal production efficiency by input-output studies of how much energy they acquire and how they use it. We can do similar metabolic balances on man. In extending energy analyses to human activities such as agriculture, one encounters a number of difficulties because man has learned to harness other sources of energy, and to use carbon energy sources in other ways. Like other animals, we derive our food energy directly or indirectly from green plants. But we have found that we can greatly amplify our metabolic ability for work through machines. By hand, one man can till only a small plot of land each day; with machines, the possibility for work can be increased hundreds of times.

Most of our expenditures of fossil energy accomplish some sort of work. In manufacturing a steel filing cabinet, energy is used to refine the metal and organize it into a useable object. But the product does not represent a further source of energy to man. In fact, with most uses of fossil fuels, the energy is dissipated to the environment or embodied in the form of the manufactured good. Two major exceptions exist—fuel production in coal mining and agriculture. Each has outputs which can be described in energy terms and used as sources of energy. Agricultural products, for example, are organic and the heat energy of their combustion can be measured.

In recent years, there have been many such input-output studies in agriculture. Among the best known are those of David Pimentel and his colleagues at Cornell University, and Vashek Cervinka and his associates in California. Pimentel's analysis of corn production trends in Iowa since 1950 shows a steady increase in the fossil fuel use per acre of land. This is related to steady increases in the industrialization and intensification of agriculture, and includes the energy-equivalent costs of fertilizer, pesticides, machinery manufacture, operation and maintenance, and other essential inputs, including human labor. The resulting ratio of 3 calories output as corn grain for each calorie of input remained essentially unchanged between 1950 and 1980. Crop yields have increased in step with energy inputs. The ratio increases to 6 if the total crop (leaves, stems and grain) is considered.

A basic flaw in such analyses is that agriculture is an economic activity and is not really concerned with heat-of-combustion yields, any more than we would ask about the work content of formed metal. Efficiency in conversion, digestibility, and nutritional content are important factors in food production—but even more important may be dietary and social preferences as reflected in the marketplace. The manufacturing of beer, for example, involves considerable refrigeration. Add to that the cost of cans and distribution, and beer is seen to be very costly of energy per digestible calorie—yet it remains a favored item and a multi-million dollar industry.

Energy Use and Conservation
About 5 percent of our state's overall use of energy goes to the production and processing of agricultural commodities. The on-farm use is only about 1 percent (not including maintenance of farm buildings and machinery), thus relatively more energy is used in food distribution, marketing, and home preparation. Nationwide, all food-system energy uses amount to about 11 percent of total use. Worldwide, about 5 percent of energy use is in agricultural production.

It is difficult to grasp the full meaning of those percentages. In absolute terms, an amount of energy equivalent to about 40 gallons of gasoline per capita is used each year in crop production in California. For comparison, consider a single round trip between Sacramento and Los Angeles by jet airliner, at 25 gallons per passenger, or by private car, at 50 gallons per car. Basically, Americans and others are using far more energy in pursuits other than food production.

There is more room for energy conservation downstream from the farm than in the production processes themselves. Processing, packaging, distribution, marketing and home use are all relatively energy intensive. With a sharp increase in energy costs, we may see trends back to greater use of locally grown, in-season fruits and vegetables, as well as to less meat in the diet (and fewer pets in the home).

But what will happen to food production as petroleum and natural gas become less available? Will modern agriculture collapse? Will we return to horses and mules for motive power? Neither is likely. Conversion back to horses would require an enormous increase in labor, with lessened timeliness and quality in operations. In addition, perhaps a quarter of our farmland would have to be used for production of feed. (Pleasure horses alone now use the production from about 3 percent of California farmland.) But there is room for conservation. There is now more use of fuel-efficient diesel tractors. Considerable land is being put to "minimum tillage," in which the land is not plowed prior to planting. This saves large amounts of tractor fuel while requiring only a small increase in energy used for herbicides. Earlier maturing grain varieties also conserve on drying costs—though at lower average yields.

There also is considerable hope that through research we might find ways to improve nitrogen fixation by legumes and thus reduce energy use in fertilizer manufacture. Legumes are very good at providing fixed nitrogen, but—even plowed under as green manures—they don't leave sufficient nitrogen in the soil to insure maximum production from subsequent crops. Even worse, we may consume more fossil energy in cultivating legumes than is required to manufacture the same amount of nitrogen, and the proportion of acreage which could be planted to non-legume would be reduced. Since grain crops provide about 2/3 of the world's dietary energy and protein, grain-legume rotations would have to be markedly improved in order to meet the food needs of the present world population without help from fertilizer nitrogen.

Intensification

The best way to conserve energy use in agriculture may be to produce more from less land. That can be done by increasing both energy and labor inputs to achieve higher yields. This approach has been illustrated in Figure 1. Perry Stout of U.C. Davis and his colleagues found that certain "fixed" costs such as seed bed preparation, irrigation, and harvesting are more or less the same, regardless of the yield level achieved. So if small additional amounts of energy expended for nutrient supply, weed and pest control, and careful management result in significant yield increases, then the energy cost per unit of production decreases. In Stout's study, energy efficiency was greatly improved by increasing nitrogen fertilization up to the amount necessary for near-maximum grain yield (and less land is required to meet society's needs). Dutch scientists have come to similar conclusions and are now exploring possibilities for a two-tiered farming system: intensive agriculture on the best lands to produce the basic food supply, supplemented by an extensive, low-energy agriculture practiced on marginal lands.

The optimum route to an intensive, high-yield agriculture depends upon the social and technological context within which the farming is practiced. C.T. de Wit's view of this is illustrated in Figure 2. De Wit arrived at this relationship by assuming that labor and energy inputs in agriculture are to some extent substitutable. We can, for example, till the land by hand or with a mechanical beast. (But substitution is not complete—we don't have robots which could handle all decisions and operations; and it is difficult or impossible to maintain a high fertility by labor alone.) The yield responses to additional labor and energy both follow diminishing returns (plateaus), like the fertilizer response shown in Figure 1. This led de Wit to the yield relations shown in Figure 2 where the two inputs interact. The two data points represent the approximate present states of agriculture in Indonesia and The Netherlands. The position of Indonesia shows that it can scarcely expect to increase production through further applications of labor, unless it somehow supplies more of the limiting nutrients.

This is a system where small amounts of fossil energy expended on nutrient cycling can give very great increases in production. Their position corresponds to the low-yield, low-nutrient end of Figure 1 where the response to added nutrients may be as great as 50 pounds of grain per pound nitrogen, in contrast to a ratio of 5 to 10 pounds per pound as one approaches maximum yield. The Netherlands—and California—are closer to optimum yield and further increases are more likely through improved management.

Much, therefore, depends on the relative values of labor and energy. Thus far, the value of labor has inflated as fast or faster than that of energy. Despite recent increases, fossil energy is still about as cheap as it has ever been, though that will change as petroleum and natural gas supplies decline further. Availability of fuels is not the problem—at the present rate of use, 1,000 to 2,000 years' supply remains—but coal, oil shale and tar sands will not provide cheap sources of liquid fuel. Still, agriculture will be able to compete for that fuel. Given the choice between essential food or travel and living conveniences, people will choose food. Thus, the relative cost of food and the proportion of income spent on it can be expected to increase.

Energy Sources and the Environment

We do need to conserve certain critical energy resources for agriculture. In theory, because the conversion of dinitrogen gas (N_2) to nitrate (NO_3^-) is an exothermic reaction, we should be able to burn the N_2 found in air to nitrate, thus fertilizing our crops without much energy cost; but we don't yet have a catalytic process to do so without large inputs of energy. Nor do bacteria—they use even greater amounts of energy for nitrogen fixation. (A Nobel Prize is assured for anyone who breaks this riddle!) Meanwhile, the most energy-efficient industrial process for fertilizer production involves natural gas as a feedstuff for the production of ammonia (NH_3). Natural gas, therefore, has a special use in agriculture. In California, however, we have chosen to ameliorate our air pollution problems by consuming vast amounts of natural gas in electrical generation and in industrial processes such as steel making. Perhaps in the long run we will judge that decision ill-advised.

The use of fossil fuels is attended by two potentially serious environmental problems for agriculture. Coal and oil generally contain sulfur.

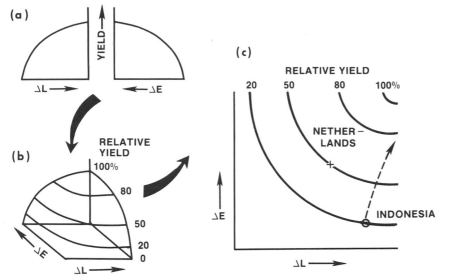

FIGURE 22.2 Agricultural yield dependence on labor and energy. Application of either more labor or more energy will tend to increase agricultural yield (a); we may visualize both inputs interacting in (b), which represents potential yield by contour lines at 20, 50, 80 and 100 percent; yields in any given farming systems will be dependent on the particular combination of labor and energy used (c). Conceptual points are shown for rice farming in Indonesia and arable agriculture in The Netherlands. The imaginary track (dotted line) for improvements of rice yields in Indonesia would depend mainly on additions of energy (fertilizer) since they have reached diminishing returns for labor. California's position is similar to The Netherlands. *Adapted from C. T. de Wit, Neth. J. Agri. Sci. 23:145–162, 1975.*

Burning them may release considerable sulfur dioxide as an air pollutant unless elaborate control devices are used. Also, all fossil fuels release carbon dioxide to the atmosphere. That appears to be the principal cause of the rising content of that gas in our air. Since carbon dioxide is a strong absorber of infrared radiation, its increasing content in the air may lead to changes in earth's climate. At the latitude of California, research to date suggests that temperature would increase and rainfall would decrease. However, it is difficult, in the face of a rapidly increasing population, to see how conservation and solar and geothermal alternatives alone could slow the rise in carbon dioxide.

The Farm as a Source of Energy

Considerable research has been given to agriculture as a solar converter for a source of fuels. Historically, woodlots provided fuel for the farmhouse, and forages and feed grains for horses and mules provided motive power for both farm and town. So why not establish a system for generating liquid fuels, petrochemical substrates, and electricity from crops? There are several major problems. First, plants are poor converters of solar energy. Second, it is impossible to avoid competition with food crops for available land—so a high price must be set for the energy. Problems of nutrient recycling or replacement will also be greatly increased. And the yield of energy will be small in both relative and absolute terms.

If we were to use *all* of California's 10 million acres of cropland for energy production, about 100 million tons of biomass with an energy content equivalent to 220 million bbl of petroleum could be produced. That production level, however, is greater than the average we now achieve with conventional crops, and would require even more nutrients and water. There also would be large energy costs in transport and processing. But the main problem is in achieving an efficient conversion to useable forms of energy. A large proportion of biomass is cellulose and lignin, which is best converted by burning to generate electrical energy with an efficiency of only about 25 percent. The much publicized production of ethanol from corn grain (a *favorable* substrate) actually uses more fossil energy in the process than is realized in ethanol. Whatever the conversion method, if we could obtain as much as 50 percent of the energy in the 100 million tons of biomass, the resulting equivalent of 110 million bbl of petroleum would represent only 11 percent of California's current energy use of 1 billion bbl equivalents per year.

While higher biomass yields are possible, conventional agriculture promises much greater returns to man than biomass production—at a much lower cost of scarce resources of land, water, and nutrients.

The Information Content of Agriculture

"Information" can be defined very broadly. In an ecological perspective, it includes the genetic traits of the species, as well as our own technological and scientific knowledge of crops and livestock and their culture, and of soils, climates and pests. Considering the multiple alternatives in choice of crops, varieties, and methods and timing of cultural operations, farming has become a very complex task in systems analysis.

Information Systems

Modern agriculture involves more than just good scientific and technological skills; increasingly, it depends on knowledge and ability in business management. Perhaps the most dramatic change in American agriculture during the past 30 years is the increased need for farmers to respond to information from beyond the farm itself. This includes knowledge of markets, financial and tax management, environmental regulation and employee welfare. Hedging production in the futures market, once left to middlemen, is now a standard part of farm management. The proportion of a farmer's time given to management as compared to operations is much greater than in the past.

The larger the farm, the more it seems to benefit from specialized services in accounting, tax consulting, marketing and pest control. While larger farm size is made possible by increases in labor productivity, information requirements seem to be another force that threatens the survival of the small farm and promotes the success of larger ones.

Some hints about future trends in this area are available. The needs for environmental protection and the urban dominance of California's legislative interests are among the factors suggesting continued expansion of governmental regulation of farming. Trends in banking towards highly integrated, full-spectrum services, on the other hand, suggest a simplification of management for the farmer.

We can also look for changes in more technological areas such as engineering, pest management and long range weather forecasting. Efforts in integrated pest management, for example, include detailed monitoring of pests and their natural controls with the view towards min-

imal use of pesticides on a selective, critical-timing basis. These services involve considerable increases in nonfarm technicians servicing agriculture. But the approach brings the possibility for field-by-field recommendations for best management. Updated pest status, the best advice on current methods of control, and recommendations for action could be transmitted to the farm office console from a central computer. If successful, such a program would greatly simplify the farm manager's decision making while improving environmental protection.

California has lagged in the development of such information systems. South Dakota has cable systems providing up-to-the-minute market and weather information to the farm. Cumulative evapotranspiration data have been a part of English farm radio services for many years but only recently have they been introduced in California.

California farmers have, however, been leaders in some types of computer-based services. Herd records such as those run by the Dairy Herd Improvement Association are one example. "Least-cost" rationing is another. California feedlots typically base their feed purchases and ration formulation on computer calculations which minimize the feed cost per unit of animal gain. The extension of such concepts to crop production would involve historical records of field characteristics and performance as well as calculation of growth patterns.

The future should see considerable expansion of such information and computing services to agriculture. Cable services from central computers probably will be more useful than on-farm computers since farmers may not have the time for detailed programming and data entry. The cable service has potential to meet record-keeping and computing needs as well as to provide encyclopedic information. Our future farmers will need computer training to help them manage this new technology.

Genetic Information

In a biological sense, the genetic coding of the genes in plants and animals is also information, apparent in the differences in adaptation and performances of different organisms. Bits of good genetic information stacked together result in a cultivar having superior performance in agriculture.

Concerns are aired from time to time that the genetic information base of agriculture is becoming too narrow. The idea is that our food system rests on only a few species (16 species of plants provide 80 percent of world food production). Improved cultivars of these species are adopted widely and the fear is that older cultivars carrying useful genetic information may be lost.

There are several answers to this problem. Clearly we must continue to collect and conserve as wide a collection aspossible of cultivated and wild-neighbor genotypes of crop plants. For the present, gene banks are living collections and thus subject to further evolution. In the future, dry or frozen storage of isolated genetic material may suffice. It is also clear that agriculture, at least for some species, now depends upon the activities of plant breeders just to maintain the *status quo*. New strains of pests and diseases occur continuously, requiring the genetic modification of present genotypes and the release of new cultivars, or a return to older ones. This may be termed "maintenance research." It is an essential need of modern agriculture, though often little understood or appreciated by those who fund and administer agricultural research. But the rewards to society are immense. One can calculate that the tax value alone of returns from one virus-resistant barley cultivar developed by a breeder at U.C. Davis is worth millions of dollars each year to the state of California.

In another sense, genetic diversity is not an emergency concern. Our crop plants have been with us for thousands of years and show high stability in production. Evolution has accommodated them to agriculture and time and again to perturbations of environment and pests. Only a few species such as corn and wheat have actually been the subject of intensive breeding efforts; nearly all of the fruits and vegetables have evolved through simple selection techniques. In an emergency, we have possibilities for substitution of cultivars, or of species (e.g., wheat for barley). World collections of crop plant resources are large—not as large as they might be, but larger than the available plant breeders can explore. The real shortage is of breeders rather than germ plasm.

Recent advances in molecular biology have led to the possibility of DNA transfer and thus genetic transfer from one species to another. The possibilities are considerable. We can arrange for pharmaceutical production of a wide range of important substances from bacteria. The current explosion in efforts at "genetic engineering" is a result. The potential for applications with higher plants and animals, however, remains uncertain. The performance of such complex organisms depends upon integration of a multitude of cellular traits. The genetic controls of integration are complex and largely unknown, but are easily manipulated by conventional methods.

New Science and New Technology

New science and technology can have large effects on agriculture. Some of the new information comes through innovative efforts of farmers and their close community, and some from institutional research. The track-laying tractor developed for Delta farming, for example, gave birth to the Caterpillar Company and later the military tank. Recent advances in laser-guided land leveling were developed elsewhere but eventually found an application in agriculture. Rural electrification, barbed wire fencing, understanding of vitamin requirements of animals, and the findings of David Hoagland's plant nutrition group at Berkeley—such developments had enormous social and technological impacts in agriculture.

Agricultural research conducted by the U.S. Department of Agriculture and state experiment stations plays a special role in fostering scientific and technological change. This remarkable public effort had its roots in the British agricultural advances of 1650 to 1850. During the transition from feudal farming systems, the new yeoman farmers paid great attention to animal breeding, manuring and crop rotation. They met in agricultural societies and fairs to exchange information, an idea which later spread to America. In a short span of time, Hereford and Angus cattle and many other new breeds of animals emerged. And the first formal agricultural research station was established at Rothamsted.

In America, research at the experiment stations has focused on how plants, animals and soils function and how farming methods might be improved. Agricultural scientists have pioneered in subjects ranging from clay minerology, statistics, virology and plant physiology to animal nutrition, microclimatology and watershed management. If Max Kleiber's classic work at U.C. Davis on animal digestion and energetics had been available a quarter of a century sooner, for example, the tragedy of starvation in Robert Scott's 1912 polar expedition could have been avoided.

As one looks around, it is truly remarkable how much of our present scientific knowledge resulted from the need for answers in agriculture.

But such research has had curious consequences for agriculture. On one hand, agriculture has become more productive, safer, and more efficient in its use of land and scarce resources. Research has helped bring food of ever better quality at less relative expense to our tables. Clearly, research has been a strong force for positive change. In another sense, it has been exploitive of farmers for the public good. Farmers must stay abreast of new information and adjust their practices to remain competitive with their fellow entrepreneurs. They must be better educated than in the old days, but they also provide more adequate food supply and better stewardship of our land resources than in the past.

Overview

Land, resources, energy and information are taken here as themes of concern and future change. One might wish to add other topics such as pesticides, air pollution and climate change. Pesticides and air pollution are under enormous scrutiny and regulation, however, and should disappear as public issues if they are brought to manageable proportions. Other less controllable issues may be more alarming. Climate change can come from the use of fossil fuels or from cyclic changes in sun-spot activity, sun-earth-planet distances, and long-term oscillations of ocean-atmosphere equilibria. A natural return to glacial climates or a sharp decline in rainfall would have far-reaching consequences. While not beyond human understanding, these may be beyond our abilities for management.

So the topics chosen have been those about which we can do something—where both understanding and action or reaction are possible. They are set forth as opinions from one student of agriculture as the really critical issues facing agriculture during the next generation. What we do in the future nevertheless depends to a great extent on the larger society. Agriculturalists are only one small voice in the complex modern world. If they are to have a favorable environment for their professional pursuits, the public must have a better understanding of their tasks. And that, after all, is what this book has been about.

References

Cox, G.W., and Atkins, M.D. *Agricultural Ecology.* San Francisco: W.H. Freeman Co., 1979.

Jenny, H. *The Soil Resource.* Ecological Studies, Vol. 37. New York: Springer-Verlag, 1980.

Loomis, R.S. "Agricultural Ecosystems." *Scientific American.* 235(1976):99–105.

Loomis, R.S. "Agricultural Systems," in *Physiological Plant Ecology: Productivity and Ecosystems Processes.* Eds. O.L. Lange, P.S. Nobel, C.B. Osmond and H. Ziegler. Encyclopedia of Plant Physiology, New Series, Vol. 13D (in press), 1982.

Contributors

REUBEN ALBAUGH
Extension Animal Scientist, Emeritus
University of California, Davis

ROY BAINER
Professor of Agricultural Engineering,
    Emeritus
University of California, Davis

ROBERT B. BALL
Extension Specialist
University of California, Davis

JAMES A. BEUTEL
Extension Pomologist
University of California, Davis

ROBERT A. BRENDLER
Extension Farm Advisor
Ventura County

ROYCE S. BRINGHURST
Professor of Pomology
University of California, Davis

DILLON BROWN
Professor of Pomology, Emeritus
University of California, Davis

HAROLD O. CARTER
Professor of Agricultural Economics
University of California, Davis

VASHEK CERVINKA
Senior Agriculture Systems Analyst
California Department of Food and
    Agriculture

FRED CONTE
Extension Aquaculture Specialist
University of California, Davis

JULIAN C. CRANE
Professor of Pomology
University of California, Davis

CONSTANT DELWICHE
Professor of Geobiology
University of California, Davis

JOHN R. DUNBAR
Extension Animal Scientist
University of California, Davis

KENNETH W. ELLIS
Extension Animal Scientist
University of California, Davis

RALPH A. ERNST
Extension Poultry Specialist
University of California, Davis

DELWORTH GARDNER
Professor of Agricultural Economics
University of California, Davis

LEON GAROYAN
Extension Marketing Economist
University of California, Davis

MELVIN R. GEORGE
Extension Agronomist
University of California, Davis

HUDSON T. HARTMANN
Professor of Pomology, Emeritus
University of California, Davis

RAYMOND F. HASEK
Extension Environmental Horticulturist
University of California, Davis

HAROLD HEADY
Professor of Range Management
University of California, Berkeley

F. JAMES HILLS
Extension Agronomist
University of California, Davis

KARL H. INGEBRETSEN
Extension Agronomist
University of California, Davis

HUNTER JOHNSON, JR.
Extension Vegetable Specialist
University of California, riverside

WARREN E. JOHNSTON
Professor of Agricultural Economics
University of California, Davis

DESMOND A. JOLLY
Extension Consumer Economist
University of California, Davis

PAULDEN F. KNOWLES
Professor of Agronomy
University of California, Davis

SYLVIA LANE
Professor of Agricultural Economics
University of California, Davis

JAMES H. LARUE
Extension Farm Advisor
Tulare County

A. STARKER LEOPOLD
Professor of Zoology and Forestry,
   Emeritus
University of California, Berkeley

ROBERT S. LOOMIS
Professor of Agronomy
University of California, Davis

OSCAR A. LORENZ
Professor of Vegetable Crops
University of California, Davis

KEITH S. MAYBERRY
Extension Farm Advisor
Imperial County

NORMAN F. MCCALLEY
Extension Farm Advisor
Monterey and Santa Cruz Counties

CHESTER O. MCCORKLE
Professor of Agricultural Economics
University of California, Davis

ROLAND D. MEYER
Extension Soils Specialist
University of California, Davis

WARREN C. MICKE
Extension Pomologist
University of California, Davis

WALTER MINGER
Senior Vice President
Bank of America, San Francisco

DUANE S. MIKKELSEN
Professor of Agronomy
University of California, Davis

F. GORDON MITCHELL
Extension Pomologist
University of California, Davis

EMIL M. MRAK
Professor of Food Science and
   Technology, Emeritus
University of California, Davis

FRANK D. MURRILL
Extension Dairy Scientist
University of California, Davis

ERIC C. MUSSEN
Extension Apiculturist
University of California, DAvis

CAROLE A. NUCKTON
Staff Research Associate
University of California, Davis

JOSEPH W. OSGOOD
Extension Farm Advisor
Tehama County

CLEMENT L. PELISSIER
Extension Dairy Scientist
University of California, Davis

ROBERT G. PLATT
Subtropical Horticulture Specialist,
  Emeritus
University of California, Riverside

LINDA PRATO
Senior Research Assistant
University of California, Davis

DAVID E. RAMOS
Extension Pomologist
University of California, Davis

REFUGIO I. ROCHIN
Associate Professor of Agricultural
  Economics
University of California, Davis

DONALD ROUGH
Extension Farm Advisor
San Joaquin County

LOY L. SAMMET
Professor of Agricultural Economics,
  Emeritus
University of California, Berkeley

CHARLES W. SCHALLER
Professor of Agronomy
University of California, Davis

ANN SCHEURING
Editor
University of California, Davis

ROBERT W. SCHEUERMAN
Extension Farm Advisor
Merced County

BERNARD S. SCHWEIGERT
Professor of Food Science and
  Technology
University of California, Davis

R. HENRY SCIARONI
Extension County Director
San Mateo-San Francisco Counties

WILLIAM L. SIMS
Extension Vegetable Specialist
University of California, Davis

JERRY SCRIBNER
Deputy Director
California Department of Food and
  Agriculture

J. HERBERT SNYDER
Director, Water Resources Center
University of California, Davis

MARVIN J. SNYDER
Extension Farm Advisor
San Luis Obispo County

MICHAEL STIMMANN
Extension Pesticide Coordinator
University of California, Davis

LARRY R. TEUBER
Assistant Professor of Agronomy
University of California, Davis

CARL L. TUCKER
Specialist, Bean Breeding
University of California, Davis

KENT B. TYLER
Extension Vegetable Specialist
San Joaquin Valley Field Station

HENRY VAUX, SR.
Professor of Forestry, Emeritus
University of California, Berkeley

A. DINSMOOR WEBB
Professor of Enology
University of California, Davis

EDWARD YEARY
Extension Farm Advisor
San Joaquin Valley Field Station

PAUL J. ZINKE
Professor of Soils and Forestry Influences
University of California, Berkeley

JOHN A. ZIVNUSKA
Professor of Forest Economics and Policy
University of California, Berkeley

Index